T0270956

CRITICAL DYNAMICS

Introducing a unified framework for describing and understanding complex interacting systems common in physics, chemistry, biology, ecology, and the social
sciences, this comprehensive overview of dynamic critical phenomena covers
the description of systems at thermal equilibrium, quantum systems, and non-
equilibrium systems.

- Powerful mathematical techniques for dealing with complex dynamic systems
  are carefully introduced, including field-theoretic tools and the perturbative
  dynamical renormalization group approach, rapidly building up a mathematical toolbox of relevant skills.
- Heuristic and qualitative arguments outlining the essential theory behind each
  type of system are introduced at the start of each chapter, alongside real-world
  numerical and experimental data, firmly linking new mathematical techniques
  to their practical applications.
- Each chapter is supported by carefully tailored problems for solution, and comprehensive suggestions for further reading, making this an excellent introduction
  to critical dynamics for graduate students and researchers across many disciplines
  within the physical and life sciences.

UWE C. TÄUBER is a Professor in the Department of Physics at Virginia Tech,
where his research focuses on the characterization of phase transitions and generic
scale invariance in non-equilibrium systems.

CRITICAL DYNAMICS

A Field Theory Approach to Equilibrium and Non-Equilibrium Scaling Behavior

UWE C. TÄUBER

Virginia Tech

CAMBRIDGE
UNIVERSITY PRESS

CAMBRIDGE
UNIVERSITY PRESS

University Printing House, Cambridge CB2 8BS, United Kingdom

One Liberty Plaza, 20th Floor, New York, NY 10006, USA

477 Williamstown Road, Port Melbourne, VIC 3207, Australia

4843/24, 2nd Floor, Ansari Road, Daryaganj, Delhi - 110002, India

79 Anson Road, #06-04/06, Singapore 079906

Cambridge University Press is part of the University of Cambridge.

It furthers the University's mission by disseminating knowledge in the pursuit of education, learning and research at the highest international levels of excellence.

www.cambridge.org
Information on this title: www.cambridge.org/9780521842235

First published 2014

A catalogue record for this publication is available from the British Library

ISBN 978-0-521-84223-5 Hardback

To Karin, with love

Contents

Preface

The goal of this advanced graduate-level textbook is to provide a description of the field-theoretic renormalization group approach for the study of time-dependent phenomena in systems either close to a critical point, or displaying generic scale invariance. Its general aim is a unifying treatment of classical near-equilibrium, as well as quantum and non-equilibrium systems, providing the reader with a thorough grasp of the fundamental principles and physical ideas underlying the subject.

Scaling ideas and the renormalization group philosophy and its various mathematical formulations were developed in the 1960s and early 1970s. In the realm of statistical physics, they led to a profound understanding of critical singularities near continuous phase transitions in thermal equilibrium. Beginning in the late 1960s, these concepts were subsequently generalized and applied to dynamic critical phenomena. By the mid-1980s, when I began my research career, critical dynamics had become a mature but still exciting field with many novel applications. Specifically, extensions to quantum critical points and to systems either driven or initialized far away from thermal equilibrium opened fertile new areas for in-depth analytical and numerical investigations.

By now there exists a fair sample of excellent textbooks that provide profound expositions of the renormalization group method for static critical phenomena, adequately introducing statistical field theory as the basic tool, and properly connecting it with its parent, quantum field theory. However, novice researchers who wish to familiarize themselves with the basic techniques and results in the study of dynamic critical phenomena still must resort largely to the original literature, supplemented with a number of very good review articles. Over the years, various outstanding texts have emerged that address specific subfields such as one-dimensional systems, numerical methods, equilibrium statistical and quantum field theory, quantum phase transitions, non-equilibrium quantum dynamics, growth models, pattern formation, and the glass transition, to list just a few. Yet there has been no published attempt at a comprehensive introductory overview of field-theoretic tools and perturbative renormalization group methods applied to stochastic non-linear systems, near-equilibrium critical dynamics, non-equilibrium phase transitions, and driven systems displaying generic scale invariance.

Given this startling gap in the literature, I immediately and gladly consented when in 1997 John Cardy proposed to me to coauthor precisely such a text. John had recently completed his brilliant, concise book *Scaling and Renormalization in Statistical Physics*; and I had just delivered a brief lecture course on *Dynamic Critical Phenomena* at the University of Oxford, where I was a postdoctoral research associate at the time. Our initial, in retrospect rather naive, vision was that based on our lecture notes and research papers, we should be able to compile a comprehensive introduction to the field of critical dynamics within about a year or two. That task, however, turned out to be much more daunting, and progress excruciatingly slower than we had anticipated. As time passed, only very few chapters had materialized, and John's principal research interest increasingly deviated from classical non-equilibrium phenomena, he decided to withdraw from the project. Following a year-long hiatus, I eventually endeavored to revive our book plans in 2003; and it has still taken me another decade to write this present and considerably evolved version of the text John and I had originally envisaged.

It is the aim of this book to give a thorough account, in a conceptually unified manner, of time-dependent behavior of classical systems close to an equilibrium critical point, dynamical critical phenomena in quantum systems, and scaling properties of systems far from equilibrium, which either exhibit a phase transition between different non-equilibrium stationary states as some parameter is varied, or whose dynamics naturally leads to a scale-invariant critical state. I wish to emphasize that this text does not intend to provide a comprehensive review of this immense field, but instead to present an accessible introduction to the main and basic results that have accumulated through the combined efforts of many accomplished researchers, along with the fundamental analytical tools of field-theoretic representations for stochastic dynamics and the renormalization group approach. Although the full power of the field-theoretic method will be developed, an effort has been made to keep the text from becoming too technical, always trying to refer back to the physical interpretation of the formalism as it evolves. In particular, introductory material in each section describes more heuristic physical arguments before plunging into the mathematics, and various illustrations, from both real and numerical experiments, that have been included. I confess that maintaining this delicate balance between intuitive physical insight and more rigorous mathematical treatment has been one of my major challenges in writing this treatise.

The first two chapters are intended to give a rather quick yet still self-contained introduction and overview of the theory of thermodynamic phase transitions, and of various formulations of stochastic dynamics. Along with the first exposition of dynamical scaling and critical dynamics in Chapter 3, and the various less technical introductions in later chapters, these should be readily usable in graduate-level lecture courses; indeed, I have employed them myself in advanced courses on *Statistical Mechanics* and *Solid State Physics*. Much of the remainder of this text, however, will probably be covered less frequently in standard course sequences. Instead, the bulk of this book may hopefully serve as an adequate foundation for self or independent group study. I trust, therefore, that the collections of problems and exercises at the end of each chapter will entice diligent readers to delve deeper into the technical and mathematical issues involved, and provide additional opportunities to practice newly acquired concepts and techniques.

Field-theoretic methods generally provide an enormously powerful toolbox, and the renormalization group is definitely indispensible in establishing universal features. However, there are indisputably other prominent theoretical approaches that researchers working in this field are well-advised to consult and employ. Naturally, exact solutions, though typically only feasible for one-dimensional 'simple' model systems, are invaluable cornerstones of any theoretical development, and may quite obviously serve as crucial test cases for approximate analysis. For example, various very clever approximation schemes to deal with stochastic master and Langevin equations have been developed and brought to fruition for many specific applications. And, last but certainly not least, numerical algorithms and computer simulations have become fantastically efficient over the past years, and often provide detailed insights into complex behavior emergent in stochastic interacting systems. Data obtained from careful and properly analyzed Monte Carlo simulations in fact have to a large extent replaced real laboratory experiments in establishing the basic facts that any successful theory needs to reproduce or predict. But this development has not been entirely beneficial to the field, and I fervently hope that the near future will see a revival of truly quantitative and high-quality experimental investigations of non-equilibrium systems; it appears as if perhaps new data will originate predominantly from a biological context.

Since there already exist outstanding expositions of exactly solvable models and their mathematical treatment, as well as extensive introductions to modern computer simulation algorithms and data analysis, these topics are only barely discussed in this book. Inevitably, I found that I had to make tough decisions on which topics to include. Naturally, the final choice of material is strongly biased towards my own expertise and personal research trajectory. I most sincerely apologize, therefore, to any readers and distinguished colleagues who find their own important contributions inadequately represented or improperly acknowledged, and stress again that a complete and comprehensive survey is not what I have strived to achieve. My perhaps most notable omissions include the powerful Keldysh formalism for non-equilibrium quantum dynamics; non-perturbative numerical renormalization group approaches that have enjoyed a tremendously successful comeback in the past decade; turbulence and scale invariance in fluid dynamics; the depinning transition of manifolds driven through random media; and dynamic scaling in structural and spin glasses. Each chapter of this text lists the most important references that I have regularly consulted, along with a bibliography of suggested original or more advanced literature for further in-depth studies. Once again, I do not claim these lists to be complete or exhaustive, but I nevertheless trust them to be useful.

As a student and young researcher, I learned a substantial fraction of the material presented in this text from wonderful lecture courses delivered by Franz Schwabl, Wolfgang Götze, Wilhelm Brenig, and Reinhard Folk at the Technische Universität München; David Nelson and Daniel Fisher at Harvard University; John Cardy and John Chalker at the University of Oxford; and from the remarkable lecture notes by Hans-Karl Janssen at the Heinrich–Heine-Universität Düsseldorf. My mentors will inevitably find parts of their excellent lectures reflected in various chapters of this book. Yet while I owe a tremendous amount of knowledge and insight to their outstanding didactic efforts, naturally any errors and misconceptions in this text and other inadequacies are solely mine and should not be blamed on these exceptional scholars.

I am moreover deeply indebted to my research collaborators in various topics in statistical physics over many years that are at least in part represented in this volume. These include my esteemed colleagues as well as current and past students Vamsi Akkineni, Jeremy Allam, Timo Aspelmeier, Hiba Assi, Christian Baumgärtel, Paul Bourgine, Michael Bulenda, Thomas Bullard, John Cardy, Harshwardhan Chaturvedi, George Daquila, Jayajit Das, Oliviér Deloubrière, Hans-Werner Diehl, Sebastian Diehl, Ulrich Dobramysl, Barbara Drossel, Reinhard Eckl, Vlad Elgart, Kimberley Forsten-Williams, Erwin Frey, Wilmut Gasser, Ivan Georgiev, Yadin Goldschmidt, Manoj Gopalakrishnan, Bertrand Halperin, Qian He, Malte Henkel, Henk Hilhorst, Haye Hinrichsen, Martin Howard, Terence Hwa, Hans-Karl Janssen, Swapnil Jawkar, Alex Kamenev, Bernhard Kaufmann, Thananart Klongcheongsan, Vivien Lecomte, Jerôme Magnin, Mauro Mobilia, Thomas Nattermann, David Nelson, Klaus Oerding, Michel Pleimling, Gunnar Prüssner, Zoltán Rácz, Leo Radzihovsky, Matthew Raum, Beth Reid, Andrew Rutenberg, Jaime Santos, Beate Schmittmann, Alfred Schorgg, Gunter Schütz, Franz Schwabl, Matthew Shimer, Robin Stinchcombe, Steffen Trimper, Benjamin Vollmayr-Lee, Mark Washenberger, Carsten Wengel, Frédéric van Wijland, and Royce Zia. In addition to John Cardy, who originally incited this endeavor, many of the colleagues listed above have in fact crucially contributed to the contents of this text, directly through their invaluable contributions to the field, via the many insightful discussions we enjoyed during our collaborations, by providing me with corrections, constructive criticism, and feedback, or by cheerfully encouraging me to pursue and persist with this daunting enterprise. I would especially like to thank Ulrich Dobramysl for generating the colorful book cover image, George Daquila for his drawing of Fig. 11.1; and Michel Alba and Jeremy Allam for their kind permissions to peruse their experimental data (Figs. 3.7 and 9.8) prior to peer-reviewed publication.

In the long course of writing this book, several academic institutions have provided me with considerable financial and administrative support. During my postdoctoral years, I was fortunate to hold research positions at Harvard University, the University of Oxford, and the Technische Universität München, in part funded through the Deutsche Forschungsgemeinschaft and the European Union's Marie Curie Fellowship program. Since 1999, Virginia Tech has been my scientific home, and my colleagues there and our students have continuously generated an inspiring and stimulating environment for my academic teaching and research program. The National Science Foundation's Division of Materials Research, the US Department of Energy's Office of Basic Energy Sciences, and Bank of America's Jeffress Memorial Trust have crucially supported my group's research during the past decade. I also wish to specifically acknowledge the friendly hospitality of the Université de Paris-Sud Orsay and the University of Oxford during my first sabbatical in 2005, and the Institut des Systèmes Complexes – Paris Île-de-France (ISC-PIF) during my most recent research leave in the fall of 2012, during which several chapters of this text were completed.

I emphatically profess that this book could and would not have been written without the firm and committed support of my beloved family. My wife Karin has been absolutely crucial to the success of this project through her indefatigable encouragement and deeply rooted trust in my ability to finally complete this task even at the various instances when I almost faltered. It is therefore only apt and just that this volume be dedicated to her, the love of my life. My daughters Lilian and Judith have always been curious what this strange book was all about that caused their father so much anxiety, and took his precious time away from them. Lilian even won our ancient bet that I would not have finished the manuscript by the time she left for college; my only consolation is that the text will at least be published prior to her attaining a postgraduate degree, and Judith's graduation from high school. My mother Gertraud Täuber has always been unwavering in her love and understanding, and throughout my entire life laid the foundations of my career and accomplishments. I hope they all will find some joy and satisfaction too in this finally completed product.

Last but not least, I am very grateful to a series of capable and experienced editors and staff at Cambridge University Press, namely Lindsay Barnes, Laura Clark, Graham Hart, Sarah Matthews, Rebecca Mikulin, Rufus Neal,

Antonaeta Ouzounova, Fiona Saunders, Eoin O'Sullivan, Megan Waddington, Emily Yossarian; and most prominently Simon Capelin, who served as my responsible editor both at the beginning and happy ending of this enterprise, as well as Vania Cunha, Elizabeth Horne, Frances Nex, and Zoë Pruce, whose kind and expert assistance in its final stages were invaluable. I sincerely appreciate their professional expertise and helpful advice over more than a decade, and most of all, their seemingly indestructible patience.

Part I

Near-equilibrium critical dynamics

Introduction

Originally, the term *'dynamic critical phenomena'* was coined for time-dependent properties near second-order phase transitions in thermal *equilibrium*. The kinetics of phase transitions in magnets, at the gas–liquid transition, and at the normal-to superfluid phase transition in helium 4 were among the prominent examples investigated already in the 1960s. The *dynamic scaling hypothesis*, generalizing the scaling ansatz for the static correlation function and introducing an additional dynamic critical exponent, successfully described a variety of these experiments. Yet only the development of the systematic *renormalization group* (RG) approach for critical phenomena in the subsequent decade provided a solid conceptual foundation for phenomenological scaling theories. Supplemented with exact solutions for certain idealized model systems, and guided by invaluable input from computer simulations in addition to experimental data, the renormalization group now provides a general framework to explore not only the static and dynamic properties near a critical point, but also the large-scale and low-frequency response in stable thermodynamic phases. Scaling concepts and the renormalization group have also been successfully applied to phase transitions at zero temperature driven by quantum rather than thermal fluctuations. It is to be hoped that RG methods may help to classify the strikingly rich phenomena encountered in far-from-equilibrium systems as well. Recent advances in studies of simple reaction-diffusion systems, active to absorbing state phase transitions, driven lattice gases, and scaling properties of moving interfaces and growing surfaces, among others, appear promising in this respect.

The first part of this book focuses on equilibrium critical phenomena, dominated by strong *thermal fluctuations* near a thermodynamic instability. Here we introduce most of the fundamental concepts and analytical tools needed also for the analysis of quantum and non-equilibrium critical dynamics. We begin with a review of thermodynamic singularities and the behavior of the order parameter correlations near a critical point. In addition to simple mean-field theory as encompassed in the generic Landau–Ginzburg approach, we briefly survey Wilson's momentum shell RG method. Next, we introduce the basic principles that will allow us to study inherently dynamic fluctuation phenomena. Chapter 2 covers linear response theory as well as a discussion of master, Fokker–Planck, and Langevin equations that capture stochastic kinetics on different description levels. We specifically highlight the restrictive detailed-balance conditions required to reach thermal equilibrium in the long-time limit. In the subsequent Chapter 3, we turn our attention to critical slowing down and the implications of the dynamic scaling hypothesis. The crucial influence of 'slow' diffusive conserved quantities besides the order parameter is elucidated.

Thereby equipped with a basic understanding of the relevant physical picture, we venture into more formal chapters on dynamic perturbation theory and the field-theoretic variant of the renormalization group procedure. These central chapters provide a powerful general analytic framework for many of the subsequent discussions. Chapter 4 contains a detailed exposition of the perturbation expansion for purely relaxational kinetics with either non-conserved or conserved order parameter. In Chapter 5, the renormalization program is explicitly carried through for these time-dependent Landau–Ginzburg models. The critical exponents are computed to lowest non-trivial loop order, where a small parameter for the perturbation series is given by the deviation $\epsilon = d_c - d$ from the (upper) critical dimension $d_c = 4$. We also explain how the emergence of massless Goldstone modes leads to generic scale invariance in the entire low-temperature phase of systems with broken continuous symmetry.

Chapter 6 explores the effect of additional conserved hydrodynamic modes and reversible non-linear mode couplings. Exploiting underlying symmetries, we derive scaling relations for the dynamic critical exponents associated with, e.g., the equilibrium critical dynamics of magnetic systems, superfluid helium 4, and binary liquids. A link is provided to self-consistent mode-coupling theory, a valuable tool for the calculation of scaling functions and quantitative comparison with experiments. Finally, Chapter 7 is concerned with phase transitions in quantum systems and quantum-critical phenomena. Here, we introduce coherent-state path integrals for bosonic and fermionic many-particle systems, and discuss several illustrative examples, most prominently the properties of boson superfluids and crossover features in quantum antiferromagnets.

1

Equilibrium critical phenomena

To set the stage for our subsequent thorough discussion of dynamic critical phenomena, we first review the theoretical description of second-order equilibrium
phase transitions. (Readers already well acquainted with this material may readily
move on to Chapter 2.) To this end, we compare the critical exponents following from the van-der-Waals equation of state for weakly interacting gases with
the results from the Curie–Weiss mean-field approximation for the ferromagnetic
Ising model. We then provide a unifying description in terms of Landau–Ginzburg
theory, i.e., a long-wavelength expansion of the effective free energy with respect
to the order parameter. The Gaussian model is analyzed, and a quantitative criterion
is established that defines the circumstances when non-linear fluctuations need to
be taken into account properly. Thereby we identify $d_c = 4$ as the upper critical
dimension for generic continuous phase transitions in thermal equilibrium. The
most characteristic feature of a critical point turns out to be the divergence of the
correlation length that renders microscopic details oblivious. As a consequence,
not only the correlation functions, but remarkably the thermodynamics as well of
a critical system are governed by an emergent unusual symmetry: scale invariance.
A simple scaling ansatz is capable of linking different critical exponents; as an
application, we introduce the basic elements of finite-size scaling. Finally, a brief
sketch of Wilson's momentum shell renormalization group method is presented,
intended as a pedagogical preview of the fundamental RG ideas. Exploiting the
scale invariance properties at the critical point, the scaling forms of the free energy
and the order parameter correlation function are derived. The critical exponents
are computed perturbatively to first order in $\epsilon = 4 - d$. Beginning with Chapter 4,
we shall later venture into a more formal discussion of both static and dynamic
critical phenomena, utilizing the framework of renormalized field theory based on
non-linear Langevin stochastic equations of motion.

1.1 Mean-field theory

We begin our review with a brief discussion of the classic *mean-field theories* for the gas–liquid and para-/ferromagnetic equilibrium critical point.[1] The common feature of such mean-field approaches is that spatial fluctuations of the thermodynamic variable serving as the order parameter for the transition are neglected. Ordinarily, in a macroscopic system with $N \gg 1$ degrees of freedom, arguments akin to the central-limit theorem can be safely applied to any extensive thermodynamic quantity, whereupon one would expect its fluctuations relative to its mean ($\sim N$) to be of order $N^{-1/2}$. However, in the vicinity of a critical point, the thermodynamic response function associated with the order parameter diverges. This indicates that its mean-square thermal fluctuations become of order N^2 (in mean-field approximation) rather than N, and cannot generally be disregarded in the computation of thermodynamic potentials and correlation functions. Nevertheless, mean-field theories often provide qualitatively correct pictures of the essential physics and phase diagrams.

1.1.1 Van-der-Waals equation of state

The *van-der-Waals equation of state* for a weakly interacting gas,

$$P(T, V, N) = \frac{N k_{\mathrm{B}} T}{V - Nb} - \frac{N^2 a}{V^2} , \tag{1.1}$$

relates its pressure P to the volume V, temperature T, and particle number N. The parameter $b > 0$ stems from the short-range repulsions between the gas molecules, and corresponds to the average excluded volume per hard-core particle. This excluded volume repulsion naturally increases the gas pressure P. Attractive two-body molecular forces, on the other hand, will reduce the pressure by a term $\propto -(N/V)^2$. For weak long-range pair interactions, e.g., of the van-der-Waals form $V(r) \propto -1/r^6$ originating in fluctuating electric dipole moments, essentially only the radially averaged pair potential $\sim a$ matters, which leads to the second term in Eq. (1.1).

Typical van-der-Waals isotherms $P(v)$ at $T = $ const. are sketched in Fig. 1.1 as function of the reduced volume (inverse particle density) $v = V/N = 1/n$. At high temperatures $T \gg T_{\mathrm{c}}$, they approach the *universal* ideal gas law $PV = N k_{\mathrm{B}} T$, independent of the microscopic interaction parameters a and b. Upon lowering T, the isotherms develop an inflection point with decreasing slope. As we shall see, the critical temperature $T = T_{\mathrm{c}}$ is reached when this inflection point turns into

---

[1] More detailed descriptions can be found in most modern graduate level textbooks on equilibrium statistical mechanics, e.g.: Chaikin and Lubensky (1995), Pathria (1996), Cowan (2005), Schwabl (2006), Kardar (2007a, b), Reichl (2009), and Van Vliet (2010).

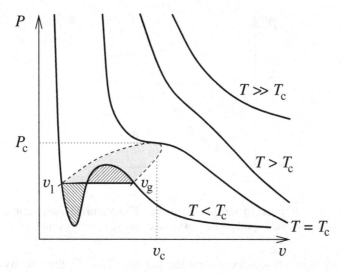

Fig. 1.1 Sketch of van-der-Waals isotherms in different temperature regimes. The critical point (v_c, P_c) (dotted) and the coexistence curve (dashed), which encloses the coexistence area (grey), are indicated. For $T < T_c$, Maxwell's construction demands that the shaded areas delimited by v_l and v_g be equal.

a saddle point with horizontal tangent. For $T < T_c$, the function $P(v)$ has two extrema. However, the van-der-Waals equation becomes unphysical in the region where $(\partial P / \partial v)_T > 0$, because thermodynamic stability requires the *isothermal compressibility* $\kappa_T = -v^{-1}(\partial v / \partial P)_T$ to be positive in thermal equilibrium. In fact, in the grand-canonical ensemble it is related to the mean-square particle number fluctuations,

$$\kappa_T = \frac{v(\Delta N)^2}{N k_B T} > 0 . \tag{1.2}$$

We interpret the instability for $T < T_c$ to indicate *phase separation* between a more dilute *gaseous* (g) state and a denser *liquid* (l) phase. In equilibrium, these two phases are not only at the same temperature and pressure, but must have identical chemical potentials $\mu_g = \mu_l$. Following Maxwell's construction, we thus replace the van-der-Waals isotherm for $v_l < v < v_g$ with a line of constant pressure $P_0(T)$. Employing the Gibbs–Duhem relation $\mu = G/N$ and the differential of the free enthalpy (Gibbs free energy) $dG = -S dT + V dP + \mu dN$, we have for $T, N = $ const.:

$$0 = \mu_g - \mu_l = \int_{v_l}^{v_g} v \, dP , \tag{1.3}$$

which represents the oriented area under the $v(P)$ curve. Therefore v_l and v_g are uniquely determined by the condition that the shaded areas in Fig. 1.1 be equal.

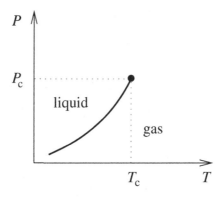

Fig. 1.2 Phase diagram in the (T, P) plane. The vapor pressure curve $P_0(T)$ separating the gas and liquid phases terminates at the critical point (T_c, P_c).

As the pressure is increased at constant temperature $T < T_c$, the density *discontinuously* jumps from $n_g = 1/v_g$ to $n_l = 1/v_l$; the associated change in entropy per particle s releases the *latent heat* $q_L = T(s_g - s_l)$. Thus, for $T < T_c$, the van-der-Waals equation describes a *first-order* gas–liquid phase transition.

The *phase coexistence region* (the surface bounded by the dashed line in Fig. 1.1) does not extend beyond T_c. Correspondingly, the *vapor pressure curve* $P_0(T)$, defined as the projection of the coexistence surface onto the (T, P) plane, terminates at the *critical point* (Fig. 1.2). While the vapor pressure curve marks a first-order transition line, at T_c the latent heat vanishes (see Problem 1.1), and the gas–liquid phase transition becomes of *second order*. This very special critical point in the phase diagram is defined by the conditions $(\partial P/\partial v)_{T_c, v_c} = 0 = (\partial^2 P/\partial v^2)_{T_c, v_c}$, and $P_c = P(T_c, v_c)$. Notice that the first equation implies the divergence of κ_T in the thermodynamic limit $N, V \to \infty$ (with n held fixed). Inserting the van-der-Waals equation of state, one readily identifies

$$v_c = 3b , \quad k_B T_c = \frac{8a}{27b} , \quad P_c = \frac{a}{27b^2} , \qquad (1.4)$$

and hence $P_c v_c / k_B T_c = 3/8$ universally for *all* fluids. In fact, upon rescaling to quantities measured relative to their critical values, $\varphi = (v - v_c)/v_c$, $\tau = (T - T_c)/T_c$, and $p = (P - P_c)/P_c$, the microscopic parameters a and b disappear from the equation of state (1.1) entirely:

$$1 + p = \frac{4(1 + \tau)}{1 + \frac{3}{2}\varphi} - \frac{3}{(1 + \varphi)^2} . \qquad (1.5)$$

According to this *law of corresponding states*, when expressed in terms of φ, τ, and p, the equations of states for *all* weakly interacting fluids should, at least approximately, be described by the same *universal* relation.

We may now derive the properties in the vicinity of the critical point by expanding Eq. (1.5) for small φ, retaining only the lowest-order non-vanishing terms. This yields

$$p \approx 4\tau - 6\tau\varphi - \frac{3}{2}\varphi^3 , \qquad (1.6)$$

wherefrom we immediately infer the cubic *critical isotherm* ($\tau = 0$),

$$-p \approx \frac{3}{2}\varphi^3 . \qquad (1.7)$$

For the *vapor pressure curve*, because of the antisymmetry near v_c, we may simply put $\varphi = 0$, and obtain $p_0 \approx 4\tau$ ($\tau < 0$). This sets the reference point for determining the spontaneous specific volume change at the phase transition. The *coexistence curve* is defined as the projection onto the (v, P) plane; hence $-6\tau\varphi = \frac{3}{2}\varphi^3$, or

$$\varphi_g = -\varphi_l \approx (-4\tau)^{1/2} \quad (\tau < 0) \qquad (1.8)$$

on the gas and liquid side, respectively (compare the parabolic form of the dashed curve in Fig. 1.1). The specific volume difference $\varphi_g - \varphi_l = 2\varphi_g$, setting on *continuously* at $T = T_c$, may serve as the phase separation *order parameter*. We proceed to compute the isothermal compressibility,

$$\kappa_T P_c = -\left(\frac{\partial\varphi}{\partial p}\right)_\tau \approx \frac{1}{6\tau + \frac{9}{2}\varphi^2} \approx \begin{cases} 1/6\tau , & \tau > 0, \\ 1/12|\tau| , & \tau < 0. \end{cases} \qquad (1.9)$$

Hence, $\kappa_T = \kappa_\pm |\tau|^{-1}$ diverges on both sides of the critical point, but with different amplitudes, $\kappa_+/\kappa_- = 2$. According to Eq. (1.2), this implies very large particle number or density fluctuations. These cause strong light scattering near the critical point, a phenomenon known as *critical opalescence* (see Section 1.1.3). Finally, employing the caloric equation of state, one finds that the specific heat C_v displays a discontinuity $\Delta C_v = \frac{9}{2} N k_B$ at T_c (Problem 1.1).

1.1.2 Mean-field theory for the ferromagnetic Ising model

In the theory of magnetism, and for the understanding of phase transitions and critical phenomena, the *(Lenz–)Ising model* has played a pivotal role. Its degrees of freedom are N discrete spin variables $\sigma_i = \pm 1$ on d-dimensional lattice sites x_i. Subject to an external magnetic field h (in units of energy), the energy of a given configuration of $\{\sigma_i\}$ is given by the Ising Hamiltonian

$$H(\{\sigma_i\}) = -\frac{1}{2} \sum_{i,j=1}^{N} J_{ij}\sigma_i\sigma_j - h \sum_{i=1}^{N} \sigma_i , \qquad (1.10)$$

with *exchange couplings* J_{ij}. Notice that in the absence of h, this Hamiltonian is symmetric with respect to sign inversion $\sigma_i \to -\sigma_i \; \forall i$. Henceforth, we shall assume *ferromagnetic* interactions, favoring parallel spin alignment, and require translational invariance, $J_{ij} = J(x_i - x_j) \geq 0$. We may then perform a discrete Fourier transform[2]

$$J(q) = \sum_i J(x_i) e^{-iq \cdot x_i} . \tag{1.11}$$

In equilibrium statistical mechanics, the task now is to evaluate the *canonical partition sum* over all possible spin configurations,

$$Z(T, h, N) = \sum_{\{\sigma_i = \pm 1\}} e^{-H(\{\sigma_i\})/k_B T} . \tag{1.12}$$

Thermodynamic properties are then given as averages of appropriate functions of the binary spin variables σ_i,

$$\langle A(\{\sigma_i\}) \rangle = \frac{1}{Z} \sum_{\{\sigma_i = \pm 1\}} A(\{\sigma_i\}) e^{-H(\{\sigma_i\})/k_B T} , \tag{1.13}$$

and may often be obtained via appropriate partial derivatives with respect to temperature T or field h. In one dimension, this program can be easily carried through explicitly, see Problem 1.2. Even in two dimensions, the Ising model may be solved exactly, albeit with considerably greater effort. In higher dimensions, it has, however, eluded any such attempts, and one must resort to approximations.

In Curie–Weiss mean-field theory, essentially the effective 'local' field

$$h_{\text{eff},i} = -\frac{\partial H}{\partial \sigma_i} = h + \sum_j J_{ij} \sigma_j \tag{1.14}$$

is replaced with its average $\langle h_{\text{eff}} \rangle = h + \tilde{J}m$, where $\tilde{J} = J(q = 0) = \sum_i J(x_i)$, and $m = \langle \sigma_i \rangle = M/N$ denotes the magnetization per spin. We would clearly expect this approximation to work best when the exchange interactions are long-range, and their effect on any site roughly uniform; or in high dimensions, when the interactions with many neighboring sites average out local fluctuations. The Ising Hamiltonian thus becomes that of a simple two-state paramagnet, subject to the combined external and internal field $\langle h_{\text{eff}} \rangle$. The magnetization m can then be determined self-consistently. More precisely, we decompose the local spin into its average and fluctuation, $\sigma_i = m + (\sigma_i - \langle \sigma_i \rangle)$, whence $\sigma_i \sigma_j = m^2 + m(\sigma_i - \langle \sigma_i \rangle + \sigma_j - \langle \sigma_j \rangle) + (\sigma_i - \langle \sigma_i \rangle)(\sigma_j - \langle \sigma_j \rangle)$. Upon

---

[2] Throughout this book, vector quantities will not be specifically indicated through arrows or boldface typing. Neither will distinct symbols be used for the Fourier transforms of scalar or vector fields; if required to avoid confusion, rather the arguments will be noted explicit.

neglecting the last contribution, whose average would yield the spin correlations C_{ij}, the ensuing mean-field Hamiltonian decouples into N independent terms *linear* in the spin variables,

$$H_{\mathrm{mf}}(\{\sigma_i\}) = N \frac{m^2}{2} \tilde{J} - (h + \tilde{J}m) \sum_{i=1}^{N} \sigma_i \, . \tag{1.15}$$

The partition function is then readily evaluated,

$$Z_{\mathrm{mf}}(T, h, N) = \mathrm{e}^{-Nm^2 \tilde{J}/2k_{\mathrm{B}}T} \left(2 \cosh \frac{h + \tilde{J}m}{k_{\mathrm{B}}T} \right)^N , \tag{1.16}$$

wherefrom we obtain the (Helmholtz) *free energy*

$$F_{\mathrm{mf}}(T, h, N) = -k_{\mathrm{B}}T \ln Z_{\mathrm{mf}}(T, h, N) \, , \tag{1.17}$$

and thence the *Curie–Weiss equation of state*

$$m(T, h) = -\frac{1}{N} \left(\frac{\partial F_{\mathrm{mf}}}{\partial h} \right)_{T,N} = \tanh \frac{h + \tilde{J}m(T, h)}{k_{\mathrm{B}}T} \, . \tag{1.18}$$

We see that any non-vanishing field induces a non-zero magnetization. It always reaches its maximum value $|m| = 1$ at $T = 0$, where all spins are aligned, while for $T > 0$ entropy favors at least some random spin orientations. At zero field $h = 0$, the self-consistent equation for the *spontaneous magnetization* becomes $m = \tanh(\tilde{J}m/k_{\mathrm{B}}T)$. Aside from the trivial solution $m = 0$, describing the *paramagnetic phase*, because the function $\tanh x$ is concave and saturates at 1 as $x \to \infty$, the existence of any non-zero solution m requires the slope of the hyperbolic tangent for small arguments to be larger than 1, i.e.: $\tilde{J} > k_{\mathrm{B}}T$. Within the mean-field approximation, we therefore identify the *critical temperature*

$$k_{\mathrm{B}}T_{\mathrm{c}} = \tilde{J} = \sum_i J(x_i) \, . \tag{1.19}$$

For example, for a hypercubic lattice in d dimensions with only nearest-neighbor uniform couplings J, $k_{\mathrm{B}}T_{\mathrm{c}} = 2dJ$. For $T < T_{\mathrm{c}}$, independent of the lattice dimension, mean-field theory predicts a *ferromagnetic phase* with $m \neq 0$. Nature then has to choose between the two degenerate values for the ferromagnetic order parameter $\pm|m|$, and the original Ising Z_2 symmetry is *spontaneously broken*. The corresponding *coexistence lines* $m(T)$ are sketched in Fig. 1.3(a), as well as the equation of state for $h \neq 0$. As the external magnetic field is ramped up or down for $T > T_{\mathrm{c}}$, the magnetization follows continuously, as is typical for paramagnetic behavior. However, for constant $T < T_{\mathrm{c}}$, the induced magnetization jumps between the values $\pm|m(T)|$ upon varying h, indicating a *first-order* phase transition. As

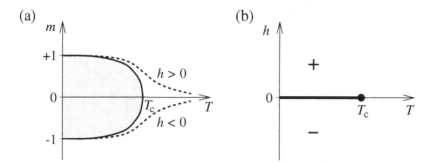

Fig. 1.3 Phase diagrams for the ferromagnetic Ising model in mean-field approximation: (a) magnetization $m(T, h)$ for both $h = 0$ (coexistence lines, full) and $h \neq 0$ (dashed); (b) phase diagram in the (T, h) plane. The first-order transition line ends at the critical point $(T_c, 0)$.

shown in the (T, h) phase diagram Fig. 1.3(b), the phase coexistence region at $h = 0$ terminates at the critical temperature T_c.

In order to obtain the critical properties near T_c and $h_c = 0$, we expand the equation of state (1.18) for $|\tau| = |T - T_c|/T_c \ll 1$, $h \ll \tilde{J}$, and hence $|m| \ll 1$: $m \approx h/k_B T + m T_c/T - \frac{1}{3}(m T_c/T)^3$, which leads to

$$\frac{h}{k_B T_c} \approx \tau m + \frac{m^3}{3} . \tag{1.20}$$

The *critical isotherm* is thus given by

$$h \approx \frac{k_B T_c}{3} m^3 , \tag{1.21}$$

compare with Eq. (1.7) for the van-der-Waals gas, and the *coexistence curve* below T_c becomes

$$m \approx \pm(-3\tau)^{1/2} \quad (\tau < 0) , \tag{1.22}$$

similar to Eq. (1.8), while the spontaneous magnetization $m(h = 0)$ vanishes for $T > T_c$. To find the *isothermal magnetic susceptibility* $\chi_T = N(\partial m/\partial h)_T$, we take the derivative of Eq. (1.20) with respect to the external field, which yields $1/k_B T_c \approx (\tau + m^2)\chi_T/N$, or

$$\chi_T \approx \frac{N}{k_B T_c} \frac{1}{\tau + m^2} \approx \frac{N}{k_B T_c} \begin{cases} 1/\tau, & \tau > 0, \\ 1/2|\tau|, & \tau < 0; \end{cases} \tag{1.23}$$

i.e., $\chi_T = \chi_\pm |\tau|^{-1}$ diverges from both sides as the critical point is approached, as depicted in Fig. 1.4(a), with the characteristic amplitude ratio $\chi_+/\chi_- = 2$,

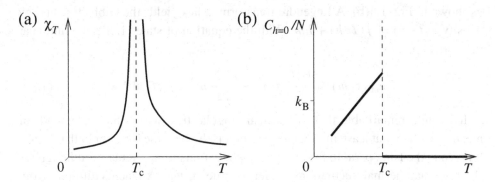

Fig. 1.4 Critical singularities for the Ising model in mean-field approximation: (a) divergence of the isothermal magnetic susceptibility χ_T at T_c; and (b) discontinuity of the specific heat $C_{h=0}$.

again as in Eq. (1.9). Because of the equilibrium relation with the magnetization fluctuations,

$$\chi_T = \frac{(\Delta M)^2}{k_B T} , \qquad (1.24)$$

the latter must be expected to contribute strongly to the magnet's thermodynamics in the vicinity of T_c, which renders the mean-field approximation rather questionable in the critical region.

With $\cosh x = (1 - \tanh^2 x)^{-1/2}$, and using Eqs. (1.18) and (1.19), the *free energy* per spin $f = F/N$ may be written as

$$f_{\mathrm{mf}}(T, h) = \frac{k_B T_c}{2} m^2 + \frac{k_B T}{2} \ln(1 - m^2) - k_B T \ln 2 . \qquad (1.25)$$

For $h = 0$ and $\tau > 0$, $f_{\mathrm{mf}}(T, 0) \approx -k_B T \ln 2$, the usual high-temperature limit for a paramagnet. In the critical region, we may expand for $|m| \ll 1$:

$$f_{\mathrm{mf}}(\tau, 0) \approx -\frac{k_B T_c}{2} \tau m^2 - \frac{k_B T_c}{4} m^4 - k_B T_c \ln 2 . \qquad (1.26)$$

On the paramagnetic side, $f_{\mathrm{mf}}(\tau, 0) \approx -k_B T_c \ln 2$. In the ferromagnetic phase, on the other hand, we need to insert (1.22), whereupon

$$f_{\mathrm{mf}}(\tau, 0) \approx -\frac{3k_B T_c}{4} \tau^2 - k_B T_c \ln 2 . \qquad (1.27)$$

This result also establishes the stability of the solution (1.22) for $\tau < 0$ as compared to $m = 0$. Moreover, we may now compute the *specific heat discontinuity* at T_c from $C_{h=0}/N = -T(\partial^2 f/\partial T^2)_{h=0}$, which yields

$$\frac{C_{h=0}}{N} \approx \begin{cases} 0, & \tau > 0, \\ 3k_B T/2T_c, & \tau < 0, \end{cases} \qquad (1.28)$$

as shown in Fig. 1.4(b). A Legendre transform at last yields the Gibbs free energy density $g(T, m) = f(T, h) + hm$. With the equation of state (1.20), we find near the critical point

$$g_{mf}(\tau, m) \approx \frac{k_B T_c}{2} \tau m^2 + \frac{k_B T_c}{12} m^4 - k_B T_c \ln 2 . \tag{1.29}$$

It is quite remarkable that the critical singularities, within the framework of mean-field theory at least, turn out to be of precisely the same form for both the gas–liquid and the Ising para- to ferromagnetic phase transitions. Actually, although we shall see that thermal fluctuations render the true critical exponents different from the numbers obtained in the mean-field approximation, they remain identical for all systems with a scalar order parameter subject to short-range interaction potentials – a phenomenon termed *universality* – despite the fact that the microscopic forces driving the phase transition itself can be vastly different. Indeed, beyond the mean-field approximation, phase separation in binary fluids can be captured through the *Ising lattice gas* model, which follows from the Hamiltonian (1.10) via the substitution

$$\sigma_i = 2n_i - 1 , \quad n_i = \frac{1 + \sigma_i}{2} = \begin{cases} 0, & \sigma_i = -1, \\ 1, & \sigma_i = +1. \end{cases} \tag{1.30}$$

The binary variables $n_i = 0, 1$ may be interpreted as site occupation numbers. A local dominance of empty/filled sites respectively corresponds to the dilute/dense phase. The ferromagnetic interaction then maps to an attractive force between either two 'particles' or two 'holes', whereas particles and holes repel each other. Upon imposing the global constraint $2 \sum_i n_i = N + M = $ const., phase separation of dense and dilute regions ensues for $T < T_c \approx \tilde{J}/k_B$.

1.1.3 Ornstein–Zernicke correlation function

Concluding this section on mean-field theory, we compute the order parameter correlation function near the critical point. We shall use the language of the Ising ferromagnet. Yet it should be clear from the preceding comments that the results apply to the gas–liquid phase transition as well. It is convenient to formally introduce a local field h_i, thus replacing the paramagnetic term in Eq. (1.10) with $-\sum_i h_i \sigma_i$. The partition function associated with this modified Hamiltonian then serves as *generating function* for local moments and correlations; e.g.,

$$m_i = \langle \sigma_i \rangle = k_B T \frac{\partial \ln Z}{\partial h_i} = -\frac{\partial F}{\partial h_i} , \tag{1.31}$$

which now depends on the site index i.

Next we define the *local* response function

$$\chi_{ij} = \frac{\partial m_i}{\partial h_j} = k_B T \frac{\partial^2 \ln Z}{\partial h_i \partial h_j} = -\frac{\partial^2 F}{\partial h_i \partial h_j} = \chi_{ji} \,. \tag{1.32}$$

Inserting the definition of $\langle \sigma_i \rangle$ in the canonical ensemble (1.13), a straightforward calculation yields the identity

$$\chi_{ij} = \frac{1}{k_B T} \left(\langle \sigma_i \sigma_j \rangle - \langle \sigma_i \rangle \langle \sigma_j \rangle \right) = \frac{1}{k_B T} C_{ij} \,, \tag{1.33}$$

where $C_{ij} = \langle (\sigma_i - \langle \sigma_i \rangle)(\sigma_j - \langle \sigma_j \rangle) \rangle = \langle \sigma_i \sigma_j \rangle - \langle \sigma_i \rangle \langle \sigma_j \rangle$ represents the (two-point) spin *correlation function*. The *fluctuation-response* theorem (1.33) establishes that the local response is intimately related to spatial correlations. After performing the partial derivatives with respect to the h_i, we may let $h_i \to h$ again, whereupon translational invariance is restored. Then both χ_{ij} and C_{ij} become functions of $x_i - x_j$ only, and we find for their Fourier transforms, cf. Eq. (1.11),

$$\chi(q) = \frac{1}{k_B T} C(q) \,. \tag{1.34}$$

The thermodynamic susceptibility follows by taking $q \to 0$,

$$\chi_T = \sum_{i,j=1}^{N} \chi_{ij} = N \chi(q = 0) = \frac{N}{k_B T} \sum_i C(x_i) \to \frac{N}{k_B T v_d} \int d^d x \, C(x) \tag{1.35}$$

in the *continuum limit* (where v_d denotes the volume of the d-dimensional unit cell). This is of course just another form of Eq. (1.24).

For the Ising model in mean-field approximation, m_i is given by the equation of state (1.18) with $h \to h_i$ and $\tilde{J}m \to \sum_k J_{ik} m_k$. Upon taking the derivative with respect to h_j, we arrive at an implicit equation for the local susceptibility in the limit $h_i \to 0$ (and thus $m_i \to m$),

$$\chi_{ij} = \frac{\delta_{ij} + \sum_k J_{ik} \chi_{kj}}{k_B T \left[\cosh(\tilde{J}m/k_B T) \right]^2} = \frac{1 - m^2}{k_B T} \left(\delta_{ij} + \sum_k J_{ik} \chi_{kj} \right) \,.$$

Here we have applied the equation of state again. After Fourier transformation and applying the convolution theorem, this is solved by

$$\chi(q)^{-1} = \frac{k_B T}{1 - m^2} - J(q) \,. \tag{1.36}$$

We next employ a long-wavelength expansion,

$$J(q) \approx \sum_i J(x_i) \left[1 - iq \cdot x_i - \frac{(q \cdot x_i)^2}{2} + \cdots \right] \approx \tilde{J} - J'q^2 \,,$$

where we have assumed lattice inversion symmetry and isotropy, and introduced $J' = \sum_i x_i^2 J(x_i)/2d$. In the vicinity of $T_c = \tilde{J}/k_B$ and for $h = 0$, $|m| \ll 1$, and the correlation function takes the *Ornstein–Zernicke* form

$$C(q) \approx \frac{k_B T_c}{k_B(T - T_c) + k_B T_c m^2 + J' q^2} = \frac{k_B T_c}{J'} \frac{1}{\xi^{-2} + q^2}. \qquad (1.37)$$

The characteristic length scale ξ is called the *correlation length*. In the high- and low-temperature phases, respectively,

$$\xi \approx \left(\frac{J'}{k_B T_c}\right)^{1/2} \begin{cases} 1/\tau^{1/2}, & \tau > 0, \\ 1/|2\tau|^{1/2}, & \tau < 0. \end{cases} \qquad (1.38)$$

At the critical point, extremely long-range correlations develop, as indicated by the divergence of $\xi = \xi_\pm |\tau|^{-1/2}$ on both sides of the phase transition, with the ampli- tude ratio $\xi_+/\xi_- = \sqrt{2}$. Then $C(q) \approx k_B T_c/J' q^2$ becomes very large for small wavevectors q. The correlation function $C(q)$ is measured directly in scattering experiments, where $q = q_f - q_i$ denotes the change in wavevector of the scattered particles. Near T_c, Eq. (1.38) therefore predicts strong forward scattering $\sim q^{-2}$, a phenomenon known as *critical opalescence* in light scattering at the gas–liquid critical point. For $T \neq T_c$, the correlation length is finite, and the spatial correlations decay exponentially, $C(x) \sim |x|^{2-d} e^{-|x|/\xi}$ for $d > 2$; e.g., in three dimensions

$$C(x) \approx \frac{k_B T_c}{J'} \frac{e^{-|x|/\xi}}{4\pi |x|}. \qquad (1.39)$$

The integral in Eq. (1.35) then roughly contributes a factor $\sim \xi^d$, and χ_T/N is finite in the thermodynamic limit. However, at T_c the exponential decay of $C(x)$ is replaced by a power law $C(x) \sim 1/|x|^{d-2}$ in $d > 2$ dimensions. These correlations of infinite range render the susceptibility integral quadratically divergent $\sim \xi^2$ as $\xi \to \infty$, or $\chi_T \sim |\tau|^{-1}$ in accord with our earlier result (1.23). We may thus view the divergent length scale ξ near a second-order phase transition as the origin of the emerging thermodynamic singularities.

1.2 Landau–Ginzburg theory

Landau's *phenomenological* approach to the statistical mechanics of phase tran- sitions sets a general framework for mean-field theories. It is based only on the underlying *symmetries* of the physical problem, and the assumption that the free energy density $f = F/V$ is an analytic function of the order parameter ϕ. There- fore, near a phase transition where $|\phi| \ll 1$, it is asserted that $f(\phi)$ may be expanded into a Taylor series. Landau–Ginzburg theory generalizes these ideas to allow for spatially inhomogeneous configurations. It turns out, though, that near a critical

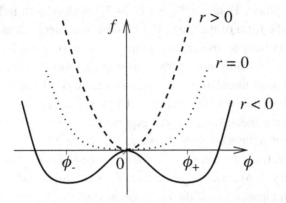

Fig. 1.5 Landau free energy density $f(\phi)$ at zero external field ($h = 0$) describing a second-order phase transition at $r = 0$ (dotted); $r > 0$ corresponds to $T > T_c$ (dashed), while $r < 0$ describes the low-temperature phase with two degenerate minima at ϕ_+ and $\phi_- = -\phi_+$ (full line).

point the free energy cannot simply be expanded into a regular power series. Nevertheless, we shall see that the Landau–Ginzburg–Wilson Hamiltonian serves as an apt starting point for the more general and powerful renormalization group approach.

1.2.1 Mean-field theory revisited: Landau expansion

Let us write down an expansion of the free energy density f for a system with discrete Ising symmetry Z_2. In the absence of the thermodynamically conjugate field h, $f(\phi)$ has to be invariant under sign change of the order parameter $\phi \to -\phi$, compare Eq. (1.29). Thus, apart from a constant,

$$f(\phi) = \frac{r}{2}\,\phi^2 + \frac{u}{4!}\,\phi^4 + \frac{v}{6!}\,\phi^6 + \cdots . \tag{1.40}$$

For $u > 0$, and $|\phi| \ll 1$, the sixth-order term may be dropped, as it merely leads to small corrections to the leading scaling behavior near the phase transition (see Problem 1.3). To lowest order in the symmetry-breaking field h therefore

$$f(\phi) = \frac{r}{2}\,\phi^2 + \frac{u}{4!}\,\phi^4 - h\,\phi , \tag{1.41}$$

which is depicted for $h = 0$ and different values of r in Fig. 1.5. For $r > 0$, the free energy as a function of the order parameter is essentially parabolic, with a single minimum at $\phi = 0$, and $f(0) = 0$. On the other hand, if $r < 0$, $f(\phi)$ is concave near the origin, and the quartic term $\propto u$ is required to stabilize the system for larger values of the order parameter. Thus, degenerate minima at ϕ_+ and $\phi_- = -\phi_+$ appear, which we naturally associate with the two states in the symmetry-broken low-temperature phase. Hence $r(T) = a(T - T_c)$ with $a > 0$; indeed, Eq. (1.41) then reproduces the form of the mean-field Gibbs free energy density for the Ising model (1.29). Precisely at the critical point $r = h = 0$, $f(\phi)$ becomes a purely quartic function. Its flat minimum with vanishing curvature at the origin signals the strong order parameter fluctuations associated with a *second-order* phase transition. As becomes obvious through straightforward dimensional analysis, Eq. (1.41) must once again yield the mean-field critical exponents: $\phi^2 \propto -r/u \sim T_c - T$ (u may be taken to be constant near T_c); furthermore $h \sim \phi^3$ at $T = T_c$.

More precisely, the equilibrium condition for the order parameter ϕ is that the free energy density $f(\phi)$ acquire a *minimum*. Thus, we first need f to be stationary, $f'(\phi) = 0$, which already yields the *equation of state*

$$h(T, \phi) = r(T)\phi + \frac{u}{6}\phi^3 \,. \tag{1.42}$$

Second, we must require $0 < f''(\phi) = r + \frac{u}{2}\phi^2$ for any ground-state candidate ϕ. If there happen to be more than one such local minima, then the one with lowest free energy will describe the equilibrium state. Notice that the inverse isothermal order parameter *susceptibility* follows from Eq. (1.42) via

$$V\chi_T^{-1} = \left(\frac{\partial h}{\partial \phi}\right)_T = r + \frac{u}{2}\phi^2 \,; \tag{1.43}$$

thermodynamic stability therefore implies $\chi_T > 0$. Setting $r = 0$ ($T = T_c$) in the equation of state, we obtain the *critical isotherm*

$$h(T_c, \phi) = \frac{u}{6}\phi^3 \,. \tag{1.44}$$

For $h = 0$, Eq. (1.42) determines the *spontaneous order parameter*. As $u > 0$, for $r > 0$ the only solution is $\phi = 0$, which becomes unstable at T_c. For $r < 0$, on the other hand, there are two additional real solutions

$$\phi_\pm = \pm\left(\frac{6|r|}{u}\right)^{1/2} \,. \tag{1.45}$$

At $T = T_c$, the function $\phi(T)$ is continuous, but non-differentiable with the typical mean-field square-root singularity, as in Fig. 1.3(a). The isothermal susceptibility (1.43) at $h = 0$ becomes

$$\frac{\chi_T}{V} = \begin{cases} 1/r, & r > 0, \\ 1/2|r|, & r < 0. \end{cases} \tag{1.46}$$

Upon approaching the transition from both above and below T_c, the response function diverges, $\chi_T \sim |T - T_c|^{-1}$, with the same amplitude ratio as in the van-der-Waals and Curie–Weiss mean-field theories, compare with Fig. 1.4(a). For

$T < T_c$, we find the free energy

$$f(\phi_\pm) = \frac{r}{4} \phi_\pm^2 = -\frac{3r^2}{2u} < 0 , \tag{1.47}$$

while $f(0) = 0$ and hence $C_{h=0} = -VT(\partial^2 f/\partial T^2)_{h=0} = 0$ in the high-temperature phase. Below T_c, the *specific heat* becomes

$$C_{h=0} = VT \frac{3a^2}{u} > 0 , \tag{1.48}$$

displaying a discontinuity at T_c, as depicted in Fig. 1.4(b).

The free energy density (1.41) with positive quartic term thus generically describes physical properties near a second-order phase transition occurring when the quadratic coefficient r vanishes. If $u < 0$, however, a positive sixth-order term $\propto v$ is required for thermodynamic stability. The Landau model (1.40) then yields a *first-order* phase transition for $h = 0$ with an order parameter discontinuity at $r_d = 5u^2/8v$. For a non-zero field h, two additional critical lines appear in parameter space, which merge with the previous second-order line ($u > 0$, $h = 0$) at the *tricritical point* $r = h = u = 0$. For $u = 0$, simple scaling yields that $\phi^4 \propto -r/v$ and $h \sim \phi^5$: The tricritical point singularities are characterized by different exponents from those near ordinary critical lines. (For more details, see Problem 1.3.)

1.2.2 Fluctuations in Gaussian approximation

The divergence of the order parameter response (e.g., the isothermal compressibility for a fluid or the magnetic susceptibility for a ferromagnet) at a critical point indicates the appearance of strong fluctuations. We must therefore expect simple mean-field theory, which neglects such fluctuations entirely, to possibly produce wrong results near a continuous phase transition. Our goal thus must be to include the order parameter fluctuations and spatial correlations appropriately in the statistical mechanics of a critical system. The Landau expansion (1.41) for the free energy density may guide us in the construction of an effective Hamiltonian for the critical degrees of freedom. To this end, we replace the order parameter ϕ with a space-dependent field $\phi(x)$, which represents a coarse-grained *order parameter density*. In both the ordered ($\langle \phi(x) \rangle \neq 0$) and disordered ($\langle \phi(x) \rangle = 0$) phases, spatial variations should be energetically suppressed. In a gradient or long-wavelength expansion, it should therefore be sufficient to retain only the first term compatible with the presumed spatial inversion symmetry of the system, namely $c(\nabla\phi)^2$, with $c > 0$. The positive constant c can then be absorbed into a rescaled order parameter field $S(x) = \sqrt{c}\,\phi(x)$, which upon neglecting higher-order terms leads us to the

Landau–Ginzburg–Wilson Hamiltonian

$$\mathcal{H}[S] = \int d^d x \left[\frac{r}{2} S(x)^2 + \frac{1}{2} [\nabla S(x)]^2 + \frac{u}{4!} S(x)^4 - h(x)S(x) \right]. \quad (1.49)$$

This energy functional reflects the underlying Z_2 Ising symmetry, i.e., invariance with respect to sign inversion $S(x) \to -S(x)$.

In order to obtain the associated canonical partition function, we must sum over all possible configurations $S(x)$, weighted with the appropriate Boltzmann factor,

$$Z[h] = \int \mathcal{D}[S] e^{-\mathcal{H}[S]/k_B T}. \quad (1.50)$$

As explained in more detail in the appendix (Section 1.5), the functional integral measure here is defined through a discretization $x \to x_i$, and then integrating over all values of the order parameter field $S(x_i)$ at each lattice site x_i, $\mathcal{D}[S] = \prod_i dS(x_i)$. In the end, the continuum limit is to be taken again. The probability of finding a configuration $S(x)$ realized is then given by

$$\mathcal{P}_{st}[S] = \frac{1}{Z[h]} e^{-\mathcal{H}[S]/k_B T}. \quad (1.51)$$

In general, evaluating the functional integral (1.50) is a very difficult task, and one must resort to approximations. Outside the critical regime, the exponential probability distribution (1.51) is usually sharply peaked, and the thermodynamics of the system is dominated by the most likely configuration. The latter is given as a minimum of the Hamiltonian, i.e., through setting the variation of $\mathcal{H}[S]$ with respect to the order parameter field $S(x)$ to zero (stationarity condition). This functional derivative (Section 1.5) then yields the *Landau–Ginzburg* (or *classical field*) *equation*

$$0 = \frac{\delta \mathcal{H}[S]}{\delta S(x)} = \left[r - \nabla^2 + \frac{u}{6} S(x)^2 \right] S(x) - h(x). \quad (1.52)$$

For homogeneous solutions $S(x) = \text{const.} = \phi$, one recovers the equation of state (1.42), and therefore $r = a(T - T_c^0)$, where T_c^0 denotes the critical temperature in mean-field approximation. In the low-temperature phase, the solution to Eq. (1.52) for $h = 0$ with the boundary conditions $S_K(x) \to \phi_\pm$ as $|x_1| \to \infty$ in one space direction describes an Ising *domain wall* (kink), extending over the range of the correlation length (Problem 1.4). Linearizing about the constant stationary configurations for $h = 0$, i.e., writing $S(x) = \phi + \delta S(x)$ and omitting all quadratic and higher-order terms in δS, one finds

$$\delta h(x) \approx \left(r - \nabla^2 + \frac{u}{2} \phi^2 \right) \delta S(x). \quad (1.53)$$

After Fourier transformation, this yields the order parameter *response function* in the linear approximation,

$$\chi_0(q) = \frac{1}{r + \frac{u}{2}\phi^2 + q^2} = \frac{1}{\xi^{-2} + q^2}, \tag{1.54}$$

where we have defined the *correlation lengths*

$$\xi = \begin{cases} 1/r^{1/2}, & r > 0, \\ 1/|2r|^{1/2}, & r < 0, \end{cases} \tag{1.55}$$

in the high- and low-temperature phases, compare with Eqs. (1.37) and (1.38).

In order to see under which circumstances fluctuations may alter the mean-field critical exponents, we evaluate the partition function (1.50) in the *Gaussian approximation*. In the high-temperature phase, we simply neglect the non-linearity $\propto u$. The resulting Hamiltonian $\mathcal{H}_0$ is bilinear in the real integration variables $S(x)$, and therefore readily diagonalized through a Fourier transform (in d dimensions),

$$S(x) = \frac{1}{V} \sum_q S(q) \, e^{iq \cdot x} \rightarrow \int \frac{d^d q}{(2\pi)^d} S(q) \, e^{iq \cdot x}, \tag{1.56}$$

where the latter integral applies in the continuum limit, and $S(-q) = S^*(q)$. Thus, the functional integral measure becomes explicitly

$$\mathcal{D}[S] = \prod_q \frac{dS(q)}{V} = \prod_{q,q_1>0} \frac{d\,\mathrm{Re}\,S(q)\,d\,\mathrm{Im}\,S(q)}{V}, \tag{1.57}$$

where we have indicated that the product in the last expression comprises only the positive Cartesian wavevector components $q_1 > 0$ (with the q_1 direction chosen arbitrarily). Then

$$\mathcal{H}_0[S] = \frac{1}{V} \sum_q \left[\frac{1}{2} \left(r + q^2\right) |S(q)|^2 - h(q)S(-q) \right], \tag{1.58}$$

and upon applying the linear transformation $\widetilde{S}(q) = S(q) - h(q)/(r + q^2)$, we obtain

$$Z_0[h] = \int \mathcal{D}[S] \, e^{-\mathcal{H}_0[S]/k_B T}$$

$$= \exp\left(\frac{1}{2k_B T V} \sum_q \frac{|h(q)|^2}{r + q^2} \right) \int \mathcal{D}[\widetilde{S}] \exp\left(-\sum_q \frac{r + q^2}{2k_B T V} |\widetilde{S}(q)|^2 \right).$$

$$\tag{1.59}$$

First of all, we thereby recover our previous result (1.54) for the zero-field two-point correlation function $C_0(q) = k_B T \, \chi_0(q)$,

$$\langle S(q)S(q')\rangle_0 = \frac{(k_B T V)^2}{Z_0[h]} \frac{\partial^2 Z_0[h]}{\partial h(-q)\partial h(-q')}\Big|_{h=0} = \frac{k_B T}{r+q^2} V \delta_{q,-q'}$$

$$\rightarrow \frac{(k_B T)^2}{Z_0[h]} \frac{(2\pi)^{2d}}{\delta h(-q)\delta h(-q')}\frac{\delta^2 Z_0[h]}{\delta h(-q)\delta h(-q')}\Big|_{h=0} = C_0(q)(2\pi)^d \delta(q+q') . \quad (1.60)$$

The origin of the delta function in the continuum limit of the second line here is of course the assumed translational invariance of the system. Next, after performing the Gaussian integrals with the measure (1.57), Eq. (1.59) yields the free energy

$$F_0[h] = -k_B T \ln Z_0[h] = -\frac{1}{2V} \sum_q \left(\frac{|h(q)|^2}{r+q^2} + k_B T V \ln \frac{2\pi \, k_B T}{r+q^2} \right) . \quad (1.61)$$

At zero external field, we obtain the *leading singularity* for the specific heat in the continuum limit,

$$C_{h=0} = -T \left(\frac{\partial^2 F_0}{\partial T^2} \right)_{h=0} \approx \frac{V k_B \, (a T_c^0)^2}{2} \int \frac{d^d q}{(2\pi)^d} \frac{1}{(r+q^2)^2} , \quad (1.62)$$

omitting less singular contributions. With $k = q/\sqrt{r} = q\xi$, and the surface area of the d-dimensional unit sphere $K_d = 2\pi^{d/2}/\Gamma(d/2)$, this becomes

$$C_{\text{sing}} \approx \frac{V k_B (a T_c^0)^2 \xi^{4-d}}{2^d \pi^{d/2} \Gamma(d/2)} \int_0^\infty \frac{k^{d-1}}{(1+k^2)^2} \, dk . \quad (1.63)$$

We must now distinguish different regimes, depending on the space dimension d. In high dimensions $d > 4$, the integral in (1.63) diverges as $k \to \infty$. This singularity can be regularized by means of an ultraviolet (UV) cutoff Λ, which in a crystal is naturally provided by the Brillouin zone boundary:

$$\int_0^{\Lambda\xi} \left[\frac{k^{d-1}}{(1+k^2)^2} - k^{d-5} + k^{d-5} \right] dk = -\int_0^{\Lambda\xi} \frac{k^{d-5} + 2k^{d-3}}{(1+k^2)^2} \, dk + \frac{(\Lambda\xi)^{d-4}}{d-4} .$$

The remaining integral is now finite, while in the second term the dependence on ξ cancels. In addition to the Landau discontinuity $\Delta C_{h=0} = 3V T_c^0 a^2/u$, the specific heat thus develops a non-differentiable cusp of the form $B - A(T - T_c^0)^{(d-4)/2}$. In four dimensions, which clearly represents a marginal case here,

$$\int_0^{\Lambda\xi} \frac{k^3}{(1+k^2)^2} \, dk \sim \ln(\Lambda\xi)$$

diverges logarithmically as either $\Lambda \to \infty$ or $T \to T_c^0$. Finally, for $d < 4$, the integral in (1.63) is finite in the ultraviolet, and the specific heat diverges upon approaching the critical point according to the power law

$$C_{\text{sing}} \sim \left|T - T_c^0\right|^{-\alpha}, \quad \alpha = \frac{4 - d}{2} \tag{1.64}$$

(for the behavior in the low-temperature phase, see Problem 1.5).

We have just found that above the *upper critical dimension* $d_c = 4$, Gaussian fluctuations at least do not modify the form of the leading mean-field singularities. However, for $d \le d_c = 4$, the fluctuation contributions in fact dominate over the mean-field singularities, and need to be properly accounted for. It is already obvious from the emerging infrared singularities in the integrals that a purely perturbational approach is bound to fail. Yet well outside the asymptotic critical region, the Landau–Ginzburg approximation is expected to provide at least qualitatively correct answers. More quantitatively, the *Ginzburg–Levanyuk criterion* states that fluctuation effects are essentially negligible as long as the specific heat discontinuity surpasses the Gaussian contributions. Introducing the quantity $\xi_0 = (aT_c^0)^{-1/2}$, according to this criterion, the mean-field results should be valid provided

$$\frac{\Delta C_{h=0}}{k_B} \frac{\xi_0^d}{V} \left|\tau\right|^{(4-d)/2} \gg 1 . \tag{1.65}$$

1.3 Scaling theory

Our analysis of the Gaussian model indicates that for dimensions $d < d_c = 4$, strong fluctuations in the vicinity of T_c will in general lead to values for the critical exponents that differ from those predicted by mean-field theory. Yet we may assert (and later confirm by means of the renormalization group) that the crucial general scaling features near a critical point remain valid. Thus the free energy and order parameter correlation function should become generalized homogeneous functions in the critical regime, albeit with the mean-field or Gaussian critical exponents replaced with different numbers. This hypothesis leads us to *scaling relations* between the critical exponents, and to the discovery that at a generic second-order phase transition, actually only *two* of these are independent. As Section 1.4 will reveal, their values are determined solely by global properties such as the system's dimension and underlying symmetries, as reflected in the number of order parameter components. These define the *universality class* for the critical scaling properties.

1.3.1 Scaling hypothesis: free energy, correlation function

The *scaling hypothesis* for the *singular* part of the free energy density near a continuous phase transition states

$$f_{\text{sing}}(\tau, h) = |\tau|^{2-\alpha} \, \hat{f}_{\pm}(h/|\tau|^{\Delta}) \, , \qquad (1.66)$$

where $\tau = (T - T_c)/T_c$, and h is the thermodynamic field conjugate to the order parameter. $\hat{f}_+$ and $\hat{f}_-$ denote two different *scaling functions* in the high- and low-temperature phase, respectively, that are both regular for small scaling variable arguments $x = h/|\tau|^{\Delta}$. With Eq. (1.45), the Landau free energy (1.41) clearly satisfies this relation with $\alpha = 0$ and $\Delta = 3/2$. As in the Gaussian approximation (1.64), the exponent α characterizes the divergence of the *specific heat* at the phase transition. For at vanishing external field $h = 0$, we have $f_{\text{sing}}(\tau, 0) = |\tau|^{2-\alpha} \, \hat{f}_{\pm}(0)$, and consequently

$$C_{h=0} = -\frac{VT}{T_c^2} \left(\frac{\partial^2 f_{\text{sing}}}{\partial \tau^2} \right)_{h=0} = C_{\pm} |\tau|^{-\alpha} \, , \qquad (1.67)$$

diverging with the same exponent α on both sides of the transition as $\tau \to 0$. Notice that this simple power law implies that as a function of temperature, the specific heat becomes *self-similar* near the critical point, i.e., it looks the same on different temperature scales. A striking confirmation for this remarkable property is depicted in Fig. 1.6, which shows experimental data for the specific heat of liquid helium 4 near its normal- to superfluid 'lambda' transition (actually named after the shape of these curves).

Next, we obtain the *equation of state*,

$$\phi(\tau, h) = -\left(\frac{\partial f_{\text{sing}}}{\partial h} \right)_{\tau} = -|\tau|^{2-\alpha-\Delta} \, \hat{f}'_{\pm}(h/|\tau|^{\Delta}) \, . \qquad (1.68)$$

Upon multiplying by the factor $|\tau|^{\alpha+\Delta-2}$, the equation of state becomes a function of the *single* variable $x = h/|\tau|^{\Delta}$, rather than τ and h independently. Equation (1.68) demands that in the vicinity of a critical point, the equation of state is in essence captured by the two distinct scaling functions $\hat{f}'_{\pm}(x)$ for the high- and low-temperature phases. Thus, when measured values for $\phi(\tau, h) |\tau|^{\alpha+\Delta-2}$ are plotted against the variable x, the data in either phase should collapse onto a single *master curve*.

The simple scaling form (1.68) moreover allows us to obtain non-trivial relationships between different critical exponents. At $h = 0$ and $\tau < 0$, we find for the *coexistence line*

$$\phi(\tau, 0) = -|\tau|^{2-\alpha-\Delta} \, \hat{f}'_{-}(0) \sim |\tau|^{\beta} \, , \quad \beta = 2 - \alpha - \Delta \, . \qquad (1.69)$$

Indeed, with $\alpha = 0$ and $\Delta = 3/2$, we recover the mean-field square-root singularity $\beta = 1/2$. For the *critical isotherm*, on the other hand, the temperature

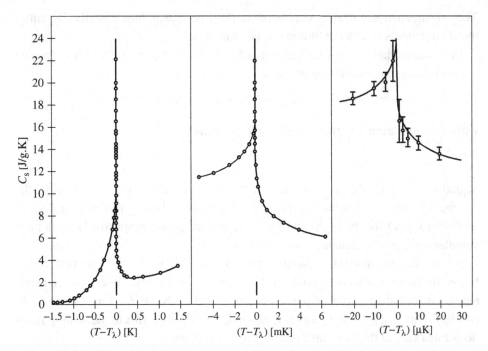

Fig. 1.6 Temperature dependence of the specific heat near the normal- to superfluid transition of helium 4, shown in successively reduced temperature scales. Close to the phase transition, the different graphs look remarkably alike. [Figure reproduced with permission from M. J. Buckingham and W. M. Fairbank, in: *Progress in Low Temperature Physics*, Vol. III, C. J. Gorter (ed.), 80–112, North-Holland (Amsterdam, 1961); copyright (1961) by Elsevier Ltd.]

dependence in the scaling functions $\hat{f}'_{\pm}$ must cancel the algebraic prefactor, i.e., $\hat{f}'_{\pm}(x) \sim x^{(2-\alpha-\Delta)/\Delta}$ as $x \to \infty$. Consequently for $\tau = 0$,

$$\phi(0, h) \sim h^{(2-\alpha-\Delta)/\Delta} = h^{1/\delta} , \qquad \delta = \Delta/\beta . \qquad (1.70)$$

In mean-field approximation, this yields $\delta = 3$, consistent with Eq. (1.44); moreover, $2 - \alpha = \beta(1 + \delta)$. Finally, we find for the isothermal *susceptibility*

$$\frac{\chi_\tau}{V} = \left(\frac{\partial \phi}{\partial h} \right)_{\tau, h=0} = \chi_\pm |\tau|^{-\gamma} , \qquad \gamma = \alpha + 2(\Delta - 1) , \qquad (1.71)$$

which assumes the correct value $\gamma = 1$ in mean-field theory. We may now eliminate Δ: $\gamma + 2\beta = \alpha + 2(\Delta - 1) + 2(2 - \alpha - \Delta) = 2 - \alpha$. With the earlier scaling relation we then obtain $\gamma + \beta = \Delta = \beta\delta$. This completes the set of relationships for the thermodynamic critical exponents α, β, γ, δ, and Δ,

$$\Delta = \beta\delta , \qquad \alpha + \beta(1 + \delta) = 2 = \alpha + 2\beta + \gamma , \qquad \gamma = \beta(\delta - 1) . \qquad (1.72)$$

The *two* independent scaling exponents of the free energy thus describe the entire thermodynamics near a continuous phase transition.

In the same manner, we may generalize Eq. (1.37) to the following scaling ansatz for the order parameter two-point correlation function,

$$C(\tau, q) = |q|^{-2+\eta} \hat{C}_{\pm}(q\xi) , \tag{1.73}$$

with the correlation length diverging at T_c according to

$$\xi = \xi_{\pm} |\tau|^{-\nu} . \tag{1.74}$$

Equation (1.73) implies that when data for $C(\tau, q) |q|^{2-\eta}$ are plotted against the scaling variable $y = q\xi = q\xi_{\pm}|\tau|^{-\nu}$, they should collapse onto unique scaling functions $\hat{C}_{\pm}(y)$ for the high- and low-temperature phase, respectively, both nonsingular for *large* arguments $y \to \infty$.

Within the mean-field approximation, $\eta = 0$ and $\nu = 1/2$. The two correlation function exponents can be related to the thermodynamic critical exponents via the fluctuation-response theorem (1.35). Near T_c, this yields $\chi_{\tau} \approx (N/k_B T_c) \lim_{q \to 0} |q|^{-2+\eta} \hat{C}_{\pm}(q\xi)$; as $q \to 0$, the leading terms in the scaling functions must cancel the singular prefactor, and therefore

$$\chi_{\tau} \sim \xi^{2-\eta} \sim |\tau|^{-\nu(2-\eta)} = |\tau|^{-\gamma} , \quad \gamma = \nu(2 - \eta) . \tag{1.75}$$

Applying the same reasoning for the spatial correlations as $|x| \to \infty$, we see that

$$C(\tau, x) = |x|^{-(d-2+\eta)} \tilde{C}_{\pm}(x/\xi) \sim \xi^{-(d-2+\eta)} \sim |\tau|^{\nu(d-2+\eta)} . \tag{1.76}$$

On the other hand, in the ordered phase for $T < T_c$ and at large distances $|x|$, $\langle S(x)S(0) \rangle \to \langle S \rangle^2 = \phi^2 \sim (-\tau)^{2\beta}$, which suggests the additional identities

$$\beta = \frac{\nu}{2} (d - 2 + \eta) , \quad 2 - \alpha = d\nu . \tag{1.77}$$

These contain the space dimension d, and are usually referred to as *hyperscaling* relations. In this form, they are valid only for $d \leq d_c = 4$; in order to correctly reproduce the mean-field values, one must insert $d = 4$. Equations (1.75) and (1.77) establish that the thermodynamic critical singularities are induced by the divergence of the correlation length governed by ν and the algebraic behavior of $C(x)$ at T_c, which is characterized by the *Fisher exponent* η. We note that a non-zero value of η implies that the spatial structure of correlated regions becomes *fractal*.

1.3.2 Finite-size scaling

Scaling theory is also very useful in considering *finite* systems. Thermodynamic singularities can only develop in the thermodynamic limit of infinitely many degrees of freedom, keeping all densities of extensive quantities fixed. For the extraction of critical exponents from numerical simulations, which are of course always performed in finite systems, it is therefore crucial to know how to correctly extrapolate to infinite system size. Let us assume that a typical system extension in any direction is given by L (e.g., the periodicity scale in a simulation on a hypercubic lattice with periodic boundary conditions), and let ξ denote the correlation length of the corresponding *infinite* system. We further consider short-range interactions only, such that near a critical point we can reach the regime of universal scaling behavior, when ξ has become large compared to any microscopic length scales. The system extension clearly represents an additional relevant scale: for $\xi \ll L$, we still expect to observe the thermodynamic singularities of the infinite system. If the system is finite in d_L directions, but infinite in the others, the limit $\xi \gg L$ will render the system effectively $(d - d_L)$-dimensional. For a truly finite system ($d_L = d$), the apparent singularities will become rounded as the critical point is approached.

These observations can be subsumed into a generalized scaling ansatz for the thermodynamic variables. For example, consider the latter case of a system bounded in all directions. In the scaling regime the order parameter susceptibility should take on the form

$$\chi_\tau(\tau, L) = |\tau|^{-\gamma}\, \hat{\chi}_\pm(\xi/L)\,, \tag{1.78}$$

with the finite-size scaling functions $\hat{\chi}_\pm(y) \to \chi_\pm$ as $y \to 0\,(L \to \infty)$. On the other hand, for $\xi \sim |\tau|^{-\nu} \gg L$ we infer $\hat{\chi}_\pm(y) \sim y^{-\gamma/\nu}$ in order for the singularities to cancel as $\tau \to 0$. We thus arrive at the equivalent statement

$$\chi_\tau(\tau, L) = L^{\gamma/\nu}\, \tilde{\chi}_\pm(|\tau|\,(L/\xi_\pm)^{1/\nu})\,, \tag{1.79}$$

where the new scaling functions $\tilde{\chi}_\pm$ depend on the geometry of the system and the choice of boundary conditions, but are regular for any finite $L < \infty$. As a function of temperature, $\tilde{\chi}_\pm$ should display pronounced maxima in the scaling limit, yet not necessarily at $\tau = 0$, the critical point of the infinite system. Rather the location of the rounded singularity will be shifted by an amount that should scale as $\sim L^{1/\nu}$; moreover, its amplitude is predicted to grow $\sim L^{\gamma/\nu}$ as the system size L increases. Thereby, the two independent critical exponents ν and γ can be obtained by studying the finite-size scaling behavior in computer simulations (or in actual experiments on small samples), despite the fact that the truly asymptotic limit cannot be reached. Yet it is important to remember that relations such as Eq. (1.79) are valid only *provided* there are no additional dangerously irrelevant variables, and that the universal scaling regime has been reached already for the system sizes under consideration.

The dependence of certain quantities on the system size may also illuminate universal features in *ordered phases*. For example, one would expect the free

energy cost for creating a domain wall in a system with broken symmetry to scale
with its linear dimension L as

$$\Sigma = \sigma L^{\Theta} , \qquad (1.80)$$

where σ denotes the interfacial tension. At non-zero temperature, domain walls will
be created spontaneously with probability $\propto \exp(-\Sigma/k_B T)$. A positive *stiffness
exponent* $\Theta > 0$ implies that large domain walls are statistically suppressed, and the
ordered phase remains stable in the thermodynamic limit. However, for $\Theta < 0$, thermally created domain walls will tend to grow and proliferate, eventually destroying
long-range order. We should expect such fluctuations to be stronger in low dimensions. If Θ therefore changes sign as function of d, then the condition $\Theta(d_{lc}) = 0$
determines the *lower critical dimension* of the system: long-range order is only
possible at finite temperatures for $d > d_{lc}$. In the Ising model, or the corresponding
Landau–Ginzburg–Wilson Hamiltonian (1.49), domain walls are essentially one-
dimensional objects, and therefore $\Theta = d - 1$ (see Problem 1.4). Thus, $d_{lc} = 1$ for
systems with Ising symmetry, and there is no long-range order at $T > 0$ in one
dimension, as shown explicitly in Problem 1.2. For the Heisenberg model with continuous spin rotation symmetry and short-range interactions, the domain walls with
lowest free energy extend into two space dimensions, and consequently $\Theta = d - 2$,
implying the lower critical dimension $d_{lc} = 2$ (*Mermin–Wagner–Hohenberg theorem*). Indeed, we shall see in Section 5.4 that the assumption of spontaneous
long-range order leads to inconsistencies in the ensuing renormalization group
flow equations in two dimensions.

1.4 Momentum shell renormalization group

The challenge to a successful theory of critical phenomena is to *derive* the scaling
laws, *explain* the emergence of universality, and to provide a *systematic* method
for computing critical exponents. Yet precisely because of the strongly cooperative
behavior and the predominance of fluctuations in the critical regime, the usual
classical approximation schemes of equilibrium statistical mechanics are of very
limited value only. For example, the marked correlations render low-order virial,
cluster, or cumulant expansions inadequate. Simple perturbation theory fails as
well, because of the prominence of infrared (IR) singularities in low dimensions,
as we shall discuss in detail in Section 5.1. The crucial observation which led to
a breakthrough in the study of second-order phase transitions is that the critical
power laws reflect an underlying *scale invariance* in the system, the origin of which
can be traced to the diverging correlation length ξ. This special symmetry may then
be exploited through studying the scale dependence of the interaction parameters
in detail by means of *renormalization group* (RG) transformations.

In order to construct this renormalization program, we notice that often a partial partition sum can be readily carried out over those degrees of freedom that vary on short length scales. If necessary, one may even proceed perturbatively here, as the dangerous infrared problems only occur on length scales comparable to or larger than the correlation length, or for Fourier modes with $q\xi \ll 1$. After integrating out these short-wavelength modes, and rescaling to the original system extension, one obtains recursion relations that systematically connect the effective couplings on different length scales. Clearly, two such consecutive operations yield another one of the same type, which establishes the *semi-group* character of the renormalization transformation (there is in general no unique inverse element).

There are then three overall possibilities. A specific effective coupling either becomes enhanced or is diminished under each renormalization step; it is then respectively called a *relevant* or *irrelevant* variable. Third, a *marginal* parameter is not affected markedly at all by scale transformations. Certain non-linear couplings such as the quartic u in the Landau–Ginzburg–Wilson Hamiltonian (1.49) require specific attention. As we demonstrate below, the RG transformations leads u to zero for $d \geq d_c = 4$. Nevertheless, we may not simply delete it from our model, as its very existence in fact drives the phase transition; u in this case represents a *dangerously irrelevant* variable. In $d < 4$ dimensions, and for $T \to T_c$, u actually approaches a *universal* value after many RG steps, independent of its initial value that characterizes a given physical system. Such a renormalization group *fixed point* manifestly describes a scale-invariant regime, where the interaction parameters remain unchanged under scale transformations. Its stability obviously demands that all relevant control parameters be set to zero initially: this defines the *critical surface* in parameter space (e.g., $T = T_c$ and $h = 0$ for the Ising model). In contrast, irrelevant quantities disappear automatically under successive renormalizations.

This is the origin of universality: all physical systems located in parameter space within the region of attraction of an RG fixed point are characterized by identical large-scale properties, and form a *universality class*. The eigenvalues of the linearized recursion relations near the respective RG fixed points eventually yield the *scaling exponents* of the distinct thermodynamic phases, and of the phase transitions separating them. Notice that once in a scale-invariant regime, via successive renormalization steps one may infer the system's desired infrared properties from a systematic study of its scaling behavior at *small* wavelengths, i.e., in the ultraviolet (UV) region.

1.4.1 RG transformation; Gaussian model

We now pursue the above ideas to obtain the critical properties of the Landau–Ginzburg–Wilson Hamiltonian (1.49) for the Ising model. In this section, we follow

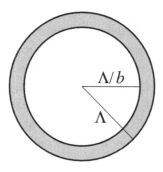

Fig. 1.7 Momentum shell renormalization group transformation: a partial partition function integral is carried out for all Fourier modes $S(q)$ with wavevectors in the interval $\Lambda/b \leq |q| \leq \Lambda$.

Wilson's intuitive approach and perform a partial partition function integral (1.50) in a momentum shell of short-wavelength degrees of freedom $S(q)$.[3] As we expect the interesting physics to emerge from the fluctuations at small wavevectors, we may simply apply a spherical cutoff in momentum space, i.e., mimic the Brillouin zone by a d-dimensional sphere of radius Λ. Wilson's momentum shell renormalization group transformation then consists of the following two steps:

(1) Carry out the partition integral over all Fourier components $S(q)$ with wavevectors $\Lambda/b \leq |q| \leq \Lambda$, where $b > 1$, see Fig. 1.7. Thereby these short-wavelength modes become eliminated.
(2) Apply a scale transformation (with the same scale parameter b that defined the momentum shell):

$$x \rightarrow x' = x/b \,, \quad q \rightarrow q' = b\,q \,. \tag{1.81}$$

Accordingly, we also need to rescale the fields

$$S(x) \rightarrow S'(x') = b^{\zeta} S(x) \,, \quad S(q) \rightarrow S'(q') = b^{\zeta-d} S(q) \,, \tag{1.82}$$

see Eq. (1.56). By means of an appropriate choice of ζ, the ensuing rescaled Hamiltonian may then be cast into the same functional form as the original one. Upon comparison, scale-dependent effective couplings can be constructed, and their dependence on b analyzed.

Let us first apply these concepts to the Gaussian model (1.58). It is diagonal in the Fourier modes $S(q)$, and therefore the two regimes $|q| < \Lambda/b$ and $\Lambda/b \leq |q| \leq \Lambda$

---

[3] This section essentially follows the concise exposition in Schwabl and Täuber (1995).

simply split additively,

$$\mathcal{H}_0[S] = \mathcal{H}_0[S_<] + \mathcal{H}_0[S_>]$$

$$= \left(\int_q^< + \int_q^> \right) \left[\frac{1}{2}(r + q^2) |S(q)|^2 - h(q)S(-q) \right]. \quad (1.83)$$

Here we work in the continuum limit, and introduce the short-hand notations

$$\int_q^< \cdots = \int_{|q|<\Lambda/b} \frac{\mathrm{d}^d q}{(2\pi)^d} \cdots, \quad \int_q^> \cdots = \int_{\Lambda/b \leq |q| \leq \Lambda} \frac{\mathrm{d}^d q}{(2\pi)^d} \cdots. \quad (1.84)$$

The partition integral over the momentum shell degrees of freedom is then readily performed, see Eqs. (1.59) and (1.61),

$$Z_0[h] = \int \mathcal{D}[S_<] \, e^{-\mathcal{H}_0[S_<]/k_B T}$$

$$\times \exp \left(\frac{1}{2k_B T} \int_q^> \left[\frac{|h(q)|^2}{r + q^2} + k_B T V \ln \frac{2\pi k_B T}{r + q^2} \right] \right). \quad (1.85)$$

Of course, eliminating the Fourier modes $S_>(q)$ yields the appropriate additive contribution to the free energy $F_0[h]$. Rescaling according to (1.81) and (1.82) then renders the Hamiltonian $\mathcal{H}_0[S_<] \to \mathcal{H}'_0[S'_<]$ of the same form as the original $\mathcal{H}_0[S]$, provided we choose $\zeta = (d-2)/2$. As a consequence, we obtain the rescaled temperature variable

$$r \to r' = b^2 r, \quad (1.86)$$

and field

$$h(q) \to h'(q') = b^{-\zeta} h(q), \quad h(x) \to h'(x') = b^{d-\zeta} h(x). \quad (1.87)$$

Both r and $h(x)$ are *relevant* scaling variables; the critical surface is therefore defined as $r = h = 0$, as is to be expected. The renormalization group transformation for the temperature variable has three fixed points: $r = \pm\infty$ correspond to the high- and low-temperature phases, respectively, while $r = 0$ describes the phase transition. In the vicinity of this critical fixed point, Eq. (1.86) already constitutes the linearized transformation, and as the correlation length, like any distance, must scale as $\xi \to \xi' = \xi/b$, we infer that $r'\xi'^2 = r\xi^2$, or $\xi \propto r^{-1/2}$. Similarly, from (1.82) we obtain for the correlation function

$$C'(x') = b^{2\zeta} C(x), \quad (1.88)$$

and consequently $C(x) \propto 1/|x|^{d-2}$ at criticality. These observations once again establish the Gaussian scaling exponents $\nu = 1/2$ and $\eta = 0$.

Yet we can now also investigate the influence of perturbations of the Gaussian model. Let us begin with higher-order terms in the gradient expansion for the Landau–Ginzburg–Wilson Hamiltonian. If the first such contribution, namely $c_2 \int d^d x \, (\nabla^2 S)^2$, were to remain invariant under the RG transformation, the coefficient c must be rescaled according to $c_2 \rightarrow c_2' = b^{d-4-2\zeta} c_2 = b^{-2} c_2$. Consequently, under subsequent iterations with $b > 1$ eventually $c_2 \rightarrow 0$. This perturbation is irrelevant with respect to the renormalization group, and we shall see that it therefore does not affect the leading critical behavior. More generally, the couplings for terms of the form $c_s \int d^d x \, (\nabla^s S)^2$ rescale according to $c_s \rightarrow c_s' = b^{d-2s-2\zeta} c_s = b^{-2(s-1)} c_s$ and hence all represent irrelevant perturbations to the Gaussian model.

Next we explore the arbitrary non-linearities $u_{2p} \int d^d x \, S(x)^{2p}$ allowed under the assumed Z_2 order parameter symmetry, which includes the quartic coupling $u = u_4$ present in the functional (1.49). Upon rescaling, we find $u_{2p} \rightarrow u_{2p}' = b^{d-2p\zeta} u_{2p} = b^{2p-(p-1)d} u_{2p}$, which indicates that these couplings become marginal at

$$d_c(p) = \frac{2p}{p-1} , \qquad (1.89)$$

will be relevant for $d < d_c(p)$, and irrelevant for $d > d_c(p)$. Specifically, for $d < d_c(2) = 4$, u is relevant, and non-linear fluctuations will lead to deviations of the critical exponents from their Gaussian values. In four dimensions, u is marginal; for $d > 4$, u becomes irrelevant, and the Gaussian critical exponents should apply. However, one needs to be careful here: u may *not* be simply set to zero, as the non-linearity is crucial for the very existence of the phase transition. Above four dimensions, u is a proto-typical example for a dangerously irrelevant parameter. For a sixth-order non-linearity, $v \int d^d x \, S(x)^6$, for example, we obtain $v \rightarrow v' = b^{6-2d} v$ with $d_c(3) = 3$. Yet near $d_c = 4$ the coupling v is irrelevant: $v' = b^{-2} v$. At least in the vicinity of the upper critical dimension, the influence of all possible non-linear terms is thus under control, with the quartic one representing the only really decisive contribution. Note that the ratio $u_{2(p+1)}/u_{2p} \rightarrow u_{2(p+1)}'/u_{2p}' = b^{2-d} u_{2(p+1)}/u_{2p}$ becomes irrelevant in any dimension $d > 2$. Indeed, $\zeta = 0$ for $d = 2$, and thus all powers of $S(x)$ rescale identically, which renders the Landau–Ginzburg–Wilson expansion inapplicable in two dimensions (see Section 7.4).

In general, near a critical fixed point the above considerations, specifically the relation (1.88), lead us to the identification

$$\zeta = \frac{d-2+\eta}{2} , \qquad (1.90)$$

and we may obtain the *two independent* critical exponents from the RG transformation for the two *relevant* scaling variables,

$$\tau \rightarrow \tau' = b^{1/\nu} \tau , \qquad h \rightarrow h' = b^{(d+2-\eta)/2} h . \qquad (1.91)$$

Combining these results immediately produces the scaling laws of Section 1.3.1. For example, the order parameter susceptibility is predicted to scale according to $\chi'_\tau = b^{2\zeta-d}\chi_\tau = b^{-2+\eta}\chi_\tau$, whence with $b = \xi/\xi'$: $\chi_\tau \sim \xi^{2-\eta}$, which yields Eq. (1.75). Or, even more directly, $\phi = \langle S \rangle \sim \xi^{-\zeta}$, equivalent to the first hyperscaling relation in (1.77). Lastly, if we eliminate the scale factor b from (1.82) and (1.87), we obtain $\phi^{d-\zeta} \sim h^\zeta$, i.e.,

$$\delta = \frac{d-\zeta}{\zeta} = \frac{d+2-\eta}{d-2+\eta}. \tag{1.92}$$

Let us now assume that the system under consideration has only two relevant control parameters, τ and h. In addition, there may be some irrelevant couplings $v_i \rightarrow v'_i = b^{-y_i}v_i$, with $y_i > 0$, and a few marginal variables which we assume to approach *stable fixed points* u_i^*, i.e., $u_i \rightarrow u'_i = u_i^* + b^{-x_i}u_i$ in the vicinity of u_i^*, with $x_i > 0$. Then, after one renormalization step, we expect the free energy density to transform as $f \rightarrow f' = b^d f + \text{const.}$; we therefore obtain for its singular part

$$f_{\text{sing}}(\tau, h, \{u_i\}, \{v_i\}) = b^{-d} f_{\text{sing}}(b^{1/\nu}\tau, b^{d-\zeta}h, \{u_i^* + b^{-x_i}u_i\}, \{b^{-y_i}v_i\}). \tag{1.93}$$

After sufficiently many, say l, iterations of the renormalization procedure, the irrelevant variables will all have scaled to zero, while the marginal ones assume their fixed point values, $u_i \rightarrow u_i^*$:

$$f_{\text{sing}}(\tau, h, \{u_i\}, \{v_i\}) = b^{-ld} f_{\text{sing}}(b^{l/\nu}\tau, b^{l(d+2-\eta)/2}h, \{u_i^*\}, \{0\}). \tag{1.94}$$

We may now choose the scale parameter b conveniently. Upon setting $b^l|\tau|^\nu = 1$, we are led to

$$f_{\text{sing}}(\tau, h) = |\tau|^{d\nu} \hat{f}_\pm(h/|\tau|^{\nu(d+2-\eta)/2}), \tag{1.95}$$

where $\hat{f}_\pm(x) = f_{\text{sing}}(\pm 1, x, \{u_i^*\}, \{0\})$. With the exponent relations (1.77) and (1.92), this establishes the scaling form (1.66) for the singular part of the free energy. Similarly, after l renormalization steps we find for the order parameter correlation function at $h = 0$:

$$C(\tau, x, \{u_i\}, \{v_i\}) = b^{-2l\zeta} C(b^{l/\nu}\tau, x/b^l, \{u_i^*\}, \{0\}). \tag{1.96}$$

Applying the *matching condition* $b^l = |x|/\xi_\pm = |\tau|^{-\nu}|x|/\xi$, we arrive at the desired scaling form (1.76).

We have therefore *derived* the scaling laws for the free energy and the correlation function, *provided* that there exists an infrared-stable RG fixed point $\{u_i^*\}$, and that there are only two relevant variables, namely the temperature and the external field. On the critical surface $\tau = 0$ and $h = 0$, the renormalization group transformations then lead into the fixed point $u_i \rightarrow u_i^*$, $v_i \rightarrow 0$ as $l \rightarrow \infty$, independent of the initial values for these parameters which describe a specific physical system. Thus *universality* emerges: all systems belonging to the region of attraction for the RG fixed point $\{u_i^*\}$ are asymptotically described by identical singularities on and near the critical surface, and constitute a universality class. Moreover, we see that the sub-leading power laws describing *corrections* to the universal leading scaling behavior are determined by the exponents νx_i and νy_i.

1.4.2 Perturbation expansion and Wick's theorem

Henceforth, we shall set $k_B T = 1$. For critical properties, we are interested in the vicinity of T_c, and thus $k_B T \approx k_B T_c$ is a constant, which can be absorbed in the definition of the overall energy scale. The non-linear 'interaction' term in the Landau–Ginzburg–Wilson Hamiltonian $\mathcal{H}$ then reads in Fourier space

$$\mathcal{H}_{\text{int}}[S] = \frac{u}{4!} \int_{|q_i|<\Lambda} S(q_1)S(q_2)S(q_3)S(-q_1 - q_2 - q_3) . \tag{1.97}$$

Furthermore, we notice that the full partition function (1.50) may be rewritten in terms of purely Gaussian averages as

$$Z[h] = Z_0[h] \langle e^{-\mathcal{H}_{\text{int}}[S]} \rangle_0 . \tag{1.98}$$

The *N-point correlation functions* are then readily obtained from the generating functional $Z[h]$ by taking N functional derivatives with respect to the source fields h; e.g., in Fourier space,

$$\begin{aligned}
\langle S(q_1) \cdots S(q_N) \rangle &= \frac{1}{Z[h]} \frac{(2\pi)^{Nd} \delta^N Z[h]}{\delta h(-q_1) \cdots \delta h(-q_N)} \\
&= \frac{1}{Z[h]} \int \mathcal{D}[S] \, S(q_1) \cdots S(q_N) e^{-\mathcal{H}[S]} \\
&= \frac{\langle S(q_1) \cdots S(q_N) e^{-\mathcal{H}_{\text{int}}[S]} \rangle_0}{\langle e^{-\mathcal{H}_{\text{int}}[S]} \rangle_0} .
\end{aligned} \tag{1.99}$$

We may now treat the effects of the non-linear term (1.97) perturbatively by expanding the exponentials in Eq. (1.99) with respect to u. The task is then reduced to computing N-point correlation functions in the *Gaussian ensemble* with quadratic Hamiltonian (1.58). In the high-temperature phase, and for vanishing external fields $h = 0$, any correlation of an *odd* number of fields clearly vanishes for symmetry reasons, while all the *even* moments simply factorize into a sum of all possible combinations of products of two-point correlation functions. In order to see how this important property, called *Wick's theorem*, comes about, let us consider the four-point correlations in a Gaussian ensemble. Using Eq. (1.59), i.e., $\ln Z_0[h] = \frac{1}{2} \int_q h(-q) C_0(q) h(q) + \text{const.}$, where

$$C_0(q) = \frac{1}{r + q^2} \qquad (1.100)$$

is the harmonic zero-field two-point function (1.60), also referred to as the *propagator*, we obtain for $h = 0$

$$
\begin{aligned}
\langle S(q_1) S(q_2) S(q_3) S(q_4) \rangle_0 &= \frac{1}{Z_0[h]} \frac{(2\pi)^{4d} \, \delta^4 Z_0[h]}{\delta h(-q_1) \, \delta h(-q_2) \, \delta h(-q_3) \, \delta h(-q_4)} \bigg|_{h=0} \\
&= C_0(q_1) (2\pi)^d \delta(q_1 + q_2) \, C_0(q_3) (2\pi)^d \delta(q_3 + q_4) \\
&\quad + C_0(q_1) (2\pi)^d \delta(q_1 + q_3) \, C_0(q_2) (2\pi)^d \delta(q_2 + q_4) \\
&\quad + C_0(q_1) (2\pi)^d \delta(q_1 + q_4) \, C_0(q_2) (2\pi)^d \delta(q_2 + q_3) .
\end{aligned}
$$

$$(1.101)$$

The three different terms in this sum correspond to the three possible ways in which the four fields $S(q_i)$ can be paired up or 'contracted' to two-point correlation functions $C_0(q_i)$. We may represent Eq. (1.101) symbolically as

$$
\langle S(q_1) S(q_2) S(q_3) S(q_4) \rangle_0 = \underline{S(q_1) S(q_2)} \; \underline{S(q_3) S(q_4)} + \underline{S(q_1) S(q_3)} \; \underline{S(q_2) S(q_4)}
$$
$$
+ \underline{S(q_1) S(q_4)} \; \underline{S(q_2) S(q_3)} , \qquad (1.102)
$$

where

$$\underline{S(q) S(q')} = \langle S(q) S(q') \rangle_0 = C_0(q) (2\pi)^d \delta(q + q') . \qquad (1.103)$$

The result (1.101) for the four-point correlation function represents an example for Wick's theorem, which states that while for a purely quadratic Hamiltonian all correlations of an odd number of fields vanish, any even N-point function can be written as a sum of products of two-point correlators C_0, which result from all possible complete pairings of the N fields:

$$
\langle S(q_1) S(q_2) \cdots S(q_{N-1}) S(q_N) \rangle_0
$$
$$
= \sum_{\substack{\text{permutations} \\ i_1(1) \ldots i_N(N)}} \underline{S(q_{i_1(1)}) S(q_{i_2(2)})} \cdots \underline{S(q_{i_{N-1}(N-1)}) S(q_{i_N(N)})} . \qquad (1.104)
$$

The general proof of this fundamental property of Gaussian ensembles proceeds through successive functional derivatives of the generating functional $Z_0[h]$, and then setting $h = 0$.

As an example for the perturbation expansion, consider the zero-field two-point correlation function $\langle S(q) S(q') \rangle = C(q) (2\pi)^d \delta(q + q')$. To first order in u, the numerator in Eq. (1.99) for this quantity becomes

$$\left\langle S(q)S(q') \left[1 - \frac{u}{4!} \int_{|q_i|<\Lambda} S(q_1)S(q_2)S(q_3)S(-q_1 - q_2 - q_3) \right] \right\rangle_0 .$$

We now apply Wick's theorem. A first set of contractions contains $S(q)S(q')$, multiplied by precisely the first-order contributions to the denominator of Eq. (1.99); this leaves us simply with $\langle S(q)S(q')\rangle_0$. Any corrections to the Gaussian correlation function require therefore that we contract the *external* fields $S(q)$ and $S(q')$ with any of the *internal* fields $S(q_i)$ inside the momentum integrals. One such term reads

$$\int_{|q_i|<\Lambda} S(q)S(q_1)\,S(q_2)S(q_3)\,S(-q_1 - q_2 - q_3)S(q')$$

$$= C_0(q)^2\,(2\pi)^d\delta(q + q')\int_{|p|<\Lambda} C_0(p) ,$$

after exploiting the delta functions in (1.103). Clearly the other terms to this order, which stem from mere permutations of the internal wavevector labels, yield the same contribution. Straightforward combinatorics tells us that there are twelve such identical terms. We have four choices of fields to contract with $S(q)$, times three possibilities for the remaining ones to connect to $S(q')$. The first-order perturbation expansion to the two-point function thus yields

$$\langle S(q)S(q')\rangle = C_0(q)\,(2\pi)^d\delta(q + q')\left[1 - \frac{u}{2}\,C_0(q)\int_{|p|<\Lambda} C_0(p) + O(u^2) \right] .$$

$$(1.105)$$

We interpret this result as the first term of the geometric series representing the corresponding Dyson equation (see Section 4.4.1), whereupon we find

$$C(q)^{-1} = r + q^2 + \frac{u}{2}\int_{|p|<\Lambda} \frac{1}{r + p^2} + O(u^2) . \qquad (1.106)$$

Thus, to order u there is no modification of the momentum dependence of the two-point correlation function; only its long-wavelength limit, i.e., the 'mass' $r = C(q = 0)^{-1}$ becomes altered through fluctuations.

1.4.3 RG recursion relations and critical exponents

We now return to the renormalization group transformations, where first the short-wavelength degrees of freedom are eliminated, followed by a scale transformation. Upon separating the order parameter field variables into their parts in the outer ($S_>$) and inner momentum shell ($S_<$), respectively, to first order in u one arrives at terms of the following (symbolically written) forms: (i) $u \int S_<^4\,e^{-\mathcal{H}_0[S]}$ just needs to be re-exponentiated, since these degrees of freedom are not eliminated; (ii) any terms that contain an odd numbers of $S_<$ or $S_>$ such as $u \int S_<^3 S_> \, e^{-\mathcal{H}_0[S]}$ vanish at zero external field; (iii) $u \int S_>^4 \, e^{-\mathcal{H}_0[S]}$ yields a constant, contributing to the free energy; and finally (iv) $u \int S_<^2 S_>^2 \, e^{-\mathcal{H}_0}$, for which the Gaussian integral over the outer shell components $S_>$ may be performed using the two-point function (propagator) (1.60). Quite generally, according to Wick's theorem, expressions of the form $\langle \prod_i^m S_>(q_i) \rangle_0$ factorize into a sum of products of all possible pairs $\langle S_>(q) S_>(-q) \rangle_0$, if m is even, and vanish otherwise.

To first order in u, we saw that there is no correction to the wavevector dependence of the two-point correlation function. Therefore, we may still choose $\zeta = (d - 2)/2$, i.e., Eq. (1.90) with $\eta = 0$. Furthermore, replacing the integral in (1.106) with its momentum shell counterpart, and then rescaling appropriately, we are led to

$$r' = b^2 \left[r + \frac{u}{2} A(r) + O(u^2) \right] = b^2 \left[r + \frac{u}{2} S_d \int_{\Lambda/b}^{\Lambda} \frac{p^{d-1}}{r + p^2} \, dp + O(u^2) \right],$$
(1.107)

with a geometric factor originating from the change to spherical coordinates

$$S_d = \frac{K_d}{(2\pi)^d} = \frac{1}{2^{d-1} \pi^{d/2} \, \Gamma(d/2)}$$
(1.108)

that contains the surface area K_d of the d-dimensional unit sphere. Similarly, but with a little more work, we may obtain the recursion relation for u to lowest non-trivial order (Problem 1.6):

$$u' = b^{4-d} u \left[1 - \frac{3}{2} u \, B(r) + O(u^2) \right]$$

$$= b^{4-d} u \left[1 - \frac{3}{2} u \, S_d \int_{\Lambda/b}^{\Lambda} \frac{p^{d-1}}{(r + p^2)^2} \, dp + O(u^2) \right].$$
(1.109)

For large $r \gg 1$, i.e., far away from criticality, we see that the fluctuation contributions disappear, and the Gaussian results are recovered. In the critical regime, we may evaluate the wavenumber integrals in a series expansion with respect to $r \ll 1$:

$$A(r) = S_d \Lambda^{d-2} \frac{1 - b^{2-d}}{d - 2} - r S_d \, \Lambda^{d-4} \frac{1 - b^{4-d}}{d - 4} + O(r^2),$$
(1.110)

$$B(r) = S_d \Lambda^{d-4} \frac{1 - b^{4-d}}{d - 4} + O(r).$$
(1.111)

Inspection of Eq. (1.109) shows again that for $d < d_c = 4$ fluctuations become relevant, and any initial $u > 0$ grows under renormalization. In order to obtain the scaling behavior in this case, we have to search for a non-trivial, finite fixed point. This is most readily done by introducing a *differential RG flow*, via $b = e^{\delta l}$ with $\delta l \to 0$, whereby the number of RG steps effectively becomes a continuous variable. The discrete recursions (1.107) and (1.109) are then transformed into the following differential relations for the *running couplings* $\tilde{r}(l)$ and $\tilde{u}(l)$,

$$
\frac{d\tilde{r}(l)}{dl} = 2\tilde{r}(l) + \frac{1}{2}\tilde{u}(l)\,S_d\Lambda^{d-2} - \frac{1}{2}\tilde{r}(l)\tilde{u}(l)\,S_d\Lambda^{d-4}
$$
$$
+ O(\tilde{u}\tilde{r}^2, \tilde{u}^2)\,, \tag{1.112}
$$
$$
\frac{d\tilde{u}(l)}{dl} = (4-d)\,\tilde{u}(l) - \frac{3}{2}\tilde{u}(l)^2 S_d\Lambda^{d-4} + O(\tilde{u}\tilde{r}, \tilde{u}^2)\,. \tag{1.113}
$$

Here we have used Eqs. (1.110) and (1.111); alternatively, to this order we could also have simply replaced $\int_{\Lambda(1-\delta l)}^{\Lambda} f(p)dp \approx f(\Lambda)\Lambda\delta l$, and at last expanded with respect to r. A *renormalization group fixed point* is now defined by the condition that $d\tilde{r}(l)/dl = 0 = d\tilde{u}(l)/dl$. Indeed, for any non-zero initial value of u, the flow equation (1.113) asymptotically drives $\tilde{u}(l)$ towards the non-trivial *Ising fixed point*

$$
u_I^* = \frac{2}{3}(4-d)\,S_d^{-1}\Lambda^{4-d}\,. \tag{1.114}
$$

Its stability is readily seen by linearizing with respect to $\delta\tilde{u}(l) = \tilde{u}(l) - u_I^*$: $d\delta\tilde{u}(l)/dl \approx (d-4)\,\delta\tilde{u}(l)$, which is solved by $\delta\tilde{u}(l) \approx \delta\tilde{u}(0)\,e^{(d-4)l}$, approaching zero for $d < 4$ and $l \to \infty$. The fixed point u_I^* describes scale-invariant behavior and should govern the universal critical properties of the Ising model at its continuous phase transition at $h = 0$, $T = T_c$.

The original assumption that u be small, justifying a perturbation expansion, now means that the expansion parameter here effectively becomes the deviation $\epsilon = 4 - d$ from the upper critical dimension $d_c = 4$. Thereby the non-trivial Ising fixed point is also seen to emerge continuously from the Gaussian fixed point $u_0^* = 0$ in $d < 4$ dimensions. Inserting into Eq. (1.112) and collecting contributions of order ϵ, one finds

$$
r_I^* = -\frac{1}{4}u_I^* S_d\Lambda^{d-2} = -\frac{1}{6}\epsilon\Lambda^2\,. \tag{1.115}
$$

Physically, we may interpret this fixed point value as representing a *downward shift* of the transition temperature due to thermal fluctuations. In terms of the parameter $\epsilon = 4 - d$, a systematic perturbation series may be constructed for the Ising fixed point (r_I^*, u_I^*). Notice that the renormalization procedure generates additional interactions not present in the original Hamiltonian; for instance, terms $\propto S^6$, $\nabla^2 S^4$ etc. are produced, which in successive steps couple back into the recursion relations for r and u. However, it turns out that to order ϵ^3 all these corrections may be disregarded.

We now introduce the deviation from the actual transition temperature $\tau = r - r_I^* \propto T - T_c$. The differential flow equation for this variable becomes

$$\frac{d\tilde{\tau}(l)}{dl} = \tilde{\tau}(l)\left[2 - \frac{1}{2}\tilde{u}(l)\,S_d\Lambda^{d-4}\right].$$ (1.116)

In the vicinity of the fixed point (1.114), this is solved by

$$\tilde{\tau}(l) = \tilde{\tau}(0)\,e^{(2-\epsilon/3)l}.$$ (1.117)

On the other hand, the correlation length scales as $\tilde{\xi}(l) = \xi(0)\,e^{-l}$; through comparison, we infer $\xi \sim \tau^{-\nu}$ with $\nu^{-1} = 2 - \epsilon/3$. Consistently to first order in ϵ, therefore,

$$\nu = \frac{1}{2} + \frac{\epsilon}{12} + O(\epsilon^2)\,, \qquad \eta = 0 + O(\epsilon^2)\,.$$ (1.118)

The universality of these lowest-order results for the Ising critical exponents becomes manifest by the fact that they solely depend on the space dimension d, but not on the original 'microscopic' Landau–Ginzburg parameters (a, T_c^0, and u). We finally note that at the upper critical dimension $d_c = 4$, the solution of Eq. (1.113) becomes algebraic rather than exponential,

$$\tilde{u}(l) = \frac{\tilde{u}(0)}{1 + 3\tilde{u}(0)\,l/16\pi^2}\,,$$ (1.119)

inducing *logarithmic corrections* to the mean-field exponents (see Section 5.3.3). For generalizations of the RG analysis to lowest non-trivial order to models with n order parameter components, either with rotational or cubic symmetry, see Problems 1.6 and 1.7.

1.5 Appendix: **Functional derivatives and integration**

Consider a *functional* $\mathcal{H}[S(x)]$. Its functional *derivative* is formally introduced by perturbing its argument $S(x)$ through adding an infinitesimal function $\sigma(x)$:

$$\mathcal{H}[S(x) + \sigma(x)] = \mathcal{H}[S(x)] + \int \frac{\delta\mathcal{H}[S(x)]}{\delta S(x')}\,\sigma(x')\,dx' + O(\sigma^2)\,.$$ (1.120)

We may view this definition as the continuum generalization of the first-order term in the Taylor expansion of a multi-variable function $S(\{x_i\})$,

$$S(\{x_i + \delta x_i\}) = S(\{x_i\}) + \sum_i \frac{\partial S(\{x_i\})}{\partial x_i}\,\delta x_i + \frac{1}{2}\sum_{i,j}\frac{\partial^2 S(\{x_i\})}{\partial x_i\,\partial x_j}\,\delta x_i\,\delta x_j + \cdots\,.$$

In analogy with $\partial x_i / \partial x_j = \delta_{ij}$, we have

$$\frac{\delta S(x)}{\delta S(x')} = \delta(x - x') . \qquad (1.121)$$

By means of the definition (1.120), one easily verifies the *product rule*

$$\frac{\delta(\mathcal{G}[S(x)] \, \mathcal{H}[S(x)])}{\delta S(x')} = \frac{\delta \mathcal{G}[S(x)]}{\delta S(x')} \, \mathcal{H}[S(x)] + \mathcal{G}[S(x)] \, \frac{\delta \mathcal{H}[S(x)]}{\delta S(x')} , \qquad (1.122)$$

and the *chain rule*

$$\frac{\delta \mathcal{H}[\mathcal{G}[S(x)]]}{\delta S(x')} = \int dy \, \frac{\delta \mathcal{H}[\mathcal{G}(x)]}{\delta \mathcal{G}(y)} \, \frac{\delta \mathcal{G}[S(y)]}{\delta S(x')} . \qquad (1.123)$$

In Landau–Ginzburg theory, but also in Lagrangian classical mechanics, the functional is of the form

$$\mathcal{H}[S] = \int f[S(x), \nabla S(x)] \, dx . \qquad (1.124)$$

Thus, to first order in σ,

$$\mathcal{H}[S(x) + \sigma(x)] - \mathcal{H}[S(x)] = \int \left[\frac{\partial f(S, \nabla S)}{\partial S(x)} \sigma(x) + \frac{\partial f(S, \nabla S)}{\partial \nabla S(x)} \cdot \nabla \sigma(x) \right] dx ,$$

whence, after integration by parts applied to the second term, and assuming that $\sigma(x)$ vanishes for $|x| \to \infty$,

$$\frac{\delta \mathcal{H}[S]}{\delta S(x)} = \frac{\partial f(S, \nabla S)}{\partial S(x)} - \nabla \cdot \frac{\partial f(S, \nabla S)}{\partial \nabla S(x)} . \qquad (1.125)$$

As mentioned in Section 1.2, the functional integral (1.50) is to be understood as the continuum limit of the multi-dimensional discretization

$$Z = \int \prod_i dS(x_i) \, e^{-\mathcal{H}[S(x_i)]/k_B T} . \qquad (1.126)$$

Thus at each lattice point x_i, the integration extends over all possible values of $S_i = S(x_i)$, weighted here by the associated Boltzmann factor. In the limit of vanishing lattice constant, this entails integration over all possible configurations or 'paths' $S(x)$. For example, in the case of decoupled harmonic oscillators with local elastic constants $C_i = C(x_i)$, i.e., $\mathcal{H}(S_i) = \frac{1}{2} \sum_i C_i S_i^2$, the independent Gaussian integrations over the real variables S_i yield $Z = \prod_i \sqrt{2\pi k_B T / C_i}$. Note that in computations of thermal averages, similar products will appear, canceling those terms in the partition function that have no well-defined continuum limit. The free energy becomes $F = -\frac{1}{2} k_B T \sum_i \ln(2\pi k_B T / C_i)$, compare with Eq. (1.61). Here we may directly perform the continuum limit, whence

$$F = \frac{k_B T}{2} \int d^d x \, \ln \frac{C(x)}{2\pi k_B T} .$$

For uniform $C(x) = C$, the free energy becomes properly extensive, $F = \frac{1}{2} k_B T\, V \ln(C/2\pi\, k_B T)$. Quadratic but non-local couplings are readily diagonalized via Fourier transformation, as demonstrated in Section 1.2.2 for the Gaussian model.

Problems

1.1 *Specific heat discontinuity and latent heat for a van-der-Waals gas*
  (a) The *caloric equation of state* for a van-der-Waals gas in d dimensions with dispersion relation $\epsilon(p) \propto p^s$ reads

$$E(T, V, N) = \frac{d}{s} Nk_B T - \frac{N^2 a}{V}.$$

  Compute the jump in the *specific heat* $C_v = (\partial E/\partial T)_{V,N}$ at the critical point T_c.

  (b) Using the *Clausius–Clapeyron equation* for the vapor pressure curve,

$$\frac{dP_0(T)}{dT} = \frac{s_g - s_l}{v_g - v_l},$$

  show that the *latent heat* per particle vanishes as $q_L \propto (T_c - T)^{1/2}$ upon approaching T_c.

1.2 *Ising model in one dimension*
  (a) Evaluate the partition sum $Z(T, N)$ for a one-dimensional open *Ising chain* with N spins $\sigma_i = \pm 1$,

$$H_N(\{\sigma_i\}) = -\sum_{i=1}^{N-1} J_i \sigma_i \sigma_{i+1}.$$

  Hint: derive the recursion $Z(T, N) = 2Z(T, N-1)\cosh(J_{N-1}/k_B T)$.

  (b) Compute the two-spin correlation function

$$G_{i,n} = \langle \sigma_i \sigma_{i+n} \rangle = \prod_{k=i}^{i+n-1} \tanh(J_k/k_B T).$$

  For uniform $J_i = J\ \forall i$, define the correlation length ξ via $G_n = e^{-n/\xi}$, and demonstrate that it diverges *exponentially* as $T \to 0$.

  (c) Calculate the isothermal magnetic susceptibility

$$\chi_T = \frac{1}{k_B T} \sum_{i,j=1}^{N} \langle \sigma_i \sigma_j \rangle = \frac{N}{k_B T}\left(\frac{1+\alpha}{1-\alpha} - \frac{2\alpha}{N}\frac{1-\alpha^N}{(1-\alpha)^2} \right)$$

  for uniform exchange couplings, where $\alpha = \tanh(J/k_B T)$.

Hint: count the number of terms with $|i - j| = n$ in the sums.
Show that as $T \to 0$, $\chi_T/N \propto \xi$ in the thermodynamic limit.

1.3 *Landau theory for the ϕ^6 model*

Consider the effective free energy density

$$f(\phi) = \frac{r}{2}\phi^2 + \frac{u}{4!}\phi^4 + \frac{v}{6!}\phi^6 - h\phi \ .$$

Here, $r = a(T - T_0)$, $v > 0$, and h denotes an external field.

(a) Show that for $u > 0$, there is a *second-order* phase transition at $h = 0$ and $T = T_0$ with the usual mean-field critical exponents β, γ, δ, and α. Why can v be neglected near the critical point?

(b) Compute β_t, γ_t, δ_t, and α_t at the *tricritical point* $u = 0$.

(c) Now assume $u = -|u| < 0$ and $h = 0$. Show that there is a *first-order* phase transition at $r_d = 5u^2/8v$, and calculate the jump in the order parameter and the associated free energy barrier.

(d) For a non-zero external field $h \neq 0$ and $u < 0$, find parametric equations $r_c(|u|, v)$ and $h_c(|u|, v)$ for two additional *second-order* transition lines, with all three continuous phase boundaries merging at the tricritical point $u = 0$, $h = 0$.

1.4 *Ising domain wall*

(a) Solve the Landau–Ginzburg equation (1.52) for $h = 0$ and $r < 0$ with the boundary conditions $S_K(x) \to \phi_\pm$ as $x_1 \to \pm\infty$ (where x_1 denotes the coordinate direction perpendicular to the domain wall):

$$S_K(x) = \phi_+ \tanh \frac{x_1 - x_0}{2\xi} \ .$$

(b) Compute the free energy cost Σ for a one-dimensional domain wall ('kink') in a system of size L^d ($L \gg \xi$) relative to $F(\phi_\pm)$.

1.5 *Gaussian approximation for the Heisenberg model*

Isotropic magnets with continuous rotational spin symmetry are described by the *Heisenberg model*. The corresponding effective Landau–Ginzburg–Wilson Hamiltonian reads

$$\mathcal{H}[S] = \int d^dx \sum_{\alpha=1}^{n} \left(\frac{r}{2} [S^\alpha(x)]^2 + \frac{1}{2} [\nabla S^\alpha(x)]^2 \right.$$

$$\left. + \frac{u}{4!} \sum_{\beta=1}^{n} [S^\alpha(x)]^2 [S^\beta(x)]^2 - h^\alpha(x) S^\alpha(x) \right),$$

where $S^\alpha(x)$ is an n-component order parameter vector field.

(a) Determine the two-point correlation functions in the high- and low-temperature phases in harmonic (Gaussian) approximation.

Notice: for $T < T_c$, it is useful to expand about the spontaneous magnetization: e.g., $S^\alpha(x) = \pi^\alpha(x)$ for $\alpha = 1, \ldots, n-1$, and $S^n(x) = \phi + \sigma(x)$; then $\langle \pi^\alpha \rangle = 0 = \langle \sigma \rangle$. The components along and perpendicular to ϕ must be carefully distinguished (see Section 5.4).

(b) For $d < d_c = 4$, compute the specific heat in Gaussian approximation on both sides of the phase transition, and show that $C_{h=0} = C_\pm |\tau|^{-\alpha}$, with the critical exponent given in Eq. (1.64), and compute the *universal amplitude ratio* $C_+/C_- = 2^{-d/2}n$.

1.6 *First-order recursion relations for the Heisenberg model*

For the n-component *Heisenberg model* of Problem 1.5, derive the renormalization group recursion relations

$$r' = b^2 \left[r + \frac{n+2}{6} u \, A(r) \right], \quad u' = b^{4-d} u \left[1 - \frac{n+8}{6} u \, B(r) \right].$$

Determine the associated RG fixed points and discuss their stability. Compute the critical exponent ν to first order in $\epsilon = 4 - d$.

1.7 *RG flow equations for the n-vector model with cubic anisotropy*

The $O(n)$ rotational invariance of the Hamiltonian in Problem 1.5 is broken by additional quartic terms with cubic symmetry,

$$\Delta \mathcal{H}[S] = \int d^d x \sum_{\alpha=1}^{n} \frac{v}{4!} [S^\alpha(x)]^4 .$$

(a) Derive the differential RG flow equations for the running couplings $\tilde{r}(l)$, $\tilde{u}(l)$, and $\tilde{v}(l)$ that generalize Eqs. (1.112) and (1.113).

(b) Discuss the ensuing RG fixed points and their stability as a function of the number n of order parameter components, and compute the associated correlation length critical exponents ν.

References

Buckingham, M. J. and W. M. Fairbank, 1961, in: *Progress in Low Temperature Physics*, Vol. III, ed. C. J. Gorter, 80–112, Amsterdam: North-Holland.

Chaikin, P. M. and T. C. Lubensky, 1995, *Principles of Condensed Matter Physics*, Cambridge: Cambridge University Press, chapters 4, 5.

Cowan, B., 2005, *Topics in Statistical Mechanics*, London: Imperial College Press, chapters 3, 4.

Kardar, M., 2007a, *Statistical Physics of Particles*, Cambridge: Cambridge University Press, chapter 5.

Kardar, M., 2007b, *Statistical Physics of Fields*, Cambridge: Cambridge University Press, chapters 1–5.

Pathria, R. K., 1996, *Statistical Mechanics*, Oxford: Butterworth–Heinemann, 2nd edn., chapters 11–13.
Reichl, L. E., 2009, *A Modern Course in Statistical Physics*, Weinheim: Wiley–VCH, 3rd edn., chapters 4, 5.
Schwabl, F., 2006, *Statistical Mechanics*, Berlin: Springer, 2nd edn., chapters 5–7.
Schwabl, F. and U. C. Täuber, 1995, Phase transitions, renormalization, and scaling, in: *Encyclopedia of Applied Physics*, ed. G. L. Trigg, Vol. 13, 343–371, New York: VCH.
Van Vliet, C. M., 2010, *Equilibrium and Non-equilibrium Statistical Mechanics*, New Jersey: World Scientific, 2nd edn., chapters 9, 10.

Further reading

Binney, J. J., N. J. Dowrick, A. J. Fisher, and M. E. J. Newman, 1993, *The Theory of Critical Phenomena*, Oxford: Oxford University Press, 2nd edn.
Brézin, E., 2010, *Introduction to Statistical Field Theory*, New York: Cambridge University Press.
Cardy, J., 1996, *Scaling and Renormalization in Statistical Physics*, Cambridge: Cambridge University Press.
Domb, C. and M. S. Green (eds.), 1972–1976, *Phase Transitions and Critical Phenomena*, Vols. 1–6, London: Academic Press.
Fisher, M. E., 1967, The theory of equilibrium critical phenomena, *Rep. Prog. Phys.* **30**, 615–730.
Fisher, M. E., 1974, The renormalization group in the theory of critical behavior, *Rev. Mod. Phys.* **46**, 597–616.
Goldenfeld, N., 1992, *Lectures on Phase Transitions and the Renormalization Group*, Reading: Addison–Wesley.
Henkel, M., 1999, *Conformal Invariance and Critical Phenomena*, Berlin: Springer.
Ma, S.-k., 1976, *Modern Theory of Critical Phenomena*, Reading: Benjamin–Cummings.
Mazenko, G. F., 2003, *Fluctuations, Order, and Defects*, Hoboken: Wiley–Interscience.
McComb, W. D., 2004, *Renormalization Methods: a Guide for Beginners*, Oxford: Oxford University Press.
Patashinskii, A. Z. and V. L. Pokrovskii, 1979, *Fluctuation Theory of Phase Transitions*, New York: Pergamon Press.
Privman, V. (ed.), 1990, *Finite Size Scaling and Numerical Simulation of Statistical Systems*, Singapore: World Scientific.
Stanley, H. E., 1971, *Introduction to Phase Transitions and Critical Phenomena*, Oxford: Clarendon Press.
Uzunov, D. I., 2010, *Introduction to the Theory of Critical Phenomena: Mean Field, Fluctuations and Renormalization*, Singapore: World Scientific, 2nd edn.
Wilson, K. G. and J. Kogut, 1974, The renormalization group and the ϵ expansion, *Phys. Rep.* **12 C**, 75–200.
Wilson, K. G., 1975, The renormalization group: critical phenomena and the Kondo problem, *Rev. Mod. Phys.* **47**, 773–840.
Yeomans, J. M., 1991, *Statistical Mechanics of Phase Transitions*, Oxford: Clarendon Press.

2

Stochastic dynamics

In this chapter, we develop the basic tools for our study of dynamic critical phenomena. We introduce dynamic correlation, response, and relaxation functions, and explore their general features. In the linear response regime, these quantities can be expressed in terms of equilibrium properties. A fluctuation-dissipation theorem then relates dynamic response and correlation functions. Under more general non-equilibrium conditions, we must resort to the theory of stochastic processes. The probability $P_1(x, t)$ of finding a certain physical configuration x at time t is governed by a master equation. On the level of such a 'microscopic' description, we discuss the detailed-balance conditions which guarantee that $P_1(x, t)$ approaches the probability distribution of an equilibrium statistical ensemble as $t \to \infty$. Taking the continuum limit for the variable(s) x, we are led to the Kramers–Moyal expansion, which often reduces to a Fokker–Planck equation. Three important examples elucidate these concepts further, and also serve to introduce some calculational methods; these are biased one-dimensional random walks, a simple population dynamics model, and kinetic Ising systems. We then venture towards a more 'mesoscopic' viewpoint which focuses on the long-time dynamics of certain characteristic, 'relevant' quantities. Assuming an appropriate separation of time scales, the remaining 'fast' degrees of freedom are treated as stochastic noise. As an introduction to these concepts, the Langevin–Einstein theory of free Brownian motion is reviewed, and the associated Fokker–Planck equation is solved explicitly. Our considerations are then generalized to Brownian particles in an external potential, leading to the Smoluchowski equation for the probability distribution. Finally, some general properties of Langevin-type stochastic differential equations are listed. Specifically, we note the sufficient conditions for the system to asymptotically approach thermal equilibrium, among them Einstein's relation for the relaxation constant and the strength of the thermal noise.

2.1 Dynamic response and correlation functions

As yet, there exists no unifying description of systems far from thermal equilibrium in terms of macroscopic thermodynamic variables akin to the spectacularly successful statistical ensemble approach in the equilibrium theory. For such inherently dynamical situations, we generally need either to solve the complete set of microscopic equations of motion, which is rarely feasible, or to try and obtain an at least mesoscopic description through appropriate coarse-graining. However, the situation much improves in the *linear response* regime, where an equilibrium system is only slightly perturbed by a weak time-dependent potential. Macroscopic averages in the system then deviate from their stationary values only linearly in the external field, and the corresponding dynamic response functions are fully characterized by its *equilibrium* properties. Moreover, the dynamic correlation functions and susceptibilities are related through a *fluctuation-dissipation theorem*. Linear response theory thus sets the equilibrium baseline for the discussion of dynamic correlations under more general non-equilibrium conditions. Since our aim is to cover quantum critical phenomena as well, we employ the language and formalism of quantum statistical mechanics in this section. In fact, some derivations are actually facilitated in the quantum formalism.[1]

2.1.1 Dynamic correlation functions

In *quantum statistical mechanics*, $\langle A \rangle = \mathrm{Tr}(\rho A)$ denotes the ensemble average of an observable A with respect to a normalized density matrix (statistical operator) $\rho = \sum_j p_j |\psi_j\rangle\langle\psi_j|$, $\mathrm{Tr}\,\rho = \sum_j p_j = 1$. The orthonormal quantum states $|\psi_j\rangle$, $\langle\psi_i|\psi_j\rangle = \delta_{ij}$, are subject to *Schrödinger's equation* with the system's Hamiltonian H,

$$i\hbar \frac{\partial |\psi_j(t)\rangle}{\partial t} = H(t) |\psi_j(t)\rangle \,. \tag{2.1}$$

We write its formal solution as $|\psi_j(t)\rangle = U(t, t_0)|\psi_j(t_0)\rangle$, introducing the unitary time evolution operator $U(t, t_0)$, $U(t, t_0)^\dagger = U(t, t_0)^{-1}$. It obviously satisfies the Schrödinger equation as well,

$$i\hbar \frac{\partial U(t, t_0)}{\partial t} = H(t)\, U(t, t_0) \,, \tag{2.2}$$

with the initial condition $U(t_0, t_0) = 1$. Thus, both the density matrix $\rho(t)$ and the average $\langle A(t) \rangle = \mathrm{Tr}[\rho(t)A]$ are in general time-dependent quantities. Alternatively,

---

[1] This section builds on the expositions in Chaikin and Lubensky (1995), Cowan (2005), Schwabl (2008), and Van Vliet (2010).

one may employ the *Heisenberg picture* with time-dependent operators $A(t) = U(t, t_0)^\dagger A(t_0) U(t, t_0)$. Using the cyclic invariance of the trace, we then obtain $\langle A(t) \rangle = \text{Tr}[\rho(t_0) A(t)]$. For conservative systems with a stationary Hamiltonian H, Eq. (2.2) is solved by $U(t, t_0) = e^{-iH(t-t_0)/\hbar}$, and $A(t)$ obeys *Heisenberg's equation of motion*

$$\frac{dA(t)}{dt} = \frac{i}{\hbar} [H, A(t)] + \frac{\partial A(t)}{\partial t} , \qquad (2.3)$$

where the last term stems from any explicit time dependence of the operator A in the Schrödinger representation, transformed into the Heisenberg picture. We can now readily proceed to define a *correlation function* of two observables A and B at different times t and t':

$$C_{AB}(t, t') = \langle A(t) B(t') \rangle = \text{Tr}[\rho(t_0) A(t) B(t')] . \qquad (2.4)$$

In *classical* statistical mechanics, ρ is to be interpreted as a phase-space trajectory density, and the trace becomes an integral over all generalized coordinates and conjugate momenta. The commutator in Eq. (2.3) is to be replaced with the Poisson bracket, $\frac{i}{\hbar} [H, A] \rightarrow \{H, A\}$. In any case, in *thermal equilibrium* $\rho(H)$ becomes a function of the Hamiltonian only (and perhaps additional conserved quantities), and therefore commutes with $U(t, 0) = e^{-iHt/\hbar}$. Upon invoking the cyclic invariance of the trace once again, we find (setting $t_0 = 0$)

$$\begin{aligned}
C_{AB}(t, t') &= \text{Tr}(\rho(H) e^{iHt/\hbar} A e^{-iH(t-t')/\hbar} B e^{-iHt'/\hbar}) \\
&= \text{Tr}(\rho(H) e^{iH(t-t')/\hbar} A e^{-iH(t-t')/\hbar} B) = C_{AB}(t - t', 0) \\
&= \text{Tr}(\rho(H) A e^{-iH(t-t')/\hbar} B e^{iH(t-t')/\hbar}) = C_{AB}(0, t' - t) . \qquad (2.5)
\end{aligned}$$

Consequently, upon defining the temporal Fourier transform via

$$A(t) = \frac{1}{2\pi} \int A(\omega) e^{-i\omega t} d\omega , \qquad A(\omega) = \int A(t) e^{i\omega t} dt , \qquad (2.6)$$

we obtain

$$\langle A(\omega) B(\omega') \rangle = C_{AB}(\omega) 2\pi \delta(\omega + \omega') , \qquad (2.7)$$

where $C_{AB}(\omega)$ is the Fourier transform of $C_{AB}(t) = C_{AB}(t, 0)$. The relation (2.7) is valid whenever *time translation invariance* holds. In this case, obviously $\langle A(t) \rangle = \langle A \rangle$ is independent of t, and $\langle A(\omega) \rangle = \langle A \rangle 2\pi \delta(\omega)$.

In the equilibrium canonical ensemble,

$$\rho(H) = \frac{1}{Z(T)} e^{-H/k_B T} , \qquad Z(T) = \text{Tr} \, e^{-H/k_B T} , \qquad (2.8)$$

we may write

$$C_{AB}(t) = \frac{1}{Z(T)} \, \text{Tr}(e^{-H/k_B T} \, e^{iHt/\hbar} A \, e^{-iHt/\hbar} B)$$

$$= \frac{1}{Z(T)} \sum_{n,m} e^{-E_n/k_B T} \, e^{i(E_n - E_m)t/\hbar} \, \langle n|A|m \rangle \langle m|B|n \rangle , \qquad (2.9)$$

where we have inserted a complete set of energy eigenstates $|n\rangle$ with eigenvalues E_n: $H|n\rangle = E_n|n\rangle$, $\sum_n |n\rangle\langle n| = 1$, and used the same basis for performing the trace, $\text{Tr} \, A = \sum_n \langle n|A|n \rangle$. A Fourier transform then yields the *spectral representation*

$$C_{AB}(\omega) = \frac{2\pi\hbar}{Z(T)} \sum_{n,m} e^{-E_n/k_B T} \langle n|A|m \rangle \langle m|B|n \rangle \, \delta(E_n - E_m + \hbar\omega) . \qquad (2.10)$$

Upon exchanging the summation indices $m \leftrightarrow n$ and exploiting the delta function, we see that

$$C_{BA}(-\omega) = C_{AB}(\omega) e^{-\hbar\omega/k_B T} . \qquad (2.11)$$

Scattering experiments directly probe dynamic correlation functions. For example, in inelastic light scattering, the coherent scattering cross section is proportional to $S_c(q, \omega)$, the Fourier transform of the normalized *density–density correlation function*

$$S_c(q, t) = \frac{1}{N} \langle n(q, t)n(-q, 0) \rangle . \qquad (2.12)$$

Here, $n(x, t) = \sum_{i=1}^{N} \delta(x - x_i(t))$ denotes the density operator for an N-particle system, and $n(q, t) = \int n(x, t) e^{-iqx} d^d x = \sum_{i=1}^{N} e^{-iqx_i(t)}$ its spatial Fourier transform. Equation (2.11) now reads $S_c(-q, -\omega) = S_c(q, \omega) e^{-\hbar\omega/k_B T}$ and has an immediate physical implication. The intensity of the 'anti-Stokes' emission lines ($\omega < 0$) is suppressed by a temperature-dependent detailed-balance factor as compared with the 'Stokes' absorption lines ($\omega > 0$). Indeed, as $T \to 0$, the system is in the ground state, and no emission is possible at all. In the classical limit $k_B T \gg \hbar\omega$, the absorption and emission line strengths become equal. The equal-time correlation

$$S_c(q) = S_c(q, t = 0) = \frac{1}{N} \langle |n(q, 0)|^2 \rangle \qquad (2.13)$$

is called the static *structure factor*. Notice that $S_c(q = 0, \omega) = N \, 2\pi \delta(\omega)$.

2.1.2 Dynamic susceptibilities and relaxation functions

Another means to investigate dynamic properties of a physical system is a *relaxation* experiment. Such a situation may be described theoretically by adding to the stationary Hamiltonian H_0 a time-dependent term of the form $H'(t) = -F(t)B$, where the function $F(t)$ denotes an external 'force' that couples to the system via the Hermitean operator $B = B^\dagger$. This interaction induces a deviation $\delta A(t) = \langle A(t) \rangle - \langle A \rangle_0$ of the ensemble average for the observable A from its time-independent unperturbed value $\langle A \rangle_0$, which is determined by the equilibrium density matrix $\rho(H_0)$. This change can be formally expanded in terms of powers of $F(t)$,

$$\delta A(t) = \int \chi_{AB}(t - t')F(t')dt' + \frac{1}{2} \int \chi^{(2)}_{ABB}(t - t', t - t'')F(t')F(t'')dt'dt'' + \cdots ,$$

$$(2.14)$$

which defines the (linear) *dynamic susceptibility* (*response function*)

$$\chi_{AB}(t - t') = \frac{\delta \langle A(t) \rangle}{\delta F(t')}\bigg|_{F=0} , \qquad (2.15)$$

and the (second-order) *non-linear response function*

$$\chi^{(2)}_{ABB}(t - t', t - t'') = \frac{\delta^2 \langle A(t) \rangle}{\delta F(t')\,\delta F(t'')}\bigg|_{F=0} , \qquad (2.16)$$

respectively. Higher-order non-linear response functions, perhaps involving couplings to additional operators, follow by means of straightforward generalization. The stationarity of H_0 implies that the susceptibilities must be functions of the time differences $t - t'$, $t - t''$, etc. only.

It is important to realize that the dynamic response functions are fully determined by the *equilibrium* properties of the system. For a weak field $F(t)$, we expect $|\delta A(t)/\langle A \rangle_0| \ll 1$, and taking into account only the linear response should provide an adequate description. If the external perturbation is *instantaneous*, i.e., represented by a delta peak at time $t = 0$: $F(t) = F\delta(t)$, the linear response simply becomes $\delta A(t) = 0$ for $t < 0$, and $\delta A(t) = F\chi_{AB}(t)$ for $t \geq 0$, see Fig. 2.1(a). Generally, as a consequence of *causality* (effects cannot precede their causes), the susceptibilities should be proportional to the product of Heaviside step functions of their arguments, $\chi_{AB}(t - t') \propto \Theta(t - t')$, $\chi^{(2)}_{ABB}(t - t', t - t'') \propto \Theta(t - t')\Theta(t - t'')$, and so forth. Thus, we may write for the linear response

$$\delta A(t) = \int_{-\infty}^{t} \chi_{AB}(t - t')F(t')\,dt' = \int_{0}^{\infty} \chi_{AB}(s)F(t - s)\,ds , \qquad (2.17)$$

(a)   (b)

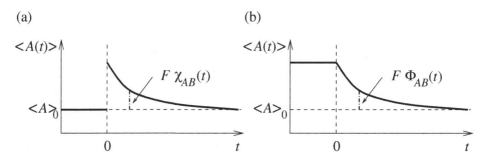

Fig. 2.1 Relaxation of the average $\langle A(t) \rangle$ towards its equilibrium value $\langle A \rangle_0$: (a) following an instantaneous perturbation F at $t = 0$, and (b) after switching off the external field F at $t = 0$; the deviation from $\langle A \rangle_0$ in these situations is given by the dynamic response and Kubo relaxation functions, respectively.

where $s = t - t'$. The convolution theorem yields for the Fourier transform of Eq. (2.14): $\langle A(\omega) \rangle = \langle A \rangle_0\, 2\pi\, \delta(\omega) + \chi_{AB}(\omega) F(\omega) + O(F^2)$, or

$$\chi_{AB}(\omega) = \int_0^\infty \chi_{AB}(t)\, e^{i\omega t}\, dt = \left. \frac{\partial \langle A(\omega) \rangle}{\partial F(\omega)} \right|_{F=0}. \qquad (2.18)$$

Hence, the Laplace transform of the dynamic susceptibility $\chi_{AB}(\omega)$ directly provides the response of the Fourier component $\langle A(\omega) \rangle$ to the external perturbation at the same frequency.

Another typical situation is that a perturbation existing for all previous times is switched off at $t = 0$, $F(t) = F\,\Theta(-t)$. The system's linear response may then be written as $\delta A(t) = F\,\Phi_{AB}(t)$, see Fig. 2.1(b), which defines *Kubo's relaxation function* $\Phi_{AB}(t)$. Comparison with Eq. (2.17) immediately gives the relation

$$\Phi_{AB}(t) = \begin{cases} \int_0^\infty \chi_{AB}(s)\, ds = \chi_{AB}(\omega = 0), & t \leq 0, \\ \int_t^\infty \chi_{AB}(s)\, ds, & t > 0. \end{cases} \qquad (2.19)$$

Thus, the Kubo relaxation function and linear dynamic susceptibility are by no means independent quantities, but $\chi_{AB}(t) = -d\Phi_{AB}/dt$ for $t > 0$, with the boundary conditions $\Phi_{AB}(t = 0) = \chi_{AB}(\omega = 0)$ and $\Phi_{AB}(t \to \infty) = 0$. Integration by parts then yields for the Laplace transform

$$\Phi_{AB}(\omega) = \int_0^\infty \Phi_{AB}(t)\, e^{i\omega t}\, dt = \frac{1}{i\omega}[\chi_{AB}(\omega) - \chi_{AB}(\omega = 0)]. \qquad (2.20)$$

It therefore suffices to discuss the properties of either the linear response function $\chi_{AB}(\omega)$ or the relaxation function $\Phi_{AB}(\omega)$ (see Problems 2.1 and 2.2).

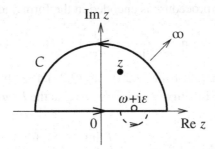

Fig. 2.2 Integration contour C in the complex frequency plane for the derivation of the Kramers–Kronig relations for the dynamic susceptibility $\chi_{AB}(z)$. The dashed semicircle represents the required contour deformation as the frequency $z = \omega + i\varepsilon$ approaches the real axis ($\varepsilon \downarrow 0$).

Let us now consider the analytic continuation of $\chi_{AB}(\omega)$ to complex frequencies z,

$$\chi_{AB}(z) = \int \chi_{AB}(t)\, e^{izt}\, dt \; . \tag{2.21}$$

Causality demands $\chi_{AB}(t)$ to vanish for $t < 0$. Therefore, $\chi_{AB}(z)$ has to be analytic in the upper complex half-plane (Im $z > 0$). For any integration contour C in the analytic region, Cauchy's integral theorem states that

$$\chi_{AB}(z) = \frac{1}{2\pi i} \int_C \frac{\chi_{AB}(z')}{z' - z}\, dz' \; . \tag{2.22}$$

We choose the integration contour as depicted in Fig. 2.2, and assume that $\chi_{AB}(z)$ decays sufficiently fast as $|z| \gg 1$ for the line integral along the semicircle in the upper half-plane to vanish when its radius is pushed to infinity. Equation (2.22) then reduces to an integral over the entire real axis,

$$\chi_{AB}(z) = \frac{1}{2\pi i} \int \frac{\chi_{AB}(\omega')}{\omega' - z}\, d\omega' \; . \tag{2.23}$$

Upon approaching real frequencies $z = \omega + i\varepsilon$, $\varepsilon \downarrow 0$, we need to deform the integration contour as indicated by the dashed path in Fig. 2.2. The integral now becomes a sum of the Cauchy principal value and the contour integral of the dashed infinitesimal semicircle, which contributes precisely one half of the residue of the pole at $\omega' = \omega$:

$$\lim_{\varepsilon \downarrow 0} \chi_{AB}(\omega + i\varepsilon) = \frac{1}{2\pi i} \mathcal{P} \int \frac{\chi_{AB}(\omega')}{\omega' - \omega}\, d\omega' + \frac{1}{2} \chi_{AB}(\omega) \; ,$$

or

$$\chi_{AB}(\omega) = \frac{1}{i\pi} \mathcal{P} \int \frac{\chi_{AB}(\omega')}{\omega' - \omega}\, d\omega' \; . \tag{2.24}$$

The preceding limiting procedure is encoded in the formal identity

$$\lim_{\varepsilon \downarrow 0} \frac{1}{x \mp i\varepsilon} = \mathcal{P} \frac{1}{x} \pm i\pi\, \delta(x) .\tag{2.25}$$

Setting $x = \omega' - \omega$, and inserting (2.25) into (2.23) directly results in Eq. (2.24). Separating into real and imaginary parts, we arrive at the *Kramers–Kronig relations*

$$\mathrm{Re}\, \chi_{AB}(\omega) = \frac{1}{\pi} \mathcal{P} \int \frac{\mathrm{Im}\, \chi_{AB}(\omega')}{\omega' - \omega}\, d\omega' ,\tag{2.26}$$

$$\mathrm{Im}\, \chi_{AB}(\omega) = -\frac{1}{\pi} \mathcal{P} \int \frac{\mathrm{Re}\, \chi_{AB}(\omega')}{\omega' - \omega}\, d\omega' .\tag{2.27}$$

As a consequence of causality, the real and imaginary parts of the dynamic susceptibility $\chi_{AB}(\omega)$ are intimately connected, and either can be computed if the other happens to be known at all (real) frequencies.

2.1.3 Linear response and fluctuation-dissipation theorem

We now proceed to express the linear susceptibility in terms of a dynamic correlation function. Within the framework of *linear* response, we merely need to apply first-order time-dependent perturbation theory. The total Hamiltonian is $H(t) = H_0 + H'(t) = H_0 - F(t)B$, where we assume that $F(t) = 0$ for $t \le t_0$. As we presume to know the solution with the unperturbed stationary Hamiltonian H_0, it is beneficial to transform to the *interaction representation* via $U(t, t_0) = e^{-iH_0(t-t_0)/\hbar}\, U'(t, t_0)$. By means of Eq. (2.2), we see that $U'(t, t_0)$ obeys the equation of motion

$$i\hbar \frac{\partial U'(t, t_0)}{\partial t} = e^{iH_0(t-t_0)/\hbar} \left[i\hbar \frac{\partial U(t, t_0)}{\partial t} - H_0\, U(t, t_0) \right]$$

$$= e^{iH_0(t-t_0)/\hbar}\, H'(t)\, e^{-iH_0(t-t_0)/\hbar}\, U'(t, t_0)$$

$$= H'_I(t)\, U'(t, t_0) ,\tag{2.28}$$

where $H'_I(t)$ denotes the perturbing Hamiltonian in the interaction representation. The equivalent integral equation

$$U'(t, t_0) = 1 + \frac{1}{i\hbar} \int_{t_0}^{t} dt'\, H'_I(t')\, U'(t', t_0)\tag{2.29}$$

is readily solved iteratively,

$$U'(t, t_0) = 1 + \frac{1}{i\hbar} \int_{t_0}^{t} dt'\, H'_I(t') + \frac{1}{(i\hbar)^2} \int_{t_0}^{t} dt' \int_{t_0}^{t'} dt''\, H'_I(t')\, H_I(t'') \cdots .\tag{2.30}$$

This expression may be nicely rewritten by means of a time-ordered product. However, we just require the first-order contribution, which we insert into

$$\langle A(t) \rangle = \text{Tr}[\rho_0(t_0)\, A(t)] = \text{Tr}[\rho_0(t_0)\, U(t, t_0)^\dagger A(t_0)\, U(t, t_0)]$$

$$= \text{Tr}[\rho_0(t_0)\, e^{iH_0(t-t_0)/\hbar}\, A(t_0)\, e^{-iH_0(t-t_0)/\hbar}]$$

$$+ \frac{1}{i\hbar} \int_{t_0}^{t} dt'\, \text{Tr}\left(\rho_0(t_0)[e^{iH_0(t-t_0)/\hbar}\, A(t_0)\, e^{-iH_0(t-t_0)/\hbar}\, ,\, H_I'(t')]\right).$$

The first term represents the unperturbed average $\langle A \rangle_0$. For the second term we recall Eq. (2.28), whereupon we arrive at

$$\delta A(t) = \frac{i}{\hbar} \int_{t_0}^{t} dt'\, \big\langle \big[e^{iH_0(t-t_0)/\hbar}\, A(t_0)\, e^{-iH_0(t-t_0)/\hbar} \,,$$

$$e^{iH_0(t'-t_0)/\hbar}\, B(t_0)\, e^{-iH_0(t'-t_0)/\hbar} \big] \big\rangle_0 F(t'). \tag{2.31}$$

To this order, the commutator is just $[A(t), B(t')]$. Taking the limit $t_0 \to -\infty$, we at last find upon comparison with the definition in Eq. (2.14) that the linear susceptibility is given by a retarded commutator,

$$\chi_{AB}(t - t') = \frac{i}{\hbar}\, \langle [A(t), B(t')] \rangle_0 \Theta(t - t'), \tag{2.32}$$

where the ensemble average is taken with respect to the *equilibrium* Hamiltonian H_0 (we shall henceforth drop the index '0').

This fundamental result contains the antisymmetric combination

$$\chi_{AB}''(t) = \frac{1}{2\hbar}\, \langle [A(t), B(0)] \rangle = \frac{1}{2\hbar}\, [C_{AB}(t) - C_{BA}(-t)] = -\chi_{BA}''(-t), \tag{2.33}$$

whence $\chi_{AB}''(\omega) = -\chi_{BA}''(-\omega)$. The relation (2.11) now yields the important quantum-mechanical *fluctuation-dissipation theorem* (FDT)

$$\chi_{AB}''(\omega) = \frac{1}{2\hbar}\left(1 - e^{-\hbar\omega/k_B T}\right) C_{AB}(\omega). \tag{2.34}$$

In the *classical limit* $\hbar\omega/k_B T \ll 1$, we may expand the exponential to obtain

$$\chi_{AB}''(\omega) = \frac{\omega}{2k_B T}\, C_{AB}(\omega), \tag{2.35}$$

$$\chi_{AB}(t) = 2i\, \chi_{AB}''(t)\, \Theta(t) = -\frac{\Theta(t)}{k_B T}\, \frac{dC_{AB}(t)}{dt}. \tag{2.36}$$

In order to compute its Fourier transform, we need the integral representation of the Heaviside step function,

$$\Theta(t) = \lim_{\varepsilon \downarrow 0} \int \frac{d\omega}{2\pi} \frac{i\,e^{-i\omega t}}{\omega + i\varepsilon} . \qquad (2.37)$$

Indeed, for $t > 0$, convergence considerations enforce the closure of the integration path in the lower complex frequency half-plane, and the residue theorem yields $\Theta(t) = 1$, while for $t < 0$, the contour is to be closed in the upper half-plane where the integrand is analytic, and the integral vanishes. By means of the convolution theorem then

$$\chi_{AB}(\omega) = \lim_{\varepsilon \downarrow 0} \frac{1}{\pi} \int \frac{\chi''_{AB}(\omega')}{\omega' - \omega - i\varepsilon} \, d\omega' , \qquad (2.38)$$

and upon inserting Eqs. (2.33) and (2.10), we arrive at the spectral representation for the dynamic response function,

$$\chi_{AB}(\omega) = \lim_{\varepsilon \downarrow 0} \frac{1}{Z(T)} \sum_{n,m} \langle n|A|m\rangle \langle m|B|n\rangle \frac{e^{-E_n/k_B T} - e^{-E_m/k_B T}}{E_m - E_n - \hbar(\omega + i\varepsilon)} . \qquad (2.39)$$

In the classical limit, Eq. (2.38) reads

$$\chi_{AB}(\omega) = \lim_{\varepsilon \downarrow 0} \frac{1}{2\pi k_B T} \int \frac{\omega' \, C_{AB}(\omega')}{\omega' - \omega - i\varepsilon} \, d\omega' , \qquad (2.40)$$

and the *thermodynamic susceptibility* becomes

$$\chi_{AB} = \chi_{AB}(\omega = 0) = \frac{1}{k_B T} \int \frac{d\omega}{2\pi} C_{AB}(\omega) = \frac{C_{AB}(t = 0)}{k_B T} = \frac{\langle AB\rangle}{k_B T} . \qquad (2.41)$$

Thus, we recover the classical *fluctuation-response theorem*, see Eq. (1.33). Applying the identity (2.25) to Eq. (2.38), we furthermore obtain

$$\chi_{AB}(\omega) = \chi'_{AB}(\omega) + i\,\chi''_{AB}(\omega) , \qquad (2.42)$$

where

$$\chi'_{AB}(\omega) = \frac{1}{\pi} \mathcal{P} \int \frac{\chi''_{AB}(\omega')}{\omega' - \omega} \, d\omega' = \chi'_{BA}(-\omega) . \qquad (2.43)$$

If $\chi''_{AB}(\omega)$ is real, this just represents the decomposition into real and imaginary parts and the Kramers–Kronig relation (2.26). This is certainly true in the case $B = A^\dagger$. Notice that $\chi'_{AA^\dagger}(\omega)$ and $\chi''_{AA^\dagger}(\omega)$ are symmetric and antisymmetric functions of ω, respectively. It is instructive to discuss the case of a periodic external perturbation with $F(t) = F \cos \omega t$. According to *Fermi's golden rule* of first-order time-dependent perturbation theory, the transition rate (as $t \to \infty$) from energy eigenstates $|n\rangle$ to $|m\rangle$ is

$$\Gamma_{n \to m}(\omega) = \frac{2\pi}{\hbar} F^2 |\langle m|A|n\rangle|^2 [\delta(E_m - E_n - \hbar\omega) + \delta(E_m - E_n + \hbar\omega)].$$

(2.44)

For $\omega > 0$, the two terms here correspond to absorption and emission of an energy quantum $\hbar\omega$, respectively. In order to compute the total *dissipated power* at frequency ω, we have to multiply by the energy transfer $E_m - E_n$, and sum over all possible initial and final states, weighted with the canonical probability distribution for the initial states:

$$P(\omega) = \sum_{n,m} \frac{e^{-E_n/k_B T}}{Z(T)} \Gamma_{n \to m}(\omega)(E_m - E_n) = 2\pi\omega F^2$$

$$\times \sum_{n,m} \frac{e^{-E_n/k_B T}}{Z(T)} |\langle m|A|n\rangle|^2 [\delta(E_m - E_n - \hbar\omega) - \delta(E_m - E_n + \hbar\omega)].$$

(2.45)

Comparing with the spectral representation (2.10), we find

$$P(\omega) = \frac{\omega}{\hbar} F^2 [C_{AA^\dagger}(\omega) - C_{A^\dagger A}(-\omega)] = 2\omega F^2 \chi''_{AA^\dagger}(\omega).$$ (2.46)

Therefore, $\chi''_{AA^\dagger}(\omega)$ describes the *dissipative* response to an external perturbation at frequency ω, which explains the term fluctuation-dissipation theorem for Eq. (2.34), while the real part of the dynamic susceptibility $\chi'_{AA^\dagger}(\omega)$ gives the *reactive* response. Inserting $F(t) = \frac{F}{2}(e^{i\omega t} + e^{-i\omega t})$ into Eq. (2.17), separating $\chi_{AA^\dagger}(\omega)$ into its real and imaginary parts, and exploiting their symmetries results in

$$\delta A(t) = \frac{F}{2} [e^{-i\omega t} \chi_{AA^\dagger}(\omega) + e^{i\omega t} \chi_{AA^\dagger}(-\omega)]$$

$$= F [\cos \omega t \, \chi'_{AA^\dagger}(\omega) + \sin \omega t \, \chi''_{AA^\dagger}(\omega)].$$ (2.47)

Both the reactive and dissipative response occur at the applied frequency, but the dissipative part acquires a phase shift of $\frac{\pi}{2}$.

Finally, we take n derivatives of Eq. (2.33) with respect to time:

$$\frac{d^n \chi''_{AB}(t)}{dt^n} = \int \frac{d\omega}{2\pi} \chi''_{AB}(\omega)(-i\omega)^n e^{-i\omega t} = \frac{1}{2\hbar} \left\langle \left[\frac{d^n A(t)}{dt^n}, B(0) \right] \right\rangle.$$

(2.48)

Via Heisenberg's equation of motion (2.3), the right-hand side can be recast in terms of n commutators with the Hamiltonian H. Setting $t = 0$ yields the *sum rules*

$$\int \frac{d\omega}{\pi} \omega^n \chi''_{AB}(\omega) = \frac{i^n}{\hbar} \left\langle \left[\frac{d^n A(t)}{dt^n}, B(0) \right] \Big|_{t=0} \right\rangle$$

$$= \frac{1}{\hbar^{n+1}} \langle [[\ldots [A, H], \ldots, H], B] \rangle , \qquad (2.49)$$

which provide *exact* relations for the frequency moments of the dissipative response (applications for the density response are treated in Problem 2.3).

2.2 Stochastic processes

In situations far from thermal equilibrium, especially outside the linear response regime, the probability distribution for a physical system's accessible microstates will in general be a non-trivial function of time t. With prescribed time-independent boundary conditions, e.g., with a fixed particle or energy current running through the system, one expects it to reach a *non-equilibrium stationary state*, after a sufficiently long time has elapsed since its preparation. The identification and characterization of the ensuing special stationary probability distributions and currents, as well as a quantitative description of the associated relaxation phenomena, are among the central research goals of current non-equilibrium statistical mechanics.

To this end, we could in principle take recourse to a fully microscopic description, and solve the quantum-mechanical *von-Neumann equation*

$$\frac{\partial \rho(t)}{\partial t} = -\frac{i}{\hbar} [H, \rho] , \qquad (2.50)$$

which follows immediately from the definition of the density matrix ρ and Schrödinger's equation (2.1), or its classical counterpart, namely *Liouville's equation* for the phase space density,

$$\frac{\partial \rho(t)}{\partial t} = -\{H, \rho\} , \qquad (2.51)$$

where the Poisson bracket replaces the quantum-mechanical commutator. Notice that in both cases, the total time derivative (in the Heisenberg picture within the quantum framework) becomes $d\rho(t)/dt = 0$. Classically, this is just Liouville's theorem, and corresponds to overall probability conservation. However, such a microscopic approach is rarely feasible in practice. Instead, we may utilize our knowledge of the possible transitions between the microstates, and apply the mathematical theory of stochastic processes.[2] By means of coarse-graining, one might

[2] Van Kampen (1981); see also Van Vliet (2010).

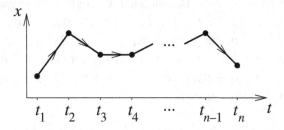

Fig. 2.3 A stochastic process as ordered time sequence of random variables $\{x(t_i)\}$.

hope to subsequently arrive at equations for the time evolution of appropriate meso- or macroscopic variables.

2.2.1 Time-dependent probability distributions

Let us consider a random variable x, dependent on a real parameter t we call 'time'. We then refer to the ordered sequence $\{x(t_1), x(t_2), \dots, x(t_n)\}$, with $t_1 \le t_2 \le \cdots \le t_n$, as a *stochastic process* (Fig. 2.3). Next we introduce the n-point probability distribution P_n. The joint probability for the random variable x to assume a value in the interval $[x_1, x_1 + dx_1]$ at time t_1, a value in the interval $[x_2, x_2 + dx_2]$ at time t_2, etc., is given by $P_n(x_1, t_1; x_2, t_2; \dots; x_n, t_n)\, dx_1 \cdots dx_n$. Its obvious properties are positivity: $P_n(x_1, t_1; x_2, t_2; \dots; x_n, t_n) \ge 0$, and $P_n(\dots x_i, t_i; \dots; x_j, t_j; \dots) = P_n(\dots x_j, t_j; \dots; x_i, t_i; \dots)$, i.e., the ordering of the arguments is irrelevant. In addition, if we demand $\int dx_1\, P_1(x_1, t_1) = 1$ and prescribe the *hierarchy rule*

$$\int dx_n\, P_n(x_1, t_1; \dots; x_{n-1}, t_{n-1}; x_n, t_n) = P_{n-1}(x_1, t_1; \dots; x_{n-1}, t_{n-1}),$$

(2.52)

we ensure proper normalization for all n-point probability distributions,

$$\int dx_1 \cdots \int dx_n\, P_n(x_1, t_1; \dots; x_n, t_n) = 1.$$ (2.53)

Time-dependent averages, moments, and correlations are defined via

$$\langle x(t_1) \cdots x(t_n) \rangle = \int dx_1 \cdots \int dx_n\, P_n(x_1, t_1; \dots; x_n, t_n)\, x_1 \cdots x_n.$$ (2.54)

We call a stochastic process *stationary*, if time translation invariance holds, $P_n(x_1, t_1 + \tau; \dots; x_n, t_n + \tau) = P_n(x_1, t_1; \dots; x_n, t_n)$. Setting $\tau = -t_1$, we note $P_n(x_1, t_1; x_2, t_2; \dots; x_n, t_n) = P_n(x_1, 0; x_2, t_2 - t_1; \dots; x_n, t_n - t_1)$ in this case, whence $P_1(x)$ and $\langle x \rangle$ become time-independent, while for the n-point correlations $\langle x_1(t_1)\, x_2(t_2) \cdots x_n(t_n) \rangle = \langle x_1(0)\, x_2(t_2 - t_1) \cdots x_n(t_n - t_1) \rangle$.

Causal sequences can now be encoded through *conditional* probabilities. Let $P_{m|k}(x_{k+1}, t_{k+1}; \ldots; x_{k+m}, t_{k+m}|x_1, t_1; \ldots; x_k, t_k)\,dx_{k+1}\cdots dx_{k+m}$ indicate the probability of finding x in the intervals $[x_{k+1}, x_{k+1} + dx_{k+1}], \ldots, [x_{k+m}, x_{k+m} + dx_{k+m}]$ at the instances $t_{k+1}, \ldots, t_{k+m}$, respectively, *provided* it already appeared in $[x_1, x_1 + dx_1], \ldots, [x_k, x_k + dx_k]$ at $t_1, \ldots, t_k$. In terms of the n-point distributions, the conditional probabilities become

$$P_{m|k}(x_{k+1}, t_{k+1}; \ldots; x_{k+m}, t_{k+m}|x_1, t_1; \ldots; x_k, t_k)$$

$$= \frac{P_{k+m}(x_1, t_1; \ldots; x_{k+m}, t_{k+m})}{P_k(x_1, t_1; \ldots; x_k, t_k)}, \qquad (2.55)$$

and with Eq. (2.52) we obtain their normalization

$$\int dx_{k+1} \cdots dx_{k+m}\, P_{m|k}(x_{k+1}, t_{k+1}; \ldots; x_{k+m}, t_{k+m}|x_1, t_1; \ldots; x_k, t_k) = 1 .$$

$$(2.56)$$

If the first k outcomes $\{x_1(t_1), \ldots, x_k(t_k)\}$ do not influence the subsequent events $\{x_{k+1}(t_{k+1}), \ldots, x_{k+m}(t_{k+m})\}$, i.e.,

$$P_{m|k}(x_{k+1}, t_{k+1}; \ldots; x_{k+m}, t_{k+m}|x_1, t_1; \ldots; x_k, t_k)$$

$$= P_m(x_{k+1}, t_{k+1}; \ldots; x_{k+m}, t_{k+m}) , \qquad (2.57)$$

the event sequence $x_{k+1}, \ldots, x_{k+m}$ is said to be *statistically independent* of the preceding $x_1, \ldots, x_k$, and we infer

$$P_{k+m}(x_1, t_1; \ldots; x_{k+m}, t_{k+m})$$

$$= P_k(x_1, t_1; \ldots; x_k, t_k)\, P_m(x_{k+1}, t_{k+1}; \ldots; x_{k+m}, t_{k+m}) . \qquad (2.58)$$

Thus, the random variables sets $\{x_1, \ldots, x_k\}$ and $\{x_{k+1}, \ldots, x_{k+m}\}$ are *not correlated*. For a *fully uncorrelated* stochastic process, with no memory at all of previous events, the n-point distribution factorizes completely, $P_n(x_1, t_1; \ldots; x_n, t_n) = \prod_{j=1}^n P_1(x_j, t_j)$.

Markov chains represent another special situation. Here, the value x_n of the random variable x at time t_n depends *only* on its values at the preceding instant t_{n-1}; i.e., there is short-term memory, but any recording of x wipes out the effects of the entire earlier history. Mathematically, the Markovian character means that

$$P_{1|n-1}(x_n, t_n|x_1, t_1; \ldots; x_{n-1}, t_{n-1}) = P_{1|1}(x_n, t_n|x_{n-1}, t_{n-1}) . \qquad (2.59)$$

It is intuitively clear that a Markov process is fully determined by the initial configuration given by $P_1(x_1, t_1)$ and the sequence of intermediate *transition probabilities*

$P_{1|1}(x_{j+1}, t_{j+1}|x_j, t_j)$. Indeed, using Eq. (2.55) and the Markovian property, one readily proves

$$P_n(x_1, t_1; \ldots; x_n, t_n) = P_{1|1}(x_n, t_n|x_{n-1}, t_{n-1})$$

$$\times P_{1|1}(x_{n-1}, t_{n-1}|x_{n-2}, t_{n-2}) \cdots P_{1|1}(x_2, t_2|x_1, t_1) \, P_1(x_1, t_1).$$

$$(2.60)$$

For example, for $t \leq \bar{t} \leq t'$ we can write

$$P_2(x, t; x', t') = \int d\bar{x} \, P_3(x, t; \bar{x}, \bar{t}; x', t')$$

$$= \int d\bar{x} \, P_{1|1}(x', t'|\bar{x}, \bar{t}) \, P_{1|1}(\bar{x}, \bar{t}|x, t) \, P_1(x, t) \, ;$$

dividing with $P_1(x, t)$ then yields the *Chapman–Kolmogorov equation*

$$P_{1|1}(x', t'|x, t) = \int d\bar{x} \, P_{1|1}(x', t'|\bar{x}, \bar{t}) \, P_{1|1}(\bar{x}, \bar{t}|x, t) \, . \qquad (2.61)$$

Its content is that the transition from x to x' can be split into intermediate steps $x \to \bar{x}$ and $\bar{x} \to x'$, and the second transition is independent of the origin of the first step. The associated probability is obtained by multiplying the separate transition probabilities, and integrating over all possible intermediate values $\bar{x}$.

A *Gaussian* stochastic process is fully characterized by the second moments, Eq. (2.54) for $n = 2$, because all higher correlations factorize, compare with Eq. (1.104). One may then show that the factorization property of the increments $\langle [x(t + \tau) - x(t)][x(t' + \tau) - x(t')] \rangle = \langle x(t + \tau) - x(t) \rangle \langle x(t' + \tau) - x(t') \rangle$ for $t \neq t'$ is equivalent to the Markovian character of the process (Problem 2.4).

2.2.2 Master equation

Our goal is to construct an equation of motion for the single-time probability distribution $P_1(x, t)$. To this end, we begin with the identity

$$P_1(x, t') = \int dx' \, P_2(x', t; x, t') = \int dx' \, P_{1|1}(x, t'|x', t) \, P_1(x', t) \, . \qquad (2.62)$$

Setting $t' = t + \tau$, we may now take the continuous-time limit:[3]

$$\frac{\partial P_1(x, t)}{\partial t} = \lim_{\tau \to 0} \frac{P_1(x, t + \tau) - P_1(x, t)}{\tau}$$

$$= \int dx' \, P_1(x', t) \lim_{\tau \to 0} \frac{P_{1|1}(x, t + \tau|x', t) - P_{1|1}(x, t|x', t)}{\tau} \, . \qquad (2.63)$$

---

[3] Notice that we have defined the time derivative via *forward* discretization here; we shall follow this convention throughout this book.

For $\tau = 0$, naturally $P_{1|1}(x, t|x', t) = \delta(x - x')$, which indeed solves (2.62) for $t = t'$. To first order in τ, we try the ansatz $P_{1|1}(x, t + \tau|x', t) = A(\tau)\delta(x - x') + \tau W(x' \to x, t)$, where $W(x' \to x, t)$ represents the *transition rate* from the random variable value x' to x, which we henceforth also interpret as states or configurations of a physical system, during the time interval $[t, t + \tau]$. From the normalization of $P_{1|1}$ we infer $1 = A(\tau) + \tau \int dx\, W(x' \to x, t) + O(\tau^2)$, whence to order τ:

$$P_{1|1}(x, t + \tau|x', t) = \left[1 - \tau \int d\bar{x}\, W(x' \to \bar{x}, t)\right]\delta(x - x') + \tau\, W(x' \to x, t).$$

$$(2.64)$$

Inserting into Eq. (2.63), and renaming integration variables, we finally obtain the *master equation*

$$\frac{\partial P_1(x, t)}{\partial t} = \int dx'\, [P_1(x', t)\, W(x' \to x, t) - P_1(x, t)\, W(x \to x', t)]. \quad (2.65)$$

This fundamental temporal evolution equation balances *gain* and *loss* terms for $P_1(x, t)$ owing to transitions from and to other states $x' \neq x$, respectively. While the master equation (2.65) is valid for any stochastic process in the continuous-time limit, a complete characterization of the kinetics requires full knowledge of the generally time- and implicitly history-dependent transition rates $W(x \to x', t)$.

Clearly, both positivity and normalization of $P_1(x, t)$ are preserved under the time evolution (2.65); e.g.,

$$\frac{\partial}{\partial t} \int dx\, P_1(x, t) = \int dx \int dx'\, [P_1(x', t)\, W(x' \to x, t) - P_1(x, t)\, W(x \to x', t)]$$

$$= 0 \qquad\qquad (2.66)$$

after exchanging integration variables in the second term. We may then define the associated time-dependent *entropy* as

$$S(t) = -k_B \langle \ln P_1(x, t) \rangle = -k_B \int dx\, P_1(x, t) \ln P_1(x, t), \qquad (2.67)$$

and obtain for its temporal evolution

$$\frac{\partial S(t)}{\partial t} = -k_B \int dx\, [\ln P_1(x, t) + 1] \frac{\partial P_1(x, t)}{\partial t}$$

$$= \frac{k_B}{2} \int dx \int dx'\, [P_1(x, t)W(x \to x', t) - P_1(x', t)W(x' \to x, t)]$$

$$\times \ln \frac{P_1(x, t)}{P_1(x', t)} = \sigma(t) - \langle J_S(x, t) \rangle. \qquad (2.68)$$

The expression in the second line follows after using Eq. (2.65) and symmetrizing. Eq. (2.68) introduces the net *entropy flux* to configuration x as the average of

$$J_S(x, t) = k_B \int dx' \, W(x \rightarrow x', t) \, \ln \frac{W(x \rightarrow x', t)}{W(x' \rightarrow x, t)} , \tag{2.69}$$

and the non-negative *entropy production rate*

$$\sigma(t) = \frac{k_B}{2} \int dx \int dx' [P_1(x, t) \, W(x \rightarrow x', t) - P_1(x', t) \, W(x' \rightarrow x, t)]$$

$$\times \ln \frac{P_1(x, t) \, W(x \rightarrow x', t)}{P_1(x', t) \, W(x' \rightarrow x, t)} \geq 0 , \tag{2.70}$$

where the final inequality is a consequence of the convexity of the logarithm function: $(x - x')(\ln x - \ln x') \geq 0$.

In the special case of *time-independent* transition rates, $\partial W(x \rightarrow x')/\partial t = 0$, one can show that there exists at least one *stationary solution* $P_{st}(x)$ with $\partial P_{st}(x)/\partial t = 0$ (provided x is confined to a finite interval). Moreover, if $P_{st}(x)$ is unique, it is also stable and $\lim_{t \rightarrow \infty} P_1(x, t) = P_{st}(x)$. A *sufficient* condition for the existence of such a stable stationary state can be read off from the master equation (2.65), as well as Eqs. (2.68) and (2.70). Specifically, the 'in' and 'out' terms precisely balance each other for all pairs of states x, x', provided the transition rates fulfill the *detailed balance* relation

$$P_{st}(x') \, W(x' \rightarrow x) = P_{st}(x) \, W(x \rightarrow x') . \tag{2.71}$$

Notice that $\sigma(t) = 0$ if and only if (2.71) holds. The approach to the stationary solution $P_{st}(x)$ represents an *irreversible* process which terminates when $P_1(x, t) = P_{st}(x)$ for all states x. Yet in order to check the detailed balance requirement (2.71), one already needs to know the stationary distribution $P_{st}(x)$. A criterion based on the complete set of transition rates only is obviously preferable. In fact, a necessary and sufficient condition for the existence of a detailed balance solution to the master equation (2.65) is that for *all* possible cycles of arbitrary length N (Fig. 2.4), with not necessarily distinct intermediate states $\{x_i\}$, the forward and backward processes be equally likely (*Kolmogorov criterion*),

$$W(x_0 \rightarrow x_1) \, W(x_1 \rightarrow x_2) \cdots W(x_{N-1} \rightarrow x_0)$$

$$= W(x_0 \rightarrow x_{N-1}) \cdots W(x_2 \rightarrow x_1) \, W(x_1 \rightarrow x_0) . \tag{2.72}$$

Markov processes are fully characterized by $P_1(x, t)$ and the transition probabilities $P_{1|1}(x', t'|x, t)$. Inserting (2.64) with $t' = t + \tau$ into the Chapman–Kolmogorov

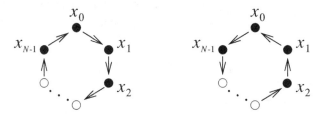

Fig. 2.4 Kolmogorov's condition for the existence of a detailed balance solution: for *any* cycle, the products of forward and backward transition rates must be equal.

equation (2.61) with initial state x_0 at time t_0 yields

$$\frac{\partial P_{1|1}(x, t|x_0, t_0)}{\partial t} = \int dx' \, [P_{1|1}(x', t|x_0, t_0) \, W(x' \to x, t)$$

$$- P_{1|1}(x, t|x_0, t_0) \, W(x \to x', t)] \qquad (2.73)$$

(Problem 2.5). Hence, for Markov processes, both $P_1(x, t)$ and $P_{1|1}(x, t|x_0, t_0)$ obey an identical master equation, which therefore provides a *complete* description of the stochastic dynamics.

In the case of a *finite* number of discrete random variables $\{n\}$, the master equation (2.65) can be written in matrix form,

$$\frac{\partial P_n(t)}{\partial t} = \sum_{n'} [P_{n'}(t) \, W_{n' \to n}(t) - P_n(t) \, W_{n \to n'}(t)]$$

$$= - \sum_{n'} L_{nn'}(t) \, P_{n'}(t) . \qquad (2.74)$$

If we collect the configuration probabilities $P_n(t)$ in a *state vector* $|P(t)\rangle$, this reads in compact notation akin to an 'imaginary-time' Schrödinger equation

$$\frac{\partial}{\partial t} |P(t)\rangle = -\mathcal{L}(t) |P(t)\rangle , \qquad (2.75)$$

where the *Liouville operator* $\mathcal{L}(t)$, which generates the temporal evolution, has the matrix elements

$$L_{nn'}(t) = -W_{n' \to n}(t) + \delta_{nn'} \sum_{\bar{n}} W_{n \to \bar{n}}(t) . \qquad (2.76)$$

Notice that $\sum_n L_{nn'}(t) = 0$, i.e., summing over the matrix elements of $\mathcal{L}(t)$ in each column must yield zero. According to Eq. (2.74), this simply expresses total probability conservation, $\sum_n P_n(t) = 1$ for all t. Formally, this can be written as the inner product $\langle 1|P(t)\rangle = 1$ of the *projection state* vector $\langle 1| = (1, 1, \ldots, 1)$

with $|P(t)\rangle$. Upon integrating Eq. (2.75) over an infinitesimal time step, we see again that conservation of probability implies $\langle 1|\mathcal{L}(t) = \sum_n L_{nn'}(t) = 0$.

When the transition rates and thus $\mathcal{L}$ are time-independent, Eq. (2.75) is formally solved by

$$|P(t)\rangle = e^{-\mathcal{L}t}|P(0)\rangle . \qquad (2.77)$$

This simply reflects that (2.74) represents a set of coupled first-order linear differential equations with constant coefficients. The ansatz $P_n(t) = e^{-\lambda t}\varphi_n$ then leads to an eigenvalue problem $\mathcal{L}|\alpha\rangle = \lambda^{(\alpha)}|\alpha\rangle$, or explicitly

$$\sum_{n'} L_{nn'}\varphi_{n'}^{(\alpha)} = \lambda^{(\alpha)}\varphi_n^{(\alpha)} . \qquad (2.78)$$

But as in general the real matrix $L_{nn'}$ is not symmetric, these right eigenvectors differ from the left eigenvectors, which are determined as the solutions of $\langle\beta|\mathcal{L} = \lambda^{(\beta)}\langle\beta|$, i.e.,

$$\sum_{n'} \psi_{n'}^{(\beta)} L_{n'n} = \lambda^{(\beta)}\psi_n^{(\beta)} . \qquad (2.79)$$

The eigenvectors form a bi-orthonormal set, $\langle\beta|\alpha\rangle = \sum_n \psi_n^{(\beta)}\varphi_n^{(\alpha)} = \delta^{\alpha\beta}$, and the matrix elements (2.76) may be written as $L_{nn'} = \sum_\alpha \lambda^{(\alpha)}\varphi_n^{(\alpha)}\psi_{n'}^{(\alpha)}$. The eigenvalues $\lambda^{(\alpha)}$ of the Liouville operator generally come in *complex*-conjugate pairs with $\mathrm{Re}\,\lambda^{(\alpha)} \geq 0$, and determine the oscillatory and relaxational behavior of the general solution to Eq. (2.74),

$$P_n(t) = \sum_\alpha c^{(\alpha)} e^{-\lambda^{(\alpha)}t}\varphi_n^{(\alpha)} . \qquad (2.80)$$

Since $\langle 1|\mathcal{L}(t) = 0$, there is always at least one left eigenstate with eigenvalue 0, representing the stationary state.

If *detailed balance* holds, i.e., $P_{n'}W_{n'\to n} = P_n W_{n\to n'}$ with the stationary solution $P_n = \lim_{t\to\infty} P_n(t)$, the time evolution matrix may be symmetrized,

$$\tilde{L}_{nn'} = \tilde{L}_{n'n} = \begin{cases} \sum_{\bar{n}} W_{n\to\bar{n}} , & n = n' , \\ -W_{n'\to n}\sqrt{P_{n'}/P_n} = -W_{n\to n'}\sqrt{P_n/P_{n'}} , & n \neq n' . \end{cases} \qquad (2.81)$$

For $\phi_n^{(\alpha)} = \varphi_n^{(\alpha)}/\sqrt{P_n} = \psi_n^{(\alpha)}\sqrt{P_n}$, both Eqs. (2.78) and (2.79) read

$$\sum_{n'} \tilde{L}_{nn'}\phi_{n'}^{(\alpha)} = \lambda^{(\alpha)}\phi_n^{(\alpha)} , \qquad (2.82)$$

with *real*, positive eigenvalues $\lambda^{(\alpha)} \geq 0$. The lowest non-zero eigenvalue λ_m clearly dominates at sufficiently long times, when $P_n(t)$ relaxes towards the stationary solution, namely the eigenstate with zero eigenvalue.

We can construct P_{n_N} explicitly from any starting state P_{n_0} through a sequence of intermediate states with non-vanishing transition rates,

$$P_{n_N} = P_{n_{N-1}} \frac{W_{n_{N-1} \to n_N}}{W_{n_N \to n_{N-1}}} = \cdots = P_{n_0} \prod_{j=0}^{N-1} \frac{W_{n_j \to n_{j+1}}}{W_{n_{j+1} \to n_j}} .$$

Exploiting the condition (2.72), this becomes

$$P_{n_N} = P_{n_0} \frac{W_{n_0 \to n_N}}{W_{n_N \to n_0}} , \tag{2.83}$$

independent of the actual path $n_0 \to \cdots \to n_N$. One may then introduce a *potential* function Φ_n via

$$P_n = \frac{1}{Z} e^{-\Phi_n} , \qquad Z = \sum_n e^{-\Phi_n} , \tag{2.84}$$

and Eq. (2.83) becomes

$$\Phi_{n_N} - \Phi_{n_0} = \sum_{j=0}^{N-1} \ln \frac{W_{n_{j+1} \to n_j}}{W_{n_j \to n_{j+1}}} = \ln \frac{W_{n_N \to n_0}}{W_{n_0 \to n_N}} . \tag{2.85}$$

We may identify the potential with an effective Hamiltonian (with energy measured in units of some fixed temperature $k_B T$), and Z with the associated canonical partition function. Thus, a stochastic process satisfying the detailed-balance condition may be viewed as describing a physical system relaxing towards thermal equilibrium. However, when the transition rates violate Eq. (2.72), there exists a non-vanishing probability current between the different configurations, and the system is inherently out of equilibrium.

In the *micro-canonical* equilibrium ensemble, the system is held at fixed total energy. But for all accessible microstates x on the energy shell *micro-reversibility* should hold, i.e., $W(x \to x') = W(x' \to x)$, compare with Fermi's golden rule (2.44). The detailed-balance condition (2.71) then implies $P_{st}(x) = P_{st}(x')$, i.e., in thermal equilibrium the system will be found with equal probability in any of the microstates compatible with the fixed total energy constraint. In the *canonical* ensemble, the physical system described by a Hamiltonian $H(x)$ is in thermal contact with a heat bath, which provides a constant temperature T. The associated stationary probabilities are given by the canonical distribution (Boltzmann factors),

$$P_{st}(x) = \frac{1}{Z(T)} e^{-H(x)/k_B T} . \tag{2.86}$$

Transition rates satisfying detailed balance must hence obey the condition

$$\frac{W(x \to x')}{W(x' \to x)} = e^{-[H(x')-H(x)]/k_B T} . \tag{2.87}$$

This is actually the microscopic kinetic background for the relation (2.11) between the absorptive and emissive branches of dynamic correlation functions, which led to the fluctuation-dissipation theorem (2.34).

A Monte Carlo computer simulation may be interpreted as a numerical solution of a master equation. One first defines the possible states of a system, e.g., for each site on a discrete lattice. Next certain rules, corresponding to the transition rates, are established according to which the configurations evolve. In order to measure physical quantities and correlations, one performs ensemble averages, over many runs, and/or temporal averages. In the master equation language, this information is encoded in the probabilities $P_1(x, t)$. If one wants to ensure that the simulation eventually reaches a stationary state such that equilibrium properties can be accessed, the assigned transition rates should satisfy detailed balance. For example, the widely employed *Metropolis algorithm* uses the rates (in units of computer time)

$$W(x \to x') = \min\left(1, e^{-[H(x')-H(x)]/k_B T}\right) . \tag{2.88}$$

The move from configuration x to x' is always performed if the associated energy decreases, while such an update is done with a finite probability when $H(x') > H(x)$, i.e., is exponentially suppressed for large energy differences.

2.2.3 *Kramers–Moyal expansion, Fokker–Planck equation*

In the following, we assume that the possible values of the random variables x (characterizing a physical configuration) are in a continuous set. With $x' = x - \xi$ and $t' = t + \tau$, Eq. (2.62) becomes

$$P_1(x, t + \tau) = \int d\xi \, P_{1|1}(x, t + \tau | x - \xi, t) \, P_1(x - \xi, t) . \tag{2.89}$$

Next, provided the transition probabilities from $x - \xi$ to x are invariant with respect to time translations, and therefore $P_{1|1}(x, t + \tau | x - \xi, t) = P_{tr}(x - \xi, \xi, \tau)$ does not depend on the initial time t, we may expand with respect to the small increment ξ:

$$P_1(x, t + \tau) = \int d\xi \left(P_{tr}(x, \xi, \tau) P_1(x, t) - \xi \frac{\partial}{\partial x} [P_{tr}(x, \xi, \tau) P_1(x, t)] \right.$$

$$\left. + \cdots + \frac{(-\xi)^k}{k!} \frac{\partial^k}{\partial x^k} [P_{tr}(x, \xi, \tau) P_1(x, t)] + \cdots \right) .$$

Upon defining the following limit for the kth increment moment of the transition probability,

$$\alpha_k(x) = \lim_{\tau \to 0} \frac{1}{\tau} \int d\xi \, \xi^k \, P_{tr}(x, \xi, \tau) , \qquad (2.90)$$

and noticing that $\int d\xi \, P_{tr}(x, \xi, \tau) = 1$, we arrive at the *Kramers–Moyal expansion*

$$\frac{\partial P_1(x, t)}{\partial t} = \lim_{\tau \to 0} \frac{P_1(x, t + \tau) - P_1(x, t)}{\tau}$$

$$= \sum_{k=1}^{\infty} \frac{(-1)^k}{k!} \frac{\partial^k}{\partial x^k} [\alpha_k(x) \, P_1(x, t)] . \qquad (2.91)$$

Specifically for Markov processes, starting from the Chapman–Kolmogorov equation (2.61) and following the same procedure one readily derives a Kramers–Moyal expansion for the transition probability $P_{1|1}(x, t|x_0, t_0)$ itself (Problem 2.5).

In many instances, only the first two moments contribute to the formal expansion (2.91), and the remaining Kramers–Moyal coefficients vanish. Whereas $\int d\xi \, \xi \, P_{tr}(x, \xi, \tau) = \tau \alpha_1(x)$ and $\int d\xi \, \xi^2 \, P_{tr}(x, \xi, \tau) = \tau \alpha_2(x)$, in this case $\int d\xi \, \xi^k \, P_{tr}(x, \xi, \tau) = O(\tau^2)$, whence $\alpha_k = 0$ for $k \geq 3$. Equation (2.91) then reduces to the *Fokker–Planck equation*

$$\frac{\partial P_1(x, t)}{\partial t} = -\frac{\partial J_1(x, t)}{\partial x} = -\frac{\partial}{\partial x} [\alpha_1(x) \, P_1(x, t)] + \frac{1}{2} \frac{\partial^2}{\partial x^2} [\alpha_2(x) \, P_1(x, t)] ,$$

$$(2.92)$$

which has the form of a continuity equation for the probability density $P_1(x, t)$, with the *probability current*

$$J_1(x, t) = \alpha_1(x) \, P_1(x, t) - \frac{1}{2} \frac{\partial}{\partial x} [\alpha_2(x) \, P_1(x, t)] . \qquad (2.93)$$

If the random variable x is a vector quantity, i.e., more than one label is required to characterize the configuration, $\partial/\partial x$ is of course to be interpreted as the divergence operation. The Kramers–Moyal coefficient $\alpha_1(x)$ then becomes a vector as well, and $\alpha_2(x)$ a matrix (second-rank tensor). The first and second contributions on the right-hand side of (2.92) are referred to as the *drift* and *diffusion* terms, respectively (see also Section 2.3.1).

In fact, *Pawula's theorem* states that for non-vanishing transition probabilities $P_{tr}(x, \xi, \tau) > 0$, the Kramers–Moyal expansion (2.91) either may terminate after the first or second term, which then yields the Fokker–Planck equation (2.92), or must contain *infinitely* many terms. In order to prove this remarkable statement, we introduce the scalar product $(f|g) = \int d\xi \, f(\xi) g(\xi) \, P_{tr}(x, \xi, \tau)$, and exploit the associated generalized Schwarz inequality $(f|g)^2 \le (f|f)(g|g)$, i.e.,

$$\left[\int d\xi\ f(\xi)g(\xi)\,P_{\mathrm{tr}}(x,\xi,\tau)\right]^2 \le \int d\xi\ f(\xi)^2\,P_{\mathrm{tr}}(x,\xi,\tau)$$

$$\times \int d\xi\ g(\xi)^2\,P_{\mathrm{tr}}(x,\xi,\tau)\,. \qquad (2.94)$$

We assume $\alpha_k \neq 0$, but all $\alpha_{k'} = 0$ for $k' > k$, and lead this assertion to contradictions for $k > 2$. For k odd, we set $f(\xi) = \xi^{(k+1)/2}$ and $g(\xi) = \xi^{(k-1)/2}$. Inserting into the inequality (2.94), dividing by τ^2, and taking $\tau \to 0$, we obtain for $k \ge 3$: $0 < \alpha_k(x)^2 \le \alpha_{k+1}(x)\alpha_{k-1}(x)$. But then $\alpha_{k+1} = 0$ would imply $\alpha_k = 0$ as well, which contradicts our assumption. Yet for $k = 1$, we find instead $\tau\,\alpha_1(x)^2 \le \alpha_2(x) = 0$, perfectly compatible with the limit $\tau \to 0$. For even k, on the other hand, the choices $f(\xi) = \xi^{(k+2)/2}$ and $g(\xi) = \xi^{(k-2)/2}$ similarly lead to $0 < \alpha_k(x)^2 \le \alpha_{k+2}(x)\alpha_{k-2}(x) = 0$ for $k \ge 4$, while for $k = 2$ merely $\tau\,\alpha_2(x)^2 \le \alpha_4(x)$. If the Kramers–Moyal expansion does not terminate at $k = 2$, we observe that certainly all even coefficients $\alpha_{2k} \neq 0$, whereas some of the odd ones might vanish.

2.3 Three examples

This concludes our brief exposition of stochastic processes. In this section, we discuss three hopefully instructive examples. Aside from providing interesting physical applications and illustrations for the general concepts introduced in Section 2.2, the important systems to be explored here closely relate to the non-equilibrium chapters on driven diffusive (Chapter 11) and reaction-diffusion systems (Chapter 9), and to the three subsequent chapters on equilibrium critical dynamics.

2.3.1 One-dimensional random walk and biased diffusion

As a first example, we consider a random walk on a one-dimensional chain consisting of N discrete lattice sites j, with lattice constant a_0. The possible configurations of the system can be labeled by the walker's position $x_j = ja_0$ at time t. After each time step of duration τ, the particle may either hop to the right, with probability $0 \le p \le 1$, or to the left, with probability $1 - p$. The corresponding Markovian stochastic process is then defined by specifying the non-vanishing time-independent transition rates. For $2 \le j \le N - 1$, we have $W_{j \to j+1} = p/\tau$ and $W_{j \to j-1} = (1 - p)/\tau$. At this point, we need to specify the boundary conditions at the chain ends. For *periodic* boundary conditions, upon identifying the sites $N + 1$ with 1 and 0 with N, these rules can simply be maintained for $j = N$

and $j = 1$. In the case of *reflecting* boundary conditions, we set $W_{1 \to 2} = p/\tau$, and $W_{N \to N-1} = (1 - p)/\tau$, while allowing no other transitions from sites 1 and N. The associated master equation reads

$$\frac{\partial P_j(t)}{\partial t} = \frac{1}{\tau} \left[p P_{j-1}(t) - P_j(t) + (1 - p) P_{j+1}(t) \right], \tag{2.95}$$

valid for all sites for periodic boundary conditions but only for $2 \le j \le N - 1$ for an open chain, in which case we must supplement Eqs. (2.95) with

$$\frac{\partial P_1(t)}{\partial t} = \frac{1}{\tau} \left[-p P_1(t) + (1 - p) P_2(t) \right],$$

$$\frac{\partial P_N(t)}{\partial t} = \frac{1}{\tau} \left[p P_{N-1}(t) - (1 - p) P_N(t) \right]. \tag{2.96}$$

The choice of boundary conditions turns out to be crucial if we seek stationary solutions P_j to (2.95) that satisfy detailed balance, i.e.,

$$p \, P_j = (1 - p) \, P_{j+1} \tag{2.97}$$

for all sites j and either set of boundary conditions. For the closed chain, a detailed-balance solution thus exists *only* in the unbiased case with $p = 1/2$, when obviously $P_j = 1/N$, and independent random walkers become uniformly distributed as $t \to \infty$. Indeed, Kolmogorov's criterion (2.72) is clearly violated for a biased random walk with $p \ne 1/2$ if we choose the cycle to be the entire system: $p^N \ne (1 - p)^N$. The system is therefore genuinely out of equilibrium, and in the stationary state, a macroscopic particle current runs through the ring. In fact, the simple uniform probability distribution $P_j = 1/N$ satisfies Eq. (2.95). Yet for a proper characterization of this non-equilibrium stochastic process one must also specify the uniform current $\propto (2p - 1)a_0/\tau$.

In the case of *reflecting* boundary conditions, on the other hand, the normalization $1 = \sum_{j=1}^{N} P_j = P_1 \sum_{j=0}^{N-1} [p/(1 - p)]^j$ fixes (for $p \ne 1/2$)

$$P_1 = \frac{1 - 2p}{(1 - p)[1 - p^N/(1 - p)^N]}, \tag{2.98}$$

and thence all other $P_j = [p/(1 - p)]^{j-1} P_1$ through the geometric progression (2.97). This yields a stationary probability distribution that, in the continuum limit, corresponds to a barometric height formula for the particle density, as shown in Fig. 2.5. In the fully directed limit with $p = 0$ one finds $P_1 = 1$, $P_j = 0$ for $2 \le j \le N$; as expected, the particles then accumulate at the left boundary. In the unbiased case $p = 1/2$, we again recover the flat distribution $P_j = 1/N$. In equilibrium statistical mechanics language, these two extreme cases correspond,

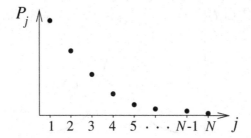

Fig. 2.5 Sketch of the stationary probability distribution for a one-dimensional biased random walk with $0 < p < 1/2$ and reflecting boundary conditions.

for a fixed bias potential $U > 0$, to zero and infinite temperature, respectively: $p = 1/(1 + e^{U/k_B T})$.

In order to take the *continuum limit*, we rewrite Eq. (2.95)

$$\frac{\partial P_j(t)}{\partial t} = \frac{1 - 2p}{2\tau}\left[P_{j+1}(t) - P_{j-1}(t)\right] + \frac{1}{2\tau}\left[P_{j+1}(t) - 2P_j(t) + P_{j-1}(t)\right],$$

(2.99)

neglecting the boundaries. With $x_j = ja_0$, and letting the lattice constant $a_0 \to 0$, we have $P_j(t) \to P_1(x, t)$, and the two terms in square brackets become $2a_0\, \partial P_1(x, t)/\partial x$ and $a_0^2\, \partial^2 P_1(x, t)/\partial x^2$. Thus we directly obtain the Fokker–Planck equation, which takes the form of a *drift-diffusion* equation:

$$\frac{\partial P_1(x, t)}{\partial t} = -v\frac{\partial P_1(x, t)}{\partial x} + D\frac{\partial^2 P_1(x, t)}{\partial x^2},$$

(2.100)

and we may identify the constant Kramers–Moyal coefficients with the *drift velocity* and *diffusion constant*,

$$v = \frac{2a_0}{\tau}\left(p - \frac{1}{2}\right), \quad D = \frac{a_0^2}{2\tau}.$$

(2.101)

For this Markov process, an identical Fokker–Planck equation holds for the transition probabilities $P_{1|1}(x, t|x_0, 0)$.

We proceed to solve the partial differential equation (2.100) for an infinite chain $L = (N - 1)a_0 \to \infty$, and initial particle location $x_0 = 0$: $P_1(x, t = 0) = \delta(x)$ (which is also the proper initial condition for the transition probability). A Galilean transformation $P_1(x, t) = G(x - vt, t)$ eliminates the drift term, and the resulting pure diffusion equation for $G(x, t)$ reads in Fourier space

$$\frac{\partial G(q, t)}{\partial t} = -Dq^2\, G(q, t) + \delta(t)$$

(2.102)

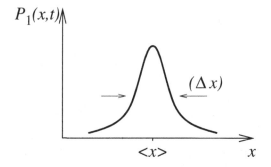

Fig. 2.6 Sketch of the solution (2.104) to the continuum drift-diffusion equation.

for $t \geq 0$, since $G(q = 0, t) = \Theta(t)$. Its solution is just the diffusion *Green function*

$$G(q, t) = e^{-Dq^2 t}\, \Theta(t) \,. \tag{2.103}$$

In coordinate space, this is a Gaussian as well, and we obtain finally

$$P_1(x, t) = G(x - vt, t) = \frac{1}{\sqrt{4\pi Dt}}\, e^{-(x-vt)^2/4Dt}\, \Theta(t) \,, \tag{2.104}$$

see Fig. 2.6. Its first moments are $\langle x \rangle = vt$ and $(\Delta x)^2 = \langle x(t)^2 \rangle - \langle x \rangle^2 = 2Dt$, which is *Fick's diffusion law* for the mean-square displacement. Since the random hopping process is Markovian, the transition probability satisfies the Fokker–Planck equation (2.100) as well. All higher moments factorize for the ensuing Gaussian distribution (2.104), and are therefore at least of order t^2, which confirms that the Kramers–Moyal expansion indeed terminates after the second term. Finally, with the aid of the Green function (2.104), the solution for any arbitrary initial distribution $P_1(x, 0)$ can be constructed:

$$P_1(x, t) = \int dx'\, G(x - x' - vt, t)\, P_1(x', 0) \,, \tag{2.105}$$

which of course is just Eq. (2.89), since we may identify $G(x - x' - vt, t) = P_{1|1}(x, t | x', 0)$.

2.3.2 Population dynamics

Our second example represents a very simplified zero-dimensional model for population dynamics that is devoid of any spatial structure. The possible configurations are indicated by the integer number n of particles or individuals A in a population present at time t. Let us consider offspring production $A \rightarrow A + A$, with rate σ, and spontaneous death, $A \to \emptyset$, with rate κ, which both constitute 'first-order' reactions. The ensuing Markovian dynamics is implemented by defining the corresponding transition rates for the branching and decay processes, namely $W_{n \to n+1} = \sigma n$ and $W_{n \to n-1} = \kappa n$, both of which are proportional to the population number. Notice that the configuration $n = 0$ represents an *absorbing state* in the following sense: once reached, the stochastic process stops there, and there are no fluctuations that allow the system to leave this empty configuration. Quite obviously then, detailed balance cannot be satisfied, and we are dealing with a genuine non-equilibrium system. The corresponding master equations for the time-dependent probabilities $P_n(t)$ are readily written down:

$$\frac{\partial P_n(t)}{\partial t} = \sigma(n-1)P_{n-1}(t) - (\sigma+\kappa)nP_n(t) + \kappa(n+1)P_{n+1}(t) . \qquad (2.106)$$

In order to solve the coupled set of (infinitely many) differential equations (2.106), we introduce the *generating function*

$$g(x, t) = \sum_{n=0}^{\infty} x^n P_n(t) . \qquad (2.107)$$

The desired probabilities $P_n(t)$ are just the nth Taylor coefficients of $g(x, t)$ in the auxiliary variable x. For example, the probability for reaching the empty, absorbing state at time t is $P_0(t) = g(0, t)$, whence the *survival probability* becomes

$$P_a(t) = 1 - g(0, t) . \qquad (2.108)$$

Furthermore, averages are given by appropriate derivatives of $g(x, t)$, e.g., the mean population size is

$$\langle n(t) \rangle = \sum_{n=1}^{\infty} n P_n(t) = \left. \frac{\partial g(x, t)}{\partial x} \right|_{x=1} , \qquad (2.109)$$

and

$$\langle n(t)^2 \rangle = \sum_{n=1}^{\infty} n^2 P_n(t) = \left. \frac{\partial}{\partial x} \left(x \frac{\partial g(x, t)}{\partial x} \right) \right|_{x=1}$$

$$= \left. \frac{\partial g(x, t)}{\partial x} \right|_{x=1} + \left. \frac{\partial^2 g(x, t)}{\partial x^2} \right|_{x=1} , \qquad (2.110)$$

etc. In addition, if initially there are n_0 particles present, i.e., $P_n(t = 0) = \delta_{nn_0}$, the generating function satisfies $g(x, 0) = x^{n_0}$; normalization finally gives

$$g(1, t) = \sum_{n=0}^{\infty} P_n(t) = 1 . \qquad (2.111)$$

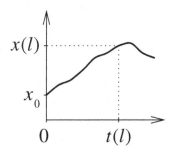

Fig. 2.7 Method of characteristics: The goal is to find a curve parametrization $x(l)$, $t(l)$ such that $x(0) = x_0$, $t(0) = 0$, and $dg/dl = 0$.

The crucial point, however, is that the set of infinitely many coupled master equations (2.106) is contained in the *single* partial differential equation

$$\frac{\partial g(x, t)}{\partial t} = \sum_{n=1}^{\infty} [\sigma x^{n+1} - (\sigma + \kappa)x^n + \kappa x^{n-1}]n P_n(t)$$

$$= \kappa(1 - x)\left(1 - \frac{\sigma}{\kappa} x\right) \frac{\partial g(x, t)}{\partial x} , \qquad (2.112)$$

subject to the boundary condition (2.111). In order to solve (2.112), we employ the *method of characteristics*. The idea is to view the solution as a curve in (x, t) space (Fig. 2.7) and find a parametrization $x(l)$, $t(l)$ such that

$$\frac{dg(x(l), t(l))}{dl} = \frac{\partial g(x, t)}{\partial x} \frac{dx(l)}{dl} + \frac{\partial g(x, t)}{\partial t} \frac{dt(l)}{dl} = 0 , \qquad (2.113)$$

because then

$$g(x(l), t(l)) = g(x(0) = x_0, t(0) = 0) = x_0^{n_0} \qquad (2.114)$$

according to the initial condition. Comparing Eqs. (2.112) and (2.113), we see that we may simply choose time itself as the curve parameter, $t(l) = l$, which leaves us with a first-order ordinary differential equation

$$\frac{dx(t)}{dt} = -\kappa [1 - x(t)]\left[1 - \frac{\sigma}{\kappa} x(t)\right] . \qquad (2.115)$$

Straightforward integration yields

$$-\kappa t = \int_{x_0}^{x(t)} \frac{dx'}{(1 - x')(1 - \sigma x'/\kappa)} = \frac{\kappa}{\kappa - \sigma} \ln\left[\frac{\kappa - \sigma x(t)}{\kappa - \sigma x_0} \frac{1 - x_0}{1 - x(t)}\right]$$

for $\sigma \neq \kappa$, whereas $-\kappa t = [1 - x(t)]^{-1} - [1 - x_0]^{-1}$ for equal rates $\sigma = \kappa$. Solving for $x_0(x, t)$ and inserting into (2.114) one arrives at

$$g(x, t) = x_0(x, t)^{n_0} = \left[\frac{(\kappa - \sigma x)\, e^{(\kappa - \sigma)t} - \kappa(1 - x)}{(\kappa - \sigma x)\, e^{(\kappa - \sigma)t} - \sigma(1 - x)} \right]^{n_0} \qquad (2.116)$$

if $\sigma \neq \kappa$, and

$$g(x, t) = \left[\frac{1 + (1 - x)(\kappa t - 1)}{1 + (1 - x)\kappa t} \right]^{n_0} \qquad (2.117)$$

in the degenerate case $\sigma = \kappa$. It is easily checked that the solutions (2.116) and (2.117) satisfy the normalization condition (2.111).

From Eqs. (2.116) and (2.117), we obtain the extinction probability

$$P_0(t) = g(0, t) = \begin{cases} \left([e^{(\kappa - \sigma)t} - 1] / [e^{(\kappa - \sigma)t} - \frac{\sigma}{\kappa}] \right)^{n_0}, & \sigma \neq \kappa, \\ [\kappa t / (1 + \kappa t)]^{n_0}, & \sigma = \kappa. \end{cases} \qquad (2.118)$$

The extinction probability grows (decreases) exponentially in time if $\kappa > \sigma$ ($\kappa < \sigma$), with a characteristic time scale $\tau_c = 1/|\kappa - \sigma|$. As the control parameter $\kappa/\sigma \to 1$, τ_c diverges, and $P_0(t)$ initially follows a power law. In this sense, our zero-dimensional model displays features that resemble critical behavior near a continuous phase transition. As a possible order parameter, one may use the asymptotic survival probability

$$P_a = 1 - \lim_{t \to \infty} P_0(t) = \begin{cases} 1 - (\kappa/\sigma)^{n_0}, & \kappa < \sigma, \\ 0, & \kappa \geq \sigma, \end{cases} \qquad (2.119)$$

plotted in Fig. 2.8. Lastly, we determine the mean population size using Eq. (2.109), $\langle n(t) \rangle = n_0\, e^{(\sigma - \kappa)t}$. Consequently, $n_\infty = \lim_{t \to \infty} \langle n(t) \rangle$ vanishes in the *inactive* state ($\kappa > \sigma$), diverges in the *active* phase ($\kappa < \sigma$), whereas $\langle n(t) \rangle = n_0$ precisely at the extinction threshold. A straightforward calculation (Problem 2.6) yields that the mean-square particle number fluctuations grow linearly in time at this 'critical point': $(\Delta n)^2 = 2n_0\kappa t$. With balancing production and decay rates, the stochastic process $\{n(t)\}$ may be viewed as a one-dimensional unbiased random walk, starting at n_0, and with effective 'diffusion' constant $n_0\kappa$. We remark that if we replace the branching process with spontaneous particle creation $\emptyset \to A$ with rate $\tau > 0$, i.e., $W_{n \to n+1} = \tau$ (independent of n), the system always resides in the active state with finite particle density $n_\infty = \tau/\kappa$ (see Problem 2.7).

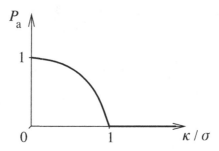

Fig. 2.8 Asymptotic survival probability for the population model as a function of the ratio κ/σ of the decay and branching rates.

2.3.3 Kinetic Ising models

As a final example, we construct *equilibrium* kinetics for the ferromagnetic Ising model (1.10). Its possible configurations are given by the set of spin values $\{\sigma_i = \pm 1\}$ at each lattice site i. We take the thermodynamic limit, and will not worry about boundary conditions here. Starting from an arbitrary initial state, the simplest Markovian dynamics consists of allowing local *spin flips* $\sigma_i \to -\sigma_i$ with certain rates $W(\sigma_i)$. This is usually referred to as *Glauber* kinetics. In order to ensure that the system eventually relaxes to the canonical distribution, we impose detailed balance, which we may conveniently write in terms of the effective local fields (1.14) as

$$e^{h_{\text{eff},i}\sigma_i/k_B T}\, W(\sigma_i) = e^{-h_{\text{eff},i}\sigma_i/k_B T}\, W(-\sigma_i)\,. \tag{2.120}$$

As $\sigma_i^2 = 1$, we have the identity

$$e^{\pm h_{\text{eff},i}\sigma_i/k_B T} = \left(1 \pm \sigma_i \tanh \frac{h_{\text{eff},i}}{k_B T}\right) \cosh \frac{h_{\text{eff},i}}{k_B T}\,,$$

and a simple choice for the transition rates that satisfies Eq. (2.120) is

$$W(\sigma_i) = \Gamma_G \left(1 - \sigma_i \tanh \frac{h_{\text{eff},i}}{k_B T}\right) \tag{2.121}$$

with a constant flip rate Γ_G. Notice that for $T \to \infty$, the energetics becomes irrelevant, and all rates equal to Γ_G. For $T = 0$, on the other hand, $W(\sigma_i) = \Gamma_G(1 - \sigma_i \operatorname{sgn} h_{\text{eff},i})$. The spin at site i may only flip, with rate $2\Gamma_G$, if the net effective field there points in the opposite direction.

In one dimension, and with nearest-neighbor exchange couplings only, $h_{\text{eff},i} = h + J(\sigma_{i-1} + \sigma_{i+1})$, and the rates $W(\sigma_i)$ depend just on the values of the two adjacent spins. This yields the following list of possible processes and respective

Glauber transition rates:

$$\uparrow\uparrow\uparrow \;\;\rightarrow\;\; \uparrow\downarrow\uparrow \qquad \Gamma_G \left(1 - \tanh \frac{h+2J}{k_B T}\right) ,$$

$$\uparrow\uparrow\downarrow \;\;\rightarrow\;\; \uparrow\downarrow\downarrow \qquad \Gamma_G \left(1 - \tanh \frac{h}{k_B T}\right) ,$$

$$\downarrow\uparrow\uparrow \;\;\rightarrow\;\; \downarrow\downarrow\uparrow \qquad \Gamma_G \left(1 - \tanh \frac{h}{k_B T}\right) ,$$

$$\downarrow\uparrow\downarrow \;\;\rightarrow\;\; \downarrow\downarrow\downarrow \qquad \Gamma_G \left(1 - \tanh \frac{h-2J}{k_B T}\right) ,$$

$$\uparrow\downarrow\uparrow \;\;\rightarrow\;\; \uparrow\uparrow\uparrow \qquad \Gamma_G \left(1 + \tanh \frac{h+2J}{k_B T}\right) ,$$

$$\uparrow\downarrow\downarrow \;\;\rightarrow\;\; \uparrow\uparrow\downarrow \qquad \Gamma_G \left(1 + \tanh \frac{h}{k_B T}\right) ,$$

$$\downarrow\downarrow\uparrow \;\;\rightarrow\;\; \downarrow\uparrow\uparrow \qquad \Gamma_G \left(1 + \tanh \frac{h}{k_B T}\right) ,$$

$$\downarrow\downarrow\downarrow \;\;\rightarrow\;\; \downarrow\uparrow\downarrow \qquad \Gamma_G \left(1 + \tanh \frac{h-2J}{k_B T}\right) . \qquad (2.122)$$

Of course, the external field h favors spin flips in its direction. For $h = 0$, this bias disappears, and the first four of the above processes become degenerate with their $\uparrow \leftrightarrow \downarrow$ 'mirror' images. We may then employ the *domain wall* representation of the Ising system, where a pair of opposite spins is labeled as a 'particle' A, while a pair of parallel spins is viewed as an empty space $\emptyset$. In this language, the four elementary Glauber processes translate to pair creation, unbiased hopping, and pair annihilation of domain walls, respectively, with the following rates:

$$\emptyset\emptyset \;\rightarrow\;\; A\,A \qquad \Gamma_G \left(1 - \tanh \frac{2J}{k_B T}\right) ,$$

$$\emptyset\,A \;\rightarrow\;\; A\,\emptyset \qquad \Gamma_G ,$$

$$A\,\emptyset \;\rightarrow\;\; \emptyset\,A \qquad \Gamma_G ,$$

$$A\,A \;\rightarrow\;\; \emptyset\emptyset \qquad \Gamma_G \left(1 + \tanh \frac{2J}{k_B T}\right) . \qquad (2.123)$$

As the temperature decreases to zero, domain wall creation is rendered impossible, and the annihilation rate becomes twice the hopping rate. The kinetic Ising system approaches the ordered state through expelling the interfaces. Within the mean-field approximation, we may infer the overall density $n(t)$ of domain walls by means of an approximate rate equation that balances the spontaneous production and pair annihilation contributions,

$$\frac{1}{\Gamma_G}\frac{dn(t)}{dt} = \left(1 - \tanh\frac{2J}{k_B T}\right)[1 - n(t)]^2 - \left(1 + \tanh\frac{2J}{k_B T}\right)n(t)^2 . \quad (2.124)$$

In the stationary state, we thus find the domain wall density

$$n_s = \left(1 + \sqrt{\frac{1 + \tanh(2J/k_B T)}{1 - \tanh(2J/k_B T)}}\right)^{-1} = \frac{1}{1 + e^{2J/k_B T}} , \quad (2.125)$$

which vanishes exponentially as $T \to 0$, proportional to the inverse correlation length (compare with Problem 1.2), whereas $n_s \to 1/2$ as $T \to \infty$.

Under Glauber kinetics, the total magnetization $M = \sum_i \sigma_i$ is not conserved. However, for the Ising model without additional anisotropies, M is actually a *conserved* quantity, its fixed value being determined by the external field h. A more appropriate microscopic relaxation mechanism therefore consists of *spin exchanges* $\sigma_i \leftrightarrow \sigma_j \neq \sigma_i$, called *Kawasaki* dynamics, with transition rates $W(\sigma_i \leftrightarrow \sigma_j) \propto \Gamma_K \frac{1}{2}(1 - \sigma_i \sigma_j)$. In the lattice gas representation (1.30), this simply describes particle transfers to empty sites; h is then related to the chemical potential that fixes the overall particle number $\sum_i n_i = \frac{1}{2}(N + M)$. According to detailed balance, the exchange rates $W(\sigma_i \leftrightarrow \sigma_j)$ will again be determined by the energetics, but become independent of the external field h. In one dimension and for purely nearest-neighbor interactions, we have the following elementary processes, both in the spin and domain wall picture, and corresponding rates:

$$
\begin{array}{ccccccc}
\uparrow\uparrow\downarrow\uparrow & \to & \uparrow\downarrow\uparrow\uparrow & \emptyset\, A\, A & \to & A\, A\, \emptyset & \Gamma_K , \\
\uparrow\uparrow\downarrow\downarrow & \to & \uparrow\downarrow\uparrow\downarrow & \emptyset\, A\, \emptyset & \to & A\, A\, A & \Gamma_b , \\
\downarrow\uparrow\downarrow\uparrow & \to & \downarrow\downarrow\uparrow\uparrow & A\, A\, A & \to & \emptyset\, A\, \emptyset & \Gamma_a , \\
\downarrow\uparrow\downarrow\downarrow & \to & \downarrow\downarrow\uparrow\downarrow & A\, A\, \emptyset & \to & \emptyset\, A\, A & \Gamma_K . \quad (2.126)
\end{array}
$$

In particle language, the first and last process represent hopping, the second one branching, and the third one fusion. As $T \to \infty$, all rates must become equal, while for $T \to 0$ one should expect $\Gamma_b \to 0$ and $\Gamma_a \to 2\Gamma_K$. In general, $\Gamma_a + \Gamma_b = 2\Gamma_K$; in addition, the detailed balance conditions imply

$$\Gamma_b = e^{-4J/k_B T}\Gamma_a \quad (2.127)$$

for the second and third of the above spin exchange processes and their reverse, whence

$$\Gamma_a = \frac{2\Gamma_K}{1 + e^{-4J/k_BT}} = \Gamma_K \left(1 + \tanh \frac{2J}{k_BT} \right) ,$$

$$\Gamma_b = \frac{2\Gamma_K}{1 + e^{4J/k_BT}} = \Gamma_K \left(1 - \tanh \frac{2J}{k_BT} \right) . \tag{2.128}$$

The mean-field rate equation for the average domain wall density now reads

$$\frac{1}{\Gamma_K} \frac{dn(t)}{dt} = \left(1 - \tanh \frac{2J}{k_BT} \right) n(t)[1 - n(t)]^2 - \left(1 + \tanh \frac{2J}{k_BT} \right) n(t)^3 , \tag{2.129}$$

with the same stationary solution (2.125) as for the Glauber model.

2.4 Langevin equations

As mentioned above, solving the fully microscopic equations of motion for stochastic dynamical systems is rarely feasible. Neither is it really desired: a description in terms of a few *mesoscopic* degrees of freedom, whose averages yield macroscopic observables, is much preferable. This requires some sort of *coarse-graining*. Moreover, we are typically interested in the long-time behavior of certain characteristic quantities, and not in their complete short-time kinetics. Provided an appropriate *separation of time scales* applies, we may attempt to formulate dynamic equations for the relevant 'slow' mesoscopic degrees of freedom. The 'fast' microscopic variables then act on the former as random forces, and can be viewed as system-inherent stochastic *noise*. This leads to *Langevin* stochastic differential equations, which have proven quite useful in the study of dynamic critical phenomena, both in and away from thermal equilibrium.[4]

2.4.1 Langevin–Einstein theory of Brownian motion

We begin with a brief study of *Brownian motion*. A large, heavy particle, with mass m, moves with velocity v (a d-dimensional vector) in a fluid consisting of many small, light particles. The impacts with the fluid particles are modeled through

---

[4] Good introductions to Langevin dynamics can also be found in Reif (1985), Chaikin and Lubensky (1995), Pathria (1996), Cowan (2005), Schwabl (2006), Kardar (2007), Reichl (2009), and Van Vliet (2010).

Stochastic dynamics

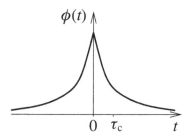

$\phi(t)$

$0 \quad \tau_c \quad t$

Fig. 2.9 Typical correlation function for the stochastic forces.

a stochastic force $f(t)$. In addition, the random collisions on average generate a *friction* force $-m\zeta v$. Newton's equation of motion for the Brownian particle thus becomes a *stochastic differential equation*

$$m \frac{\partial v(t)}{\partial t} = -m\zeta\, v(t) + f(t) . \tag{2.130}$$

In order to solve the *Langevin equation* (2.130), we need to specify the statistical properties of the random force $f(t)$. As we have already taken care of the mean effect of the collisions through the friction term, the time average of the stochastic force on a given large particle, or equivalently, the ensemble average over a large set of alike Brownian particles, vanishes,

$$\langle f(t) \rangle = 0 . \tag{2.131}$$

For the second moment, we assume time translation invariance, and take the d spatial force components to be uncorrelated:

$$\langle f_i(t) f_j(t') \rangle = \phi(t - t') \delta_{ij} . \tag{2.132}$$

In the spirit of the central-limit theorem we demand that the specification of the first two moments (2.131) and (2.132) should suffice to fully characterize this stochastic process, which we hence presume to be *Gaussian*.

We expect the stochastic forces to lose any correlations beyond a typical time scale τ_c, roughly the duration of a collision with a fluid particle; e.g.,

$$\phi(t) = \frac{\lambda}{2\tau_c}\, e^{-|t|/\tau_c} . \tag{2.133}$$

In general, $\phi(t) = \phi(-t)$ is a symmetric function, with $\phi(0) = \langle f(t)^2 \rangle / d \geq 0$, and we expect $\lim_{|t-t'| \to \infty} \phi(t - t') \delta_{ij} = \langle f_i(t) \rangle \langle f_j(t') \rangle = 0$. Because of $0 \leq \langle [f(t) \pm f(t')]^2 \rangle = 2d\, [\phi(0) \pm \phi(t - t')]$, furthermore $|\phi(t)| \leq \phi(0)$; see Fig. 2.9. The Fourier transform of the memory function $\phi(t)$ is symmetric as well,

$\phi(\omega) = \phi(-\omega)$, and satisfies the *Wiener–Khintchine theorem*

$$\phi(t = 0) = \frac{1}{d}\langle f(t)^2\rangle = \frac{1}{\pi}\int_0^\infty \phi(\omega)\,d\omega\,. \tag{2.134}$$

For example, the Fourier transform of Eq. (2.133) is a Lorentzian with line width $1/\tau_c$,

$$\phi(\omega) = \frac{\lambda/\tau_c^2}{\omega^2 + 1/\tau_c^2}\,. \tag{2.135}$$

If we are interested in the long-time limit $t \gg \tau_c$ only, we may effectively replace (2.133) with a delta function,

$$\phi(t) \to \lambda\,\delta(t)\,, \tag{2.136}$$

or equivalently $\phi(\omega) \to \phi(\omega = 0) = \lambda$, yielding uncorrelated *white noise*.

In this limit, the associated Gaussian probability distribution for the stochastic forces becomes

$$P[f] = C\exp\left[-\frac{1}{2\lambda}\int_{t_0}^{t_f} f(t)^2\,dt\right] \tag{2.137}$$

for processes starting at time t_0 and ending at t_f. Averages over random force histories may then be computed via functional integration $\langle A[f]\rangle = \int \mathcal{D}[f]\,A[f]\,P[f]$. The integral over all possible 'paths' $f(t)$ is constructed through temporal discretization with M time steps of length $\tau = (t_f - t_0)/M$, $t_l = t_0 + l\tau$, $l = 0,\ldots,M$, where we identify $t_M = t_f$. Upon defining the functional integration measure explicitly via

$$\mathcal{D}[f] = \lim_{\tau\to 0}\prod_{l=0}^{M-1}\left(\frac{\tau}{2\pi\lambda}\right)^{d/2} d^d f(t_l)\,, \tag{2.138}$$

we obtain

$$\int \mathcal{D}[f]\,P[f] = \lim_{\tau\to 0}\int\prod_{l=0}^{M-1}\left(\frac{\tau}{2\pi\lambda}\right)^{d/2} d^d f(t_l)\exp\left[-\frac{\tau}{2\lambda}\sum_{l=0}^{M-1} f(t_l)^2\right] = 1\,,$$

whereupon we can set the normalization constant $C = 1$. As a check, we also compute

$$\langle f_i(t_l)f_j(t_{l'})\rangle = \lambda\frac{\delta_{ll'}}{\tau}\delta_{ij}\,,$$

i.e., the appropriately discretized version of (2.132) and (2.136).

In order to proceed with the analysis of the Langevin equation (2.130), we apply the Green function technique. The associated differential equation $\partial G(t)/\partial t + \zeta G(t) = \delta(t)$ is solved by $G(t) = e^{-\zeta t} \Theta(t)$. With $v(t=0) = v_0$, the solution of the homogeneous part of Eq. (2.130) reads $v(t) = v_0 e^{-\zeta t}$, whence that of the full, inhomogeneous equation becomes

$$v(t) = v_0 e^{-\zeta t} + \frac{1}{m} \int_0^\infty G(t-t') f(t') \, dt'$$

$$= e^{-\zeta t} \left(v_0 + \frac{1}{m} \int_0^t e^{\zeta t'} f(t') \, dt' \right) . \tag{2.139}$$

Let us now consider the velocity correlation function

$$\langle v_i(t) v_j(t') \rangle = e^{-\zeta(t+t')} \left[v_{0i} v_{0j} + \frac{\delta_{ij}}{m^2} \int_0^t d\tau \int_0^{t'} d\tau' \, e^{\zeta(\tau+\tau')} \phi(\tau - \tau') \right],$$

where we have inserted the moments (2.131) and (2.132). For $\tau_c \to 0$, we may use Eq. (2.136), whereupon the double integral here reduces to $\lambda(e^{2\zeta t} - 1)/2\zeta$ for $t < t'$, and thus

$$\langle v_i(t) v_j(t') \rangle = \left(v_{0i} v_{0j} - \frac{\lambda}{2\zeta m^2} \delta_{ij} \right) e^{-\zeta(t+t')} + \frac{\lambda}{2\zeta m^2} \delta_{ij} e^{-\zeta |t-t'|} . \tag{2.140}$$

Asymptotically for $t, t' \gg 1/\zeta$, only the last term survives. If we assume that the Brownian particle eventually equilibrates with the fluid at temperature T, we may employ the classical equipartition theorem, $\frac{d}{2} k_B T = \frac{m}{2} \langle v(t)^2 \rangle = d\lambda/4\zeta m$, to obtain *Einstein's relation*

$$\lambda = 2\zeta m k_B T . \tag{2.141}$$

Hence, in thermal *equilibrium*, the relaxation coefficient ζ is determined by the strength of the noise correlations λ and $k_B T$. Notice that a *double* separation of time scales has been applied here, namely $\tau_c \ll 1/\zeta \ll t$. If we relax the first condition, we just need to replace λ with $\phi(\omega = 0)$,

$$\zeta = \frac{\phi(\omega = 0)}{2m k_B T} = \frac{1}{2dm k_B T} \int_{-\infty}^\infty \langle f(t) \cdot f(0) \rangle \, dt . \tag{2.142}$$

This relation is called the *fluctuation-dissipation theorem of the second kind*. It ensures that the kinetic energy dissipation in the Langevin equation (2.130) is on average balanced by the fluctuating force input, see Problem 2.8, and that the system eventually relaxes to thermal equilibrium.

The Brownian particle's mean-square displacement follows from $x(t) = \int_0^t v(\tau) \, d\tau$ and our previous result (2.140):

$$
\begin{aligned}
\langle x(t)^2 \rangle &= \int_0^t d\tau \int_0^t d\tau' \, \langle v(\tau) \cdot v(\tau') \rangle \\
&= \left(v_0^2 - \frac{d\lambda}{2\zeta m^2} \right) \left(\int_0^t e^{-\zeta\tau} \, d\tau \right)^2 + \frac{d\lambda}{2\zeta m^2} \int_0^t d\tau \int_0^t d\tau' \, e^{-\zeta|\tau-\tau'|} \\
&= \left(v_0^2 - \frac{d\lambda}{2\zeta m^2} \right) \frac{1}{\zeta^2} (1 - e^{-\zeta t})^2 + \frac{d\lambda}{\zeta^2 m^2} \left[t - \frac{1}{\zeta} (1 - e^{-\zeta t}) \right]
\end{aligned}
$$

(2.143)

after straightforward integration. At short times $\tau_c \ll t \ll 1/\zeta$, a Taylor expansion to order $(\zeta t)^2$ yields *ballistic* motion with the initial velocity: $\langle x(t)^2 \rangle \approx v_0^2 t^2$. For $t \gg 1/\zeta$, on the other hand, we may neglect $e^{-\zeta t}$, and the result becomes independent of the initial condition. With $\langle x(t) \rangle = v_0 (1 - e^{-\zeta t})/\zeta$, Eq. (2.143) then yields *Fick's diffusion law* $(\Delta x)^2 \approx 2Dt$, with the diffusion coefficient

$$
D = \frac{d\lambda}{2\zeta^2 m^2} = \frac{d}{\zeta m} k_{\rm B} T \; .
$$

(2.144)

In the last equation, we have inserted the Einstein relation (2.141).

2.4.2 *Fokker–Planck equation for free Brownian motion*

The solution $\{v(t)\}$ of the Langevin equation (2.130) for a free Brownian particle can be viewed as a stochastic process. For the associated Kramers–Moyal expansion, we need the moments of the velocity increments $\xi(t) = v(t + \tau) - v(t)$,

$$
\alpha_{i_1 \ldots i_k}(v) = \lim_{\tau \to 0} \frac{1}{\tau} \int d^d\xi \, \xi_{i_1} \cdots \xi_{i_k} \, P_{\rm tr}(v, \xi, \tau) \; .
$$

(2.145)

We utilize time translation invariance, and obtain from the explicit solution (2.139): $\langle v(\tau) - v_0 \rangle = (e^{-\zeta\tau} - 1)v_0 \to -\zeta\tau v_0$ as $\tau \to 0$. Hence the first Kramers–Moyal coefficient becomes $\alpha_1(v) = -\zeta v$. For the second moment, we compute with the aid of Eq. (2.140)

$$
\langle [v_i(\tau) - v_{i0}][v_j(\tau) - v_{j0}] \rangle = v_{0i} v_{0j} (e^{-\zeta\tau} - 1)^2 - \frac{\lambda}{2\zeta m^2} \delta_{ij} (e^{-2\zeta\tau} - 1) \, ,
$$

whence $\alpha_{ij} = \lambda \delta_{ij}/m^2$, independent of v. In order to determine the remaining moments, we use $\zeta\tau \ll 1$ right away, apply forward integration to Eq. (2.130),

$$
v(t + \tau) - v(t) \approx -\zeta\tau \, v(t) + \frac{1}{m} \int_t^{t+\tau} f(t') \, dt' \, ,
$$

(2.146)

and exploit the properties of the Gaussian distribution (2.137). Thus,

$$\langle \xi_i \xi_j \xi_k \rangle \approx (-\zeta\tau)^3 v_i v_j v_k - \zeta\tau(v_i\delta_{jk} + v_j\delta_{ik} + v_k\delta_{ij})$$

$$\times \int_t^{t+\tau} dt' \int_t^{t+\tau} dt'' \frac{\lambda}{m^2} \delta(t - t') ,$$

which is of order τ^2, as the double integral in the second term yields $\lambda\tau/m^2$. Similarly, we obtain

$$\langle \xi_i \xi_j \xi_k \xi_l \rangle \approx (-\zeta\tau)^4 v_i v_j v_k v_l + (\zeta\tau)^2 (v_i v_j \delta_{kl} + 5 \text{ permutations}) \frac{\lambda}{m^2} \tau$$

$$+ (\delta_{ij}\delta_{kl} + \delta_{ik}\delta_{jl} + \delta_{il}\delta_{jk}) \left(\frac{\lambda}{m^2} \tau\right)^2 .$$

Consequently, $\alpha_{ijk}(v) = 0 = \alpha_{ijkl}(v)$, and according to Pawula's theorem, all higher Kramers–Moyal coefficients vanish as well. The probability distribution $P_1(v, t)$ for free Brownian motion therefore obeys the Fokker–Planck equation

$$\frac{\partial P_1(v, t)}{\partial t} = \zeta \frac{\partial}{\partial v} \cdot [v P_1(v, t)] + \frac{\lambda}{2m^2} \frac{\partial^2 P_1(v, t)}{\partial v^2} . \qquad (2.147)$$

It is instructive to derive Eq. (2.147) directly from the identity

$$P_1(v, t) = \langle \delta(v - v(t)) \rangle . \qquad (2.148)$$

Upon inserting Eq. (2.130) one finds

$$\frac{\partial P_1(v, t)}{\partial t} = -\frac{\partial}{\partial v} \cdot \left\langle \delta(v - v(t)) \left[-\zeta v(t) + \frac{1}{m} f(t)\right]\right\rangle .$$

In the first term, we may replace $v(t)$ with v, which yields the drift term in the Fokker–Planck equation. For the second contribution, we recall the definition of averages with the Gaussian probability distribution (2.137), and integrate by parts

$$\langle \delta(v - v(t)) f(t) \rangle = \int \mathcal{D}[f] \delta(v - v(t)) f(t) \exp\left[-\frac{1}{2\lambda} \int f(t')^2 dt'\right]$$

$$= -\lambda \int \mathcal{D}[f] \delta(v - v(t)) \frac{\delta}{\delta f(t)} \exp\left[-\frac{1}{2\lambda} \int f(t')^2 dt'\right]$$

$$= \lambda \left\langle \frac{\delta}{\delta f(t)} \delta(v - v(t)) \right\rangle = -\lambda \frac{\partial}{\partial v} \left\langle \delta(v - v(t)) \frac{\delta v(t)}{\delta f(t)} \right\rangle .$$

$$(2.149)$$

Lastly we obtain from the explicit solution (2.139)

$$\frac{\delta v(t)}{\delta f(t)} = \frac{e^{-\zeta t}}{m} \int_0^t e^{\zeta t'} \delta(t - t') dt' = \frac{1}{2m} , \qquad (2.150)$$

and collecting all contributions, we are led to the diffusion term in (2.147).

Writing the Fokker–Planck equation in the form of a continuity equation as in Eq. (2.92), we find for the probability current

$$J_1(v, t) = -\zeta v P_1(v, t) - \frac{\lambda}{2m^2} \frac{\partial P_1(v, t)}{\partial v} . \tag{2.151}$$

For a stationary solution $P_{st}(v)$, $J_{st}(v) = 0$. Provided the Einstein relation (2.141) holds, this condition is indeed satisfied by the classical *Maxwell–Boltzmann velocity distribution*

$$P_{st}(v) = \left(\frac{m}{2\pi k_B T}\right)^{d/2} e^{-mv^2/2k_B T} . \tag{2.152}$$

This is confirmed by an explicit solution of the Fokker–Planck equation (2.147). To this end, we substitute $\rho(v, t) = v\, e^{\zeta t}$ into $P_1(v, t) = Y(\rho, t)$, whence

$$\frac{\partial Y(\rho, t)}{\partial t} = d\zeta\, Y(\rho, t) + \frac{\lambda}{2m^2} e^{2\zeta t} \frac{\partial^2 Y(\rho, t)}{\partial \rho^2} .$$

The ansatz $Y(\rho, t) = X(\rho, t)\, e^{d\zeta t}$ then eliminates the homogeneous term, leaving us with a diffusion equation with time-dependent diffusion coefficient $D(t)$. The latter is solved simply through replacing Dt in the standard results with the integral $\int_0^t D(t')\, dt'$. Here, this amounts to transforming to a new time variable $\theta(t) = (e^{2\zeta t} - 1)/2\zeta$ with $\theta(0) = 0$, which yields the normal diffusion equation

$$\frac{\partial X(\rho, \theta)}{\partial \theta} = \frac{\lambda}{2m^2} \frac{\partial^2 X(\rho, \theta)}{\partial \rho^2} .$$

Notice that with the initial condition $v(t = 0) = v_0$, we have $P_1(v, t) = P_{1|1}(v, t|v_0, 0)$. As we shall see below, in the case of uncorrelated noise (2.136) the Langevin equation indeed describes a Markovian stochastic process, and the transition probability satisfies the very same Fokker–Planck equation (2.147), see Problem 2.5. For $\rho(0) = v_0$, we find with the straightforward generalization of Eq. (2.104) to d space dimensions

$$X_{1|1}\left(\rho(v, t), \theta(t)|v_0, 0\right) = \left(\frac{m^2}{2\pi \lambda\, \theta(t)}\right)^{d/2} e^{-m^2[\rho(v,t)-v_0]^2/2\lambda\theta(t)} ,$$

and hence in terms of the original variables

$$P_{1|1}(v, t|v_0, 0) = \left[\frac{\zeta m^2}{\pi \lambda (1 - e^{-2\zeta t})}\right]^{d/2} \exp\left[-\frac{\zeta m^2 (v - v_0 e^{-\zeta t})^2}{\lambda(1 - e^{-2\zeta t})}\right] . \tag{2.153}$$

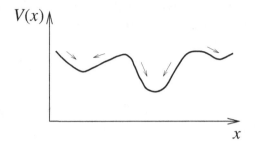

Fig. 2.10 Brownian motion in an external potential $V(x)$.

Initially, this becomes $P_{1|1}(v, 0|v_0, 0) = \delta(v - v_0)$, as it should, while asymptotically

$$P_{1|1}(v, t \to \infty|v_0, 0) = P_{st}(v) = \left(\frac{\zeta m^2}{\pi \lambda}\right)^{d/2} e^{-\zeta m^2 v^2/\lambda} \,, \tag{2.154}$$

independent of v_0. For $\lambda = 2\zeta m \, k_B T$, this is exactly the Maxwell–Boltzmann distribution (2.152). Given an arbitrary normalized initial velocity distribution $w_0(v)$ with $\int w_0(v) \, d^d v = 1$, the general solution of Eq. (2.147) reads

$$P_1(v, t) = \int P_{1|1}(v, t|v', 0) \, w_0(v') \, d^d v' \,, \tag{2.155}$$

see Eqs. (2.62) or (2.105), since $P_{1|1}(v, t|v_0, 0)$ is just the corresponding Green function. Thus $P_1(v, 0) = w_0(v)$, whereas $P_1(v, t \to \infty) = P_{st}(v)$. Quite independently of the initial conditions, the probability distribution relaxes to the same stationary limit, namely Eq. (2.154).

2.4.3 Random motion in an external potential

We now generalize the above considerations to the motion of a Brownian particle in an external potential $V(x)$, as shown in Fig. 2.10. With $v(t) = \partial x(t)/\partial t$, and the force $F(x) = -\partial V(x)/\partial x$, Newton's equation now reads

$$m \frac{\partial^2 x(t)}{\partial t^2} = -m\zeta \frac{\partial x(t)}{\partial t} + F(x) + f(t) \,, \tag{2.156}$$

where we assume the same statistical properties (2.131), (2.132) for the stochastic force as before. Specifically, we may use (2.136) in the limit of short collision times, and Einstein's relation (2.141) holds in thermal equilibrium.

In the case of strong damping, where $\zeta |v| \gg |\dot{v}|$, we may neglect the inertial term. Thus we arrive at the Langevin equation in the overdamped limit,

$$\frac{\partial x(t)}{\partial t} = -\Gamma \frac{\partial V(x)}{\partial x} + r(t) , \qquad (2.157)$$

where $\Gamma = 1/m\zeta$ and $r(t) = f(t)/m\zeta$, whence

$$\langle r(t) \rangle = 0 , \qquad \langle r_i(t) r_j(t') \rangle = 2\Gamma k_B T \, \delta_{ij} \delta(t - t') , \qquad (2.158)$$

with the associated Gaussian probability distribution

$$P[r] = \exp \left[-\frac{1}{4\Gamma k_B T} \int r(t)^2 \, dt \right] , \qquad (2.159)$$

with the functional integration measure defined as in Eq. (2.138). Under purely deterministic, overdamped dynamics, the particles will accumulate in the potential minima closest to their starting positions, as indicated in Fig. 2.10. In the presence of thermal noise, the Brownian particles may escape from these local potential minima with a finite probability, and for $t \to \infty$ the equilibrium distribution should be given by the Boltzmann weights $P_{st}(x) \propto e^{-V(x)/k_B T}$. In order to confirm this, we study the corresponding Fokker–Planck equation, called the *Smoluchowski equation* in this context. With $\xi(t) = x(t + \tau) - x(t) \approx \Gamma F(x)\tau + \int_t^{t+\tau} r(t') \, dt'$, we find $\langle \xi(t) \rangle \approx \Gamma F(x)\tau$ and $\langle \xi_i(t)\xi_j(t) \rangle \approx (\Gamma\tau)^2 F_i F_j + 2\Gamma k_B T \, \delta_{ij} \int_t^{t+\tau} dt' \int_t^{t+\tau} dt'' \delta(t - t') = 2\Gamma k_B T \, \delta_{ij} \tau + O(\tau^2)$, while all higher moments are at least of order τ^2. The only non-vanishing Kramers–Moyal coefficients are therefore $\alpha_1(x) = \Gamma F(x)$ and $\alpha_{ij}(x) = 2\Gamma k_B T \, \delta_{ij}$, and we obtain the Smoluchowski equation

$$\frac{\partial P_1(x, t)}{\partial t} = -\Gamma \frac{\partial}{\partial x} \cdot \left[F(x) P_1(x, t) - k_B T \frac{\partial P_1(x, t)}{\partial x} \right] \qquad (2.160)$$

(for a direct derivation, see Problem 2.9). Indeed, stationarity requires the probability current $J_1(x, t) = \Gamma[F(x) P_1(x, t) - k_B T \, \partial P_1(x, t)/\partial x]$ to vanish, which is satisfied by the canonical distribution

$$P_{st}(x) = \frac{e^{-V(x)/k_B T}}{\int e^{-V(x)/k_B T} \, d^d x} . \qquad (2.161)$$

For example, let us consider a harmonic potential $V(x) = \frac{f}{2} x^2$, $F(x) = -f x$. With the substitutions $x \leftrightarrow v$, $f \leftrightarrow m$, and $f\Gamma \leftrightarrow \zeta$, the associated Smoluchowski equation (2.160) becomes identical with the Fokker–Planck equation (2.147) for free Brownian motion. For the sharp initial condition $P_1(x, 0) = \delta(x - x_0)$, naturally $P_1(x, t) = P_{1|1}(x, t|x_0, 0)$, and with Eq. (2.141) we obtain from Eq. (2.153)

$$P_{1|1}(x, t|x_0, 0) = \left[\frac{f}{2\pi k_B T (1 - e^{-2f\Gamma t})} \right]^{d/2} \exp\left[-\frac{f(x - x_0 e^{-f\Gamma t})^2}{2k_B T(1 - e^{-2f\Gamma t})} \right].$$

$$(2.162)$$

The general solution for an arbitrary normalized initial distribution $w_0(x)$ reads

$$P_1(x, t) = \int P_{1|1}(x, t|x', 0) \, w_0(x') \, d^d x',$$

$$(2.163)$$

which relaxes to thermal equilibrium

$$P_{st}(x) = \left(\frac{f}{2\pi k_B T} \right)^{d/2} e^{-fx^2/2k_B T},$$

$$(2.164)$$

independent of $w_0(x)$. Upon adding an external force term $\Gamma h(t)$ and with $\gamma = \Gamma f$, the corresponding overdamped Langevin equation (2.157) becomes

$$\frac{\partial x(t)}{\partial t} = -\gamma \, x(t) + \Gamma \, h(t) + r(t).$$

$$(2.165)$$

For $h = 0$, this is essentially Eq. (2.130) again.

An alternative long-time solution for this stochastic differential equation proceeds via direct Fourier transform, which immediately yields the dynamic response function (see also Problem 2.2)

$$\chi_{ij}(\omega) = \frac{\partial \langle x_i(\omega) \rangle}{\partial h_j(\omega)} = \frac{\Gamma}{-i\omega + \gamma} \, \delta_{ij} = \chi(\omega) \, \delta_{ij}.$$

$$(2.166)$$

We can now establish the causality property $\chi(t) = 0$ for $t < 0$. To this end we analyze the Fourier integral for $\chi(t)$ by means of the residue theorem. For $t < 0$, the exponential factor in the integrand forces us to close the integration contour in the complex upper half-plane Im $\omega > 0$ (full line in Fig. 2.11). But notice that $\chi(\omega)$ is analytic in the complex upper half-plane, and therefore the integral vanishes. For $t > 0$, on the other hand, the contour lies in the lower half-plane (dashed line in Fig. 2.11), and encloses the pole at $\omega = -i\gamma$ (full circle in Fig. 2.11). The Fourier integral thus yields

$$\chi(t) = \frac{i\Gamma}{2\pi} \int \frac{e^{-i\omega t}}{\omega + i\gamma} \, d\omega = \Gamma e^{-\gamma t} \, \Theta(t).$$

$$(2.167)$$

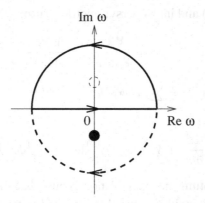

Fig. 2.11 Integration contours for the evaluation of the time-dependent response and correlation functions for an overdamped harmonic oscillator. The full and dashed circles denote the complex poles at $\omega = \mp i\gamma$, respectively.

With the aid of Eq. (2.158), we also readily compute the dynamic correlation function

$$\langle x_i(\omega)x_j(\omega')\rangle = \frac{\langle r_i(\omega)r_j(\omega')\rangle}{(-i\omega + \gamma)(-i\omega' + \gamma)} = C(\omega)\, 2\pi\, \delta(\omega + \omega')\delta_{ij}\,,$$

$$C(\omega) = \frac{2\Gamma\, k_\mathrm{B}T}{\omega^2 + \gamma^2}\,, \tag{2.168}$$

since $\langle r_i(\omega)r_j(\omega')\rangle = 2\Gamma\, k_\mathrm{B}T\, 2\pi\, \delta(\omega + \omega')\delta_{ij}$. Upon comparing with Im $\chi(\omega)$, we see that the Einstein relation (2.158) guarantees the validity of the equilibrium fluctuation-dissipation theorem (2.35). Fourier backtransform gives

$$C(t) = \frac{\Gamma\, k_\mathrm{B}T}{\pi} \int \frac{e^{-i\omega t}}{\omega^2 + \gamma^2}\, d\omega = \frac{k_\mathrm{B}T}{f}\, e^{-\gamma|t|}\,, \tag{2.169}$$

in accord with Eqs. (2.36) and (2.167). Finally, we explicitly confirm the classical equipartition theorem

$$\frac{f}{2}\,\langle x(t)^2\rangle = f\,\frac{d}{2}\, C(t=0) = \frac{d}{2}\, k_\mathrm{B}T\,. \tag{2.170}$$

Problems 2.10 and 2.11 address related applications to electrical LRC circuits and a semi-classical description of single-mode lasers.

Quite generally in one dimension, the ansatz $P_1(x, t) = \rho(x, t)\, e^{-V(x)/2k_\mathrm{B}T}$ transforms the Smoluchowski equation (2.160) to a Schrödinger equation in imaginary time (with $\tau = -i\hbar 2\Gamma\, k_\mathrm{B}T\, t$):

$$\frac{\partial\rho(x, t)}{\partial t} = 2\Gamma\, k_\mathrm{B}T \left[\frac{1}{2}\frac{\partial^2}{\partial x^2} - V^0(x)\right]\rho(x, t)\,, \tag{2.171}$$

with the potential $V^0(x)$ and its 'supersymmetric' partner $V^1(x)$ defined through

$$V^{0/1}(x) = \frac{V'(x)^2}{8(k_{\rm B}T)^2} \mp \frac{V''(x)}{4k_{\rm B}T} . \tag{2.172}$$

Variable separation $\rho(x, t) = \sum_n c_n \phi_n^{0/1}(x) e^{-2\Gamma k_{\rm B}T \epsilon_n^{0/1} t}$ leads to the eigenvalue problems

$$\left[-\frac{1}{2} \frac{\partial^2}{\partial x^2} + V^{0/1}(x) \right] \phi_n^{0/1}(x) = \epsilon_n^{0/1} \phi_n^{0/1}(x) . \tag{2.173}$$

'Supersymmetric' quantum mechanics[5] now states that the spectrum belonging to the potential $V^1(x)$ coincides with all the *excited* states for $V^0(x)$, $\epsilon_n^1 = \epsilon_n^0$ for $n > 0$. The latter has the additional normalized ground state

$$\phi_0^0(x) = \frac{e^{-V(x)/2k_{\rm B}T}}{\left(\int e^{-V(x)/k_{\rm B}T} \, dx \right)^{1/2}} , \tag{2.174}$$

with $\epsilon_0^0 = 0$, as is readily confirmed by direct calculation. Hence all $\epsilon_n^0 > 0$ for $n > 0$, and with the coefficient

$$c_0 = \int \phi_0^0(x) \rho(x, t) \, dx = \frac{\int P_1(x, t) \, dx}{\left(\int e^{-V(x)/k_{\rm B}T} \, dx \right)^{1/2}} = \frac{1}{\left(\int e^{-V(x)/k_{\rm B}T} \, dx \right)^{1/2}}$$

we obtain

$$P_1(x, t) = \frac{e^{-V(x)/k_{\rm B}T}}{\int e^{-V(x)/k_{\rm B}T} \, dx} + e^{-V(x)/2k_{\rm B}T} \sum_{n>0} c_n \phi_n^0(x) e^{-2\Gamma k_{\rm B}T \epsilon_n^0 t} , \tag{2.175}$$

which approaches the Boltzmann factor (2.161) as $t \to \infty$.

Finally, we return to the general Langevin equation (2.156). Treating $x(t)$ and $v(t)$ as independent variables, this second-order differential equation is transformed to a first-order one for the $2d$-dimensional vector (x, v),

$$\frac{\partial x(t)}{\partial t} = v(t) ,$$

$$m \frac{\partial v(t)}{\partial t} = -m\zeta \, v(t) + F(x) + f(t) . \tag{2.176}$$

At time t, the stochastic process is then characterized by the joint probability distribution

$$P_1(x, v, t) = \langle \delta(x - x(t)) \, \delta(v - v(t)) \rangle . \tag{2.177}$$

---

[5] See, e.g., Schwabl (2007), chapter 19.

With $\partial x(t)/\partial t = v(t)$,

$$\frac{\partial P_1(x, v, t)}{\partial t} = -\frac{\partial}{\partial x} \cdot \langle \delta(x - x(t)) \, \delta(v - v(t)) v(t) \rangle$$

$$-\frac{\partial}{\partial v} \cdot \left\langle \delta(x - x(t)) \, \delta(v - v(t)) \left[-\zeta v(t) + \frac{F(x)}{m} + \frac{f(t)}{m} \right] \right\rangle$$

$$= -v \cdot \frac{\partial P_1(x, v, t)}{\partial x} + \zeta \frac{\partial}{\partial v} [v \cdot P_1(x, v, t)] - \frac{F(x)}{m} \cdot \frac{\partial P_1(x, v, t)}{\partial v}$$

$$-\frac{1}{m} \frac{\partial}{\partial v} \cdot \langle \delta(x - x(t)) \, \delta(v - v(t)) f(t) \rangle \,, \qquad (2.178)$$

and the last term can be evaluated further as in Eqs. (2.149) and (2.150). The resulting general Fokker–Planck equation for Brownian motion in an external field $F(x)$ becomes

$$\frac{\partial P_1(x, v, t)}{\partial t} + v \cdot \frac{\partial P_1(x, v, t)}{\partial x} + \frac{F(x)}{m} \cdot \frac{\partial P_1(x, v, t)}{\partial v}$$

$$= \zeta \frac{\partial}{\partial v} \cdot \left[v P_1(x, v, t) + \frac{\lambda}{2\zeta m^2} \frac{\partial P_1(x, v, t)}{\partial v} \right]. \qquad (2.179)$$

The left-hand side of this partial differential equation represents the deterministic part; with no fluid of light particles present, it just represents Liouville's equation for the phase space density of classical non-interacting particles. The dissipative and stochastic terms on the right-hand side stem from the collisions with the fluid particles. Provided Einstein's relation (2.141) is fulfilled, the stationary solution to Eq. (2.179) reads

$$P_{\mathrm{st}}(x, v) = \frac{1}{Z(T)} \exp \left[-\frac{V(x)}{k_\mathrm{B} T} - \frac{mv^2}{2k_\mathrm{B} T} \right], \qquad (2.180)$$

with the classical canonical partition function

$$Z(T) = \left(\frac{2\pi k_\mathrm{B} T}{m} \right)^{d/2} \int e^{-V(x)/k_\mathrm{B} T} \, d^d x \,. \qquad (2.181)$$

2.4.4 Markovian character and equilibrium conditions

We end this introduction to stochastic differential equations with some general considerations. Langevin equations take the form

$$\frac{\partial x(t)}{\partial t} = F[x(t)] + r(t) \,, \qquad (2.182)$$

where $F(x)$ is a random function, and x may be a (possibly large) vector. Indeed, if higher time derivatives occur, we may incorporate $y(t) = \dot{x}(t)$ etc. into an enlarged vector x, with non-trivial couplings between its entries, and thus generically arrive at (2.182), compare with Eqs. (2.176). The solution $\{x(t)\}$ to this equation may be viewed as a stochastic process. It is of Markovian character, if and only if (after incorporating the average $\langle r \rangle$ into F) $r(t)$ represents Gaussian white noise,

$$\langle r(t) \rangle = 0, \quad \langle r_i(t) r_j(t') \rangle = \lambda_i \delta_{ij} \delta(t - t'). \tag{2.183}$$

Roughly, the deterministic part of Eq. (2.182) is a first-order differential equation, and $x(t + \tau)$ follows from $x(t)$ only. The delta-correlated white noise implies that the stochastic part induces no memory at all. For a more formal proof, we recall that the temporal derivatives are defined here as the limit $\tau \to 0$ of discrete time steps in forward discretization $\partial x(t)/\partial t = \lim_{\tau \to 0}[x(t + \tau) - x(t)]/\tau.$[6] For the second moment of these increments one finds

$$\frac{1}{\tau^2} \langle [x_i(t + \tau) - x_i(t)][x_j(t' + \tau) - x_j(t')] \rangle = F_i[x(t)]F_j[x(t')] + \langle r_i(t) r_j(t') \rangle,$$

$$\tag{2.184}$$

which factorizes for $t \neq t'$ (only) if (2.183) holds. The statements in Problem 2.4 then establish the Markovian character.

The noise trajectory $\{r(t)\}$ of course represents a stochastic process itself. If we take its correlations to be

$$\langle r(t) \rangle = 0, \quad \langle r_i(t) r_j(t') \rangle = \frac{\lambda_i \gamma}{2} e^{-\gamma|t - t'|} \delta_{ij}, \tag{2.185}$$

then the solution for the associated Langevin equation (2.182) represents a non-Markovian *Ornstein–Uhlenbeck process*. The stochastic process $\{r(t)\}$, on the other hand, can be obtained as the solution of the Langevin equation

$$\frac{\partial r(t)}{\partial t} = -\gamma \, r(t) + \eta(t), \tag{2.186}$$

with Gaussian white noise $\eta(t)$ precisely of the form (2.183). For, if we recall Eqs. (2.139) and (2.140), we see that the solution of Eq. (2.186) reads $r(t) = e^{-\gamma t} \int_0^t e^{\gamma t'} \eta(t') \, dt'$, leading to

$$\langle r_i(t) r_j(t') \rangle = \frac{\lambda_i \gamma}{2} \left[e^{-\gamma|t - t'|} - e^{-\gamma(t + t')} \right] \delta_{ij},$$

which reduces to (2.185) for $t, t' \gg 1/\gamma$. The two *coupled* Langevin equations (2.182) and (2.186) therefore do constitute a Markov process. However, upon eliminating $r(t)$ as an independent variable, this Markovian character is lost.

---

[6] Ambiguities can arise only in situations where the stochastic forces r or their correlators are functionals of the random variables $x(t)$ themselves; in such a situation, a more microscopic approach is called for, see Chapter 9.

Given an (effective) Hamiltonian $H(x)$, we may separate the 'reversible' forces $F_{rev}(x)$ from the dissipative terms that describe relaxation towards a minimum of H,

$$F_{rel}(x) = -\Gamma \frac{\partial H(x)}{\partial x} , \qquad (2.187)$$

where Γ denotes the corresponding Onsager coefficient. The Langevin equation (2.182) now reads

$$\frac{\partial x(t)}{\partial t} = F_{rev}[x(t)] - \Gamma \frac{\delta H[x]}{\delta x(t)} + r(t) . \qquad (2.188)$$

Furthermore, we impose the noise correlations $\langle r_i(t) r_j(t') \rangle = \phi(t - t') \delta_{ij}$. As before, the associated Kramers–Moyal coefficients are readily found:

$$\alpha_1(x) = F(x) = F_{rev}(x) - \Gamma \frac{\partial H(x)}{\partial x} , \qquad (2.189)$$

$$\alpha_{ij} = \frac{1}{\tau} \int_t^{t+\tau} dt' \int_t^{t+\tau} dt'' \, \phi(t - t') \delta_{ij} \to \phi(\omega = 0) \delta_{ij} \qquad (2.190)$$

in the limit of time scale separation, $\tau \gg \tau_c$. This yields the Fokker–Planck equation in the usual form of a continuity equation (2.92), with the probability current

$$J_1(x) = \left[F_{rev}(x) - \Gamma \frac{\partial H(x)}{\partial x} \right] P_1(x, t) - \frac{\lambda}{2} \frac{\partial P_1(x, t)}{\partial x} , \qquad (2.191)$$

where $\lambda = \phi(\omega = 0)$. For $P_1(x, t)$ to approach the stationary, canonical distribution (2.86) as $t \to \infty$, we see that there are *two* sufficient conditions, namely first the *Einstein relation*

$$\lambda = \int \phi(t) \, dt = 2\Gamma \, k_B T \qquad (2.192)$$

for the dissipative terms; and second, we need to demand that the reversible stationary probability current be *divergence-free*,

$$\frac{\partial}{\partial x} \cdot \left[F_{rev}(x) \, e^{-H(x)/k_B T} \right] = 0 . \qquad (2.193)$$

The first equilibrium condition is actually often the less stringent one; for, as long as the noise strength λ and the Onsager coefficient Γ are proportional, Eq. (2.192) may serve as the definition of an *effective* temperature T'. Yet for a genuinely non-equilibrium stationary state, at least one of these two conditions is violated, and the different contributions in (2.191) must balance each other in a highly non-trivial manner.

Problems

2.1 *Linear response functions with temporal operator derivatives*

For $\dot{A}(t) = dA(t)/dt$ and $t \geq 0$, derive the identity $\Phi_{\dot{A}B}(t) = -\chi_{AB}(t)$, provided that $A(t)$ and $B(t')$ are uncorrelated as $|t - t'| \rightarrow \infty$. As consequences, establish the important relations $\Phi_{\dot{A}B}(\omega) = -i\omega\,\Phi_{AB}(\omega) - \chi_{AB}(\omega = 0)$ and $\chi_{\dot{A}B}(\omega = 0) = -\frac{i}{\hbar}\langle[A, B]\rangle_0$.

2.2 *Dynamic response functions for a damped harmonic oscillator*

A driven classical harmonic oscillator is described by the equation of motion

$$m\left(\frac{d^2}{dt^2} + \gamma\frac{d}{dt} + \omega_0^2\right)x(t) = h(t).$$

Here, γ denotes the friction coefficient, ω_0 the eigenfrequency of the free, undamped oscillator, and $h(t)$ the external driving force. Determine the dynamic susceptibility $\chi(\omega) = \partial\langle x(\omega)\rangle/\partial h(\omega)$, its reactive and dissipative components, the relaxation function $\Phi(\omega)$, and the dynamic correlation function $C(\omega)$.

2.3 *Density response sum rules*

(a) For N particles confined to a volume V, derive the *compressibility sum rule*

$$\lim_{q\to 0} \mathcal{P}\int \frac{d\omega}{\pi}\frac{\chi_{nn}''(q, \omega)}{\omega} = \frac{N^2}{V}\kappa_T.$$

(b) For N particles of mass m, demonstrate the *f-sum rule*

$$\int \frac{d\omega}{2\pi}\,\omega\,\chi_{nn}''(q, \omega) = \frac{Nq^2}{2m}.$$

Hint: use the continuity equation with the particle current operator $j(x, t) = \frac{\hbar}{2mi}\sum_{i=1}^{N}\{\nabla_i, \delta(x - x_i(t))\}$.

2.4 *Markovian character and factorization of moments*

(a) For a Markov chain, show that $\langle[x(t + \tau) - x(t)][x(t' + \tau) - x(t')]\rangle = \langle x(t + \tau) - x(t)\rangle\langle x(t' + \tau) - x(t')\rangle$ for $t \neq t'$.

(b) Now assume the factorization of the second moments for all possible arguments for $t \neq t'$, and demonstrate that Eq. (2.60) holds for $n = 3, 4$. Thus establish the Markovian character for a *Gaussian* stochastic process with *uncorrelated* increments.

2.5 *Master and Fokker–Planck equations for Markov processes*

By means of the Chapman–Kolmogorov equation (2.61), derive

(a) the master equation (2.73), and

(b) the Fokker–Planck equation for $P_{1|1}(x, t|x_0, t_0)$

for Markovian processes with time-independent transition rates.

2.6 *Particle number fluctuations in the population dynamics model*

    (a) Compute the mean-square particle number fluctuations for the population dynamics model of Section 2.3.2 for $\sigma \neq \kappa$:

$$(\Delta n)^2 = \langle n(t)^2 \rangle - \langle n(t) \rangle^2 = n_0 \frac{\kappa + \sigma}{\kappa - \sigma} e^{(\sigma - \kappa)t} (1 - e^{(\sigma - \kappa)t}) .$$

    (b) Confirm $(\Delta n)^2 = 2n_0 \kappa t$ at the extinction threshold $\sigma = \kappa$.

2.7 *Spontaneous particle creation and death processes*

    (a) Write down the master equation for the competing processes $\emptyset \to A$ (rate τ) and $A \to \emptyset$ (rate κ), and derive the partial differential equation

$$\frac{\partial g(x, t)}{\partial t} = (1 - x) \left[\kappa \frac{\partial g(x, t)}{\partial x} - \tau g(x, t) \right]$$

    for the generating function $g(x, t)$.

    (b) Find the stationary solution $g_\infty(x) = g(x, t \to \infty)$, and subsequently determine the full time-dependent function $g(x, t)$, if initially n_0 particles are present.

    (c) Compute the survival probability $P_a(t)$ and the mean particle number $\langle n(t) \rangle$ at time t.

2.8 *Langevin equation energy balance*

For the Langevin equation (2.130), show that stationarity of the kinetic energy along with the classical equipartition theorem imply the Einstein relation (2.141).

2.9 *Direct derivation of Smoluchowski's equation*

Derive the Smoluchowski equation (2.160) starting from the identity $P_1(x, t) = \langle \delta(x - x(t)) \rangle$, using the Langevin equation (2.157).

2.10 *LRC circuit and Nyquist's theorem*

An LRC circuit consists of a resistor (resistance R), capacitor (capacitance C), and inductive coil (inductivity L).

    (a) Show that at fixed temperature T, with an external voltage $V_{\text{ext}}(t)$, and taking *thermal voltage noise* V_{th} into account, the capacitor charge $Q(t)$ obeys the Langevin equation

$$L \frac{\partial^2 Q(t)}{\partial t^2} + R \frac{\partial Q}{\partial t} + \frac{Q(t)}{C} = V_{\text{ext}}(t) + V_{\text{th}}(t) .$$

    Comparing with Problem 2.2, determine the dynamic response function $\chi(\omega) = \partial \langle Q(\omega) \rangle / \partial V_{\text{ext}}(\omega)$, and its Fourier transform $\chi(t)$.

    (b) For vanishing battery potential $V_{\text{ext}} = 0$, compute the voltage and current correlations, and confirm *Nyquist's theorem*

$$\frac{C}{2} \langle V_c(t)^2 \rangle = \frac{k_B T}{2} = \frac{L}{2} \langle I_c(t)^2 \rangle .$$

2.11 *Semi-classical description of a single-mode laser*

Semi-classically, a single-mode laser is described by a complex electric field $E(t) = |E(t)| e^{i\varphi(t)}$. Including statistical (non-thermal) field fluctuations $F(t)$, the relevant equation of motion reads

$$\frac{\partial E(t)}{\partial t} = -\kappa E(t) + [\alpha - \beta |E(t)|^2] E(t) + F(t),$$

with positive coefficients κ, α, and β. They govern, respectively, losses due to absorption, reflections, etc.; the intensity gain which is proportional to the level inversion; and the saturation at high intensities. We further assume $\langle F(t) \rangle = 0$ and $\langle F(t)F(t')^* \rangle = \Lambda \delta(t - t')$.

(a) As functions of the control parameters, determine the stable stationary solutions in the noiseless limit.

(b) Transform the associated Fokker–Planck equation to polar coordinates. Find the stationary probability distribution $P_{st}(E)$.

(c) Below the lasing threshold, i.e., for $\kappa > \alpha$, linearize both the Langevin and Fokker–Planck equations, and determine the field fluctuations as well as $P_{st}(E)$.

(d) Above threshold ($\kappa < \alpha$), similarly linearize about the non-trivial stationary solution. For the phase fluctuations, confirm that

$$\left\langle e^{i[\varphi(t) - \varphi(t')]} \right\rangle \approx e^{-\beta \Lambda |t-t'|/2(\alpha-\kappa)}.$$

Again compute and discuss the field fluctuations and $P_{st}(E)$.

References

Chaikin, P. M. and T. C. Lubensky, 1995, *Principles of Condensed Matter Physics*, Cambridge: Cambridge University Press, chapters 7, 8.

Cowan, B., 2005, *Topics in Statistical Mechanics*, London: Imperial College Press, chapter 5.

Kardar, M., 2007, *Statistical Physics of Fields*, Cambridge: Cambridge University Press, chapter 9.

Pathria, R. K., 1996, *Statistical Mechanics*, Oxford: Butterworth–Heinemann, 2nd edn., chapter 14.

Reichl, L. E., 2009, *A Modern Course in Statistical Physics*, Weinheim: Wiley–VCH, 3rd edn., chapter 7.

Reif, F., 1985, *Fundamentals of Statistical and Thermal Physics*, Singapore: McGraw–Hill, 18th edn., chapter 15.

Schwabl, F., 2006, *Statistical Mechanics*, Berlin: Springer, 2nd edn., chapter 8.

Schwabl, F., 2007, *Quantum Mechanics*, Berlin: Springer, 4th edn., chapter 19.

Schwabl, F., 2008, *Advanced Quantum Mechanics*, Berlin: Springer, 4th edn., chapter 4.

Van Kampen, N. G., 1981, *Stochastic Processes in Physics and Chemistry*, Amsterdam: North Holland.
Van Vliet, C. M., 2010, *Equilibrium and Non-equilibrium Statistical Mechanics*, New Jersey: World Scientific, 2nd edn., chapters 16, 18, 19.

Further reading

Forster, D., 1983, *Hydrodynamic Fluctuations, Broken Symmetry, and Correlation Functions*, Redwood City: Addison-Wesley, 3rd edn.
Haken, H., 1983, *Synergetics*, Berlin: Springer, 3rd edn.
Krapivsky, P. K., S. Redner, and E. Ben-Naim, 2010, *A Kinetic View of Statistical Physics*, Cambridge: Cambridge University Press.
Kubo, R., M. Toda, and N. Hashitsume, 1991, *Statistical Physics II — Nonequilibrium Statistical Mechanics*, Berlin: Springer, 2nd edn.
Lovesey, S. W., 1986, *Condensed Matter Physics: Dynamic Correlations*, Menlo Park: Benjamin–Cummings, 2nd edn.
Marro, L. and R. Dickman, 1999, *Nonequilibrium Phase Transitions in Lattice Models*, Cambridge: Cambridge University Press.
Mukamel, D., 2000, Phase transitions in nonequilibrium systems, in: *Soft and Fragile Matter: Nonequilibrium Dynamics, Metastability and Flow*, eds. M. E. Cates and M. R. Evans, Scottish Universities Summer School in Physics **53**, Bristol: Institute of Physics Publishing, 231–258.
Polettini, M., 2012, Nonequilibrium thermodynamics as a gauge theory, *EPL* **97**, 30003-1–6.
Risken, H., 1984, *The Fokker–Planck Equation*, Heidelberg: Springer.
Zwanzig, R., 2001, *Nonequilibrium Statistical Mechanics*, Oxford: Oxford University Press.

3

Dynamic scaling

In the preceding chapter, we have introduced several levels for the mathematical description of stochastic dynamics. We now use the kinetic Ising models introduced in Section 2.3.3 to formulate the dynamic scaling hypothesis which appropriately generalizes the homogeneity property of the static correlation function in the vicinity of a critical point, as established in Chapter 1. The dynamic critical exponent z is defined to characterize both the critical dispersion and the basic phenomenon of critical slowing-down. As a next step, and building on the results of Section 2.4, a continuum effective theory for the mesoscopic order parameter density, basically the dynamical analog to the Ginzburg–Landau approach, is constructed in terms of a non-linear Langevin equation. The distinction between dissipative and diffusive dynamics for the purely relaxational kinetics of either a non-conserved or conserved order parameter field, respectively, defines the universality classes A and B. Following the analysis of these models in the Gaussian approximation, they also serve to outline the construction of a dynamical perturbation theory for non-linear stochastic differential equations through direct iteration. In general, however, the order parameter alone does not suffice to fully capture the critical dynamics near a second-order phase transition. Additional hydrodynamic modes originating from conservation laws need to be accounted for as well. The simplest such situation is entailed in the relaxational models C and D, which encompass the static coupling of the order parameter to the energy density. Further scenarios emerge through reversible non-linear mode couplings in the Langevin equations of motion. The effects of such contributions are illustrated by means of the spin precession terms relevant in isotropic (Heisenberg) and planar (XY) ferromagnets, which define the dynamical universality classes J and E. In both cases, the dynamic scaling hypothesis in conjunction with the spin wave dispersion in the ordered phase fixes the values of the dynamic critical exponent z.

3.1 Dynamic scaling hypothesis

We are now equipped with all the necessary tools to generalize and extend our investigation of critical phenomena to dynamic properties. We saw in Chapter 1 that the peculiar power laws near a second-order phase transition originate from a diverging correlation length $\xi = \xi_\pm |\tau|^{-\nu}$, with $\tau \propto T - T_c$. As ever larger regions of the critical system become correlated and behave cooperatively, one expects the associated dynamical time scales to grow as well. Mathematically, this *critical slowing-down* is captured in terms of an appropriate *dynamic scaling hypothesis*.[1] Yet before we elucidate its physical consequences, let us examine a familiar case.

3.1.1 Critical slowing-down

In Section 2.3, we encountered the Glauber kinetic Ising model with spin flip dynamics (non-conserved magnetization). We saw that the dynamics can also be represented in the domain wall picture, see the mean-field rate equation (2.124) for the domain wall density n. In one dimension, domain wall boundaries between up- and down-spin regions become increasingly scarce as the temperature is lowered towards the Ising chain's degenerate critical point at $T = 0$. The contribution $\propto n^2$ can then be neglected, leaving us with

$$\frac{1}{\Gamma_G} \frac{dn(t)}{dt} \approx 2e^{-4J/k_B T} = \frac{1}{2\xi(T)^2}. \tag{3.1}$$

Thus, we find for the *characteristic time scale* governing the domain wall expulsion, i.e., spin ordering, $t_c \propto \xi(T)^2 / \Gamma_G$. Precisely at $T = 0$, Eq. (2.124) becomes $dn(t)/dt = -2\Gamma_G n(t)^2$, which is solved by

$$n(t) = \frac{n_0}{1 + 2n_0 \Gamma_G t}. \tag{3.2}$$

At criticality, the exponential approach to the saturation density n_s is replaced by the asymptotic power law $n(t) \to (2\Gamma_G t)^{-1}$, independent of the initial value n_0. Recall also the similar behavior of the survival probability for the zero-dimensional population dynamics model near the extinction threshold, Eq. (2.118).

Guided by these simple examples, we conjecture that the order parameter dynamics near a critical point is governed by a characteristic, usually relaxational, time scale that diverges upon approaching the transition as

$$t_c(\tau) \sim \xi(\tau)^z \sim |\tau|^{-z\nu}, \tag{3.3}$$

---

[1] The dynamic scaling hypothesis was independently formulated by Ferrell, Menyhàrd, Schmidt, Schwabl, and Szépfalusy (1967, 1968); Halperin and Hohenberg (1969).

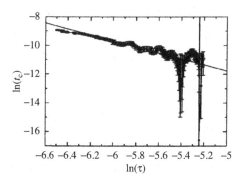

Fig. 3.1 Critical slowing-down of the relaxation time in the magnetization dynamics of a Fe bilayer grown on top of a W (110) substrate near its Curie temperature ($T_c = 453.03$ K). The experimental data for this two-dimensional Ising ferromagnet yield $zv = 2.09 \pm 0.06$. [Data and figure reproduced with permission from M. J. Dunlavy and D. Venus, *Phys. Rev. B* **71**, 144406 (2005); DOI: 10.1103/PhysRevB.71.144406; copyright (2005) by the American Physical Society.]

which defines the *dynamic critical exponent z*. For the Glauber model in mean-field approximation, we infer $z = 2$. In both real and computer experiments, the inevitable critical slowing-down causes obvious problems with equilibration that may hinder accurate determination of scaling exponents. In Monte Carlo simulations with the goal of investigating the *static* critical behavior, this inherent difficulty can be at least partially overcome through intelligent schemes, such as cluster algorithms. These employ artificial dynamical rules which aim at minimizing the effective dynamic exponent, thereby considerably accelerating the approach to thermal equilibrium.

Figure 3.1 shows experimental data for the critical slowing-down of the magnetization relaxation time in an ultrathin bilayer film of Fe atoms deposited on a W (110) surface. A power law fit to the data yields $zv = 2.09 \pm 0.06$; since in a two-dimensional Ising magnet $v = 1$ is known exactly, this implies $z = 2.09 \pm 0.06$ for the dynamic exponent, in excellent agreement with predictions for purely relaxational dynamics with non-conserved order parameter (see Section 3.2 and Chapter 5). Notice that if we view t as just one component, albeit special, of $(d + 1)$-dimensional space-time, t_c assumes the role of the correlation length in that direction with associated critical exponent $v_t = zv$; i.e., z itself represents the corresponding 'anisotropy' exponent.

In Fourier space, critical slowing-down implies that the characteristic frequency should behave as $\omega_c(\tau) \sim |\tau|^{zv}$. In order to capture the ensuing power law at T_c as well, we introduce two scaling functions $\hat{\omega}_\pm(y)$ for $T > T_c$ and $T < T_c$, respectively, with the argument $y = q\xi$, and write

$$\omega_c(\tau, q) = |q|^z \,\hat{\omega}_\pm(q\xi) . \qquad (3.4)$$

As $q \to 0$, for the Glauber model we must have $\hat{\omega}_\pm(y) \sim y^{-z}$ as $y \to 0$, whereupon (3.3) is recovered. On the other hand, $\hat{\omega}_\pm(y) \to$ const. as $y \to \infty$, yielding the *critical dispersion* $\omega_c(0, q) \sim |q|^z$.

3.1.2 Dynamic scaling functions

According to the dynamic scaling hypothesis, the moments of the order parameter all fluctuate on the scales of the correlation length ξ and characteristic time $t_c \sim \xi^z$. For example, the static power law for the spontaneous order parameter generalizes to

$$\phi(\tau, t) = |\tau|^\beta \,\hat{\phi}(t/t_c) . \qquad (3.5)$$

It is easy to see that this ansatz leads to the critical power law decay $\phi(0, t) \propto t^{-\beta/z\nu}$ as $t \to \infty$ (Problem 3.1).

In the same manner, in the critical region the frequency dependence of the dynamic order parameter susceptibility should enter only via the scaling variable $\omega/\omega_c(\tau, q)$. As a consequence of Eq. (3.4), the associated scaling function can depend only on the dimensionless variables $q\xi$ and either $\omega/D|qa_0|^z$ or alternatively $\omega\xi^z/Da_0^z$, where $1/D$ and a_0 represent some microscopic time and length scales. With the latter choice, the scaling form for the dynamic response function reads

$$\chi(\tau, q, \omega) = |q|^{-2+\eta} \,\hat{\chi}_\pm(q\xi, \omega\xi^z/Da_0^z) . \qquad (3.6)$$

In the static limit $\omega = 0$, this reduces to Eq. (1.73). By means of the classical equilibrium fluctuation-dissipation theorem (2.35), one obtains the dynamic correlation function

$$C(\tau, q, \omega) = \frac{2k_B T}{\omega} \,\mathrm{Im}\,\chi(\tau, q, \omega) = |q|^{-z-2+\eta} \,\hat{C}_\pm(q\xi, \omega\xi^z/Da_0^z) . \qquad (3.7)$$

An inverse Fourier transform then yields (Problem 3.1)

$$C(\tau, x, t) = |x|^{-(d-2+\eta)} \,\widetilde{C}_\pm(x/\xi, Da_0^z t/\xi^z) , \qquad (3.8)$$

again with the correct static limit (1.76) for $t = 0$.

In essence, therefore, the dynamics enter the description of critical phenomena through one additional scaling variable with a single new exponent z. The identification of the possible *dynamic universality classes*, the determination of the

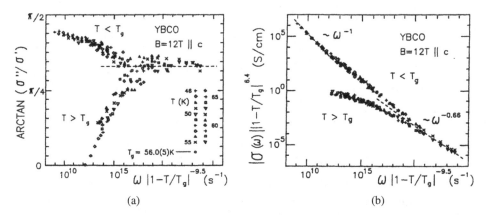

Fig. 3.2 (a) Phase and (b) scaled amplitude of the linear conductivity vs. rescaled frequency measured for the normal- to superconducting phase transition (occurring at $T_g \approx 56$ K) in an external magnetic field of $B = 12$ T in a twinned YBCO crystal. [Data and figure reproduced with permission from J. Kötzler, M. Kaufmann, G. Nakielski, R. Behr, and W. Assmus, *Phys. Rev. Lett.* **72**, 2081 (1994); DOI: 10.1103/PhysRevLett.72.2081; copyright (1994) by the American Physical Society.]

corresponding dynamic critical exponents, as well as the computation or measurement of the associated scaling functions constitute the central goals of the study of critical dynamics. In addition, successful theories should be capable of providing a solid foundation for the dynamic scaling hypothesis, and either justify the homogeneity properties (3.6) and (3.8), or extend them appropriately.

Experimentally, the consequences of these scaling laws, valid close to T_c and in the long-wavelength ($q \to 0$) and low-frequency ($\omega \to 0$) or long-time ($t \to \infty$) limit, are quite striking. For example, Eq. (3.8) entails the fact that dynamic features enter exclusively through the scaling variable $t\xi^{-z} \propto t|\tau|^{zv}$, which is occasionally referred to as the *time–temperature superposition* property. This means that dynamic correlations at t for a given value of the control parameter τ can be mapped onto those for $|\tau'| > |\tau|$ at an earlier time $t' = t(|\tau|/|\tau'|)^{zv}$. Therefore, when the quantities $|\tau|^{-\beta}\phi(\tau, t)$ and $|\tau|^{2\beta}C(\tau, t)$ (at fixed x) are plotted against $t|\tau|^{zv}$, or $|\tau|^{\gamma}\chi(\tau, \omega)$ (at fixed q) vs. $\omega|\tau|^{-zv}$, the resulting data should all collapse onto a single *master curve* in the universal scaling regime. As an example, Fig. 3.2 depicts the scaling plots for the phase (a) and appropriately scaled amplitude (b) of the linear conductivity measured in the vicinity of the normal- to superconducting phase transition in a twinned YBCO crystal (at finite external magnetic field). Notice the different scaling functions on either side of the transition; the scaling collapse indicates $zv \approx 9.5$.

3.2 Continuum theory: relaxational models

Critical phenomena are characterized by a diverging length as well as time scale. A continuum description in terms of coarse-grained variables should therefore suffice; in fact, such a mesoscopic representation will be most apt for our eventual goal to exploit the well-developed tools from field theory, and the subsequent application of renormalization group methods. To begin, we consider purely relaxational kinetics, and investigate its dynamic scaling properties in the vicinity of a critical point in the (mean-field) Gaussian approximation. We then sketch how a perturbation expansion in terms of the static non-linearity in the Landau–Ginzburg–Wilson functional can be constructed through direct iteration of the Langevin equation.[2]

3.2.1 Langevin description of critical dynamics

As we have seen in the preceding section, the characteristic time scale t_c associated with the order parameter dynamics diverges upon approaching a critical point. For the time being, let us assume there are no slow hydrodynamic modes associated with conservation laws in the system. Then, critical slowing-down provides us with a natural *separation of time scales*. Near a continuous phase transition, all other physical quantities will fluctuate much faster than the order parameter S itself. As far as the increasingly slow kinetics of the latter is concerned, the effect of all remaining dynamical variables can therefore be thought of as merely constituting a stochastic forcing on S, with short-range correlations in both space and time.

To be specific, consider again the Ising model with Glauber kinetics. Provided detailed balance holds, any deviation from the thermal equilibrium configuration is simply going to relax back towards it. This suggests a continuum representation in terms of an appropriate generalization of the overdamped Langevin equation (2.157) for the infinite set of variables $S(x, t)$. As the effective potential to provide the restoring force here, we naturally choose the Landau–Ginzburg–Wilson Hamiltonian (1.49) for a scalar order parameter with Ising symmetry,

$$\mathcal{H}[S] = \int d^d x \left[\frac{r}{2} S(x)^2 + \frac{1}{2} [\nabla S(x)]^2 + \frac{u}{4!} S(x)^4 - h(x)S(x) \right]. \quad (3.9)$$

Recall (see Chapter 1) that in an Ising ferromagnet $S(x)$ corresponds to a local coarse-grained magnetization density, and $h(x)$ denotes the external magnetic field. The quantity $r \sim T - T_c^0$ serves as the control parameter measuring the distance from the mean-field critical point at temperature T_c^0. One of the major effects of the strong fluctuations near a second-order phase transition is a *downward* shift of the

---

[2] Chapter 4 will be devoted to a systematic construction and thorough analysis of dynamic perturbation theory.

transition temperature to $T_c < T_c^0$. For models with Ising order parameter symmetry, this fluctuation-induced T_c shift becomes sufficiently strong in one dimension to push the critical point all the way to $T_c = 0$. We have furthermore assumed that the minimum free energy configuration is given by a spatially homogeneous order parameter, $\nabla S(x) = 0$. The second term in Eq. (3.9) then describes the cost in free energy associated with inhomogeneous order parameter fluctuations. It represents the lowest-order term in a gradient expansion compatible with the discrete Ising symmetry.

Summarizing the above considerations, the critical dynamics of the Glauber model should be captured by the non-linear Langevin equation

$$\frac{\partial S(x, t)}{\partial t} = -D \frac{\delta \mathcal{H}[S]}{\delta S(x, t)} + \zeta(x, t) , \tag{3.10}$$

where D denotes a constant relaxation rate. In order to guarantee that the system reaches the canonical equilibrium probability distribution

$$\mathcal{P}_{st}[S] \propto e^{-\mathcal{H}[S]/k_B T} \tag{3.11}$$

at long times, we furthermore impose the continuum analog of the Einstein relation (2.158) for the Gaussian stochastic forces $\zeta(x, t)$:

$$\langle \zeta(x, t) \rangle = 0 , \quad \langle \zeta(x, t) \zeta(x', t') \rangle = 2D \, k_B T \, \delta(x - x') \delta(t - t') . \tag{3.12}$$

The Langevin equation (3.10) with (3.12) is often referred to as the *time-dependent Ginzburg–Landau equation* or *model A* dynamics for a non-conserved order parameter.[3]

For Kawasaki dynamics, on the other hand, the total magnetization $M \propto S(q = 0)$ is *conserved*. The associated order parameter density must therefore satisfy a continuity equation

$$\frac{\partial S(x, t)}{\partial t} + \nabla \cdot J(x, t) = 0 . \tag{3.13}$$

We now need a constitutive equation for the magnetization current J. Within linear response theory, it will essentially be proportional to the gradient of the magnetization density, $J(x, t) \propto -\nabla S(x, t)$. Consequently, $S(x, t)$ relaxes according to the diffusion equation, $\partial S(x, t)/\partial t \propto \nabla^2 S(x, t)$, or $\partial S(q, t)/\partial t \propto -q^2 S(q, t)$ for its Fourier transform, which indeed satisfies $\partial S(q = 0, t)/\partial t = 0$. Accounting for the non-linearities and stochastic noise, the constitutive equation generalizes to

$$J(x, t) = -D\nabla \frac{\delta \mathcal{H}[S]}{\delta S(x, t)} + \eta(x, t) , \tag{3.14}$$

---

[3] An alphabetical classification of universality classes for critical dynamics was introduced by Hohenberg and Halperin (1977).

with vectorial Gaussian random forces $\eta(x, t)$ obeying

$$\langle \eta(x, t) \rangle = 0 , \quad \langle \eta_i(x, t)\eta_j(x', t') \rangle = 2D\, k_B T\, \delta(x - x')\delta(t - t')\, \delta_{ij} . \quad (3.15)$$

For a conserved order parameter field, one thus arrives at the stochastic equation of motion

$$\frac{\partial S(x, t)}{\partial t} = D\nabla^2 \frac{\delta \mathcal{H}[S]}{\delta S(x, t)} + \zeta(x, t) , \quad (3.16)$$

with random noise $\zeta(x, t) = -\nabla \cdot \eta(x, t)$, whence

$$\langle \zeta(x, t) \rangle = 0 , \quad \langle \zeta(x, t)\zeta(x', t') \rangle = -2D\, k_B T\, \nabla_x^2\, \delta(x - x')\delta(t - t') . \quad (3.17)$$

Equations (3.16) and (3.17) constitute the dynamical *model B* for a conserved order parameter with diffusive relaxation, and correspondingly conserved noise with correlator $\propto Dq^2$ in Fourier space. The deterministic version of Eq. (3.16) is also called the *Cahn–Hilliard* equation.

For a conserved field away from the critical point, we essentially obtain a relaxation rate $\sim q^2$, whereas non-conserved quantities are characterized by a wavevector-independent decay constant (to leading order for $q \to 0$). One may formally combine both cases and write

$$\frac{\partial S(x, t)}{\partial t} = -D(i\nabla)^a \frac{\delta \mathcal{H}[S]}{\delta S(x, t)} + \zeta(x, t) , \quad (3.18)$$

with either $a = 0$, corresponding to dissipative dynamics for a *non-conserved* field with constant relaxation rate D; or $a = 2$, which describes the relaxational dynamics of a *conserved* field with diffusion constant D. The Einstein relations (3.12) and (3.17) then read

$$\langle \zeta(x, t)\zeta(x', t') \rangle = 2k_B T\, D(i\nabla_x)^a\, \delta(x - x')\delta(t - t') . \quad (3.19)$$

In the high-temperature phase ($r > 0$ in mean-field theory), $\langle S \rangle = 0$. Sufficiently far away from the critical point, the Gaussian model

$$\mathcal{H}_0[S] = \int d^d x \left[\frac{r}{2} S(x)^2 + \frac{1}{2} [\nabla S(x)]^2 - h(x)S(x) \right] \quad (3.20)$$

with contributions up to quadratic order in S should already provide a qualitatively satisfactory model Hamiltonian. In the theory of static critical phenomena (Section 1.4), $\mathcal{H}_0$ is treated as the unperturbed Hamiltonian, and the perturbation expansion is constructed in terms of

$$\mathcal{H}_{int}[S] = \int d^d x\, \frac{u}{4!}\, S(x)^4 . \quad (3.21)$$

Through Eq. (3.18), this translates directly into a systematic perturbation theory for the critical dynamics of the relaxational models.

In principle, one would have to consider higher-order terms in an expansion in $S(x)^2$ as well, but these turn out to be irrelevant in the renormalization group sense, provided $u > 0$, as do additional terms in the gradient expansion of the harmonic part (see Sections 1.4 and 5.3). This means that these further contributions have no influence on the leading asymptotic scaling behavior at the critical point, and will henceforth be neglected. We remark that for $u \leq 0$, one would have to take into account higher even powers in S; generically, however, this leads to a first-order phase transition scenario. For $u = 0$ and with a positive sixth-order term, $r = 0$ describes a tricritical point, see Section 1.2. In the low-temperature phase ($r < 0$ in the mean-field approximation), on the other hand, $|\langle S \rangle| > 0$ as a consequence of the non-linearity $\propto u S^4$; e.g., in mean-field theory, $\langle S \rangle \approx \pm(-6r/u)^{1/2}$. Therefore the non-linear coupling u must be kept even in the harmonic approximation, i.e., an expansion about $\langle S \rangle$ up to quadratic order in $\delta S(x) = S(x) - \langle S \rangle$ (see Problem 1.5). Cubic and higher-order terms in $\delta S(x)$ may then be treated perturbatively. In this and the subsequent chapters, we shall focus on the theory in the high-temperature phase and at the critical point. We defer the explicit construction of the perturbation theory in the ordered phase to Section 5.4 (but see also Problem 3.2).

3.2.2 Models A and B in Gaussian approximation

Upon inserting the Landau–Ginzburg–Wilson Hamiltonian (3.9) into the model A/B Langevin equation (3.18), one arrives at

$$\frac{\partial S(x,t)}{\partial t} = -D(i\nabla)^a \left[(r - \nabla^2)S(x,t) + \frac{u}{6} S(x,t)^3 - h(x,t) \right] + \zeta(x,t). \quad (3.22)$$

In *dynamic perturbation theory*, we use the linearized Langevin equation, in which $\mathcal{H}$ is replaced by the Gaussian Hamiltonian $\mathcal{H}_0$, as the reference theory, and then expand with respect to the non-linear term $-D(i\nabla)^a \, \delta\mathcal{H}_{int}/\delta S = -D(i\nabla)^a u S^3/6$. This procedure is in complete analogy with the static perturbation expansion, which is of course contained in the dynamic analysis as a special case, namely in the limit $\omega \to 0$ for response functions, or $t \to 0$ for correlation functions (see Section 2.1). The ensuing unperturbed model A/B Langevin equation with $u = 0$ becomes linear in $S(x,t)$, and is readily solved via Fourier transformation, whereupon the linearized Langevin equation (3.22) becomes

$$\left[-i\omega + Dq^a(r + q^2) \right] S(q,\omega) = Dq^a h(q,\omega) + \zeta(q,\omega), \quad (3.23)$$

and the second moment of the stochastic forces reads

$$\langle \zeta(q,\omega)\zeta(q',\omega') \rangle = 2k_B T \, Dq^a \, (2\pi)^{d+1} \delta(q + q') \, \delta(\omega + \omega'). \quad (3.24)$$

Here, the Dirac delta functions originate in the spatial and temporal translational invariance of the system.

Equation (3.23) immediately yields the dynamic susceptibility for the relaxational models A and B in the Gaussian approximation,

$$\chi_0(q, \omega) = Dq^a G_0(q, \omega) = \frac{\partial \langle S(q, \omega) \rangle}{\partial h(q, \omega)}\bigg|_{h=0} = \frac{Dq^a}{-i\omega + Dq^a(r + q^2)}. \tag{3.25}$$

In Eq. (3.25), we have also defined the bare *response propagator* $G_0(q, \omega)$ by factoring out the Onsager coefficient Dq^a. We now rewrite the inverse dynamic response function in the form

$$\chi_0(r, q, \omega)^{-1} = q^2\Big[(r/q^2) + 1 - i(\omega/Dq^{2+a})\Big], \tag{3.26}$$

which may be compared directly with the general scaling form (3.6),

$$\chi(r, q, \omega)^{-1} = |q|^{2-\eta}\,\hat{\chi}_+\Big(r/|q|^{1/\nu}, \omega/D|q|^z\Big)^{-1}. \tag{3.27}$$

This leads to the identification of the *three* independent critical exponents in the Gaussian model, namely the two static exponents $\nu = 1/2$ and $\eta = 0$ for the divergent correlation length and the algebraic decay of the correlations at T_c, and the dynamic exponent z characterizing the critical slowing-down,

$$z = 2 + a = \begin{cases} 2, & \text{model A (non-conserved)}, \\ 4, & \text{model B (conserved)}. \end{cases} \tag{3.28}$$

If we (daringly) assume further that there exists no independent renormalization for the relaxation rate D, scaling suggests the simple replacement $q^2 \to |q|^{2-\eta}$ in (3.26), whereupon $z = 2 + a - \eta$. While the fluctuation correction to the mean-field value of the dynamic exponent will indeed turn out to be of order η, this relation turns out to be incorrect for model A with non-conserved order parameter. However, the conservation law for model B in fact precludes a singular behavior of D itself, thus validating the scaling relation $z = 4 - \eta$ (see Section 5.3).

Through explicit Fourier back transform, one can establish the causality property of the dynamic response function, i.e., $\chi(q, t < 0) = 0$. To this end, we proceed as in our analysis of the overdamped harmonic oscillator in Section 2.4, setting $\gamma = Dq^a(r + q^2)$ for the damping coefficient. Upon evaluating the Fourier integral for $G_0(q, t)$ by means of the residue theorem (see Fig. 2.11), we recall that for $t < 0$ we need to close the integration contour in the complex upper half-plane $\text{Im}\,\omega > 0$ (full line in Fig. 2.11); yet $G_0(q, \omega)$ is analytic in this half-plane, whence the integral vanishes. For $t > 0$, on the other hand, the integration contour must be chosen in the lower half-plane (dashed line in Fig. 2.11), enclosing the pole at

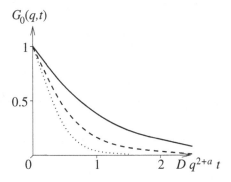

Fig. 3.3 Schematic plot of the reduced dynamic response function $G_0(q, t)$ for the Gaussian relaxational models vs. $Dq^{2+a}t$; full line: $r = 0$ ($T = T_c$), dashed: $r/q^2 = 1$, dotted: $r/q^2 = 3$.

$\omega = -iDq^a(r + q^2)$ (full circle in Fig. 2.11); thus,

$$G_0(q, t) = \int \frac{d\omega}{2\pi} \frac{1}{-i\omega + Dq^a(r + q^2)} e^{-i\omega t} = e^{-Dq^a(r+q^2)t} \Theta(t) . \qquad (3.29)$$

The function $G_0(q, t)$ decays from its initial value $G_0(q, t = 0) = 1$ to zero as $t \to \infty$, with a characteristic time scale $t_c = [Dq^a(r + q^2)]^{-1}$. For $T > T_c$, the long-wavelength dynamics ($|q| \ll r^{1/2} = \xi^{-1}$) is governed by the relaxation rate $t_c^{-1} \approx Dq^a r$, which becomes small as $r \to 0$, reflecting the phenomenon of critical slowing-down. Precisely at the critical point, it is replaced by the considerably reduced decay constant $t_c^{-1} = Dq^{2+a}$, see Fig. 3.3, which vanishes for both models A and B as $q \to 0$. A non-conserved order parameter thus becomes quasi-diffusive ($t_c^{-1} \sim Dq^2$) at a continuous transition, while a conserved order parameter acquires sub-diffusive dynamics ($t_c^{-1} \sim Dq^4$).

In order to compute the dynamic correlation function $C_0(q, \omega)$ for the relaxational models at zero external field in the Gaussian approximation, we use Eq. (3.23) in the form $S(q, \omega) = \zeta(q, \omega)/[-i\omega + Dq^a(r + q^2)]$, multiply by the corresponding expression for $S(q', \omega')$, and then take the statistical noise average with the aid of (3.24). This yields (the subscript '0' will henceforth denote correlations taken in the Gaussian ensemble)

$$\langle S(q, \omega) S(q', \omega') \rangle_0 = \frac{\langle \zeta(q, \omega) \zeta(q', \omega') \rangle}{[-i\omega + Dq^a(r + q^2)][-i\omega' + Dq'^a(r + q'^2)]}$$

$$= C_0(q, \omega) (2\pi)^{d+1} \delta(q + q') \delta(\omega + \omega') , \qquad (3.30)$$

where

$$C_0(q, \omega) = \frac{2k_B T Dq^a}{\omega^2 + [Dq^a(r + q^2)]^2} = 2k_B T Dq^a G_0(-q, -\omega) G_0(q, \omega) . \quad (3.31)$$

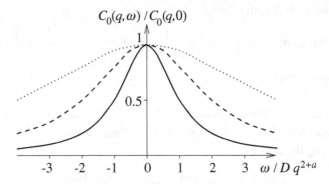

$$C_0(q,\omega)/C_0(q,0)$$

Fig. 3.4 Schematic plot of the normalized zero-field dynamic correlation function $C_0(q,\omega)/C_0(q,0)$ for the Gaussian relaxational models vs. ω/Dq^{2+a}; full line: $r=0$ ($T=T_c$), dashed: $r/q^2 = 1$, dotted: $r/q^2 = 3$.

This describes a Lorentzian curve centered at $\omega = 0$ with line width $\Gamma \sim Dq^a r$ for $T > T_c$ and $|q|\xi = |q|r^{-1/2} \ll 1$, which becomes temperature-independent and much narrower in the vicinity of the critical point, $\Gamma \sim Dq^{2+a}$ for $|q|\xi \gg 1$, as displayed in Fig. 3.4. Alternatively, we could have obtained the identical result by applying the fluctuation-dissipation theorem (2.35), $C_0(q,\omega) = 2k_B T \operatorname{Im} \chi_0(q,\omega)/\omega$. We can now rewrite Eq. (3.31) as

$$C_0(r,q,\omega)^{-1} = \frac{Dq^{4+a}}{2k_B T}\left[(r/q^2) + 1 + (\omega/Dq^{2+a})\right]^2, \qquad (3.32)$$

which is in accord with the dynamic scaling hypothesis (3.7), or

$$C(r,q,\omega)^{-1} = |q|^{z+2-\eta}\,\hat{C}_+\big(r/|q|^{1/\nu},\,\omega/D|q|^z\big)^{-1}, \qquad (3.33)$$

when the mean-field critical exponents are inserted.

In the time domain, we compute the dynamic correlation function via the Fourier integral by using the residue theorem again (see Fig. 2.11):

$$
\begin{aligned}
C_0(q,t) &= \int \frac{d\omega}{2\pi}\, \frac{2k_B T\, Dq^a}{[\omega - iDq^a(r+q^2)][\omega + iDq^a(r+q^2)]}\, e^{-i\omega t} \\
&= \frac{\mp 2\pi i}{2\pi}\, \frac{2k_B T\, Dq^a}{\mp 2i\, Dq^a(r+q^2)}\, e^{\mp Dq^a(r+q^2)t} = \frac{k_B T}{r+q^2}\, e^{-Dq^a(r+q^2)|t|}. \quad (3.34)
\end{aligned}
$$

In the second line, the upper and lower signs respectively refer to the cases $t > 0$ and $t < 0$. Direct inspection confirms the fluctuation-dissipation theorem once more, now in the form (2.36), $k_B T\, \chi_0(q,t) = -\Theta(t)\,\partial C_0(q,t)/\partial t$. The static susceptibility and equal-time correlations in the Gaussian model are given by the

Ornstein–Zernicke function (1.60):

$$\chi_0(q) = \lim_{\omega \to 0} \chi_0(q, \omega) = \frac{1}{r + q^2} = \frac{1}{k_B T} \lim_{t \to 0} C_0(q, t) = \frac{1}{k_B T} C_0(q) \,. \quad (3.35)$$

For the dynamic response and correlation functions in the ordered phase, see Problem 3.2.

Finally, we note that *Wick's theorem* (1.104) for the evaluation of averages in a Gaussian ensemble finds a straightforward generalization to dynamics. Specifically, upon introducing the abbreviations

$$\underline{k} = (q, \omega) \,, \quad \delta(\underline{k}) = (2\pi)^{d+1} \delta(q) \, \delta(\omega) \,, \quad (3.36)$$

and the fundamental contraction

$$\underline{S(\underline{k})S(\underline{k}')} = \langle S(\underline{k})S(\underline{k}')\rangle_0 = C_0(\underline{k}) \, \delta(\underline{k} + \underline{k}') \,, \quad (3.37)$$

one has for all even N-point functions

$$\langle S(\underline{k}_1)S(\underline{k}_2) \cdots S(\underline{k}_{N-1})S(\underline{k}_N)\rangle_0$$

$$= \sum_{\substack{\text{permutations} \\ i_1(1)\cdots i_N(N)}} \underline{S(\underline{k}_{i_1(1)})S(\underline{k}_{i_2(2)})} \cdots \underline{S(\underline{k}_{i_{N-1}(N-1)})S(\underline{k}_{i_N(N)})} \quad (3.38)$$

whereas any Gaussian average of an odd number of order parameter fields vanishes at zero external field in the high-temperature phase.

3.2.3 Iterative solution of non-linear Langevin equations

We can now construct a perturbative expansion in terms of the non-linearity u by iterating the stochastic Langevin equation of motion.[4] To this end, we write Eq. (3.22) in Fourier space, and multiply by the propagator $G_0(q, \omega)$ from (3.25),

$$S(q, \omega) = G_0(q, \omega) \, Dq^a \, h(q, \omega) + G_0(q, \omega) \, \zeta(q, \omega) - G_0(q, \omega) \, Dq^a$$

$$\times \frac{u}{6} \int_{q', \omega'} \int_{q'', \omega''} S(q', \omega')S(q'', \omega'')S(q - q' - q'', \omega - \omega' - \omega'') \,.$$

$$(3.39)$$

Here, the last term stems from the Fourier transform of the cubic non-linearity $-D(i\nabla)^a \, u S(x, t)^3/6$, and $\int_{q,\omega} \cdots = (2\pi)^{-(d+1)} \int d^d q \int d\omega \cdots$.

In order to compute the dynamic response function $\chi(q, \omega)$, defined via

$$\chi(q, \omega) \, (2\pi)^{d+1} \delta(q + q')\delta(\omega + \omega') = \frac{\delta\langle S(q, \omega)\rangle}{\delta h(-q', -\omega')}\bigg|_{h=0} \,, \quad (3.40)$$

---

[4] Halperin, Hohenberg, and Ma (1972).

we need to perform the functional derivative of Eq. (3.39) with respect to the external field h, and take the statistical average. To first order in perturbation theory, we may factorize the ensuing non-linear correlation functions on the right-hand side with the response and correlation functions (3.25) and (3.31) of the Gaussian model. For the response propagator or reduced dynamic susceptibility $G(q, \omega)$, defined by

$$\chi(q, \omega) = Dq^a\, G(q, \omega)\,, \qquad (3.41)$$

we thus arrive at

$$G(\underline{k})\,\delta(\underline{k} + \tilde{\underline{k}}) \approx G_0(\underline{k})\,\delta(\underline{k} + \tilde{\underline{k}})$$

$$-G_0(\underline{k})\,\frac{u}{6}\int_{\underline{k}'}\int_{\underline{k}''}\Big[Dq'^a\,G_0(\underline{k}')\,\delta(\underline{k}' + \tilde{\underline{k}})\,C_0(\underline{k}'')\,\delta(\underline{k} - \underline{k}')$$

$$+ Dq''^a\,G_0(\underline{k}'')\,\delta(\underline{k}'' + \tilde{\underline{k}})\,C_0(\underline{k}')\,\delta(\underline{k} - \underline{k}'')$$

$$+ D(q - q' - q'')^a\,G_0(-\tilde{\underline{k}})\,\delta(\underline{k} - \underline{k}' - \underline{k}'' + \tilde{\underline{k}})\,C_0(\underline{k}')\,\delta(\underline{k}' + \underline{k}'')\Big]\,,$$

where we have again employed the short-hand notation (3.36). Evaluating one set of wavevector and frequency integrals by means of the delta functions finally yields

$$G(q, \omega) \approx G_0(q, \omega)\Big[1 - 3\,G_0(q, \omega)\,Dq^a\,\frac{u}{6}\int_{q',\omega'}C_0(q', \omega')\Big]$$

$$= G_0(q, \omega) - 3\,G_0(q, \omega)\,Dq^a\,\frac{u}{6}$$

$$\times \int_{q',\omega'} G_0(q', \omega')\,2k_BT\,Dq'^a\,G_0(-q', -\omega')\,G_0(q, \omega). \quad (3.42)$$

Notice that we have already performed the remaining frequency integral in Eq. (3.34), for

$$\int \frac{d\omega}{2\pi}\,C_0(q, \omega) = C_0(q) = \frac{k_BT}{r + q^2} \qquad (3.43)$$

is just the static Ornstein–Zernicke correlation function.

Alternatively, we could have simply replaced the fields S in the integrand of the non-linear term of Eq. (3.39) by their zeroth-order counterparts from (3.23), then taken the statistical average using Eq. (3.24), and finally performed the functional derivative. This latter strategy may be applied for the computation of the dynamic correlation function

$$\langle S(q, \omega)S(q', \omega')\rangle = C(q, \omega)\,(2\pi)^{d+1}\delta(q + q')\delta(\omega + \omega')\,. \qquad (3.44)$$

We can now set $h = 0$, and to first order in u need to compute

$$C(\underline{k})\,\delta(\underline{k} + \tilde{\underline{k}}) \approx C_0(\underline{k})\,\delta(\underline{k} + \tilde{\underline{k}})$$

$$- G_0(\underline{k})\,Dq^a\,\frac{u}{6}\int_{\underline{k}'}\int_{\underline{k}''}\langle S(\tilde{\underline{k}})S(\underline{k}')S(\underline{k}'')S(\underline{k} - \underline{k}' - \underline{k}'')\rangle_0$$

$$- G_0(\tilde{\underline{k}})\,D\tilde{q}^a\,\frac{u}{6}\int_{\underline{k}'}\int_{\underline{k}''}\langle S(\underline{k})S(\underline{k}')S(\underline{k}'')S(\tilde{\underline{k}} - \underline{k}' - \underline{k}'')\rangle_0, \quad (3.45)$$

where the subscript '0' indicates again that the averages are to be taken in the Gaussian ensemble. The two four-point functions in the integrands factorize according to Wick's theorem (3.38). After evaluating one of the integrals with the aid of the ensuing delta functions, they yield three identical contributions each (Problem 3.3). Using $G_0(q, \omega) + G_0(-q, -\omega) = 2\mathrm{Re}\,G_0(q, \omega)$, the final result for the two-point correlation function to $O(u)$ becomes

$$C(q, \omega) \approx C_0(q, \omega)\left[1 - 6\,[\mathrm{Re}\,G_0(q, \omega)]\,Dq^a\,\frac{u}{6}\int_{q',\omega'}C_0(q', \omega')\right]. \quad (3.46)$$

To first order in u, we can also write

$$C(q, \omega) \approx 2k_{\mathrm{B}}T\,Dq^a\,G(-q, -\omega)\,G(q, \omega), \quad (3.47)$$

with the response propagators $G(q, \omega)$ from Eq. (3.42).

The iterative perturbation expansion for the dynamic correlation and response functions can obviously be extended to higher orders by replacing the fields S on the right-hand side of Eq. (3.39) with terms that contain successively increasing powers of u. The above examples illustrate that the ensuing analytic expressions become rather lengthy already in low orders of perturbation theory. Furthermore, a certain amount of subsequent computational manipulation appears repeatedly in identical fashion. It is therefore desirable to organize the perturbation expansion in a more transparent way. This can be achieved graphically in terms of *Feynman diagrams*. Associated Feynman graph computation rules succinctly summarize the involved standard calculations.

In order to construct a diagrammatic representation of the perturbation series, we first notice that it contains *three* different elements. First, we have the 'bare' propagator $G_0(q, \omega)$ of the Gaussian model, to which we assign a *directed line*, reflecting the causal property of the associated order parameter response function $\chi(q, t)$ in the time domain, see Eq. (3.29). We shall use the convention that time is running from right to left in our Feynman graphs, and therefore attach a left-pointed arrow to the propagator lines, linking 'incoming' and 'outgoing' order parameter fields.

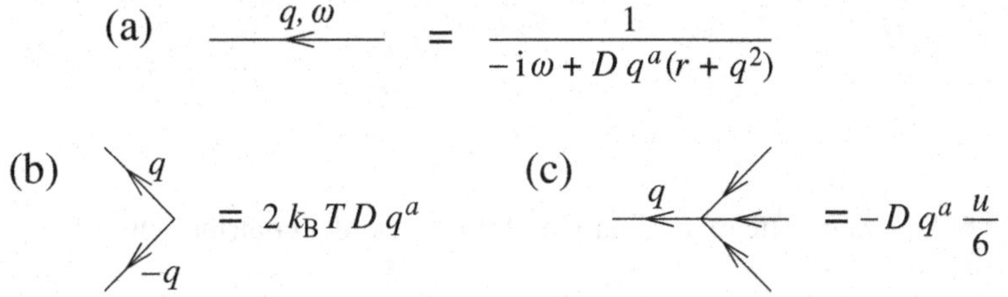

(a) $\quad \underset{q,\,\omega}{\longleftarrow} \quad = \quad \dfrac{1}{-\mathrm{i}\omega + D\,q^a(r + q^2)}$

(b) $\quad \diagdown{}^{q} \atop \diagup_{-q} \quad = 2\,k_{\mathrm{B}}T D\,q^a$

(c) $\quad \underset{q}{\longleftarrow}\kern-1em \diagup\kern-0.2em\diagdown \quad = -D\,q^a\,\dfrac{u}{6}$

Fig. 3.5 Elements of dynamic perturbation theory and their graphical representation: (a) propagator; (b) two-point noise vertex; and (c) four-point vertex.

Next, there is the random force ζ which effectively acts as a *source* for the field fluctuations $S(q, \omega)$. Eventually, when computing observable quantities, we need to average over the Gaussian noise, employing its second moment (3.24). Thus, the second basic ingredient is the noise strength $2k_{\mathrm{B}}T\,Dq^a$, which appears whenever the stochastic force averages are taken, or equivalently, when the dynamic correlation function $C_0(q, \omega)$ is expressed in terms of the product $G_0(q, -\omega)G_0(q, \omega)$ via Eq. (3.31). Thus, after averaging, the noise sources have been contracted to a *vertex* with two outgoing lines, each representing a propagator G_0.

Finally, we have the non-linear coupling $-Dq^a u/6$, linking four propagators which originate from the quartic anharmonicity in the Hamiltonian. Inspection of Eq. (3.39) suggests that the associated graph, a *four-point vertex*, should connect *one outgoing* (corresponding to the factor $G_0(q, \omega)$ in front of the integral in the iterated equation for S) and *three incoming* directed lines. The pictorial representation of these elements of perturbation theory is shown in Fig. 3.5. The arrows in Fig. 3.5(b, c) indicate how the propagators are to be attached to the vertices. We have labeled the propagator with its momentum q and frequency ω. For the vertices, it suffices to indicate the wavevectors q of the outgoing lines which determine their momentum dependence for $a = 2$, i.e., for a conserved order parameter field. As can again be seen explicitly in Eq. (3.39), at each non-linear vertex the sum of the ingoing momenta and frequencies must equal the sum of the outgoing momenta and frequencies, as a consequence of translation invariance in space and time. The noise vertex, Fig. 3.5(b), was designed in such a way that this wavevector and frequency conservation holds as well, in accord with Eqs. (3.24) and (3.31).

We saw that performing the averages over the random noise sources led to convolution integrals over 'internal' momenta and frequencies of products of Gaussian propagators and correlation functions. Graphically, these are constructed by linking up the outgoing lines of the noise vertices with incoming lines of the four-point vertices. Thereby, in the associated Feynman diagrams these integrals correspond to closed *loops* of those propagator lines which carry the *internal* wavevectors and frequencies. Notice that each of these closed loops is associated with a factor u.

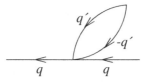

Fig. 3.6 Diagrammatic representation of the first-order contribution to $G(q, \omega)$.

Therefore, for a given correlation function, the contributions to a certain order in the perturbation series with respect to u are all characterized by an identical number of closed loops in the associated Feynman diagrams: the perturbation series diagrammatically maps onto a *loop expansion*.

The Feynman graph that represents our first-order result (3.42) for the propagator, containing one closed so-called 'Hartree' loop, is depicted in Fig. 3.6 (only the external and internal wavevectors are indicated, the frequency labeling is analogous). Notice that the combinatorial factor 3 in Eq. (3.42) is encoded in the one-loop diagram as the number of possibilities of selecting one of the three 'legs' of the vertex in Fig. 3.5(c) to join the external line; or, equivalently, as the number of possibilities of choosing two of the three incoming vertex legs for the formation of the closed loop.

An alternative diagrammatic representation of the perturbation series directly employs the dynamic correlation function $C_0(q, \omega)$ as a *second propagator*, to be pictured by a *non-directed* line. This contains the noise vertex automatically, and the elements of the ensuing Feynman graphs are now the two propagators and the non-linear vertex $\propto u$. The internal momentum and frequency integral in the first line of (3.42) then simply becomes a loop with this new propagator. Although both formulations are of course fully equivalent, we shall predominantly use the earlier representation solely in terms of the response propagator $G_0(q, \omega)$, because this renders the relationships to the dynamic field theories for quantum critical phenomena and non-equilibrium systems more transparent. The second variant has the advantage that the static limit becomes more obvious. We shall return to these issues in Chapter 4, which contains a more thorough and systematic description of the Feynman diagrams and associated computation rules, based on a field theory representation of non-linear Langevin equations.

3.3 Conserved quantities and reversible mode couplings

We have just seen that the dynamic critical behavior near a second-order phase transition is markedly different for conserved and non-conserved order parameters. Yet a complete description of the slow dynamics near a critical point requires the inclusion of *hydrodynamic modes* associated with conservation laws in addition to the order parameter field. As exemplified by the energy density in the still purely relaxational models C and D, coupling to such conserved quantities may alter the dynamic critical behavior while leaving the static critical exponents unchanged. Moreover, the fully reversible microscopic classical or quantum-mechanical equations of motion may generate entirely dynamical mode–mode couplings between the order parameter and hydrodynamic variables. As a consequence, systems with identical static critical exponents may be split up further into distinct dynamic universality classes. This will be illustrated for models J and E, describing respectively the critical dynamics of isotropic and planar (XY) ferromagnets. In each of these cases, the assumption of a *single* diverging characteristic time scale in conjunction with the dynamic scaling hypothesis will allow us to relate the dynamic exponent z to static critical exponents. We also study the hydrodynamics of these systems (in Gaussian approximation) in the ordered phase. A more detailed investigation of these models by means of the renormalization group and self-consistent mode-coupling theory will be the topic of Chapter 6.

3.3.1 Coupling to the energy density: models C and D

Any local conservation law implies a continuity equation of the form (3.13). Consequently, for the associated density $\rho(q, t)$ and current $J(q, t)$ in Fourier space, $\partial \rho(q, t)/\partial t = -iq \cdot J(q, t)$. Thus, its characteristic relaxational time scale diverges at least $\sim q^{-1}$. Typically, conserved modes will relax diffusively, $t_c(q) \sim q^{-2}$. At any rate, if we are interested in the low-frequency and long-wavelength limit, i.e., a *generalized hydrodynamic description* of a given system, we thus merely have to characterize the kinetics of its conserved modes. Near a second-order phase transition, one needs to add the slow-order parameter field (if not itself subject to a conservation law). Any coupling of the order parameter to conserved modes will certainly affect the dynamic scaling functions, but can also change the dynamic critical exponent z.

As an example, let us consider the always-conserved energy density ρ.[5] In an effective long-wavelength description near the critical point, the associated Hamiltonian should essentially be just of Gaussian form. However, we wish to allow for a non-linear coupling to the order parameter S, respecting the system's fundamental symmetries. A linear coupling $\rho(x)S(x)$ would not preserve the Ising Z_2 symmetry, for example, and the lowest-order term compatible with it is of the form $\rho(x)S(x)^2$. Thus, we add to the Landau–Ginzburg–Wilson Hamiltonian (3.9) the terms

$$\Delta \mathcal{H}[\rho, S] = \frac{1}{2C} \int d^d x \left[\rho(x)^2 + g \, \rho(x) \, S(x)^2 \right] . \qquad (3.48)$$

---

[5] Halperin, Hohenberg, and Ma (1974, 1976).

In order to grasp the physical meaning of the coupling C, we infer from (3.48) the energy correlation function $\langle \rho(x)\rho(x') \rangle = k_B T\, C$. On the other hand, the energy fluctuations in the canonical ensemble are related to the specific heat C_V via

$$(\Delta E)^2 = k_B T^2\, C_V \,, \tag{3.49}$$

whence we identify $C = C_V/VT$. Recall that $C \sim |\tau|^{-\alpha}$ in the vicinity of the phase transition.

Next we notice that the coupling g between the order parameter and the energy density does not affect the static universal critical behavior: $\rho(x)$ enters the Hamiltonian only quadratically, and upon completing the square, Gaussian integration yields

$$\int d\rho\, \exp\left[-\frac{1}{2Ck_B T}\left(\rho^2 + g\,\rho\, S^2\right)\right] \propto \exp\left(\frac{g^2}{8Ck_B T}\, S^4\right).$$

The energy density fluctuations have now been integrated out, leaving a Landau–Ginzburg–Wilson Hamiltonian for the order parameter which is precisely of the form (3.9), albeit with shifted anharmonicity strength $u \to u' = u - 3g^2/C$. Yet u asymptotically approaches a universal fixed point value u_I^* independent of its initial value (provided $u > 0$, which will be satisfied here, because $C^{-1} \sim |\tau|^{\alpha} \to 0$ as $\alpha > 0$ for a scalar order parameter). The static critical exponents are solely determined by u_I^*, and hence assume the same values as for $g = 0$.

For either non-conserved or conserved order parameter dynamics, we employ the relaxational ansatz (3.18) again, which yields

$$\frac{\partial S(x,t)}{\partial t} = -D(i\nabla)^a\left[(r - \nabla^2)S(x,t) + \frac{u}{6} S(x,t)^3 - h(x,t) + \frac{g}{C}\rho(x,t)\, S(x,t)\right]$$
$$+ \zeta(x,t) \tag{3.50}$$

supplemented by the noise correlations (3.19). Similarly, the Langevin equation for the conserved energy density $\rho(x,t)$ becomes

$$\frac{\partial \rho(x,t)}{\partial t} = \lambda\nabla^2\, \frac{\delta H[\rho, S]}{\delta\rho(x,t)} + \eta(x,t) = \frac{\lambda}{C}\nabla^2\left[\rho(x,t) + \frac{g}{2} S(x,t)^2\right] + \eta(x,t)\,, \tag{3.51}$$

with conserved noise $\eta(x,t)$ that must obey

$$\langle \eta(x,t)\rangle = 0\,, \quad \langle \eta(x,t)\eta(x',t')\rangle = -2\lambda\, k_B T\, \nabla_x^2\, \delta(x - x')\delta(t - t')\,. \tag{3.52}$$

This set of coupled Langevin equations defines the *relaxational models C and D* for a non-conserved ($a = 0$) or conserved ($a = 2$) order parameter, respectively, coupled to a conserved scalar density.

Considering the linear term in Eq. (3.51), we see that $\kappa = \lambda/C$ is just the thermal conductivity. The associated characteristic frequency is $\omega_c(\tau, q) = i\kappa(\tau)$

$q^2 \sim |\tau|^\alpha q^2$, as the relaxation constant λ of the conserved non-critical field ρ should not acquire any singular behavior at all. Let us now *assume* that there exists a single diverging characteristic time scale in the coupled order parameter and energy density system. The dynamic scaling hypothesis (3.4) then requires that $\omega_c(\tau, q) \sim |q|^z |q\xi|^{-\alpha/\nu}$ in order to match the temperature dependence in the previous relation. Upon comparing the resulting exponents for the wavenumber q, we thus infer the scaling relation

$$z = 2 + \frac{\alpha}{\nu} \qquad (3.53)$$

for the dynamic critical exponent. For model C with a non-conserved order param-eter ($z = 2$ in mean-field approximation), the coupling to the scalar energy density leads to a further slowing-down of the dynamics near the phase transition. For model D with conserved order parameter, however, the Gaussian theory gave $z = 4$, larger than the result (3.53), which therefore cannot realistically describe its critical dynamics. Rather, one would in this case expect the dynamics of the order parameter S and the energy density ρ to be governed by *different* characteristic time scales, governed by distinct dynamic exponents, namely

$$z_s = 4 - \eta , \quad z_\rho = 2 + \frac{\alpha}{\nu} . \qquad (3.54)$$

This is in fact borne out by a systematic renormalization group analysis (Sec-tion 6.1). For model C with non-conserved n-component vector order parameter, $\alpha < 0$ for $n \geq 2$, which also renders Eq. (3.53) inconsistent. Section 6.1 will be devoted to a renormalization group study of models C and D; indeed, we shall then find different dynamic scaling regimes, depending on the number n of order parameter components.

3.3.2 Critical dynamics with reversible mode couplings

Aside from purely relaxational terms in the effective long-wavelength Langevin description of dynamic critical phenomena, fully *reversible* contributions may be present, see Eq. (2.188). These represent non-linear couplings between different hydrodynamic modes and/or the order parameter that originate in the underlying microscopic dynamics.

Let us henceforth collect the set of slow variables in a vector of fields ψ^α. At zero temperature, the corresponding *microscopic* degrees of freedom obey the equations of motion

$$\frac{\partial \psi_m^\alpha(x, t)}{\partial t} = \{H[\psi_m], \psi_m^\alpha(x, t)\} , \qquad (3.55)$$

where $H[\psi_m]$ denotes the microscopic Hamiltonian, and $\{A, B\}$ the Poisson bracket between two observables A and B. In the realm of quantum mechanics, by virtue of Ehrenfest's correspondence principle we just need to replace the observables by the respective operators, and the Poisson brackets with the commutator, $\{A, B\} \leftrightarrow \frac{i}{\hbar}[A, B]$, whereupon Eq. (3.55) translates to Heisenberg's equation of motion (2.3). Upon coarse-graining, the microscopic quantities ψ_m^α and $H[\psi_m]$ become replaced by the *mesoscopic* hydrodynamic fields ψ^α and the effective Hamiltonian $\mathcal{H}[\psi]$. The slow variables ψ^α per construction provide a complete description of the low-frequency, long-wavelength limit. Hence we may expand

$$\{\mathcal{H}[\psi], \psi^\alpha(x)\} = \int d^d x' \sum_\beta \frac{\delta \mathcal{H}[\psi]}{\delta \psi^\beta(x')} \{\psi^\beta(x'), \psi^\alpha(x)\} .$$

Introducing the abbreviation

$$Q^{\alpha\beta}(x, x') = \left\{\psi^\alpha(x), \psi^\beta(x')\right\} = -Q^{\beta\alpha}(x', x) \tag{3.56}$$

for the Poisson brackets (or quantum-mechanical commutators) between the relevant slow variables, we thus find for the reversible terms in the Langevin equations of motion at $T = 0$:

$$F_{\text{rev}}^\alpha[\psi](x) = -\int d^d x' \sum_\beta Q^{\alpha\beta}(x, x') \frac{\delta \mathcal{H}[\psi]}{\delta \psi^\beta(x')} . \tag{3.57}$$

In order for these reversible forces to be compatible with the canonical equilibrium probability distribution (3.11), they need to satisfy the condition (2.193) that the corresponding probability current be divergence-free in the infinite-dimensional space spanned by the hydrodynamic fields $\psi^\alpha(x)$,

$$\int d^d x \sum_\alpha \frac{\delta}{\delta \psi^\alpha(x)} \left(F_{\text{rev}}^\alpha[\psi] \, e^{-\mathcal{H}[\psi]/k_B T} \right) = 0 . \tag{3.58}$$

This is not yet fulfilled by Eq. (3.57); in fact, one needs to supplement this ansatz with a finite-temperature term,[6]

$$F_{\text{rev}}^\alpha[\psi](x) = -\int d^d x' \sum_\beta \left[Q^{\alpha\beta}(x, x') \frac{\delta \mathcal{H}[\psi]}{\delta \psi^\beta(x')} - k_B T \frac{\delta Q^{\alpha\beta}(x, x')}{\delta \psi^\beta(x')} \right]. \tag{3.59}$$

An explicit calculation (Problem 3.4) shows that Eq. (3.58) now holds. The complete coupled set of stochastic differential equations for the slow variables finally

---

[6] This structure follows explicitly from the Mori–Zwanzig projector formalism which provides a systematic coarse-graining procedure to arrive at a mesoscopic description starting from the microscopic equations of motion. Under renormalization, further terms compatible with the system's overall symmetry properties might be generated.

reads

$$\frac{\partial \psi^\alpha(x, t)}{\partial t} = F^\alpha_{\mathrm{rev}}[\psi](x, t) - D^\alpha (i\nabla)^{a_\alpha} \frac{\delta \mathcal{H}[\psi]}{\delta \psi^\alpha(x, t)} + \zeta^\alpha(x, t) , \tag{3.60}$$

with $a_\alpha = 0$ or 2 for a non-conserved or conserved field, respectively, and

$$\langle \zeta^\alpha(x, t) \rangle = 0 , \quad \langle \zeta^\alpha(x, t)\zeta^\beta(x', t') \rangle = 2k_\mathrm{B}T \, D^\alpha (i\nabla_x)^{a_\alpha} \delta(x - x')\delta(t - t') \, \delta^{\alpha\beta} . \tag{3.61}$$

3.3.3 Isotropic ferromagnets: model J

As a first example, we consider *isotropic ferromagnets*,[7] as described microscopically by the *Heisenberg model* Hamiltonian

$$H[\{\vec{S}_j\}] = -\frac{1}{2} \sum_{j,k=1}^{N} J_{jk} \, \vec{S}_j \cdot \vec{S}_k - \vec{h} \cdot \sum_{j=1}^{N} \vec{S}_j , \tag{3.62}$$

where the three-component vectors $\vec{S}_j$ represent quantum-mechanical spin operators on each lattice site j, with the standard commutation relations

$$\left[S_j^\alpha, S_k^\beta \right] = i\hbar \sum_\gamma \epsilon^{\alpha\beta\gamma} S_j^\gamma \, \delta_{jk} . \tag{3.63}$$

Here, $\epsilon^{\alpha\beta\gamma}$ is the fully antisymmetric tensor of rank 3. We assume ferromagnetic exchange couplings, $J_{jk} = J_{kj} > 0$. For vanishing external magnetic field $\vec{h} = 0$, the Heisenberg Hamiltonian is invariant with respect to rotations in spin space; this $O(3)$ symmetry is explicitly broken by a non-zero field. Spontaneous symmetry breaking to a ferromagnetic phase with uniform magnetization occurs at sufficiently low temperatures in dimensions $d > d_{\mathrm{lc}} = 2$. Also, for $\vec{h} = 0$ the magnetization operator $\vec{M} = \sum_i \vec{S}_i$ commutes with the Hamiltonian (3.62):

$$[H, M^\alpha] = -\frac{1}{2} \sum_{ijk} J_{jk} \sum_\beta \left(S_j^\beta \left[S_k^\beta, S_i^\alpha \right] + \left[S_j^\beta, S_i^\alpha \right] S_k^\beta \right)$$

$$= \frac{i\hbar}{2} \sum_{jk} J_{jk} \sum_{\beta\gamma} \epsilon^{\alpha\beta\gamma} \left(S_j^\beta \, S_k^\gamma + S_k^\gamma \, S_j^\beta \right) = 0 \tag{3.64}$$

after summation over the antisymmetric tensor. Thus, the total magnetization is a *conserved* quantity in the $O(3)$-symmetric Heisenberg model.

Moving on to a mesoscopic description level, the microscopic lattice spin variables are replaced with components of the local magnetization density $S^\alpha(x)$, and

---

[7] Schwabl and Michel (1970), Ma and Mazenko (1974, 1975).

the quantum-mechanical commutation relations (3.63) become Poisson brackets,

$$Q^{\alpha\beta}(x, x') = \{S^\alpha(x), S^\beta(x')\} = -g \sum_\gamma \epsilon^{\alpha\beta\gamma} S^\gamma(x) \delta(x - x') . \qquad (3.65)$$

Here, we have introduced the dynamical mode coupling g which originates from the coarse-graining procedure.[8] Rotational invariance in spin space requires it to be identical for all components. As $\delta Q^{\alpha\beta}(x, x')/\delta S^\beta(x') = \sum_\gamma \epsilon^{\alpha\beta\gamma} \delta^{\beta\gamma} \delta(x - x') = 0$, we find for the reversible forces (3.59)

$$F_{\text{rev}}^\alpha[\vec{S}](x) = g \sum_{\beta\gamma} \epsilon^{\alpha\beta\gamma} S^\gamma(x) \frac{\delta\mathcal{H}[\vec{S}]}{\delta S^\beta(x)} = -g \left(\vec{S}(x) \times \frac{\delta\mathcal{H}[\vec{S}]}{\delta \vec{S}(x)} \right)_\alpha . \qquad (3.66)$$

This term may be interpreted as the precession of the magnetization density vector $\vec{S}(x)$ in the effective local field $-\delta\mathcal{H}[\vec{S}]/\delta\vec{S}(x)$, see Eq. (1.14). With the order parameter being conserved, it is to be added to a diffusive relaxational term, i.e., the vector generalizations of Eqs. (3.16) and (3.17),

$$\frac{\partial \vec{S}(x, t)}{\partial t} = -g\, \vec{S}(x, t) \times \frac{\delta\mathcal{H}[\vec{S}]}{\delta\vec{S}(x, t)} + D\nabla^2 \frac{\delta\mathcal{H}[\vec{S}]}{\delta\vec{S}(x, t)} + \vec{\zeta}(x, t) , \qquad (3.67)$$

with noise correlations

$$\langle \vec{\zeta}(x, t) \rangle = 0 , \quad \langle \zeta^\alpha(x, t)\zeta^\beta(x', t') \rangle = -2D\, k_{\mathrm{B}} T\, \nabla_x^2 \delta(x - x')\delta(t - t') \delta^{\alpha\beta} . \tag{3.68}$$

The appropriate effective potential for the Heisenberg model is the natural extension of the *Landau–Ginzburg–Wilson* Hamiltonian (3.9) to vector fields,

$$\mathcal{H}[\vec{S}] = \int d^d x \left(\frac{r}{2} \vec{S}(x)^2 + \frac{1}{2} \left[\nabla\vec{S}(x)\right]^2 + \frac{u}{4!} \left[\vec{S}(x)^2\right]^2 - \vec{h}(x) \cdot \vec{S}(x) \right) . \tag{3.69}$$

With Eqs. (3.67) and (3.68), this completes the definition of the dynamical *model J* which captures the universal aspects of the critical dynamics of isotropic ferromagnets. The reversible force term now reads explicitly $\vec{F}_{\text{rev}}[\vec{S}] = g\, \vec{S} \times \vec{h}_{\text{eff}} = g\, \vec{S} \times (\vec{h} + \nabla^2\vec{S})$, whence we arrive at

$$\frac{\partial S^\alpha(x, t)}{\partial t} = g \sum_{\beta\gamma} \epsilon^{\alpha\beta\gamma} S^\beta(x, t)\left[\nabla^2 S^\gamma(x, t) + h^\gamma(x, t)\right]$$

$$+ D\nabla^2 \left[\left(r - \nabla^2 + \frac{u}{6}\vec{S}(x, t)^2\right) S^\alpha(x, t) - h^\alpha(x, t)\right] + \zeta^\alpha(x, t). \tag{3.70}$$

---

[8] Notice that the sign of g should not matter, since upon inversion $S^\alpha \to -S^\alpha$ and $g \to -g$. In fact, we shall see in Section 6.2 that the relevant mode coupling is $\propto g^2$.

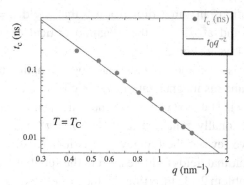

Fig. 3.7 Critical slowing-down of the characteristic relaxation time $t_c(q) = t_0 |q|^{-z}$ of the magnetic fluctuations in the three-dimensional Heisenberg ferromagnet $CdCr_2S_4$. The solid line represents a power law fit with $z = 2.47 \pm 0.02$ and $t_0 = 0.02813$ ns nm^{-z}. [Data and figure reproduced with permission from M. Alba, S. Pouget, P. Fouquet, B. Farago, and C. Pappas, Preprint arXiv:cond-mat/0703702 (2007).]

In the paramagnetic phase, linearizing this stochastic equation of motion just recovers the Gaussian model B, as the reversible additions are quadratic in the fields. It is characterized by diffusive modes with dispersion $\omega(q) = -iDq^2(r + q^2)$, called *paramagnons* in this context. Naively, we might thus expect a dynamic critical exponent close to $z = 4$; however, the *non-linear* reversible dynamics in fact alters this value in $d < 6$ dimensions.

To see this, let us apply the scaling ideas of Section 1.4. From Eq. (3.70), we infer that $g^{-1} \sim \nabla^2 S^\alpha t \sim ht$. Next we rescale $x \to x' = x/b$; in the Gaussian model, accordingly $t \to t' = t/b^z$ with $z = 4$, and the fields transform as given in Eqs. (1.82) and (1.87), whence $g \sim b^{3-d/2}$. The reversible mode coupling therefore constitutes a *relevant* perturbation in the renormalization group sense for $d < d'_c$, with a *dynamic upper critical dimension* $d'_c = 6$. Moreover, scaling theory actually allows us to express the dynamic critical exponent z in terms of the known static critical exponents for this equilibrium system. With Eqs. (1.90) or (1.91) we obtain $g \sim b^{z-(d+2-\eta)/2}$. At the transition, g (more precisely, the effective coupling g/D) approaches an RG fixed point that describes the universal properties of model J. Provided $0 < |g^*| < \infty$, the dependence on the scale b must cancel, and therefore

$$z = \frac{d + 2 - \eta}{2}, \tag{3.71}$$

where η is Fisher's exponent for the $O(3)$-symmetric Heisenberg model. In three dimensions $z \approx 5/2$, markedly smaller than predicted by mean-field theory. An experimental confirmation of this dynamic scaling law for the Heisenberg ferromagnet $CdCr_2S_4$ is shown in Fig. 3.7, giving $z = 2.47 \pm 0.02$, quite similar to numerical simulations in three dimensions[9] that yield $z = 2.49 \pm 0.09$. For $d > d_c = 4$, $\eta = 0$, and at $d'_c = 6$, the dynamic critical exponent assumes the Gaussian value $z = 4$, as it should.

It is instructive to also consider the Gaussian approximation in the ordered phase with non-zero spontaneous magnetization ϕ, which we choose in the z direction: $S^3(x, t) = \phi + \sigma(x, t)$. Thus $\langle \sigma(x, t) \rangle = 0$; the statistical average of the other two components, conventionally renamed as $S^\alpha(x, t) = \pi^\alpha(x, t)$, $\alpha = 1, 2$, remains zero as well. Next we rewrite the Langevin equations (3.70) in terms of the new transverse and longitudinal fields (with respect to the spontaneous order parameter) and linearize, see Problem 2.11. Inserting the mean-field expression $\phi = \sqrt{6|r|/u}$, we arrive at

$$\frac{\partial \pi^\alpha(x, t)}{\partial t} = -g\,\phi \sum_\beta \epsilon^{\alpha\beta3}\big[\nabla^2\pi^\beta(x, t) + h^\beta(x, t)\big]$$

$$- D\nabla^2\big[\nabla^2\pi^\alpha(x, t) + h^\alpha(x, t)\big] + \zeta^\alpha(x, t) \qquad (3.72)$$

for the transverse fluctuations, while the linearized stochastic equation of motion for the longitudinal field is just the one of model B (Problem 3.2). The transverse sector is readily diagonalized by introducing the circularly polarized modes

$$\pi^\pm(x, t) = \frac{1}{\sqrt{2}}\big[\pi^1(x, t) \pm i\,\pi^2(x, t)\big], \qquad (3.73)$$

whereupon, with similarly defined $h^\pm(x, t)$ and $\zeta^\pm(x, t)$,

$$\frac{\partial \pi^\pm(x, t)}{\partial t} = (\pm ig\,\phi - D\nabla^2)\big[\nabla^2\pi^\pm(x, t) + h^\pm(x, t)\big]. \qquad (3.74)$$

After Fourier transformation, we immediately read off the dynamic response functions in Gaussian approximation:

$$\chi_0^\pm(q, \omega) = \frac{\partial \langle \pi^\pm(q, \omega)\rangle}{\partial h^\pm(q, \omega)}\bigg|_{h=0} = \frac{\pm ig\,\phi + Dq^2}{-i(\omega \mp g\,\phi\,q^2) + Dq^4}. \qquad (3.75)$$

Either directly, or via the classical fluctuation-dissipation theorem (2.35), we find the correlation function $\langle \pi^\pm(q, \omega)\pi^\mp(q', \omega')\rangle_0 = C_0^\pm(q, \omega)\,(2\pi)^{d+1}\,\delta(q + q')\delta(\omega + \omega')$, with

$$C_0^\pm(q, \omega) = \frac{2k_B T}{\omega}\,\mathrm{Im}\,\chi_0^\pm(q, \omega) = \frac{2k_B T\,Dq^2}{\big(\omega \mp g\,\phi\,q^2\big)^2 + \big(Dq^4\big)^2}. \qquad (3.76)$$

---

[9] Murtazaev, Mutailamov, Kamilov, Khizriev, and Abuev (2003).

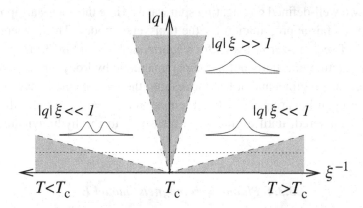

Fig. 3.8 Critical vs. hydrodynamic regions in the temperature–wavenumber plane. The (transverse) dynamic correlation function is sketched in both hydrodynamic regimes with $|q|\xi \ll 1$ (to the right for $T > T_c$, left for $T < T_c$); and in the critical region with $|q|\xi \gg 1$. The horizontal dotted line indicates a constant $|q|$ scan. [Figure adapted from Frey and Schwabl (1994).]

The poles of the transverse dynamic susceptibility (3.75) yield *propagating spin waves (magnons)* with dispersion

$$\omega_\pm(q) = \pm g\,\phi\,q^2 - iDq^4 . \tag{3.77}$$

These Goldstone modes (Section 5.4) govern the hydrodynamics of a Heisenberg ferromagnet as well as its low-temperature thermodynamics; e.g., contributing a term $\sim T^{d/2}$ to the specific heat.

Near T_c, $\mathrm{Im}\,\omega_\pm(q) \sim -|q|^z$; for the real part we notice that the Laplacians inside the square brackets of Eq. (3.74) should be interpreted as the inverse static susceptibility, i.e.: $\mathrm{Re}\,\omega_\pm(q) \sim \pm\phi\,|q|^{2-\eta}$. We may now invoke the dynamic scaling hypothesis, which demands that in the critical regime, both real and imaginary parts of the dispersion relation obey the general scaling form (3.4) for the characteristic frequency. Since $\phi \sim \xi^{-\beta/\nu}$, compatibility requires that $z = 2 - \eta + \beta/\nu$; with Eq. (1.77) we thus recover our earlier result (3.71) for the dynamic exponent of model J.

Figure 3.8 summarizes the different scenarios in the temperature–wavenumber plane, with the former parametrized by the inverse correlation length ξ^{-1}. The hydrodynamic description applies in the limit $|q|\xi \ll 1$, shown in the lower right and left of the figure for $T > T_c$ and $T < T_c$, respectively. The paramagnetic hydro-dynamic region is described well by the Gaussian model B, yielding a diffusive spin excitation. Upon approaching the critical regime where $|q|\xi \gg 1$, the paramagnon peak broadens considerably due to the critical softening and slowing-down with $\omega_c(q) \sim |q|^z$. In the ferromagnetic phase, this broad overdamped critical excitation

splits into two well-defined propagating spin waves. Here the Gaussian approximation provides a fair approximation for the transverse modes. Thus, when probing
the system at fixed non-zero wavevector q (horizontal line in Fig. 3.8), upon reducing the temperature one starts out in the paramagnetic hydrodynamic region, then
enters the critical regime, and at last moves into the ferromagnetic hydrodynamic
domain. Yet upon lowering $|q|$ at fixed temperature $T \neq T_c$, initially the critical
region is encountered, until eventually a crossover to the hydrodynamic region
occurs.

3.3.4 Planar ferromagnets: model E

Generalizing the Heisenberg model (3.62), we allow for uniaxial anisotropy in spin
space:

$$H[\{S_j^\alpha\}] = -\frac{1}{2} \sum_{j,k=1}^{N} \left[J_{jk} \left(S_j^1 S_k^1 + S_j^2 S_k^2 \right) + \tilde{J}_{jk} S_j^3 S_k^3 \right] - \vec{h} \cdot \sum_{j=1}^{N} \vec{S}_j . \quad (3.78)$$

For $J_{jk} \neq \tilde{J}_{jk}$, the $O(3)$ symmetry is broken. For vanishing external field $\vec{h} = 0$, the
Hamiltonian (3.78) remains invariant with respect to rotations in the 1-2 plane. As
a consequence only the total spin component $M^3 = \sum_i S_i^3$, which is the generator
of this reduced $O(2)$ symmetry group, is conserved,

$$\left[H, M^3 \right] = 0 , \quad (3.79)$$

whereas a brief commutator calculation akin to (3.64) yields

$$\left[H, M^{1/2} \right] = \pm \frac{i\hbar}{2} \sum_{jk} \left(J_{jk} - \tilde{J}_{jk} \right) \left(S_j^3 S_k^{2/1} + S_k^{2/1} S_j^3 \right) \neq 0 . \quad (3.80)$$

Let us first discuss the case of an *easy-axis* ferromagnet, for which $\tilde{J}_{jk} > J_{jk} >$
0. Any long-range order should then emerge in the S^3 direction. The S^1, S^2 spin
components represent non-critical *and* non-conserved degrees of freedom, and can
therefore be safely discarded in the critical region. With $S_j^3 = \frac{\hbar}{2} \sigma_j$, the Hamiltonian
(3.78) reduces to the Ising model (1.10), with a conserved order parameter. Its
critical dynamics should hence be described by the scalar model B, Eqs. (3.16) and
(3.17) with (3.9).

For $J_{jk} > \tilde{J}_{jk} > 0$, on the other hand, we encounter an *easy-plane* ferromagnet,
whose spontaneous magnetization will lie in the 1-2 plane.[10] Yet now the S^3
fluctuations may not be disregarded entirely, since $M^3 = \sum_i S_i^3$ is a conserved
quantity. The effective Hamiltonian for this system is to feature a two-component

---

[10] Halperin, Hohenberg, and Siggia (1974, 1976).

vector order parameter $\vec{S} = (S^1, S^2)$, and a decoupled non-critical scalar field, for which we may neglect any non-linear and spatially varying terms. Absorbing the prefactor of the merely harmonic term into the renamed field $S^3(x) \to M(x)$, we arrive at

$$\mathcal{H}[\vec{S}, M] = \int d^d x \left(\frac{r}{2} \vec{S}(x)^2 + \frac{1}{2} [\nabla \vec{S}(x)]^2 + \frac{u}{4!} [\vec{S}(x)^2]^2 - \vec{h}(x) \cdot \vec{S}(x) \right.$$

$$\left. + \frac{1}{2} M(x)^2 - H(x) M(x) \right) . \qquad (3.81)$$

In the coarse-grained description, the spin operator commutators (3.63) transform into the Poisson brackets

$$\{ S^1(x), S^2(x') \} = -\tilde{g} M(x) \delta(x - x') , \qquad (3.82)$$

$$\{ S^{1/2}(x), M(x') \} = \pm g \, S^{2/1}(x) \delta(x - x') . \qquad (3.83)$$

Noting that the second term in Eq. (3.59) vanishes for each component, we then compute

$$F_{\text{rev}}^{1/2}[\vec{S}, M](x) = \pm \tilde{g} \, M(x) \frac{\delta \mathcal{H}[\vec{S}, M]}{\delta S^{2/1}(x)} \mp g \, S^{2/1}(x) \frac{\delta \mathcal{H}[\vec{S}, M]}{\delta M(x)} , \qquad (3.84)$$

$$F_{\text{rev}}^M[\vec{S}, M](x) = -g \left[S^1(x) \frac{\delta \mathcal{H}[\vec{S}, M]}{\delta S^2(x)} - S^2(x) \frac{\delta \mathcal{H}[\vec{S}, M]}{\delta S^1(x)} \right] . \qquad (3.85)$$

The Langevin equation for the conserved field $M(x)$ needs to be supplemented by a relaxational term as in Eq. (3.51), whence

$$\frac{\partial M(x, t)}{\partial t} = F_{\text{rev}}^M[\vec{S}, M](x, t) + \lambda \nabla^2 \frac{\delta \mathcal{H}[\vec{S}, M]}{\delta M(x, t)} + \eta(x, t)$$

$$= g \sum_{\alpha, \beta} \epsilon^{\alpha \beta 3} S^\alpha(x, t) [\nabla^2 S^\beta(x, t) + h^\beta(x, t)]$$

$$+ \lambda \nabla^2 [M(x, t) - H(x, t)] + \eta(x, t) , \qquad (3.86)$$

with the stochastic noise correlations (3.52). In order to leave the XY model Hamiltonian (3.81) invariant under scale transformations, the conserved variable and its conjugate field $H(x)$ must rescale according to $M(x) \to M'(x') = b^{d/2} M(x)$, in contrast to the order parameter, see Eqs. (1.82) and (1.90). From Eq. (3.86) therefore $g^{-1} \sim S^1 h^2 t / M$, or $g \sim b^{z-d/2}$, consistent with the second term in (3.84) contributing to the time derivative of the order parameter, $g^{-1} \sim Mt$. In Gaussian approximation $z = 2$, from which we infer the dynamic upper critical dimension in this model to equal the static one, $d'_c = d_c = 4$. Moreover, at any finite RG fixed point $0 < |g^*| < \infty$ (as we shall see in Section 6.3, the true effective coupling is

$g^2/D\lambda$), we deduce

$$z = \frac{d}{2},$$ (3.87)

again smaller than the model A result $z = 2 + O(\epsilon)$ for $d < d_c = 4$.

For the other mode coupling, we find $\widetilde{g}^{-1} \sim M h^1 t$. In the Gaussian model, $\widetilde{g} \sim b^{1-d}$ thus constitutes an irrelevant perturbation (as the phase transition only occurs for $d > 2$). We even see that $\widetilde{g}$ remains irrelevant at any non-trivial fixed point g^* as well, because $\widetilde{g} h^1 \sim g$, and we know that the external field conjugate to the order parameter certainly is a relevant variable. For the critical dynamics of the XY model, we may therefore set $\widetilde{g} = 0$, which leaves us with the following Langevin equations for the non-conserved two-component order parameter:

$$\frac{\partial S^{1/2}(x, t)}{\partial t} = \mp g \, S^{2/1}(x, t) \frac{\delta \mathcal{H}[\vec{S}, M]}{\delta M(x, t)} - D \frac{\delta \mathcal{H}[\vec{S}, M]}{\delta S^{1/2}(x, t)} + \zeta^{1/2}(x, t)$$

$$= \mp g \, S^{2/1}(x, t) \, [M(x, t) - H(x, t)]$$ (3.88)

$$-D \left[\left(r - \nabla^2 + \frac{u}{6} \vec{S}(x, t)^2 \right) S^{1/2}(x, t) - h^{1/2}(x, t) \right] + \zeta^{1/2}(x, t),$$

with the noise correlations

$$\langle \zeta^{1/2}(x, t) \rangle = 0, \quad \langle \zeta^1(x, t) \zeta^2(x', t') \rangle = 0,$$

$$\langle \zeta^{1/2}(x, t) \zeta^{1/2}(x', t') \rangle = 2D \, k_B T \, \delta(x - x') \delta(t - t').$$ (3.89)

Eqs. (3.88), (3.89), (3.86), and (3.52) define *model E* describing the critical dynamics of planar ferromagnets. Simple scaling theory already fixed the dynamic critical exponent for this universality class to $z = d/2$ for $d < d_c = 4$, reaching the mean-field value $z = 2$ in four dimensions. This, however, tacitly assumed *strong dynamic scaling* behavior, i.e., the three components of the magnetization all fluctuate on equal time scales, $z_s = z_M$. As we shall see in Section 6.3, in the situation of *weak scaling*, Eq. (3.87) generalizes to the relation $z_s + z_M = 2$.

Following our procedure for the isotropic ferromagnet, we proceed to linearize the Langevin equations for model E in the ordered phase, setting $S^2(x, t) = \phi + \sigma(x, t)$. While the longitudinal field σ is governed by the model A equation of motion in this approximation (see Problem 3.2), the transverse order parameter fluctuation π becomes coupled to the conserved mode M,

$$\frac{\partial \pi(x, t)}{\partial t} = -g \, \phi \, [M(x, t) - H(x, t)] + D \left[\nabla^2 \pi(x, t) + h^1(x, t) \right] + \zeta^1(x, t),$$

(3.90)

$$\frac{\partial M(x, t)}{\partial t} = -g \, \phi \left[\nabla^2 \pi(x, t) + h^1(x, t) \right] + \lambda \nabla^2 \left[M(x, t) - H(x, t) \right] + \eta(x, t).$$

(3.91)

Fourier transformation now leads to two coupled linear equations for the averages of the fields π and M, which we may cast into matrix form:

$$G(q, \omega)^{-1} \begin{pmatrix} \langle \pi(q, \omega) \rangle \\ \langle M(q, \omega) \rangle \end{pmatrix} = \begin{pmatrix} D & g\,\phi \\ -g\,\phi & \lambda q^2 \end{pmatrix} \begin{pmatrix} \langle h^1(q, \omega) \rangle \\ \langle H(q, \omega) \rangle \end{pmatrix}. \tag{3.92}$$

Here

$$G(q, \omega)^{-1} = \begin{pmatrix} -i\omega + Dq^2 & g\,\phi \\ -g\,\phi q^2 & -i\omega + \lambda q^2 \end{pmatrix}, \tag{3.93}$$

which is readily inverted to

$$G(q, \omega) = \frac{1}{-[\omega - \omega_+(q)][\omega - \omega_-(q)]} \begin{pmatrix} -i\omega + \lambda q^2 & -g\,\phi \\ g\,\phi q^2 & -i\omega + Dq^2 \end{pmatrix}, \tag{3.94}$$

with $\omega_\pm(q)$ denoting the zeros of the determinant of (3.93),

$$\omega_\pm(q) = \pm \sqrt{(g\,\phi)^2 q^2 - \frac{1}{4}(D - \lambda)^2 q^4} - \frac{i}{2}(D + \lambda)q^2. \tag{3.95}$$

A simple matrix product now yields the matrix of response functions

$$\begin{pmatrix} \chi_{\pi\pi}(q, \omega) & \chi_{\pi M}(q, \omega) \\ \chi_{M\pi}(q, \omega) & \chi_{MM}(q, \omega) \end{pmatrix} = G(q, \omega) \begin{pmatrix} D & g\,\phi \\ -g\,\phi & \lambda q^2 \end{pmatrix}, \tag{3.96}$$

whence one obtains for the transverse order parameter susceptibility

$$\chi_{\pi\pi}(q, \omega) = -\frac{D(-i\omega + \lambda q^2) + (g\,\phi)^2}{[\omega - \omega_+(q)][\omega - \omega_-(q)]}, \tag{3.97}$$

the response function for the conserved variable

$$\chi_{MM}(q, \omega) = -q^2 \frac{\lambda(-i\omega + Dq^2) + (g\,\phi)^2}{[\omega - \omega_+(q)][\omega - \omega_-(q)]}, \tag{3.98}$$

and finally, the mixed susceptibility

$$\chi_{\pi M}(q, \omega) = -\chi_{M\pi}(q, \omega) = \frac{i\omega\, g\,\phi}{[\omega - \omega_+(q)][\omega - \omega_-(q)]}. \tag{3.99}$$

We thus see that the spin waves in a planar ferromagnet are characterized by the dispersion relation (3.95), i.e., $\omega_\pm(q) \approx \pm g\,\phi\,|q|$ at long wavelengths, in contrast to the quadratic dispersion (3.77) for isotropic magnets. At low temperatures, the model E spin waves thus add a contribution $\sim T^d$ to the specific heat. The spin wave damping is proportional to q^2 only, as opposed to $\sim q^4$ for model J. Note that for vanishing mode coupling $g = 0$ one recovers of course the model A result $\chi_{\pi\pi}(q, \omega)^{-1} = q^2 - i\omega/D$ (Problem 3.2), and the purely diffusive response $\chi_{MM}(q, \omega)^{-1} = 1 - i\omega/\lambda q^2$ (recall that we scaled the static susceptibility for the conserved field to unity). Model E also (approximately) describes the critical dynamics of superfluid helium 4, which is characterized by a complex (i.e., two-component real) non-conserved order parameter interacting with the conserved entropy current. The dispersion (3.95) then describes the propagation and damping of 'second sound', a mode that couples the particle and entropy densities.

In the vicinity of the critical point, the leading contributions to the real and imaginary parts of the dispersion (3.95) become $\operatorname{Re} \omega_{\pm}(q) \sim \pm \phi |q|^{1-\eta/2}$ and $\operatorname{Im} \omega_{\pm}(q) \sim -|q|^z$. This is consistent with the dynamic scaling hypothesis provided $z = 1 - \frac{\eta}{2} + \frac{\beta}{\nu} = d/2$, where we have employed Eq. (1.77) once more. The overall scenario is captured again by Fig. 3.8, including the shape of the (transverse) order parameter correlation function, which looks qualitatively as in isotropic ferromagnets. Yet, as a consequence of the reduced symmetry in a planar magnet, the XY order parameter field is no longer a conserved quantity, leading to a linear rather than quadratic spin wave dispersion, and a markedly lower value for the dynamic critical exponent z. As worked out in detail in Problem 3.5, the critical dynamics of isotropic *antiferromagnets* with staggered spin order parameter,[11] dynamically coupled to the conserved magnetization, termed *model G* in the literature, constitutes an $O(3)$ generalization of model E. In Section 6.3, we shall explore an n-component extension, the *Sasvári–Schwabl–Szépfalusy (SSS) model*, which incorporates both models E and G, by means of the dynamic renormalization group.

Problems

3.1 *Critical order parameter decay and dynamic correlation functions*
   (a) Show that the dynamic scaling ansatz (3.5) for the order parameter implies the critical decay law $\phi(0, t) \sim t^{-\beta/z\nu}$ as $t \to \infty$.
   (b) Confirm the scaling forms (3.7) and (3.8) for the dynamic correlation function.

3.2 *Relaxational models in the ordered phase in Gaussian approximation*
   Expanding on Problem 1.5, compute the dynamic response and correlation functions for the relaxational models A/B in the ordered phase, within the Gaussian approximation. (See also Problem 2.11 and Section 5.4.)

3.3 *Four-point correlation function in the Gaussian ensemble*
   Apply Wick's theorem (3.38) to the four-point correlation function $\langle S(q_1, \omega_1) S(q_2, \omega_2) \; S(q_3, \omega_3) S(q_4, \omega_4) \rangle_0$. Then insert this expression into Eq. (3.45), and thus derive Eq. (3.46) for the two-point correlation function of the relaxational models A/B to first order in u.

---

[11] Michel and Schwabl (1970); Halperin, Hohenberg, and Siggia (1974, 1976).

3.4 *Reversible mode-coupling terms and equilibrium condition*

Confirm that the ansatz (3.59) for the general form of reversible mode-coupling terms, with antisymmetric $Q^{\alpha\beta}(x, x') = -Q^{\beta\alpha}(x', x)$, satisfies the equilibrium condition (3.58).

3.5 *Critical dynamics of isotropic antiferromagnets (model G)*

Isotropic antiferromagnets are described by a Heisenberg model (3.62) with negative nearest-neighbor exchange couplings. On a bipartite lattice in $d > 2$ dimensions, at low temperatures the *staggered magnetization* $\vec{N}_i = \eta_i \, \vec{S}_i$ assumes a non-zero value, and may thus serve as the order parameter for the para- to antiferromagnetic phase transition. Here, $\eta_i = \pm 1$ labels the two sublattices, respectively.

(a) Is the total staggered magnetization $\vec{N} = \sum_i \vec{N}_i$ conserved ?

For this Heisenberg antiferromagnet, establish the Poisson brackets

$$\left\{ N^\alpha(x), N^\beta(x') \right\} = -\widetilde{g} \sum_\gamma \epsilon^{\alpha\beta\gamma} S^\gamma(x) \, \delta(x - x') \,,$$

$$\left\{ N^\alpha(x), S^\beta(x') \right\} = -g \sum_\gamma \epsilon^{\alpha\beta\gamma} N^\gamma(x) \, \delta(x - x') \,,$$

$$\left\{ S^\alpha(x), S^\beta(x') \right\} = -g \sum_\gamma \epsilon^{\alpha\beta\gamma} S^\gamma(x) \, \delta(x - x') \,,$$

and therefrom construct the reversible force terms (3.59).

(b) Perform a scaling analysis for the three mode couplings, and demonstrate that the effective dynamical *model G* describing the critical dynamics of this system is a generalization to three-component vector fields of Eqs. (3.81), (3.88), (3.89), and (3.86) for model E.

(c) Determine the dynamical upper critical dimension d_c' and the dynamic critical exponent z for this model, assuming that the components of the order parameter and the conserved magnetization all fluctuate on the same time scales, i.e., strong dynamic scaling holds.

(d) Linearize the stochastic equations of motion in the antiferromagnetically ordered phase, where $\langle N^3 \rangle = \phi$. Show that the ensuing spin waves have the identical dispersion relation (3.95) as in planar ferromagnets. What happens to the magnetization density component $S^3(x, t)$?

References

Alba, M., S. Pouget, P. Fouquet, B. Farago, and C. Pappas, 2007, Critical scattering and dynamical scaling in an Heisenberg ferromagnet: neutron spin echo versus renormalization group theory, preprint arXiv:cond-mat/0703702, 1–4.

Dunlavy, M. J. and D. Venus, 2005, Critical slowing down in the two-dimensional Ising model measured using ferromagnetic ultrathin films, *Phys. Rev. B* **71**, 144406-1–6.

Frey, E. and F. Schwabl, 1994, Critical dynamics of magnets, *Adv. Phys.* **43**, 577–683.

Ferrell, R. A., N. Menyhàrd, H. Schmidt, F. Schwabl, and P. Szépfalusy, 1967, Dispersion in second sound and anomalous heat conduction at the lambda point of liquid helium, *Phys. Rev. Lett.* **18**, 891–894.

Ferrell, R. A., N. Menyhàrd, H. Schmidt, F. Schwabl, and P. Szépfalusy, 1968, Fluctuations and lambda phase transition in liquid helium, *Ann. Phys. (NY)* **47**, 565–613.

Halperin, B. I. and P. C. Hohenberg, 1969, Scaling laws for dynamic critical phenomena, *Phys. Rev.* **177**, 952–971.

Halperin, B. I., P. C. Hohenberg, and S.-k. Ma, 1972, Calculation of dynamic critical properties using Wilson's expansion methods, *Phys. Rev. Lett.* **29**, 1548–1551.

Halperin, B. I., P. C. Hohenberg, and S.-k. Ma, 1974, Renormalization-group methods for critical dynamics: I. Recursion relations and effects of energy conservation, *Phys. Rev. B* **10**, 139–153.

Halperin, B. I., P. C. Hohenberg, and E. D. Siggia, 1974, Renormalization-group calculations of divergent transport coefficients at critical points, *Phys. Rev. Lett.* **32**, 1289–1292.

Halperin, B. I., P. C. Hohenberg, and E. D. Siggia, 1976, Renormalization-group treatment of the critical dynamics of superfluid helium, the isotropic antiferromagnet, and the easy-plane ferromagnet, *Phys. Rev. B* **13**, 1299–1328; err. *Phys. Rev. B* **21**, 2044–2045 (1980).

Halperin, B. I., P. C. Hohenberg, and S.-k. Ma, 1976, Renormalization-group methods for critical dynamics: II. Detailed analysis of the relaxational models, *Phys. Rev. B* **13**, 4119–4131.

Hohenberg, P. C. and B. I. Halperin, 1977, Theory of dynamic critical phenomena, *Rev. Mod. Phys.* **49**, 435–479.

Kötzler, J., M. Kaufmann, G. Nakielski, R. Behr, and W. Assmus, 1994, Anisotropic dynamical scaling near the vortex-glass transition of twinned $YBa_2Cu_3O_{7-\delta}$, *Phys. Rev. Lett.* **72**, 2081–2084.

Ma, S.-k. and G. F. Mazenko, 1974, Critical dynamics of ferromagnets in $6 - \epsilon$ dimensions, *Phys. Rev. Lett.* **33**, 1383–1385.

Ma, S.-k. and G. F. Mazenko, 1975, Critical dynamics of ferromagnets in $6 - \epsilon$ dimensions: general discussion and detailed calculation, *Phys. Rev. B* **11**, 4077–4100.

Michel, K. H. and F. Schwabl, 1970, On the hydrodynamics of antiferromagnets, *Z. Phys.* **240**, 354–367.

Murtazaev, A. K., V. A. Mutailamov, I. K. Kamilov, K. S. Khizriev, and Y. K. Abuev, 2003, Investigation on the critical dynamics of real magnetics models by computational physics methods, *J. Magn. Mag. Mat.* **259**, 48–50.

Schwabl, F. and K. H. Michel, 1970, Hydrodynamics of Heisenberg ferromagnets, *Phys. Rev. B* **2**, 189–205.

Further reading

Cardy, J., 1996, *Scaling and Renormalization in Statistical Physics*, Cambridge: Cambridge University Press, chapter 10.

Chaikin, P. M. and T. C. Lubensky, 1995, *Principles of Condensed Matter Physics*, Cambridge: Cambridge University Press, chapter 8.

Folk, R. and G. Moser, 2006, Critical dynamics: a field theoretical approach, *J. Phys. A: Math. Gen.* **39**, R207–R313.

Forster, D., 1983, *Hydrodynamic Fluctuations, Broken Symmetry, and Correlation Functions*, Redwood City: Addison–Wesley, 3rd edn.

Kardar, M., 2007, *Statistical Physics of Fields*, Cambridge: Cambridge University Press, chapter 9.

Landau, D. P., A. Bunker, H. G. Evertz, M. Krech, and S.-H. Tsai, 2000, Spin dynamics simulations – a powerful method for the study of critical dynamics, *Prog. Theor. Phys. Suppl.* **138**, 423–432.

Lovesey, S. W., 1986, *Condensed Matter Physics: Dynamic Correlations*, Menlo Park: Benjamin–Cummings, 2nd. edn.

Ma, S.-k., 1976, *Modern Theory of Critical Phenomena*, Reading: Benjamin–Cummings, chapters 9, 14.

Stancil, D. D. and A. Prabhakar, 2009, *Spin Waves – Theory and Applications*, New York: Springer.

4

Dynamic perturbation theory

In this chapter, we introduce and explain one of the fundamental tools in the study of dynamic critical phenomena, namely dynamic perturbation theory. Inevitably, large parts of Chapter 4 need to be rather technical. Complementary to the straightforward iterative method for the solution of non-linear Langevin equations presented in Section 3.2.3, we now describe the more elegant and efficient field-theoretic techniques. Yet both for the derivation of general properties of the perturbation series and for the sake of practical calculations, the response functional and the following elaborations on Feynman diagrams, cumulants, and vertex functions prove indispensable. As is the case with every efficient formalism, once one has become acquainted with the Janssen–De Dominicis response functional for the construction of dynamic field theory, and therefrom the perturbation expansion in terms of vertex functions, it serves to save a considerable amount of rather tedious work. For the purely relaxational $O(n)$-symmetric models A and B, we derive the fluctuation-dissipation theorem within this formalism, and then systematically construct the perturbation series and its diagrammatic representation in terms of Feynman graphs. In order to reduce the required efforts to a minimum, the generating functionals first for the cumulants, and then, motivated by Dyson's equation for the propagator, for the one-particle irreducible vertex functions are examined. As an example, the relevant vertex functions for models A and B are explicitly evaluated to two-loop order. Furthermore, alternative formulations of the perturbation expansion are discussed, and the Feynman rules are given in both the frequency and the time domain. The results of this chapter provide the foundation for the renormalization group treatment in the subsequent Chapter 5. The powerful perturbational and diagrammatic tools developed here, supplemented by straightforward extensions and generalizations, will be employed throughout this text for the analysis of non-linear stochastic dynamical systems.

4.1 Response field formalism

In Section 3.2.3, we outlined a straightforward, but rather cumbersome, method to solve non-linear Langevin equations perturbationally, through direct iteration. Instead, one may construct the perturbation series in a more transparent manner by first casting the dynamical problem into a form that is reminiscent of the field-theoretic description of equilibrium statistical mechanics, and then applying the well-established machinery of (quantum) field theory.[1] The first step in this procedure for a general set of non-linear stochastic differential equations is the subject of this section.[2] In the following Section 4.2, we will explicitly construct the perturbation expansion for the $O(n)$-symmetric relaxational models.

4.1.1 Onsager–Machlup functional

Let us consider coupled dynamical Langevin-type equations of motion for the coarse-grained stochastic field variables $\psi^\alpha(x, t)$, of the generic form

$$\frac{\partial \psi^\alpha(x, t)}{\partial t} = F^\alpha[\psi](x, t) + \zeta^\alpha(x, t) . \tag{4.1}$$

The 'systematic' forces $F^\alpha[\psi]$ here may in general contain both reversible 'mode–mode' couplings between the ψ fields, see Eq. (3.59), and irreversible relaxation terms, e.g., of the form specified in Eqs. (3.18) and (3.60). Recall that the vector ψ^α should in principle include *all* the 'slow variables', i.e., the order parameter fluctuations (denoted as S^α above), conserved quantities, broken-symmetry 'elastic' or Goldstone modes, etc. The influence of 'fast' degrees of freedom is (approximately) captured by the stochastic forces ζ^α, whose mean we take to be zero as before, $\langle \zeta^\alpha(x, t) \rangle = 0$, and whose second moment we assume to be given by diagonal white noise,

$$\langle \zeta^\alpha(x, t) \zeta^\beta(x', t') \rangle = 2L^\alpha \, \delta(x - x') \delta(t - t') \delta^{\alpha\beta} . \tag{4.2}$$

In general, the noise 'strength' L^α is an operator; for instance, for a conserved quantity ψ^α it is of the form $L^\alpha = -D^\alpha \nabla^2$ describing diffusive relaxation. If we require that the set of equations (4.1) describes stochastic processes ultimately leading to an equilibrium probability distribution, then we need to impose the integrability condition (3.58) for the reversible forces, and Einstein relations for

---

[1] For good introductions to statistical field theory, especially pertaining to this chapter, see, e.g., Ma (1976), Janssen (1979), Ramond (1981), Amit (1984), Parisi (1988), Itzykson and Drouffe (1989), Le Bellac (1991), Janssen (1992), Zinn-Justin (1993), Cardy (1996), McComb (2004), Vasil'ev (2004), and Kamenev (2011).

[2] This general formalism is based on the original work by Martin, Siggia, and Rose (1973), De Dominicis (1976), Janssen (1976), and Bausch, Janssen, and Wagner (1976).

the Onsager coefficients and the noise correlations, as embodied in (3.61), see Section 2.4. The following construction of a field theory representation does not utilize any of these additional constraints, however, and may therefore be employed for genuine non-equilibrium processes as well, provided these can be represented in terms of effective Langevin dynamics, and the corresponding noise correlations are known.

In fact, basically all we need is that the noise correlations (4.2) can be obtained from the probability distribution

$$W[\zeta] \propto \exp\left(-\frac{1}{4}\int d^d x \int dt \sum_\alpha \zeta^\alpha(x, t)[(L^\alpha)^{-1}\zeta^\alpha(x, t)]\right), \qquad (4.3)$$

as is established immediately via the ratio of the Gaussian integrals

$$\int e^{-\zeta L^{-1}\zeta/4}\, d\zeta = 2(\pi L)^{1/2}, \qquad \int \zeta^2 e^{-\zeta L^{-1}\zeta/4}\, d\zeta = 4L(\pi L)^{1/2}\,.$$

Since L^α may be an operator involving gradients, the expression $(L^\alpha)^{-1}$ should by understood as a Green function; e.g., for a conserved quantity the corresponding term in the exponent of (4.3) reads in Fourier space $-\int_{q,\omega}|\zeta^\alpha(q, \omega)|^2/4D^\alpha q^2$. We can now view the noise ζ^α as a functional of the fields ψ^α, whose relation is given by the Langevin equation (4.1), i.e., $\zeta^\alpha[\psi] = \partial\psi^\alpha/\partial t - F^\alpha[\psi]$, and thus arrive at a probability distribution $\mathcal{P}[\psi]$ for the set of slow variables of interest ψ^α,

$$W[\zeta]\mathcal{D}[\zeta] = \mathcal{P}[\psi]\mathcal{D}[\psi] \propto e^{-\mathcal{G}[\psi]}\mathcal{D}[\psi]\,. \qquad (4.4)$$

Here, the *Onsager–Machlup* functional

$$\mathcal{G}[\psi] = \frac{1}{4}\int d^d x \int dt \sum_\alpha \left(\frac{\partial\psi^\alpha(x, t)}{\partial t} - F^\alpha[\psi](x, t)\right)$$

$$\times \left[(L^\alpha)^{-1}\left(\frac{\partial\psi^\alpha(x, t)}{\partial t} - F^\alpha[\psi](x, t)\right)\right] \qquad (4.5)$$

defines the statistical weight of a certain configuration $\psi^\alpha(x, t)$, and thus formally plays a role somewhat analogous to the Hamiltonian $\mathcal{H}$ in equilibrium statistical mechanics. However, because the noise probability distribution is supposed to be properly normalized, $\int \mathcal{D}[\zeta]W[\zeta] = 1$, the corresponding 'partition function' is always unity, and the associated 'free energy' vanishes. Notice that we have not explicitly accounted for the Jacobian stemming from the non-linear variable transformation (4.4). As explained in the Appendix (Section 4.5), when the standard Itô forward discretization is applied to the Langevin equation (4.1), the functional determinant is actually just a constant and can be absorbed into the measure

$\mathcal{D}[\psi]$. Employing the time discretization $t_l = t_0 + l\,\tau$, $t_M = t_f$ of Section 2.4.1 with Eq. (2.138), assuming the specific form $L^\alpha = D^\alpha(\mathrm{i}\nabla)^{a_\alpha}$ (with $a_\alpha = 0$ or 2) for the noise operator, and therefore using a spatial Fourier mode decomposition of the fluctuating fields $\psi^\alpha(x, t_l)$ (assumed to be real), see Eqs. (1.56) and (1.57) in Section 1.2.2, we may choose for the explicit functional measure

$$\mathcal{D}[\psi] = \lim_{\tau \to 0} \prod_\alpha \prod_{l=0}^{M-1} \prod_{q,q_1 > 0} \frac{\tau\, d\,\mathrm{Re}\,\psi^\alpha(q, t_l)\, d\,\mathrm{Im}\,\psi^\alpha(q, t_l)}{4\pi\, D^\alpha q^{a_\alpha} V}, \qquad (4.6)$$

whereupon $\int \mathcal{D}[\psi]\, e^{-\mathcal{G}[\psi]} = 1$.

4.1.2 Janssen–De Dominicis response functional

In principle, Eqs. (4.4) and (4.5) already provide the desired field-theoretic representation of non-linear Langevin dynamics, and the basis for a perturbational expansion. However, the appearance of the operator L^α in the denominator of the Onsager–Machlup functional, which, e.g., leads to a singularity as $q \to 0$ in the case of conserved dynamics, and furthermore the high-order non-linearities stemming from the $F^\alpha[\psi](L^\alpha)^{-1}F^\alpha[\psi]$ contribution render this formulation rather inconvenient. It is advantageous to 'linearize' with respect to $F^\alpha[\psi]$ by means of a Gaussian (Hubbard–Stratonovich) transformation, and thereby in addition transfer the operator L^α into the numerator. Our goal is to compute averages of observables A over noise histories, $\langle A[\psi]\rangle_\zeta \propto \int \mathcal{D}[\zeta]\, A[\psi(\zeta)]\, W[\zeta]$; omitting again the Jacobian from the variable transformation $\{\zeta^\alpha\} \to \{\psi^\alpha\}$, we may impose the constraint that the fields ψ^α are subject to the Langevin dynamics (4.1) through inserting a complicatedly written unity at each space-time point,

$$1 = \int \mathcal{D}[\psi] \prod_\alpha \prod_{(x,t)} \delta\left(\frac{\partial \psi^\alpha(x, t)}{\partial t} - F^\alpha[\psi](x, t) - \zeta^\alpha(x, t)\right)$$

$$= \int \mathcal{D}[\mathrm{i}\tilde{\psi}] \int \mathcal{D}[\psi]\, \exp\left[-\int d^d x \int dt \sum_\alpha \tilde{\psi}^\alpha \left(\frac{\partial \psi^\alpha}{\partial t} - F^\alpha[\psi] - \zeta^\alpha\right)\right].$$

$$(4.7)$$

The second equation is nothing but the functional integral representation of the delta function (with factors of 2π absorbed into the functional integral measure; notice that the integral is taken along the imaginary axis), and introduces *Martin–Siggia–Rose auxiliary fields* $\tilde{\psi}$, also called *response fields* for reasons that will become clear shortly.

Employing Eq. (4.3) we thus arrive at

$$\langle A[\psi]\rangle_\zeta \propto \int \mathcal{D}[i\widetilde{\psi}] \int \mathcal{D}[\psi]\exp\left[-\int d^d x \int dt \sum_\alpha \widetilde{\psi}^\alpha \left(\frac{\partial \psi^\alpha}{\partial t} - F^\alpha[\psi]\right)\right]$$

$$\times A[\psi] \int \mathcal{D}[\zeta]\exp\left(-\int d^d x \int dt \sum_\alpha \left[\frac{1}{4}\zeta^\alpha (L^\alpha)^{-1}\zeta^\alpha - \widetilde{\psi}^\alpha \zeta^\alpha\right]\right). \quad (4.8)$$

The noise average in the second line here is now readily performed through completing the square in the Gaussian integral

$$\int e^{\widetilde{\psi}\zeta - \zeta L^{-1}\zeta/4}\, d\zeta \propto e^{\widetilde{\psi}L\widetilde{\psi}},$$

whence we arrive at the following probability distribution for the stochastic field variables ψ^α,

$$\mathcal{P}[\psi] = \mathcal{C}^{-1}\int \mathcal{D}[i\widetilde{\psi}]\, e^{-\mathcal{A}[\widetilde{\psi},\psi]}, \quad (4.9)$$

with a statistical weight determined by the dynamic *Janssen–De Dominicis response functional*

$$\mathcal{A}[\widetilde{\psi},\psi] = \int d^d x \int dt \sum_\alpha \left[\widetilde{\psi}^\alpha(x,t)\left(\frac{\partial \psi^\alpha(x,t)}{\partial t} - F^\alpha[\psi](x,t)\right)\right.$$

$$\left. - \widetilde{\psi}^\alpha(x,t)\, L^\alpha \widetilde{\psi}^\alpha(x,t)\right], \quad (4.10)$$

and the constant $\mathcal{C} = \int \mathcal{D}[i\widetilde{\psi}]\int \mathcal{D}[\psi]\, e^{-\mathcal{A}[\widetilde{\psi},\psi]}$ fixed by normalization (see Problem 4.1). Alternatively, we may again absorb the normalization into the functional measure. Employing Eq. (4.6) for $\mathcal{D}[\psi]$, and collecting factors in steps (4.7)–(4.9), one in fact finds that the simple choice

$$\mathcal{D}[i\widetilde{\psi}]\,\mathcal{D}[\psi] = \lim_{\tau \to 0}\prod_\alpha \prod_{l=0}^{M-1}\prod_q \frac{d\,\widetilde{\psi}^\alpha(q,t_l)\, d\,\psi^\alpha(q,t_l)}{2\pi i} \quad (4.11)$$

guarantees that $\int \mathcal{D}[i\widetilde{\psi}]\int \mathcal{D}[\psi]\, e^{-\mathcal{A}[\widetilde{\psi},\psi]} = 1$.

Notice that the term $-\widetilde{\psi}^\alpha L^\alpha \widetilde{\psi}^\alpha$ originates in the noise correlator (4.2), while the factor multiplying $\widetilde{\psi}^\alpha$ corresponds to the deterministic part in the Langevin equation (4.1). Of course, when we explicitly perform the functional integral over the auxiliary fields $\widetilde{\psi}^\alpha$ in Eq. (4.9) along the imaginary axis (which is required in order to ensure convergence), we recover the Onsager–Machlup functional (4.5). This follows because

$$\int e^{i(i\widetilde{\psi}\zeta[\psi]) - (i\widetilde{\psi})L(i\widetilde{\psi})}\, d(i\widetilde{\psi}) = \int e^{-\widetilde{\psi}L\widetilde{\psi}}\, d\widetilde{\psi}\ e^{-\zeta[\psi]L^{-1}\zeta[\psi]/4},$$

where $\bar{\psi} = i\tilde{\psi} - iL^{-1}\zeta/2$. As mentioned before, the Jacobian $\delta\zeta^\alpha/\delta\psi^\alpha$ has been absorbed into the functional measure, since it is actually merely a constant provided a forward (Itô) discretization is applied. As discussed in the Appendix (Section 4.5), this is equivalent to defining $\Theta(0) = 0$ for the Heaviside step function, and hence $G(q, t = 0) = 0$ for the response propagator, which automatically excludes closed response propagator loops from the perturbation theory. In general one may show that the contributions arising from the Jacobian *precisely* cancel these closed response loop diagrams, if a consistent discretization is used. We may thus omit these Jacobian terms with the understanding that *no Feynman diagrams containing closed response loops* are to be permitted in the perturbation expansion to be constructed below. Note that in this manner, *causality* becomes implemented in our field theory representation for Langevin equations of the form (4.1).

The ensuing field theory with action (4.10) can be regarded as a statistical model for the independent variables $\tilde{\psi}^\alpha(x, t)$ and $\psi^\alpha(x, t)$ in $d + 1$ dimensions, where the time-like 'direction' plays a special role. In a scale-invariant regime, one would therefore expect *anisotropic scaling*, which is the reason why the dynamic exponent relating the characteristic time and length scales at criticality according to $t_c \sim \xi^z$ is generically $z \neq 1$. These remarks conclude our general discussion.

4.2 Relaxational dynamics: systematic perturbation theory

We shall next apply the above formalism to the $O(n)$-symmetric relaxational models A and B. For this special case, we will rederive the fluctuation-dissipation theorem within the current framework of classical non-linear Langevin equations. Then we proceed to construct the perturbation expansion in terms of the static non-linear coupling u.

4.2.1 The O(n)-symmetric relaxational models

Generalizing the Landau–Ginzburg–Wilson Hamiltonians for the Ising and Heisenberg models, Eqs. (3.9) and (3.69), to n-component isotropic systems undergoing a second-order phase transition, we consider

$$\mathcal{H}[S] = \int d^d x \sum_\alpha \left(\frac{r}{2} [S^\alpha(x)]^2 + \frac{1}{2} [\nabla S^\alpha(x)]^2 \right.$$

$$\left. + \frac{u}{4!} \sum_\beta [S^\alpha(x)]^2 [S^\beta(x)]^2 - h^\alpha(x) S^\alpha(x) \right). \quad (4.12)$$

The order parameter is now an n-component vector $S^\alpha(x)$, $\alpha = 1, \ldots, n$, with the conjugate external field $h^\alpha(x)$. Notice that for $h^\alpha = 0$ the above Hamiltonian is invariant with respect to rotations in order parameter space. The discrete Z_2 reflection invariance of the Ising model (3.9) is extended to the continuous symmetry group $O(n)$. For $n = 2$, the Hamiltonian (4.12) yields the order parameter part of the XY model, whereas for $n = 3$ we recover the Heisenberg effective model.

For non-conserved ($a = 0$) or conserved ($a = 2$) purely relaxational dynamics, we may use the obvious generalization of Eq. (3.18),

$$\frac{\partial S^\alpha(x, t)}{\partial t} = -D(i\nabla)^a \frac{\delta \mathcal{H}[S]}{\delta S^\alpha(x, t)} + \zeta^\alpha(x, t)$$

$$= -D(i\nabla)^a \left[(r - \nabla^2) S^\alpha(x, t) \right.$$

$$\left. + \frac{u}{6} S^\alpha(x, t) \sum_\beta [S^\beta(x, t)]^2 - h^\alpha(x, t) \right] + \zeta^\alpha(x, t), \quad (4.13)$$

which is to be supplemented by the Einstein relation, i.e., the second moment (4.2) for the stochastic forces with $L^\alpha = D(i\nabla)^a$. Here and henceforth, we have set $k_B T = 1$: For critical dynamics, we are interested in the vicinity of T_c, and thus $k_B T \approx k_B T_c$ is a constant, which we can absorb in the definition of the overall energy scale. Notice that we require identical relaxation constants for each component $S^\alpha(x, t)$, in order to retain the rotational symmetry on the dynamic level as well. We can now identify the fields ψ^α of Section 4.1 with the n order parameter components, and thereby arrive at the response functional $\mathcal{A} = \mathcal{A}_0 + \mathcal{A}_{int}$ with the bilinear part

$$\mathcal{A}_0[\tilde{S}, S] = \int d^d x \int dt \sum_\alpha \left(\tilde{S}^\alpha(x, t) \left[\frac{\partial}{\partial t} + D(i\nabla)^a (r - \nabla^2) \right] S^\alpha(x, t) \right.$$

$$\left. - D\tilde{S}^\alpha(x, t)(i\nabla)^a \left[\tilde{S}^\alpha(x, t) + h^\alpha(x, t) \right] \right), \quad (4.14)$$

and the anharmonic contribution

$$\mathcal{A}_{int}[\tilde{S}, S] = D \frac{u}{6} \int d^d x \int dt \sum_{\alpha, \beta} \tilde{S}^\alpha(x, t)(i\nabla)^a S^\alpha(x, t) S^\beta(x, t) S^\beta(x, t). \quad (4.15)$$

Physical quantities can generally be expressed in terms of correlation functions of the fields S^α and $\tilde{S}^\alpha$, taken with the statistical weight $e^{-\mathcal{A}[\tilde{S}, S]}$,

$$\left\langle \prod_{ij} S^{\alpha_i}(x_i, t_i) \tilde{S}^{\alpha_j}(x_j, t_j) \right\rangle$$

$$= \frac{\int \mathcal{D}[i\tilde{S}] \int \mathcal{D}[S] \prod_{ij} S^{\alpha_i}(x_i, t_i) \tilde{S}^{\alpha_j}(x_j, t_j) e^{-\mathcal{A}[\tilde{S}, S]}}{\int \mathcal{D}[i\tilde{S}] \int \mathcal{D}[S] e^{-\mathcal{A}[\tilde{S}, S]}}. \quad (4.16)$$

An important example is provided by the *dynamic susceptibility*

$$\chi^{\alpha\beta}(x - x', t - t') = \frac{\delta\langle S^\alpha(x, t)\rangle}{\delta h^\beta(x', t')}\bigg|_{h=0}.$$ (4.17)

According to Eq. (4.14), the functional derivative of $e^{-A[\widetilde{S}, S]}$ with respect to the external field h^β takes down a factor $D(i\nabla)^a \widetilde{S}^\beta$, which leads to the identity

$$\chi^{\alpha\beta}(x - x', t - t') = D\langle S^\alpha(x, t)(i\nabla)^a \widetilde{S}^\beta(x', t')\rangle,$$ (4.18)

valid for the purely relaxational models A $(a = 0)$ and B $(a = 2)$. In Fourier space, therefore, by exploiting the translational and rotational invariance of the system, we can express the *response propagator* $G(q, \omega)\delta^{\alpha\beta} = \chi^{\alpha\beta}(q, \omega)/Dq^a$ as

$$G(q, \omega)(2\pi)^{d+1}\delta(q + q')\delta(\omega + \omega')\delta^{\alpha\beta} = \langle S^\alpha(q, \omega)\widetilde{S}^\beta(q', \omega')\rangle.$$ (4.19)

The identity (4.19) is the origin for the term 'response field' for the variables $\widetilde{S}^\alpha$. The $\langle S^\alpha \widetilde{S}^\beta\rangle$ correlation function is intimately related to the dynamic response function $\chi^{\alpha\beta}$. Yet for dynamical models with reversible mode couplings the computation of the susceptibility requires additional correlation functions that involve composite operators, see Problem 4.2 and Section 6.2.

We are now in a position to derive the fluctuation-dissipation theorem for the relaxational models within our formalism. To this end, we observe that the Onsager–Machlup functional (4.5) reads

$$\mathcal{G}_h[S] = \mathcal{G}_{h=0}[S]$$
$$-\frac{1}{2}\int d^dx \int dt \sum_\alpha \left[\frac{\partial S^\alpha(x, t)}{\partial t} + D(i\nabla)^a \frac{\delta\mathcal{H}_{h=0}[S]}{\delta S^\alpha(x, t)}\right] h^\alpha(x, t) + O(h^2) \quad (4.20)$$

to first order in the external field, since $F_h^\alpha[S] = D(i\nabla)^a[h^\alpha - \delta\mathcal{H}_{h=0}[S]/\delta S^\alpha]$. Upon using the statistical weight $e^{-\mathcal{G}_h}$, the dynamic susceptibility (4.17) may therefore alternatively be written as

$$\chi^{\alpha\beta}(x - x', t - t') = \frac{1}{2}\left\langle S^\alpha(x, t)\left[\frac{\partial S^\beta(x', t')}{\partial t'} + D(i\nabla)^a \frac{\delta\mathcal{H}_{h=0}[S]}{\delta S^\beta(x', t')}\right]\right\rangle.$$ (4.21)

Causality demands that $\chi^{\alpha\beta}(x - x', t - t') = 0$ for $t < t'$, see Section 2.1, whence

$$t < t' : \left\langle S^\alpha(x, t)D(i\nabla)^a \frac{\delta\mathcal{H}_{h=0}[S]}{\delta S^\beta(x', t')}\right\rangle = -\left\langle S^\alpha(x, t)\frac{\partial S^\beta(x', t')}{\partial t'}\right\rangle.$$ (4.22)

Next, we apply time inversion to Eq. (4.22), using

$$\mathcal{T}S^\alpha(x, t) = \varepsilon_S S^\alpha(x, -t),$$ (4.23)

where $\varepsilon_S = \pm 1$ denotes the time inversion 'parity' of the fields S^α. (For the relaxational models, ε_S remains unspecified, but for a true magnetization density, for instance, $\varepsilon_S = -1$.) It is now important to note that the two terms in Eq. (4.22) acquire a relative minus sign under time inversion. Furthermore, we may exploit the time translation invariance (stationarity) of our stochastic process, which enables us to switch the arguments $-t$ and $-t'$ to $+t'$ and $+t$, respectively; after relabeling $t \leftrightarrow t'$, we consequently get

$$t > t': \quad \left\langle S^\alpha(x, t) D(i\nabla)^a \frac{\delta \mathcal{H}_{h=0}[S]}{\delta S^\beta(x', t')} \right\rangle = \left\langle S^\alpha(x, t) \frac{\partial S^\beta(x', t')}{\partial t'} \right\rangle. \quad (4.24)$$

Inserting this expression into Eq. (4.21) finally yields the fluctuation-dissipation theorem in the form

$$\chi^{\alpha\beta}(x - x', t - t') = \Theta(t - t') \frac{\partial}{\partial t'} \langle S^\alpha(x, t) S^\beta(x', t') \rangle$$

$$= -\Theta(t - t') \frac{\partial}{\partial t} \langle S^\alpha(x, t) S^\beta(x', t') \rangle. \quad (4.25)$$

This relation holds in the presence of reversible mode-coupling terms as well; cf. Eq. (2.36).

4.2.2 Generating functional for correlation functions

In order to compute correlation functions of the form (4.16) involving response fields, it is useful to introduce the *generating functional*

$$\mathcal{Z}[\tilde{j}, j] = \left\langle \exp \int d^d x \int dt \sum_\alpha \left(\tilde{j}^\alpha(x, t) \, \tilde{S}^\alpha(x, t) + j^\alpha(x, t) \, S^\alpha(x, t) \right) \right\rangle, \quad (4.26)$$

from which the correlation functions are obtained via functional derivatives,

$$\left\langle \prod_{ik} S^{\alpha_i}(x_i, t_i) \tilde{S}^{\alpha_k}(x_k, t_k) \right\rangle = \prod_{ik} \frac{\delta}{\delta j^{\alpha_i}(x_i, t_i)} \frac{\delta}{\delta \tilde{j}^{\alpha_k}(x_k, t_k)} \left. \mathcal{Z}[\tilde{j}, j] \right|_{\tilde{j}=0=j}. \quad (4.27)$$

This procedure is readily followed through for the Gaussian theory with the statistical weight $e^{-\mathcal{A}_0}$. In Fourier space, we can write the harmonic action (4.14) for $h^\alpha = 0$ as

$$\mathcal{A}_0[\tilde{S}, S] = \int_q \int_\omega \sum_\alpha (\tilde{S}^\alpha(-q, -\omega)[-i\omega + Dq^a(r + q^2)] S^\alpha(q, \omega)$$

$$- Dq^a \, \tilde{S}^\alpha(-q, -\omega) \tilde{S}^\alpha(q, \omega)) \quad (4.28)$$

$$= \frac{1}{2} \int_q \int_\omega \sum_\alpha (\tilde{S}^\alpha(-q, -\omega) \, S^\alpha(-q, -\omega)) \, \mathbf{A}(q, \omega) \begin{pmatrix} \tilde{S}^\alpha(q, \omega) \\ S^\alpha(q, \omega) \end{pmatrix},$$

with the Hermitian coupling matrix

$$\mathbf{A}(q,\omega) = \begin{pmatrix} -2Dq^a & -i\omega + Dq^a(r+q^2) \\ i\omega + Dq^a(r+q^2) & 0 \end{pmatrix}. \tag{4.29}$$

With the aid of the Gaussian integrals from Section 4.1, we compute

$$\mathcal{Z}_0[\tilde{j}, j]$$

$$= \frac{\int \mathcal{D}[i\tilde{S}]\int \mathcal{D}[S]\ \exp\left[-\frac{1}{2}\int_q\int_\omega\sum_\alpha(\tilde{S}^\alpha\ S^\alpha)\mathbf{A}\binom{\tilde{S}^\alpha}{S^\alpha} + \int_q\int_\omega\sum_\alpha(\tilde{j}^\alpha\ j^\alpha)\binom{\tilde{S}^\alpha}{S^\alpha}\right]}{\int \mathcal{D}[i\tilde{S}]\int \mathcal{D}[S]\ \exp\left[-\frac{1}{2}\int_q\int_\omega\sum_\alpha(\tilde{S}^\alpha\ S^\alpha)\mathbf{A}\binom{\tilde{S}^\alpha}{S^\alpha}\right]}$$

$$= \exp\left[\frac{1}{2}\int_q\int_\omega\sum_\alpha\left(\tilde{j}^\alpha(-q,-\omega)\ j^\alpha(-q,-\omega)\right)\mathbf{A}^{-1}(q,\omega)\binom{\tilde{j}^\alpha(q,\omega)}{j^\alpha(q,\omega)}\right], \tag{4.30}$$

since the denominator $\mathcal{C}_0$ is exactly cancelled by the shifted Gaussian integral in the numerator (whose explicit evaluation is left to Problem 4.1).

From Eq. (4.30), we now directly infer the matrix of two-point correlation functions in the Gaussian ensemble

$$\begin{pmatrix} \langle \tilde{S}^\alpha(q,\omega)\ \tilde{S}^\beta(q',\omega')\rangle_0 & \langle \tilde{S}^\alpha(q,\omega)\ S^\beta(q',\omega')\rangle_0 \\ \langle S^\alpha(q,\omega)\ \tilde{S}^\beta(q',\omega')\rangle_0 & \langle S^\alpha(q,\omega)\ S^\beta(q',\omega')\rangle_0 \end{pmatrix}$$

$$= \mathbf{A}^{-1}(q,\omega)(2\pi)^{d+1}\delta(q+q')\delta(\omega+\omega')\delta^{\alpha\beta}, \tag{4.31}$$

with the inverse of the harmonic coupling matrix (4.29),

$$\mathbf{A}^{-1}(q,\omega) = \frac{1}{\omega^2 + [Dq^a(r+q^2)]^2}\begin{pmatrix} 0 & -i\omega + Dq^a(r+q^2) \\ i\omega + Dq^a(r+q^2) & 2Dq^a \end{pmatrix}. \tag{4.32}$$

From this result, we gather the Gaussian response propagator and correlation function, both of which are diagonal and independent of the order parameter components α, β in the high-temperature phase,

$$G_0(q,\omega) = \frac{1}{-i\omega + Dq^a(r+q^2)}, \tag{4.33}$$

$$C_0(q,\omega) = \frac{2Dq^a}{\omega^2 + [Dq^a(r+q^2)]^2} = 2Dq^a\,|G_0(q,\omega)|^2, \tag{4.34}$$

in accord with Eqs. (3.25) and (3.31) from our previous direct calculations. Notice that

$$\langle \tilde{S}^\alpha(q,\omega)\tilde{S}^\beta(q',\omega')\rangle_0 = 0. \tag{4.35}$$

In the harmonic field theory (4.14), we may thus define two different non-vanishing contractions, namely

$$\underline{S^\alpha(q,\omega)\widetilde{S}^\beta(q',\omega')} = \langle S^\alpha(q,\omega)\widetilde{S}^\beta(q',\omega')\rangle_0$$
$$= G_0(q,\omega)(2\pi)^{d+1}\delta(q+q')\delta(\omega+\omega')\delta^{\alpha\beta}, \quad (4.36)$$

which yields the Gaussian response propagator (4.33), and

$$\underline{S^\alpha(q,\omega)S^\beta(q',\omega')} = \langle S^\alpha(q,\omega)S^\beta(q',\omega')\rangle_0$$
$$= C_0(q,\omega)(2\pi)^{d+1}\delta(q+q')\delta(\omega+\omega')\delta^{\alpha\beta}$$
$$= 2Dq^a\, G_0(-q,-\omega)G_0(q,\omega)(2\pi)^{d+1}\delta(q+q')\delta(\omega+\omega')\delta^{\alpha\beta},$$
$$(4.37)$$

denoting the dynamic correlation function (4.34), which in turn may be expressed in terms of the response propagator $|G_0(q,\omega)|^2$ and the noise strength $2Dq^a$. In real space, the contractions read

$$\underline{S^\alpha(x,t)\widetilde{S}^\beta(x',t')} = \langle S^\alpha(x,t)\widetilde{S}^\beta(x',t')\rangle_0 = G_0(x-x',t-t')\delta^{\alpha\beta}, \quad (4.38)$$
$$\underline{S^\alpha(x,t)S^\beta(x',t')} = \langle S^\alpha(x,t)S^\beta(x',t')\rangle_0 = C_0(x-x',t-t')\delta^{\alpha\beta}, \quad (4.39)$$

where $G_0(x,t) \propto \Theta(t)$ and $C_0(x,t)$ are given by the spatial Fourier transforms of Eqs. (3.29) and (3.34), respectively (with $k_BT = 1$).

4.2.3 Perturbation expansion

We are now ready to construct the perturbation series for general correlation functions. To this end, we recast Eq. (4.16) in terms of Gaussian averages, and then only need to expand the exponential of $-\mathcal{A}_{\text{int}}[\widetilde{S}, S]$,

$$\left\langle \prod_{ij} S^{\alpha_i}(x_i,t_i)\widetilde{S}^{\alpha_j}(x_j,t_j)\right\rangle = \frac{\left\langle \prod_{ij} S^{\alpha_i}(x_i,t_i)\widetilde{S}^{\alpha_j}(x_j,t_j)e^{-\mathcal{A}_{\text{int}}[\widetilde{S},S]}\right\rangle_0}{\langle e^{-\mathcal{A}_{\text{int}}[\widetilde{S},S]}\rangle_0}$$

$$= \frac{\left\langle \prod_{ij} S^{\alpha_i}(x_i,t_i)\widetilde{S}^{\alpha_j}(x_j,t_j)\sum_{l=0}^\infty(-\mathcal{A}_{\text{int}}[\widetilde{S},S])^l\big/l!\right\rangle_0}{\left\langle\sum_{l=0}^\infty(-\mathcal{A}_{\text{int}}[\widetilde{S},S])^l\big/l!\right\rangle_0}.$$
$$(4.40)$$

According to (4.15), this amounts to a power series with respect to the non-linear coupling u, where the term for $l = 0$ yields the correlation functions of the Gaussian models. The averages on the right-hand side of (4.40) may be evaluated by means of Wick's theorem, i.e., by summing over all possible contractions (4.36) and (4.37) of the fields $\widetilde{S}^\alpha$ and S^α into Gaussian response and correlation functions. For the terms with $l > 0$, this involves integrations and summations over 'internal' degrees of freedom stemming from the integrals and sums in Eq. (4.15), as opposed to the 'external' labels x_i, t_i, α_i attached to the fields in the correlators.

For example, upon writing $\mathcal{A}_{\text{int}}[\widetilde{S}, S]$ in Fourier space, and employing the short-hand notation (3.36) and the convention that repeated indices are to be summed over, the first two terms of the denominator in Eq. (4.40) read

$$
1 - D \frac{u}{6} \int_{\underline{k}_1', \underline{k}_2', \underline{k}_3'} q_1'^a \left\langle \widetilde{S}^{\alpha'}(\underline{k}_1') S^{\alpha'}(-\underline{k}_1' - \underline{k}_2' - \underline{k}_3') S^{\beta'}(\underline{k}_2') S^{\beta'}(\underline{k}_3') \right\rangle_0
$$

$$
+ \frac{1}{2} \left(D \frac{u}{6} \right)^2 \int_{\underline{k}_1', \underline{k}_2', \underline{k}_3'} q_1'^a \int_{\underline{k}_1'', \underline{k}_2'', \underline{k}_3''} q_1''^a \left\langle \widetilde{S}^{\alpha'}(\underline{k}_1') S^{\alpha'}(-\underline{k}_1' - \underline{k}_2' - \underline{k}_3') \right.
$$

$$
\left. \times S^{\beta'}(\underline{k}_2') S^{\beta'}(\underline{k}_3') \widetilde{S}^{\alpha''}(\underline{k}_1'') S^{\alpha''}(-\underline{k}_1'' - \underline{k}_2'' - \underline{k}_3'') S^{\beta''}(\underline{k}_2'') S^{\beta''}(\underline{k}_3'') \right\rangle_0 .
$$

After contracting the fields according to

$$
\langle \widetilde{S}^{\alpha'} S^{\alpha'} S^{\beta'} S^{\beta'} \rangle_0 = \underline{S^{\alpha'} \widetilde{S}^{\alpha'}} \; \underline{S^{\beta'} S^{\beta'}} + 2 \, \underline{S^{\beta'} \widetilde{S}^{\alpha'}} \; \underline{S^{\alpha'} S^{\beta'}} , \tag{4.41}
$$

the first-order term becomes

$$
-n(n+2) \frac{u}{6} \int_{\underline{k}_1} D q_1'^a \, G_0(\underline{k}_1') \int_{\underline{k}_2} C_0(\underline{k}_2')
$$

$$
= -n(n+2) \frac{u}{6} \int_{q_1'} \chi_0(q_1', t=0) \int_{q_2'} C_0(q_2', t=0) .
$$

This expression is thus proportional to $\Theta(0) = 0$ according to our forward discretization prescription. Alternatively, it would be canceled by the contributions stemming from the functional determinant in the functional (4.10), as explained in the Appendix (Section 4.5). This is actually fortunate, as the q_1'-integral is highly divergent in the ultraviolet (for wavevector cutoff $\Lambda \to \infty$). In a similar manner, in the second-order term as well as in fact in *every* non-trivial contribution to the denominator of Eq. (4.40), there occur contractions of *internal* $\widetilde{S}^\alpha$ and S^α fields, i.e., closed response loops, which all vanish. Consequently, the denominator is simply unity, and we may write

$$
\left\langle \prod_{ij} S^{\alpha_i}(x_i, t_i) \widetilde{S}^{\alpha_j}(x_j, t_j) \right\rangle = \left\langle \prod_{ij} S^{\alpha_i}(x_i, t_i) \widetilde{S}^{\alpha_j}(x_j, t_j) \sum_{l=0}^{\infty} \frac{1}{l!} \left(-\mathcal{A}_{\text{int}}[\widetilde{S}, S] \right)^l \right\rangle_0 .
$$

$$
\tag{4.42}
$$

In field-theory language, this means that for an action of the form (4.10), there exist *no vacuum* contributions.[3]

---

[3] In quantum field theory and equilibrium statistical mechanics, the *linked cluster theorem* generally states that the vacuum contributions from the denominator of expressions such as Eq. (4.40) precisely cancel with the non-connected parts of the numerator. In our dynamic theory, this denominator containing the non-Gaussian contributions to the partition function analog is unity as a consequence of causality.

Furthermore, we may easily demonstrate that all pure response field correlations vanish,

$$\left\langle \prod_j \tilde{S}^{\alpha_j}(x_j, t_j) \right\rangle = 0 . \tag{4.43}$$

Obviously, to first order, and because of Eq. (4.35), there are not even sufficiently many internal fields $S^{\alpha'}$ present to yield a non-zero contraction. Yet even to order u^l with $l > 1$, where some of the external response fields $\tilde{S}^{\alpha_j}$ can be paired with internal fields $S^{\alpha'}$, there exist l additional internal response fields $\tilde{S}^{\alpha'}$ originating in the vertices $\propto u$. Thus, at least the outgoing response field at the vertex that appears last in a time-ordered sequence must be contracted to another internal $S^{\alpha'}$, implying the inevitable appearance of closed response loops in the correlator (4.43).

4.3 Feynman diagrams

Proceeding to higher orders in the perturbation expansion quickly leads to rather cumbersome expressions. Many of the apparently different contributions, moreover, merely represent identical mathematical terms, obtained from each other through simple permutations of summation and integration variable labels. It is thus a considerable simplification to represent the perturbation series graphically, see Section 3.2.3. This also facilitates grouping of related contributions and the physical interpretation of the various terms.

4.3.1 Response and correlation functions

Following the procedure in Section 3.2.3, we now organize the perturbation expansion more lucidly by means of Feynman diagrams, containing as their fundamental ingredients directed propagator lines and interaction vertices. For the

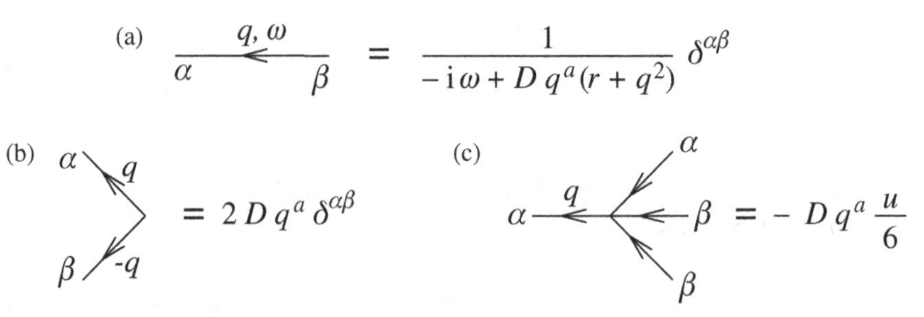

Fig. 4.1 Elements of dynamic perturbation theory for the $O(n)$-symmetric relaxational models: (a) propagator; (b) two-point noise vertex; (c) four-point vertex.

(a)   (b)   (c)

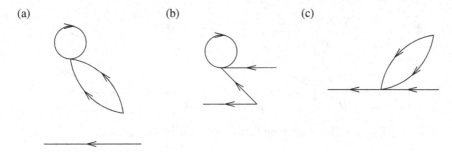

Fig. 4.2 Topology of the first-order Feynman diagrams for $G(q, \omega)$.

$O(n)$-symmetric relaxational models, the propagator, which represents the contraction (4.36), as well as the vertices carry field indices $\alpha, \beta = 1, \ldots, n$, as depicted in Fig. 4.1. As before, integrals and sums over internal degrees of freedom will be encoded in closed loops.

For example, to first order in u the response propagator reads

$$\langle S^\alpha(\underline{k})\widetilde{S}^\beta(\underline{\tilde{k}})\rangle = \langle S^\alpha(\underline{k})\widetilde{S}^\beta(\underline{\tilde{k}})\rangle_0 - D\frac{u}{6}\int_{\underline{k}_1',\underline{k}_2',\underline{k}_3'} q_1'^a$$

$$\times \langle S^\alpha(\underline{k})\widetilde{S}^\beta(\underline{\tilde{k}})\widetilde{S}^{\alpha'}(\underline{k}_1')S^{\alpha'}(-\underline{k}_1' - \underline{k}_2' - \underline{k}_3')S^{\beta'}(\underline{k}_2')S^{\beta'}(\underline{k}_3')\rangle_0 + \cdots. \quad (4.44)$$

There are twelve possible contractions for the Gaussian six-point function, which can be grouped into three classes, representing topologically different Feynman graphs (see Fig. 4.2):

$$\langle S^\alpha \widetilde{S}^\beta \ \widetilde{S}^{\alpha'} S^{\alpha'} S^{\beta'} S^{\beta'}\rangle_0 = \underline{S^\alpha \widetilde{S}^\beta}(\underline{S^{\alpha'}\widetilde{S}^{\alpha'}} \ \underline{S^{\beta'}S^{\beta'}} + 2\, \underline{S^{\beta'}\widetilde{S}^{\alpha'}} \ \underline{S^{\alpha'}S^{\beta'}})$$

$$+ 2\, \underline{S^{\alpha'}\widetilde{S}^{\alpha'}} \ \underline{S^{\beta'}\widetilde{S}^{\beta}} \ \underline{S^\alpha S^{\beta'}}$$

$$+ 2\, \underline{S^{\beta'}\widetilde{S}^{\alpha'}}(\underline{S^{\alpha'}\widetilde{S}^{\beta}} \ \underline{S^\alpha S^{\beta'}} + \underline{S^{\beta'}\widetilde{S}^{\beta}} \ \underline{S^\alpha S^{\alpha'}})$$

$$+ \underline{S^\alpha \widetilde{S}^{\alpha'}}(\underline{S^{\alpha'}\widetilde{S}^{\beta}} \ \underline{S^{\beta'}S^{\beta'}} + 2\, \underline{S^{\beta'}\widetilde{S}^{\beta}} \ \underline{S^{\alpha'}S^{\beta'}}). \quad (4.45)$$

First, we may contract the external lines, which for the primed internal fields leads exactly to the three possible vacuum contractions (4.41). The corresponding diagram, Fig. 4.2(a), splits into two disconnected pieces, namely the bare propagator and the first-order vacuum graphs with closed response loops. Were these not zero anyway, they would be canceled by the vacuum diagrams stemming from the denominator in Eq. (4.40). In the second and third lines of (4.45), the internal response field $\widetilde{S}^{\alpha'}$ has been contracted with other internal fields. This involves closed response loops again, as shown in Fig. 4.2(b).

Consequently, these six contributions vanish as well, which leaves us with the three terms described by Fig. 4.2(c), where the internal response field $\widetilde{S}^{\alpha'}$ is connected to the external field S^α. More precisely, the contractions in the fourth

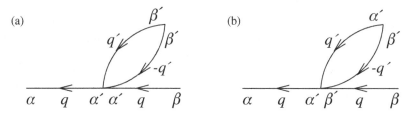

Fig. 4.3 Non-vanishing first-order contributions to $G(q, \omega)$.

line of (4.45) are represented by the labeled 'Hartree loop' diagrams of Fig. 4.3, which generalize Fig. 3.6 for the single-component relaxational models. Upon inserting the analytic expressions (4.36) and (4.37) for the contractions, we find

$$\left\langle S^\alpha(\underline{k})\tilde{S}^\beta(\tilde{\underline{k}})\tilde{S}^{\alpha'}(\underline{k}'_1)S^{\alpha'}(-\underline{k}'_1 - \underline{k}'_2 - \underline{k}'_3)S^{\beta'}(\underline{k}'_2)S^{\beta'}(\underline{k}'_3)\right\rangle_0$$

$$= G_0(\underline{k})\delta(\underline{k} + \underline{k}'_1)\delta^{\alpha\alpha'}\Big[G_0(-\tilde{\underline{k}})\delta(\tilde{\underline{k}} - \underline{k}'_1 - \underline{k}'_2 - \underline{k}'_3)\delta^{\alpha'\beta} C_0(\underline{k}'_2) \delta(\underline{k}'_2 + \underline{k}'_3)\delta^{\beta'\beta'}$$

$$+ 2\, G_0(-\tilde{\underline{k}})\delta(\tilde{\underline{k}} + \underline{k}'_3)\delta^{\beta'\beta} C_0(\underline{k}'_2)\delta(\underline{k}'_1 + \underline{k}'_3)\delta^{\alpha'\beta'}\Big], \tag{4.46}$$

and after inserting into Eq. (4.44), using $\delta^{\beta'\beta'} = n$, and renaming $\underline{k}'_2 \to \underline{k}'$,

$$G(\underline{k}) = G_0(\underline{k})\left[1 - (n+2)\,G_0(\underline{k})\,Dq^a\,\frac{u}{6}\int_{\underline{k}'}C_0(\underline{k}') + O(u^2)\right]$$

$$= G_0(\underline{k})\left[1 - (n+2)G_0(\underline{k})Dq^a\,\frac{u}{6}\int_{\underline{k}'}2Dq'^a\,G_0(-\underline{k}')G_0(\underline{k}') + O(u^2)\right]. \tag{4.47}$$

We could have obtained this first-order result directly from the graphical representation as follows. First, Fig. 4.2(c) constitutes the only possibility for connecting the relaxation vertex of Fig. 4.1(c) to one in- and one out-going propagator line, shown in Fig. 4.1(a), such that there appear no closed response loops as in Figs. 4.2(a) and (b). Analytically, Fig. 4.2(c) stands for a product of two response propagators $G_0(\underline{k})$, the vertex $-Dq^a u/6$, and the loop integral and sums over the internal degrees of freedom. Inside the loop, two response propagators are linked to the noise vertex of Fig. 4.1(b), which already carries a factor 2 for the two possible ways of assigning wavenumbers and frequencies to its outgoing legs. This yields the integral over the Gaussian correlation function $C_0(\underline{k}')$. Finally, Fig. 4.3 depicts the corresponding two possible distributions of vector field indices α among the propagator lines. Instead of explicitly performing the sum over the internal indices α' and β', we may simply take the relaxation vertex Fig. 4.1(c) and connect its outgoing leg with index α to an external line. For the $n - 1$ possible values of $\beta \neq \alpha$,

(a)   (b)

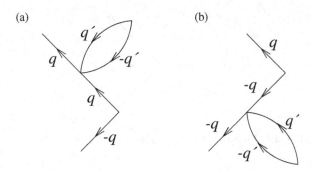

Fig. 4.4 Non-vanishing first-order contributions to $C(q, \omega)$.

there is only one way to link the incoming leg with index α to the corresponding external line, and subsequently form the correlation loop with the remaining lines carrying the index β. On the other hand, for $\beta = \alpha$ we may choose the one incoming external leg, or, equivalently, the two lines to be connected to the correlation loop, from three different possibilities. Altogether, this provides us with the correct combinatorial factor $n - 1 + 3 = n + 2$. For $n = 1$, Eq. (4.47) reduces to our earlier result (3.42), which we had obtained via direct iteration of the Langevin equation.

In a similar manner, we evaluate the first-order contributions to the two-point correlation function $\langle S^\alpha S^\beta \rangle$, for which we need

$$\langle S^\alpha S^\beta \, \widetilde{S}^{\alpha'} S^{\alpha'} S^{\beta'} S^{\beta'} \rangle_0 = \underline{S^\alpha \widetilde{S}^{\alpha'}} \, (\underline{S^\beta S^{\alpha'}} \, \underline{S^{\beta'} S^{\beta'}} + 2 \, \underline{S^\beta S^{\beta'}} \, \underline{S^{\alpha'} S^{\beta'}})$$
$$+ \, \underline{S^\beta \widetilde{S}^{\alpha'}} \, (\underline{S^\alpha S^{\alpha'}} \, \underline{S^{\beta'} S^{\beta'}} + 2 \, \underline{S^\alpha S^{\beta'}} \, \underline{S^{\alpha'} S^{\beta'}}) . \quad (4.48)$$

Here, we have only taken into account the non-vanishing contractions, where the internal response field $\widetilde{S}^{\alpha'}$ is connected to an external field S^α or S^β. The corresponding Feynman diagrams, shown in Fig. 4.4, are simply constructed by replacing each branch of $\langle S^\alpha S^\beta \rangle_0$ with a first-order propagator (4.47). Hence, to $O(u)$, we arrive at

$$C(\underline{k}) = C_0(\underline{k}) \left[1 - (n + 2) \left[G_0(\underline{k}) + G_0(-\underline{k}) \right] Dq^a \, \frac{u}{6} \int_{\underline{k'}} C_0(\underline{k'}) \right]$$

$$= 2Dq^a \, G_0(-\underline{k}) \, G_0(\underline{k}) \left[1 - (n + 2) \left[G_0(\underline{k}) + G_0(-\underline{k}) \right] \right.$$

$$\left. \times \, Dq^a \, \frac{u}{6} \int_{\underline{k'}} 2Dq'^a \, G_0(-\underline{k'}) \, G_0(\underline{k'}) \right] , \quad (4.49)$$

to be compared with Eq. (4.34) of Section 4.2. This example illustrates the power of the pictorial representation: instead of tediously evaluating the contractions

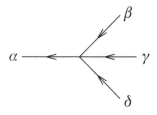

Fig. 4.5 First-order Feynman diagram for $\langle S^\alpha \widetilde{S}^\beta \widetilde{S}^\gamma \widetilde{S}^\delta \rangle$.

in Eq. (4.48), we readily construct the one-loop Feynman graphs of Fig. 4.4. Comparing with Fig. 4.2 establishes that to first order in u

$$C(\underline{k}) = 2Dq^a\, G(-\underline{k})\, G(\underline{k}) + O(u^2)\,, \qquad (4.50)$$

and inserting Eq. (4.47) here immediately leads to the result (4.49).

4.3.2 Four-point correlation functions

Before we formulate the general Feynman rules, let us examine as further examples the four-point correlation functions for the $O(n)$-symmetric relaxational models. Notice that $\langle \widetilde{S}^\alpha \widetilde{S}^\beta \widetilde{S}^\gamma \widetilde{S}^\delta \rangle = 0$, see Eq. (4.43).

Consider next $\langle S^\alpha \widetilde{S}^\beta \widetilde{S}^\gamma \widetilde{S}^\delta \rangle$; clearly, there exists no zeroth-order contribution to this correlation function. To first order in u, the only non-vanishing contractions of the ensuing Gaussian eight-point function are

$$\langle S^\alpha \widetilde{S}^\beta \widetilde{S}^\gamma \widetilde{S}^\delta \widetilde{S}^{\alpha'} S^{\alpha'} S^{\beta'} S^{\beta'} \rangle_0 = 2\, \underline{S^\alpha \widetilde{S}^{\alpha'}} (\underline{S^{\alpha'} \widetilde{S}^\beta}\ \underline{S^{\beta'} \widetilde{S}^\gamma}\ \underline{S^{\beta'} \widetilde{S}^\delta} + \underline{S^{\alpha'} \widetilde{S}^\gamma}\ \underline{S^{\beta'} \widetilde{S}^\beta}\ \underline{S^{\beta'} \widetilde{S}^\delta}$$
$$+ \underline{S^{\alpha'} \widetilde{S}^\delta}\ \underline{S^{\beta'} \widetilde{S}^\beta}\ \underline{S^{\beta'} \widetilde{S}^\gamma})\,. \qquad (4.51)$$

Inserting the analytic expressions for the contractions, this simply yields four propagators connected at a single vertex,

$$\left\langle S^\alpha(\underline{k}_\alpha)\widetilde{S}^\beta(\underline{k}_\beta)\widetilde{S}^\gamma(\underline{k}_\gamma)\widetilde{S}^\delta(\underline{k}_\delta) \right\rangle = -Dq_\alpha^a\, u\, G_0(\underline{k}_\alpha)\, G_0(\underline{k}_\beta)\, G_0(\underline{k}_\gamma)\, G_0(\underline{k}_\delta)$$

$$\times\, \delta(\underline{k}_\alpha + \underline{k}_\beta + \underline{k}_\gamma + \underline{k}_\delta)\, F^{\alpha\beta\gamma\delta} + O(u^2)\,, \quad (4.52)$$

see Fig. 4.5, with the fully symmetric tensor

$$F^{\alpha\beta\gamma\delta} = \frac{1}{3}(\delta^{\alpha\beta}\, \delta^{\gamma\delta} + \delta^{\alpha\gamma}\, \delta^{\beta\delta} + \delta^{\alpha\delta}\, \delta^{\beta\gamma})\,. \qquad (4.53)$$

Diagrammatically, the derivation of Eq. (4.52) becomes almost trivial. The graph in Fig. 4.5 merely represents the vertex $-Dq_\alpha^a u/6$ connected to one outgoing and three incoming propagators. There exist $3 \cdot 2 = 6$ different possibilities for connecting the incoming vertex legs to the external lines; the symmetrized tensor

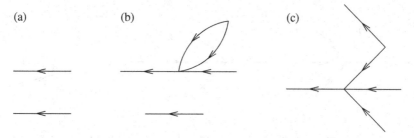

Fig. 4.6 (a) Zeroth-order, (b) disconnected first-order, and (c) connected first-order Feynman diagrams for $\langle S^\alpha S^\beta \widetilde{S}^\gamma \widetilde{S}^\delta \rangle$.

factor (4.53) then takes care of the fact that two incoming legs must carry equal vector indices. Notice that

$$\left\langle S^\alpha(\underline{k}_\alpha) \widetilde{S}^\alpha(\widetilde{\underline{k}}_\alpha) \widetilde{S}^\beta(\underline{k}_\beta) \widetilde{S}^\beta(\widetilde{\underline{k}}_\beta) \right\rangle = -D q_\alpha^a \frac{u}{3} G_0(\underline{k}_\alpha) G_0(\widetilde{\underline{k}}_\alpha) G_0(\underline{k}_\beta) G_0(\widetilde{\underline{k}}_\beta)$$

$$\times \delta(\underline{k}_\alpha + \widetilde{\underline{k}}_\alpha + \underline{k}_\beta + \widetilde{\underline{k}}_\beta) + O(u^2), \quad (4.54)$$

because there now remain only the two different possibilities for assigning the values $\underline{k}_\beta$ and $\widetilde{\underline{k}}_\beta$ to the incoming legs with index $\beta \neq \alpha$.

As a final example, we study $\langle S^\alpha S^\beta \widetilde{S}^\gamma \widetilde{S}^\delta \rangle$. As opposed to the previous case, according to Wick's theorem (3.38) the zeroth-order correlation function can be factorized into a sum of two products of two response propagators, see Fig. 4.6(a). To first order in u, one of the corresponding contractions of external fields may still be carried out, leaving a Gaussian six-point function of precisely the form (4.45) which led to the first-order corrections for the response propagator. These two types of contribution are represented by the two disconnected Feynman graphs depicted in Fig. 4.6(a) and (b). The third, connected diagram Fig. 4.6(c) finally corresponds to contractions analogous to Eq. (4.51), where all the external fields are linked to internal ones. Notice that as the internal response field $\widetilde{S}^{\alpha'}$ may now be contracted with either S^α or S^β, there emerge two such terms. To $O(u)$ one thus finds

$$\left\langle S^\alpha(\underline{k}_\alpha) S^\beta(\underline{k}_\beta) \widetilde{S}^\gamma(\underline{k}_\gamma) \widetilde{S}^\delta(\underline{k}_\delta) \right\rangle = G(\underline{k}_\alpha) G(\underline{k}_\beta)$$

$$\times \left[\delta(\underline{k}_\alpha + \underline{k}_\gamma) \delta^{\alpha\gamma} \delta(\underline{k}_\beta + \underline{k}_\delta) \delta^{\beta\delta} + \delta(\underline{k}_\alpha + \underline{k}_\delta) \delta^{\alpha\delta} \delta(\underline{k}_\beta + \underline{k}_\gamma) \delta^{\beta\gamma} \right]$$

$$- Du \left[q_\alpha^a G_0(\underline{k}_\alpha) C_0(\underline{k}_\beta) + q_\beta^a G_0(\underline{k}_\beta) C_0(\underline{k}_\alpha) \right] G_0(\underline{k}_\gamma) G_0(\underline{k}_\delta)$$

$$\times \delta(\underline{k}_\alpha + \underline{k}_\beta + \underline{k}_\gamma + \underline{k}_\delta) F^{\alpha\beta\gamma\delta} + O(u^2) \quad (4.55)$$

(Problem 4.3). As is clear from both the diagrams in Fig. 4.6 and the analytic expression (4.55), the truly new information for this four-point correlation function

(a)   (b)   (c)

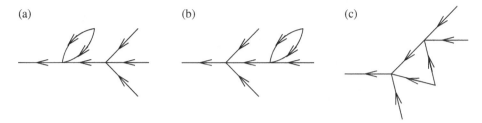

Fig. 4.7 Feynman graphs for the second-order contributions to $\langle S^\alpha \widetilde{S}^\beta \widetilde{S}^\gamma \widetilde{S}^\delta \rangle$.

is encoded in the *connected* part

$$\left\langle S^\alpha(\underline{k}_\alpha) S^\beta(\underline{k}_\beta) \widetilde{S}^\gamma(\underline{k}_\gamma) \widetilde{S}^\delta(\underline{k}_\delta) \right\rangle_c$$

$$= \left\langle S^\alpha(\underline{k}_\alpha) S^\beta(\underline{k}_\beta) \widetilde{S}^\gamma(\underline{k}_\gamma) \widetilde{S}^\delta(\underline{k}_\delta) \right\rangle - \left\langle S^\alpha(\underline{k}_\alpha) \widetilde{S}^\gamma(\underline{k}_\gamma) \right\rangle \left\langle S^\beta(\underline{k}_\beta) \widetilde{S}^\delta(\underline{k}_\delta) \right\rangle$$

$$- \left\langle S^\alpha(\underline{k}_\alpha) \widetilde{S}^\delta(\underline{k}_\delta) \right\rangle \left\langle S^\beta(\underline{k}_\beta) \widetilde{S}^\gamma(\underline{k}_\gamma) \right\rangle , \tag{4.56}$$

represented by Fig. 4.6(c). Writing the first-order result in terms of the bare response propagator only, we find

$$\left\langle S^\alpha(\underline{k}_\alpha) S^\beta(\underline{k}_\beta) \widetilde{S}^\gamma(\underline{k}_\gamma) \widetilde{S}^\delta(\underline{k}_\delta) \right\rangle_c = -2 D q_\alpha^a \, D q_\beta^a \, u$$

$$\times \left[G_0(-\underline{k}_\alpha) + G_0(-\underline{k}_\beta) \right] G_0(\underline{k}_\alpha) \, G_0(\underline{k}_\beta) \, G_0(\underline{k}_\gamma) \, G_0(\underline{k}_\delta)$$

$$\times \delta(\underline{k}_\alpha + \underline{k}_\beta + \underline{k}_\gamma + \underline{k}_\delta) \, F^{\alpha\beta\gamma\delta} + O(u^2) . \tag{4.57}$$

In a similar way, the disconnected contributions to the remaining four-point correlation functions $\langle S^\alpha S^\beta S^\gamma \widetilde{S}^\delta \rangle$ and $\langle S^\alpha S^\beta S^\gamma S^\delta \rangle$ can be subtracted easily. The generalization of Eqs. (4.52) and (4.57) for the connected correlators $\langle S^\alpha S^\beta S^\gamma \widetilde{S}^\delta \rangle_c$ and $\langle S^\alpha S^\beta S^\gamma S^\delta \rangle_c$ is a straightforward exercise (Problem 4.3).

We end the discussion of the four-point functions with a look at the second-order contributions to $\langle S^\alpha \widetilde{S}^\beta \widetilde{S}^\gamma \widetilde{S}^\delta \rangle$. In order to construct the Feynman diagrams to $O(u^2)$, we must take two relaxation vertices, and connect all their legs either to each other or to the four external lines. The resulting diagrams are displayed in Fig. 4.7. Graphs (a) and (b) here can be constructed by replacing either the outgoing or one of the incoming external lines of the first-order graph in Fig. 4.5 with the first-order contribution to the response propagator depicted in Fig. 4.2(c). The contributions of Fig. 4.7(a) and (b) hence merely constitute modifications or a 'renormalization' of the bare response propagators attached to Fig. 4.5, and the corresponding analytic expression assumes the form of Eq. (4.52), with the factors $G_0(\underline{k}_\alpha)$ etc. changed to the first-order propagators (4.47). However, Fig. 4.7(c) represents a genuinely new *vertex renormalization*.

For the four-point functions such as $\langle S^\alpha S^\beta \tilde{S}^\gamma \tilde{S}^\delta \rangle_c$ of Eq. (4.56), it is easily established that the first-order terms correspond to 'tree' diagrams like Fig. 4.5 and Fig. 4.6(c), while the Feynman graphs contributing to $O(u^2)$ contain precisely one propagator *loop* (Problem 4.5). Thus, for the *connected N*-point functions, the perturbation expansion in terms of u is encoded in a *loop expansion* for the Feynman diagrams. Furthermore, in addition to the propagator renormalization loops of Fig. 4.3, there are only two different one-loop diagrams reflecting the vertex renormalizations occurring in *all* four-point functions to this order (Problem 4.5). Even the Feynman graphs for the connected *N*-point functions still contain ample redundancies. Yet, before we analyze and eliminate these by considering the one-particle irreducible vertex functions in Section 4.4, we first establish the Feynman rules for the connected parts of the correlation functions or cumulants.

4.3.3 Feynman rules for the computation of cumulants

When the perturbation expansion is extended to higher orders in the non-linear coupling u, quite obviously the number of possible contractions to be accounted for increases rapidly, and the analytic expressions become rather lengthy and impervious. Therefore, the considerably more transparent representation in terms of Feynman graphs is an invaluable tool for keeping track of and classifying the numerous contributions to a certain order in perturbation theory. As we have just seen, *N*-point correlation functions generally contain a large amount of redundant information. It therefore suffices to study their connected parts, i.e., the *N*-point *cumulants*, which are given by functional derivatives of the logarithm of the generating functional (4.26),

$$G_{\{\alpha_i\};\{\alpha_k\}}^{(N,\tilde{N})}(\{x_i, t_i\}; \{x_k, t_k\}) = \left\langle \prod_i^N S^{\alpha_i}(x_i, t_i) \prod_k^{\tilde{N}} \tilde{S}^{\alpha_k}(x_k, t_k) \right\rangle_c$$

$$= \prod_i^N \frac{\delta}{\delta j^{\alpha_i}(x_i, t_i)} \prod_k^{\tilde{N}} \frac{\delta}{\delta \tilde{j}^{\alpha_k}(x_k, t_k)} \ln \mathcal{Z}[\tilde{j}, j] \Big|_{\tilde{j}=0=j} .$$

$$(4.58)$$

For example, we compute

$$\langle S^\alpha(x, t) S^\beta(x', t') \rangle_c = \frac{\delta}{\delta j^\beta(x', t')} \frac{1}{\mathcal{Z}[\tilde{j}, j]} \langle S^\alpha(x, t) e^{\int j^{\alpha'} S^{\alpha'}} \rangle \Big|_{j=0}$$

$$= \langle S^\alpha(x, t) S^\beta(x', t') \rangle - \langle S^\alpha(x, t) \rangle \langle S^\beta(x', t') \rangle, \quad (4.59)$$

where $\int \ldots$ symbolically stands for $\int d^d x \int dt \sum_\alpha \ldots$, and we have used $\mathcal{Z}[\tilde{j} = 0, j = 0] = 1$. Thus, above T_c where $\langle S^\alpha \rangle = 0$, the two-point cumulants are identical to the ordinary two-point correlation functions, $\langle S^\alpha S^\beta \rangle_c = \langle S^\alpha S^\beta \rangle$, and $\langle S^\alpha \tilde{S}^\beta \rangle_c = \langle S^\alpha \tilde{S}^\beta \rangle$. In the same manner, in the disordered phase one finds for the four-point cumulants (Problem 4.4)

$$\langle S^\alpha S^\beta S^\gamma S^\delta \rangle_c = \langle S^\alpha S^\beta S^\gamma S^\delta \rangle - \langle S^\alpha S^\beta \rangle \langle S^\gamma S^\delta \rangle$$

$$- \langle S^\alpha S^\gamma \rangle \langle S^\beta S^\delta \rangle - \langle S^\alpha S^\delta \rangle \langle S^\beta S^\gamma \rangle , \qquad (4.60)$$

and similarly for $\langle S^\alpha S^\beta S^\gamma \tilde{S}^\delta \rangle_c$. Because $\langle \tilde{S}^\gamma \tilde{S}^\delta \rangle = 0$, this also confirms Eq. (4.56) for $\langle S^\alpha S^\beta \tilde{S}^\gamma \tilde{S}^\delta \rangle_c$, and furthermore

$$\langle S^\alpha \tilde{S}^\beta \tilde{S}^\gamma \tilde{S}^\delta \rangle_c = \langle S^\alpha \tilde{S}^\beta \tilde{S}^\gamma \tilde{S}^\delta \rangle . \qquad (4.61)$$

In this fashion, quite generally, the cumulants $G^{(N, \tilde{N})}$ result by subtracting the disconnected parts from the full correlation functions.

We may now summarize our above considerations for the construction and properties of the perturbation expansion for the relaxational models A and B by means of the *Feynman rules* in Fourier space for the lth order term contributing to the *cumulant* of N fields S^{α_i} and $\tilde{N}$ response fields $\tilde{S}^{\alpha_j}$.

(i) Using the elements depicted in Fig. 4.1, draw *all topologically different connected Feynman graphs* with N outgoing and $\tilde{N}$ incoming lines, which contain l relaxation four-point vertices $\propto u$. Discard those diagrams that contain closed response loops.

(ii) Attach wavevectors q_i, frequencies ω_i, and vector indices α_i to all directed lines representing the propagators, such that as a consequence of translation invariance in space and time at each vertex 'momentum and energy' are conserved, i.e., the sum of all incoming q_j, ω_j equals the sum of all outgoing q_i, ω_i.

(iii) Analytically, each directed line corresponds to a response propagator $G_0(q, \omega) = [-i\omega + Dq^a(r + q^2)]^{-1}$, the two-point vertex to the noise strength $2Dq^a$, and the four-point relaxation vertex to $-Dq^a u/6$, where q and ω denote the outgoing momentum and frequency, respectively. Closed loops imply integrals over the internal wavevectors and frequencies, as well as sums over the internal vector indices.

(iv) Multiply by the combinatorial factor counting all possible ways of connecting the propagators, l relaxation vertices, and k two-point vertices leading to topologically identical graphs, including a factor $1/l!$ originating in the expansion of $\exp(-\mathcal{A}_{\text{int}}[\tilde{S}, S])$, see Eq. (4.42), and similarly a factor $1/k!$ for the noise vertices.

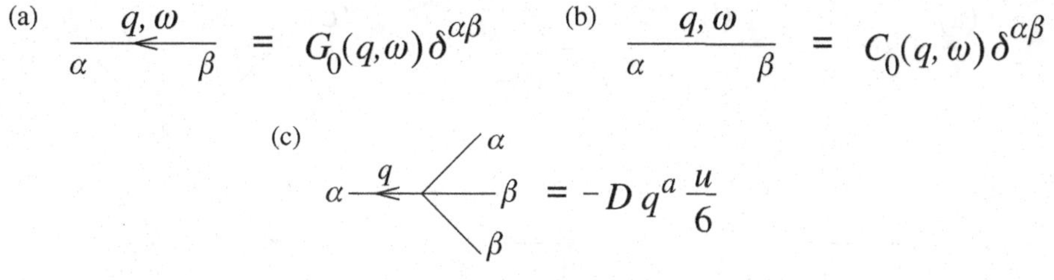

(a) $\xrightarrow[\alpha]{\quad q,\,\omega \quad}_{\beta}$ $= G_0(q,\omega)\delta^{\alpha\beta}$ (b) $\xrightarrow[\alpha]{\quad q,\,\omega \quad}_{\beta}$ $= C_0(q,\omega)\delta^{\alpha\beta}$

(c)

$$\alpha \xrightarrow{\;q\;} \overset{\alpha}{\underset{\beta}{\Big\langle}} \beta \;=\; -D\,q^a\,\frac{u}{6}$$

Fig. 4.8 Dynamic perturbation theory elements in the alternative formulation: (a) response propagator; (b) correlation propagator; and (c) relaxation vertex; compare with Fig. 4.1.

We remark that the dependence of these combinatorial factors on the number of order parameter components n can be obtained either by direct counting, or more formally by symmetrizing the four-point vertex by means of the tensor (4.53), and then contracting the indices with the Kronecker symbols $\delta^{\alpha\beta}$ attached to the propagators and noise vertices. In our earlier analysis of the two- and four-point functions, we have employed both methods. When diagrams to large order in u are evaluated, the latter approach of contracting tensor indices may sometimes be safer.

In the Feynman diagrams for the cumulants, all vertices are connected to each other through the *causal* response propagators (4.33), which in the time domain are $G_0(x, t - t') \propto \Theta(t - t')$. The arrows on the directed propagator lines therefore always point towards increasing time variables. Consequently, causality implies that the cumulants necessarily vanish if any time index on an external response field $\widetilde{S}^\alpha$ is larger than *all* time labels on the fields S^α,

$$G^{(N,\widetilde{N})}_{\{\alpha_i\};\{\alpha_j\}}(\{x_i, t_i\}; \{x_j, t_j\}) = 0 \quad \text{if any } t_j > \text{all } t_i \,. \tag{4.62}$$

Because the pure response field correlations are inevitably represented by connected graphs, Eq. (4.43) simply follows as a special case for $N = 0$ from this more general statement.

4.3.4 Correlation propagators in dynamic field theory

To be precise, the way we set up the above Feynman rules actually follows from a slight re-interpretation of the perturbation expansion as compared to the splitting of the dynamic action into (4.14) and (4.15). For, using $G_0(q, \omega)$ as the sole propagator of the theory corresponds to the harmonic action

$$\mathcal{A}_0[\widetilde{S}, S] = \int d^d x \int dt \sum_\alpha \widetilde{S}^\alpha(x, t) \left[\frac{\partial}{\partial t} + D(i\nabla)^a(r - \nabla^2) \right] S^\alpha(x, t)\,, \tag{4.63}$$

(a)   (b)   (c)

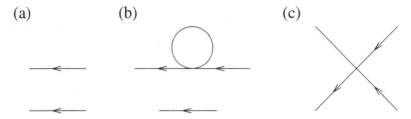

Fig. 4.9 (a) Zeroth-order, (b) disconnected first-order, and (c) connected first-order Feynman diagrams for $\langle S^\alpha S^\beta \widetilde{S}^\gamma \widetilde{S}^\delta \rangle$, using response and correlation propagators, compare with Fig. 4.6.

i.e., the noise vertex has been transferred into the anharmonic part

$$\mathcal{A}_{\text{int}}[\widetilde{S}, S] = -D \int d^d x \int dt \sum_\alpha \left[\widetilde{S}^\alpha(x, t)(i\nabla)^a \widetilde{S}^\alpha(x, t) \right.$$

$$\left. - \frac{u}{6} \widetilde{S}^\alpha(x, t)(i\nabla)^a S^\alpha(x, t) \sum_\beta S^\beta(x, t) S^\beta(x, t) \right]. \quad (4.64)$$

From $-\mathcal{A}_{\text{int}}[\widetilde{S}, S]$ we infer the two vertices of Fig. 4.1, where the factor 2 in the noise vertex already accounts for the two different ways of labeling the legs with q and $-q$, respectively. To lth order in u, we then require a certain number k of two-point vertices $\propto D$ in order to construct a specific connected Feynman diagram, which amounts to the kth order term of the expansion of $\exp(-\mathcal{A}_{\text{int}}[\widetilde{S}, S])$ with respect to the noise vertex.

Our original separation of the response functional into the Gaussian part (4.14) and the anharmonic contribution (4.15), on the other hand, led to *two* different non-vanishing contractions (4.36) and (4.37). The perturbation expansion may thus be equivalently built up from two propagators, namely the bare response and correlation functions $G_0(q, \omega)$ and $C_0(q, \omega)$, and a single non-linear vertex $\propto u$, see Fig. 4.8. With these alternative elements to construct the Feynman diagrams for the $O(n)$-symmetric relaxational models, the graphs contributing to $\langle S^\alpha S^\beta \widetilde{S}^\gamma \widetilde{S}^\delta \rangle$ to first order in u are shown in Fig. 4.9, and the diagrams to order u^2 for $\langle S^\alpha \widetilde{S}^\beta \widetilde{S}^\gamma \widetilde{S}^\delta \rangle$ are depicted in Fig. 4.10. Analytically, the equivalence of the two different formulations of perturbation theory, or the two versions of Feynman diagrams, is established via Eqs. (4.34) or (4.37), which express the correlation propagator as a product of the noise vertex and two response propagators. Indeed, this replacement immediately transforms the diagrams in Figs. 4.9 and 4.10 into those of Figs. 4.6 and 4.7.

The second formulation using both response and correlations propagators is widely used for the description of equilibrium critical dynamics. Its advantages are

(a)   (b)   (c)

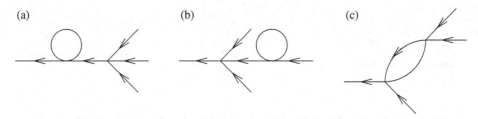

Fig. 4.10 Feynman graphs for the second-order contributions to $\langle S^\alpha \widetilde{S}^\beta \widetilde{S}^\gamma \widetilde{S}^\delta \rangle$, using response and correlation propagators, compare with Fig. 4.7.

that the perturbation expansion is directly performed with respect to the non-linear coupling u, and the static limit is manifestly encoded in the Feynman graphs. For example, the diagrams contributing to the four-point function $\langle S^\alpha S^\beta S^\gamma S^\delta \rangle$ up to order u^2 are represented by Figs. 4.9 and 4.10, if we simply replace the external directed lines with correlation propagators. The topology of these graphs is already identical to those of the static perturbation expansion for the $O(n)$-symmetric Heisenberg model Hamiltonian (4.12). Furthermore, according to Eq. (3.43) the loop in Fig. 4.9(b) yields the integral over the static correlation function $C_0(q)$. As we shall see below (Eq. (4.102)), performing the internal frequency integral in Fig. 4.10(c) in the limit of vanishing external frequency leads to a wave-vector integral over a product of two static correlation functions, precisely as in the static one-loop approximation, with the missing factors Dq^a provided by the vertices.

On the other hand, introducing a separate correlation propagator is somewhat redundant due to the identity (4.37). More importantly, the Feynman rules stated above with only a single response propagator can be more directly generalized and applied to physically quite different situations, namely for the perturbation expansion in the theory of quantum critical phenomena as well as for general non-equilibrium processes which cannot necessarily be described in terms of Langevin-type equations of motion. Bearing in mind these equivalences, we shall therefore henceforth exclusively use the earlier variant (Fig. 4.1) and the accompanying Feynman rules as stated above.

4.4 Vertex functions

Let us reconsider the general structure of the perturbation expansion. As we have seen, restricting the perturbation series to the cumulants represented by connected Feynman diagrams already eliminates some redundancies. However, we are still left with graphs such as in Figs. 4.4 and 4.7(a, b), which merely correspond to renormalized propagators attached to a bare vertex. If we are, e.g., interested

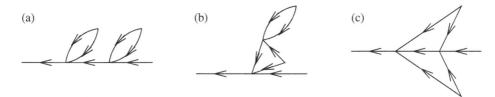

Fig. 4.11 Two-loop Feynman diagrams for the fluctuation corrections to the response propagator $G(q, \omega)$ (see text).

in the proper renormalization of the four-point vertex $\propto u$, the only relevant one-loop diagram is Fig. 4.7(c). Through the introduction of *vertex functions*, diagrammatically given by *one-particle irreducible* graphs, the perturbation expansion becomes focused on such truly essential contributions.

4.4.1 Dyson's equation and self-energy

We begin once more with an analysis of the loop expansion for the response propagator $G(q, \omega)$. The zeroth-order $G_0(q, \omega)$ is shown in Fig. 4.1(a), whereas the non-vanishing corrections to first and second order in u are depicted in Fig. 4.2(c) and Fig. 4.11. Obviously, the corresponding Feynman diagrams are all connected, i.e.,

$$G_{\alpha;\beta}^{(1,1)}(q, \omega) = G(q, \omega)\, \delta^{\alpha\beta} \qquad (4.65)$$

and to lth order in u contain precisely l closed propagator loops. Notice that Figs. 4.11(a) and (b) are readily generated from the one-loop diagram in Fig. 4.2(c) by replacing an external and internal line, respectively, with the one-loop correction to $G_0(q, \omega)$. Alternatively, the topologically different graphs in Fig. 4.11 may be generated by placing the two relaxation vertices (a) in a row, (b) on top of each other on the same side of the propagator line, and (c) on opposite sides of the propagator line.

Figure 4.11(a), stemming from the replacement of a bare external propagator line in the one-loop diagram with the first-order correction, is simply a product of one-loop contributions, and truly novel information is only contained in the other two second-order graphs. The diagrams Fig. 4.11(b) and (c) are called *one-particle irreducible* (1PI), because one cannot factorize them into lower-order graphs by severing a single internal line. Quite generally, we observe that $G(q, \omega)$ may be written as an alternating series of bare propagators $G_0(q, \omega)$ and the complete set of 1PI graphs Σ, as depicted in the first line of Fig. 4.12. We can then diagrammatically sum the entire perturbation series for $G(q, \omega)$ by combining all the terms to the

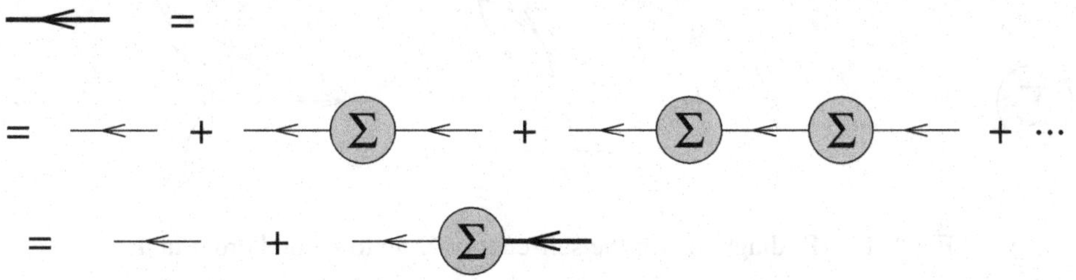

Fig. 4.12 Diagrammatic representation of Dyson's equation. The bold directed line represents the full response propagator $G(q, \omega)$, the thin one denotes $G_0(q, \omega)$.

right (or left) of the farthest left (right) 1PI insertion Σ into a full propagator line, as shown in the third line of Fig. 4.12. By simply iterating this diagrammatic equation, one reproduces the complete perturbation expansion, as illustrated in the second line.

Analytically, Fig. 4.12 becomes *Dyson's equation*,

$$G(q, \omega) = G_0(q, \omega)\Big[1 + \Sigma(q, \omega)\, G_0(q, \omega)$$

$$+ \Sigma(q, \omega)\, G_0(q, \omega)\, \Sigma(q, \omega)\, G_0(q, \omega) + \cdots \Big] \quad (4.66)$$

$$= G_0(q, \omega)[1 + \Sigma(q, \omega)\, G(q, \omega)] \,. \quad (4.67)$$

Either by summing the geometric series (4.66), or by solving Eq. (4.67) for the full response propagator, we find the *exact* expression

$$G(q, \omega) = \frac{G_0(q, \omega)}{1 - G_0(q, \omega)\, \Sigma(q, \omega)} \,, \quad (4.68)$$

or equivalently,[4]

$$G(q, \omega)^{-1} = G_0(q, \omega)^{-1} - \Sigma(q, \omega) = -i\omega + Dq^a(r + q^2) - \Sigma(q, \omega) \,. \quad (4.69)$$

As Eq. (4.68) demonstrates, by means of Dyson's equation we have effectively summed an entire geometric series of diagrams. As a consequence of this partial summation of the perturbation expansion, we are left exclusively with the 1PI graphs contributing to the *self-energy* $\Sigma(q, \omega)$, which according to Eq. (4.69) may be viewed as a renormalization of the bare dispersion relation. Thus, Dyson's equation (Fig. 4.12) exploits the general structure of the perturbation expansion to all orders, reduces the number of required graphs to a minimum, and even entails a physical interpretation of the loop corrections.

---

[4] This form of Dyson's equation holds as a matrix relation also for non-diagonal propagators.

Fig. 4.13 1PI diagrams for the self-energy Σ up to second order in u.

The contributions to $\Sigma(q, \omega)$ up to second order in u are displayed in Fig. 4.13. Notice that these 1PI Feynman diagrams are drawn with shortened external legs, as the corresponding analytic expressions do not contain the external propagators. The evaluation of the one-loop diagram proceeds precisely as before, see Fig. 4.3, and using the frequency integral (3.43) (with $k_B T = 1$), we arrive at the following explicit result to first order in u:

$$G(q, \omega)^{-1} = -i\omega + Dq^a \left[r + q^2 + \frac{n+2}{6} u \int_k \frac{1}{r + k^2} \right] + O(u^2) . \qquad (4.70)$$

Note that the one-loop contribution to the self-energy carries neither momentum nor frequency, and thus merely modifies the bare parameter r. As a consequence, both exponents η and z obviously retain their mean-field values in this approximation. To one-loop order, therefore, we shall merely find a non-trivial correlation length exponent ν (Section 5.3). However, the two-loop diagram in Fig. 4.11(c) does depend on the external wavevector q and frequency ω, and eventually leads to a second-order modification of both Fisher and dynamical exponents η and z.

4.4.2 Generating functional; two-point vertex functions

We have just seen that the perturbation series for the response propagator can be effectively reduced to the collection of one-particle irreducible (1PI) contributions by means of Dyson's equation (Fig. 4.12). In a more formal manner, this procedure can be generalized to higher cumulants as well, via the introduction of vertex functions. To this end, we define the fields

$$\tilde{\Phi}^\alpha(x, t) = \frac{\delta \ln \mathcal{Z}[\tilde{j}, j]}{\delta \tilde{j}^\alpha(x, t)} , \quad \Phi^\alpha(x, t) = \frac{\delta \ln \mathcal{Z}[\tilde{j}, j]}{\delta j^\alpha(x, t)} , \qquad (4.71)$$

which reduce to $\langle \tilde{S}^\alpha \rangle_c$ and $\langle S^\alpha \rangle_c$ at vanishing external sources $\tilde{j}^\alpha$ and j^α, and perform a Legendre transformation to these new variables,

$$\Gamma[\tilde{\Phi}, \Phi] = -\ln \mathcal{Z}[\tilde{j}, j] + \int d^d x \int dt \sum_\alpha \left(\tilde{j}^\alpha(x, t) \tilde{\Phi}^\alpha(x, t) + j^\alpha(x, t) \Phi^\alpha(x, t) \right) .$$

$$(4.72)$$

$\Gamma[\widetilde{\Phi}, \Phi]$ then serves as the *generating functional* for the $(\widetilde{N} + N)$-point vertex functions

$$\Gamma^{(\widetilde{N},N)}_{\{\alpha_i\};\{\alpha_k\}}(\{x_i, t_i\}; \{x_k, t_k\}) = \prod_i^{\widetilde{N}} \frac{\delta}{\delta\widetilde{\Phi}^{\alpha_i}(x_i, t_i)} \prod_j^{N} \frac{\delta}{\delta\Phi^{\alpha_k}(x_k, t_k)} \Gamma[\widetilde{\Phi}, \Phi]\bigg|_{\widetilde{j}=0=j},$$

(4.73)

compare with Eqs. (4.27) and (4.58) for the correlation functions and cumulants $G^{(N,\widetilde{N})}$. For example, we compute

$$\frac{\delta\Gamma[\widetilde{\Phi}, \Phi]}{\delta\widetilde{\Phi}^{\alpha}(x, t)}$$

$$= \widetilde{j}^{\alpha}(x, t) + \int d^d x' \int dt' \sum_{\alpha'} \left(\frac{\delta\widetilde{j}^{\alpha'}(x', t')}{\delta\widetilde{\Phi}^{\alpha}(x, t)} \widetilde{\Phi}^{\alpha'}(x', t') + \frac{\delta j^{\alpha'}(x', t')}{\delta\widetilde{\Phi}^{\alpha}(x, t)} \Phi^{\alpha'}(x', t') \right)$$

$$- \int d^d x' \int dt' \sum_{\alpha'} \left(\frac{\delta \ln \mathcal{Z}[\widetilde{j}, j]}{\delta\widetilde{j}^{\alpha'}(x', t')} \frac{\delta\widetilde{j}^{\alpha'}(x', t')}{\delta\widetilde{\Phi}^{\alpha}(x, t)} + \frac{\delta \ln \mathcal{Z}[\widetilde{j}, j]}{\delta j^{\alpha'}(x', t')} \frac{\delta j^{\alpha'}(x', t')}{\delta\widetilde{\Phi}^{\alpha}(x, t)} \right)$$

$$= \widetilde{j}^{\alpha}(x, t),$$

(4.74)

where we have used the chain rule for functional derivatives and the definitions (4.71). In the same manner,

$$\frac{\delta\Gamma[\widetilde{\Phi}, \Phi]}{\delta\Phi^{\alpha}(x, t)} = j^{\alpha}(x, t)$$

(4.75)

just yields the source field j^{α}, and consequently

$$\Gamma^{(1,0)}_{\alpha}(x, t) = 0 = \Gamma^{(0,1)}_{\alpha}(x, t)$$

(4.76)

in the disordered, symmetric high-temperature phase. When we interpret Eqs. (4.74) and (4.75) as the definitions of $\widetilde{j}^{\alpha}$ and j^{α}, then Eq. (4.72) provides us with the Legendre transform from $\Gamma[\widetilde{\Phi}, \Phi]$ back to $\ln \mathcal{Z}[\widetilde{j}, j]$. As in classical mechanics or thermodynamics, the descriptions in terms of either set of variables are of course equivalent, and we may choose the most convenient representation.

Via their defining Legendre transform (4.72), the vertex functions are uniquely connected with the cumulants. For example, by performing functional derivatives of Eqs. (4.74) and (4.75) with respect to the sources, one may readily derive the relations between the two-point vertex functions and cumulants. To this end, we introduce the real-space short-hand notation

$$\underline{y} = (x, t), \quad \delta(\underline{y}) = \delta(x)\,\delta(t), \quad \int_{\underline{y}} \cdots = \int d^d x \int dt \cdots$$

(4.77)

cf. Eq. (3.36) in Fourier space, and consider

$$\frac{\delta^2 \Gamma}{\delta \widetilde{\Phi}^\alpha(\underline{y}) \delta \widetilde{j}^\beta(\underline{\tilde{y}})} = \delta(\underline{y} - \underline{\tilde{y}}) \delta^{\alpha\beta}$$

$$= \int_{\underline{y'}} \left[\frac{\delta^2 \Gamma}{\delta \widetilde{\Phi}^\alpha(\underline{y}) \delta \widetilde{\Phi}^\gamma(\underline{y'})} \frac{\delta \widetilde{\Phi}^\gamma(\underline{y'})}{\delta \widetilde{j}^\beta(\underline{\tilde{y}})} + \frac{\delta^2 \Gamma}{\delta \widetilde{\Phi}^\alpha(\underline{y}) \delta \Phi^\gamma(\underline{y'})} \frac{\delta \Phi^\gamma(\underline{y'})}{\delta \widetilde{j}^\beta(\underline{\tilde{y}})} \right]$$

$$= \int_{\underline{y'}} \left[\frac{\delta^2 \Gamma}{\delta \widetilde{\Phi}^\alpha(\underline{y}) \delta \widetilde{\Phi}^\gamma(\underline{y'})} \frac{\delta^2 \ln \mathcal{Z}}{\delta \widetilde{j}^\gamma(\underline{y'}) \delta \widetilde{j}^\beta(\underline{\tilde{y}})} + \frac{\delta^2 \Gamma}{\delta \widetilde{\Phi}^\alpha(\underline{y}) \delta \Phi^\gamma(\underline{y'})} \frac{\delta^2 \ln \mathcal{Z}}{\delta j^\gamma(\underline{y'}) \delta \widetilde{j}^\beta(\underline{\tilde{y}})} \right] ,$$

$$(4.78)$$

where Eq. (4.71) was used. Taking the limit $\tilde{j}^\alpha = j^\alpha \to 0$, the functional derivative of $\ln \mathcal{Z}$ in the first term yields the vanishing cumulant $\langle \widetilde{S}^\gamma(\underline{y'}) \widetilde{S}^\beta(\underline{\tilde{y}}) \rangle_c = 0$, and in the second contribution $\langle S^\gamma(\underline{y'}) \widetilde{S}^\beta(\underline{\tilde{y}}) \rangle_c = G(x' - \tilde{x}, t' - \tilde{t}) \delta^{\beta\gamma} \propto \Theta(t' - \tilde{t})$, leading to

$$\int d^d x' \int dt' \, \Gamma^{(1,1)}_{\alpha;\beta}(x, t; x', t') \, G(x' - \tilde{x}, t' - \tilde{t}) = \delta(x - \tilde{x}) \delta(t - \tilde{t}) \delta^{\alpha\beta} . \quad (4.79)$$

Thus, $\Gamma^{(1,1)}$ is diagonal in the tensor indices; in Fourier space, the convolution in Eq. (4.79) becomes a simple product. Upon defining

$$\Gamma^{(1,1)}_{\alpha;\beta}(x, t; x', t') = \Gamma^{(1,1)}(x' - x, t' - t) \delta^{\alpha\beta} , \quad (4.80)$$

the Fourier transform of Eq. (4.79) becomes

$$\Gamma^{(1,1)}(q, \omega) = G(-q, -\omega)^{-1} = i\omega + Dq^a(r + q^2) - \Sigma(-q, -\omega) . \quad (4.81)$$

Therefore, apart from a minus sign and the inversion of the external frequency, the loop diagrams contributing to $\Gamma^{(1,1)}(q, \omega)$ are nothing but the one-particle irreducible self-energy Feynman graphs.

An explicit evaluation of the one- and two-loop diagrams of Fig. 4.13 yields (Problem 4.6)

$$\Gamma^{(1,1)}(q, \omega) = i\omega + Dq^a \left[r + q^2 + \frac{n+2}{6} u \int_k \frac{1}{r+k^2} \right.$$

$$- \left(\frac{n+2}{6} u \right)^2 \int_k \frac{1}{r+k^2} \int_{k'} \frac{1}{(r+k'^2)^2}$$

$$- \frac{n+2}{18} u^2 \int_k \frac{1}{r+k^2} \int_{k'} \frac{1}{r+k'^2} \frac{1}{r+(q-k-k')^2}$$

$$\times \left. \left(1 - \frac{i\omega}{i\omega + \Delta(k) + \Delta(k') + \Delta(q-k-k')} \right) + O(u^3) \right], \quad (4.82)$$

where we have introduced the quantity

$$\Delta(q) = Dq^a(r + q^2).$$ (4.83)

Notice that all the one- and two-loop corrections to the bare propagator are proportional to q^a, which stems from the external line on the non-linear vertex of Fig. 4.1(c). This observation obviously holds to any order in the perturbation expansion, because the outgoing line of the vertex function $\Gamma^{(1,1)}(q, \omega)$ must always be connected to a relaxation vertex. For the relaxational model B describing the critical dynamics of a conserved order parameter, therefore all the loop diagrams vanish at least $\propto q^2$ in the limit $q \to 0$, i.e.,

$$a = 2 : \quad \Gamma^{(1,1)}(q = 0, \omega) = i\omega.$$ (4.84)

This remarkable *exact* statement, valid to *all* orders of the perturbation expansion, is a consequence of the conservation law and the ensuing dependence of the non-linear vertex on the external wavevector.

Next, taking the functional derivative of Eq. (4.75) with respect to $j^\beta(\tilde{x}, \tilde{t})$ results in

$$\frac{\delta^2 \Gamma}{\delta \Phi^\alpha(\underline{y}) \delta j^\beta(\underline{\tilde{y}})} = \delta(\underline{y} - \underline{\tilde{y}}) \delta^{\alpha\beta} = \int_{\underline{y'}} \left[\frac{\delta^2 \Gamma}{\delta \Phi^\alpha(\underline{y}) \delta \widetilde{\Phi}^\gamma(\underline{y'})} \frac{\delta^2 \ln \mathcal{Z}}{\delta \tilde{j}^\gamma(\underline{y'}) \delta j^\beta(\underline{\tilde{y}})} \right.$$

$$\left. + \frac{\delta^2 \Gamma}{\delta \Phi^\alpha(\underline{y}) \delta \Phi^\gamma(\underline{y'})} \frac{\delta^2 \ln \mathcal{Z}}{\delta j^\gamma(\underline{y'}) \delta j^\beta(\underline{\tilde{y}})} \right],$$ (4.85)

whence

$$\int d^d x' \int dt' \left[\Gamma^{(1,1)}(x - x', t - t') G(\tilde{x} - x', \tilde{t} - t') \delta^{\alpha\beta} \right.$$

$$\left. + \Gamma^{(0,2)}_{\alpha\beta}(x, t; x', t') C(\tilde{x} - x', \tilde{t} - t') \right] = \delta(x - \tilde{x})\delta(t - \tilde{t}) \delta^{\alpha\beta}.$$ (4.86)

Subtracting Eq. (4.79) then immediately shows

$$\Gamma^{(0,2)}_{\alpha\beta}(x, t; x', t') = 0.$$ (4.87)

Finally, the 'mixed' functional derivative gives zero,

$$\frac{\delta^2 \Gamma}{\delta \widetilde{\Phi}^\alpha(\underline{y}) \delta j^\beta(\underline{\tilde{y}})} = 0$$

$$= \int_{\underline{y'}} \left[\frac{\delta^2 \Gamma}{\delta \widetilde{\Phi}^\alpha(\underline{y}) \delta \widetilde{\Phi}^\gamma(\underline{y'})} \frac{\delta^2 \ln \mathcal{Z}}{\delta \tilde{j}^\gamma(\underline{y'}) \delta j^\beta(\underline{\tilde{y}})} + \frac{\delta^2 \Gamma}{\delta \widetilde{\Phi}^\alpha(\underline{y}) \delta \Phi^\gamma(\underline{y'})} \frac{\delta^2 \ln \mathcal{Z}}{\delta j^\gamma(\underline{y'}) \delta j^\beta(\underline{\tilde{y}})} \right],$$

(4.88)

Fig. 4.14 Diagrammatic representation of the correlation function $C(q, \omega)$ in terms of the vertex function $\Gamma^{(2,0)}(q, \omega)$ and the full propagators $G(q, \omega)$ and $G(-q, -\omega)$.

implying

$$\int d^d x' \int dt' \Big[\Gamma^{(2,0)}_{\alpha\beta}(x, t; x', t') \, G(\tilde{x} - x', \tilde{t} - t')$$

$$+ \Gamma^{(1,1)}(x' - x, t' - t) \, C(\tilde{x} - x', \tilde{t} - t') \delta^{\alpha\beta} \Big] = 0 . \quad (4.89)$$

Using Eq. (4.81), the Fourier transform of Eq. (4.89) becomes

$$\Gamma^{(2,0)}(q, \omega) = - \frac{C(q, \omega)}{G(q, \omega) \, G(-q, -\omega)} . \quad (4.90)$$

We see that this expression is symmetric with respect to time inversion $\omega \to -\omega$, and consequently $\Gamma^{(2,0)}(q, \omega)$, like the correlation function $C(q, \omega)$, is real. In fact, the implication of Eq. (4.90) is that, apart from a minus sign again, the vertex function $\Gamma^{(2,0)}(q, \omega)$ comprises precisely the 1PI contributions to $C(q, \omega)$, which remain after the division by the full response propagators $|G(q, \omega)|^2$, see Fig. 4.14.

On the zero-loop or tree level, this just corresponds to the bare noise vertex Fig. 4.1(b), $\Gamma^{(2,0)}(q, \omega) = -2Dq^a + O(u)$. As discussed in Section 4.3, the existing one-loop graphs merely represent contributions to the external propagators emerging from the two-point vertex, see Fig. 4.4. A non-trivial renormalization of this noise vertex only appears to second order in u, namely the graph depicted in Fig. 4.15, which is readily evaluated to (Problem 4.6)

$$\Gamma^{(2,0)}(q, \omega) = -2Dq^a \Big[1 + Dq^a \, \frac{n+2}{18} \, u^2 \int_k \frac{1}{r + k^2} \int_{k'} \frac{1}{r + k'^2} \frac{1}{r + (q - k - k')^2}$$

$$\times \text{Re} \, \frac{1}{i\omega + \Delta(k) + \Delta(k') + \Delta(q - k - k')} + O(u^3) \Big] .$$

$$(4.91)$$

Notice that the second-order correction to the bare vertex contains an additional factor q^a stemming from the second vertex, see Fig. 4.15. This observation obviously holds to any order in the perturbation expansion, because for the vertex function $\Gamma^{(2,0)}(q, \omega)$ the internal loops must always be connected to *two* relaxation vertices of Fig. 4.1(c). For the relaxational model B describing the critical dynamics of a

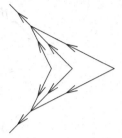

Fig. 4.15 Second-order, two-loop diagram for the vertex function $\Gamma^{(2,0)}(q, \omega)$.

conserved order parameter, therefore all the loop diagrams vanish $\propto q^4$ in the limit $q \to 0$, while the bare vertex is only $\propto q^2$. As in Eq. (4.84) for the vertex function $\Gamma^{(1,1)}(q, \omega)$, for model B we thus arrive at the exact statement

$$a = 2: \quad \frac{\partial}{\partial q^2} \Gamma^{(2,0)}(q, \omega)\bigg|_{q=0} = -2D, \tag{4.92}$$

valid to all orders in the perturbation expansion.

In the Gaussian model, the two-point vertex functions just constitute the entries of the Hermitian matrix (4.29),[5]

$$\mathbf{A}(q, \omega) = \begin{pmatrix} \Gamma^{(2,0)}(q, \omega) & \Gamma^{(1,1)}(-q, -\omega) \\ \Gamma^{(1,1)}(q, \omega) & \Gamma^{(0,2)}(q, \omega) \end{pmatrix}. \tag{4.93}$$

The full two-point vertex functions may therefore be viewed as the renormalized counterparts of the coupling coefficients appearing in the harmonic action (4.28). We can also express the full order parameter susceptibility and correlation function in terms of vertex functions. From Eqs. (4.18) and (4.81) we infer

$$\chi(q, \omega) = Dq^a \, \Gamma^{(1,1)}(-q, -\omega)^{-1}, \tag{4.94}$$

for the relaxational models A and B, and the inversion of Eq. (4.90) reads

$$C(q, \omega) = -\frac{\Gamma^{(2,0)}(q, \omega)}{|\Gamma^{(1,1)}(q, \omega)|^2}. \tag{4.95}$$

Hence, the fluctuation-dissipation theorem (2.35) connects the two-point vertex functions in a non-trivial manner,

$$\Gamma^{(2,0)}(q, \omega) = -\frac{2Dq^a}{\omega} \operatorname{Im} \Gamma^{(1,1)}(q, \omega). \tag{4.96}$$

Of course, our explicit two-loop results (4.82) and (4.91) fulfil this relation. For instance, the fact that the one-loop contributions to $\Gamma^{(1,1)}(q, \omega)$ carry no external frequency and are therefore real implies that there is no first-order correction to

---

[5] The convention (4.80) above was actually chosen in order to allow for this identification without additional frequency inversion.

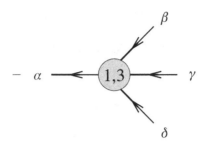

Fig. 4.16 Four-point cumulant $\langle S^\alpha \widetilde{S}^\beta \widetilde{S}^\gamma \widetilde{S}^\delta \rangle_c$, represented in terms of the vertex function $\Gamma^{(1,3)}$ and full propagators $G(q, \omega)$.

$\Gamma^{(2,0)}(q, \omega)$. Also, Eq. (4.84) for model B follows directly from Eq. (4.92) and the fluctuation-dissipation theorem (4.96) in the limit $q \to 0$.

4.4.3 Feynman rules for vertex functions

As a consequence of the $O(n)$ rotation symmetry, in the high-temperature phase all $(\widetilde{N} + N)$-point vertex functions with $\widetilde{N} + N$ odd vanish. (Formally, this can be demonstrated by means of the general Ward identity derived in Section 5.4 below.) Let us thus now explore the four-point vertex functions. By means of two additional functional derivatives of Eq. (4.78), one may readily show (Problem 4.7) that in Fourier space

$$\left\langle S^\alpha(\underline{k}_\alpha) \widetilde{S}^\beta(\underline{k}_\beta) \widetilde{S}^\gamma(\underline{k}_\gamma) \widetilde{S}^\delta(\underline{k}_\delta) \right\rangle_c = -G(\underline{k}_\alpha)\, G(\underline{k}_\beta)\, G(\underline{k}_\gamma)\, G(\underline{k}_\delta)\, \Gamma^{(1,3)}_{\alpha;\beta\gamma\delta}(\underline{k}_\alpha; \underline{k}_\beta, \underline{k}_\gamma, \underline{k}_\delta),$$

$$(4.97)$$

in full analogy with Eq. (4.90) for the two-point correlation function $C(q, \omega)$. Graphically, the relation between the four-point cumulant $\langle S^\alpha \widetilde{S}^\beta \widetilde{S}^\gamma \widetilde{S}^\delta \rangle_c$ and the vertex function $\Gamma^{(1,3)}$ is represented by Fig. 4.16, which establishes that once again the vertex function comprises the 1PI contributions to the cumulant, after truncation of the four full external response propagators.

Similar, albeit more complicated relations hold between the other four-point cumulants and vertex functions. For example, $\langle S^\alpha S^\beta \widetilde{S}^\gamma \widetilde{S}^\delta \rangle_c$ can be written in terms of $\Gamma^{(1,3)}$, $\Gamma^{(2,2)}$, and full propagators $G(q, \omega)$, $C(q, \omega)$:

$$\left\langle S^\alpha(\underline{k}_\alpha) S^\beta(\underline{k}_\beta) \widetilde{S}^\gamma(\underline{k}_\gamma) \widetilde{S}^\delta(\underline{k}_\delta) \right\rangle_c$$

$$= -G(\underline{k}_\alpha)\, C(\underline{k}_\beta)\, G(\underline{k}_\gamma)\, G(\underline{k}_\delta)\, \Gamma^{(1,3)}_{\alpha;\beta\gamma\delta}(\underline{k}_\alpha; \underline{k}_\beta, \underline{k}_\gamma, \underline{k}_\delta)$$

$$- C(\underline{k}_\alpha)\, G(\underline{k}_\beta)\, G(\underline{k}_\gamma)\, G(\underline{k}_\delta)\, \Gamma^{(1,3)}_{\beta;\alpha\gamma\delta}(\underline{k}_\beta; \underline{k}_\alpha, \underline{k}_\gamma, \underline{k}_\delta)$$

$$- G(\underline{k}_\alpha)\, G(\underline{k}_\beta)\, G(\underline{k}_\gamma)\, G(\underline{k}_\delta)\, \Gamma^{(2,2)}_{\alpha\beta;\gamma\delta}(\underline{k}_\alpha, \underline{k}_\beta; \underline{k}_\gamma, \underline{k}_\delta),$$

$$(4.98)$$

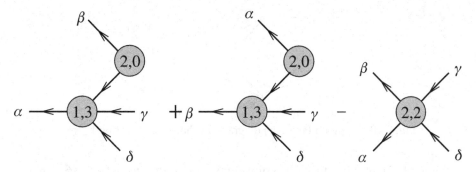

Fig. 4.17 Four-point cumulant $\langle S^\alpha S^\beta \widetilde{S}^\gamma \widetilde{S}^\delta \rangle_c$, represented in terms of the vertex functions $\Gamma^{(1,3)}$, $\Gamma^{(2,2)}$, $\Gamma^{(2,0)}$, and full response propagators $G(q, \omega)$.

see Problem 4.7. The symbolic representation of Eq. (4.98) is shown in Fig. 4.17, where Eq. (4.95) has been inserted for the correlation functions. This implies that $\Gamma^{(2,2)}$ too denotes purely 1PI contributions. Notice, however, that there is no tree-level (zero-loop) part of the vertex function $\Gamma^{(2,2)}$, for no corresponding coupling is present in the non-linear dynamic functional (4.15). Indeed, the lowest-order contribution to $\Gamma^{(2,2)}$ is given by the one-loop graph depicted in Fig. 4.18. As a consequence of rotational invariance, all the non-vanishing four-point vertex functions and cumulants are proportional to the symmetric tensor $F^{\alpha\beta\gamma\delta}$.

The generalization to higher vertex functions is straightforward, though tedious. The result simply is that all the vertex functions correspond to one-particle irreducible Feynman diagrams. A straightforward consequence of this observation is the general causality property

$$\Gamma^{(\widetilde{N},N)}_{\{\alpha_i\};\{\alpha_j\}}(\{x_i, t_i\}; \{x_j, t_j\}) = 0 \quad \text{if any } t_j > \text{all } t_i , \tag{4.99}$$

which follows from the corresponding Eq. (4.62) for the cumulants by truncating the external lines. This exchanges the role of the indices belonging to the fields and auxiliary fields, respectively. In Fourier space, this reversal leads to the replacement $\omega \to -\omega$ for the external frequencies, as in Eq. (4.81). Of course, Eq. (4.87) is just a special case of this causality requirement for $\widetilde{N} = 0$. Thus, we can write down the *Feynman rules* in Fourier space for the lth order term contributing to the *vertex function* $\Gamma^{(\widetilde{N},N)}$.

(i) Draw all topologically different, connected *one-particle irreducible graphs* with $\widetilde{N}$ outgoing and N incoming directed lines connecting l relaxation vertices $\propto u$. Do not allow closed response loops.

(ii) Attach wavevectors q_i, frequencies ω_i, and vector indices α_i to all lines, obeying 'momentum and energy' conservation at each vertex.

(iii) Analytically, each directed line corresponds to a response propagator $G_0(-q, -\omega) = [i\omega + Dq^a(r + q^2)]^{-1}$, the two-point vertex to the noise

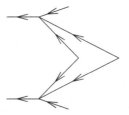

Fig. 4.18 One-loop Feynman diagram for the vertex function $\Gamma^{(2,2)}$.

strength $2Dq^a$, and the four-point relaxation vertex to $-Dq^a u/6$, where q and ω denote the outgoing momentum and frequency, respectively. Closed loops imply integrals over internal wavevectors and frequencies, as well as sums over internal vector indices.

(iv) Multiply by -1 and the combinatorial factor counting all possible ways of connecting the propagators, l relaxation vertices, and k two-point vertices leading to topologically identical graphs, including a factor $1/l!$ originating in the expansion of $\exp(-\mathcal{A}_{\mathrm{int}}[\widetilde{S}, S])$, and similarly a factor $1/k!$ for the noise vertices.

For example, translational and rotational invariance imply that the four-point vertex function $\Gamma^{(1,3)}$ is of the form

$$\Gamma^{(1,3)}_{\alpha;\beta\gamma\delta}(\underline{k}_\alpha;\underline{k}_\beta,\underline{k}_\gamma,\underline{k}_\delta) = \Gamma^{(1,3)}(\underline{k}_\alpha;\underline{k}_\beta,\underline{k}_\gamma,\underline{k}_\delta)\,\delta(\underline{k}_\alpha + \underline{k}_\beta + \underline{k}_\gamma + \underline{k}_\delta)\,F^{\alpha\beta\gamma\delta}\,.$$

(4.100)

Its tree-level contribution reads $-(-Dq^a_\alpha u/6) \cdot 3 \cdot 2 = Dq^a_\alpha u$, and the one-loop term, see Fig. 4.19(b) below, taken at symmetric external wavevectors $\underline{k}_\beta = \underline{k}_\gamma = \underline{k}_\delta = \underline{k}/2$, becomes

$$-\frac{1}{2} D \left(\frac{3}{2}q\right)^a \left(-\frac{u}{6}\right)^2 2 \cdot (n - 1 + 3 \cdot 3) \cdot 2 \cdot 2 \cdot 3$$

$$\times \int_k \int_{\omega'} \frac{D(q - k)^a}{i(\omega - \omega') + \Delta(q - k)} \frac{2Dk^a}{\omega'^2 + [Dk^a(r + k^2)]^2}$$

$$= -Dq^a \frac{n + 8}{3} u^2 \int_k \frac{1}{r + k^2} \frac{D(q - k)^a}{i\omega + \Delta(k) + \Delta(q - k)}\,.$$

(4.101)

Symmetrizing with respect to the internal wavevectors then leads to

$$\Gamma^{(1,3)}(-3\underline{k}/2; \underline{k}/2, \underline{k}/2, \underline{k}/2) = D \left(\frac{3}{2}q\right)^a u \left[1 - \frac{n + 8}{6} u\right.$$

$$\left. \times \int_k \frac{1}{r + k^2} \frac{1}{r + (q - k)^2} \left(1 - \frac{i\omega}{i\omega + \Delta(k) + \Delta(q - k)}\right) + O(u^2)\right].$$

(4.102)

In a similar manner, the one-loop contribution to the vertex function $\Gamma^{(2,2)}$, taken at symmetric external wavevectors and frequencies, reads

$$\Gamma^{(2,2)}(-\underline{k}, -\underline{k}; \underline{k}, \underline{k}) = -2Dq^{2a} \frac{n+8}{9} u^2 \int_k \frac{1}{k^a (r + k^2)^3} + O(u^3), \qquad (4.103)$$

see Problem 4.8, and is thus independent of the external frequency ω.

4.4.4 Perturbation theory in the time domain

So far, we have evaluated the Feynman graphs in Fourier space, based on the propagators $G_0(q, \omega)$. Thus, for the construction of the perturbation theory, wavevectors and frequencies followed largely analogous rules. The special role of time as reflected in the causality properties of the cumulants and vertex functions was taken into account via the representation of the response propagators through *directed* lines. Furthermore, we carried out the internal frequency integrations by means of the residue theorem, leaving us with purely momentum-dependent loop integrals.

Occasionally, however, it is useful to perform the calculations in the time domain. To this end, one merely needs to modify the above rules (ii) and (iii), and replace them with *Feynman rules* for the lth order term contributing to the *vertex function* $\Gamma^{(\tilde{N},N)}$ in the *time domain*.

(ii') Attach wavevectors q_i and vector indices α_i to all directed lines, such that 'momentum' conservation holds at each vertex. Label all vertices with time indices t_i.

(iii') Analytically, each directed line that connects the vertices labeled with t_i and t_j corresponds to a response propagator $G_0(q, t_i - t_j) = \Theta(t_i - t_j)$ $e^{-Dq^a(r+q^2)(t_i-t_j)}$, the two-point vertex to the noise strength $2Dq^a$, and the four-point relaxation vertex to $-Dq^a u/6$, where q denotes the outgoing momentum. Closed loops imply integrations over internal wavevectors and times, subject to the causality constraints imposed by the propagators, as well as summations over internal vector indices.

Temporal Fourier transforms of course then yield the same expressions as those obtained directly from the Feynman rules in frequency space.

Let us illustrate these Feynman rules in the time domain with the evaluation of the one-loop diagrams for the vertex functions $\Gamma^{(1,1)}$ and $\Gamma^{(1,3)}$, see Fig. 4.19, which is labeled according to the prescription (ii'). In the loop integral of Fig. 4.19(a) for $\Gamma^{(1,1)}$, causality enforces that $t' \leq t$. Consequently, according to rule (iii') it is given by

$$\int_k 2Dk^a \int_{-\infty}^t e^{-2Dk^a(r+k^2)(t-t')} \, dt' = \int_k \frac{1}{r + k^2},$$

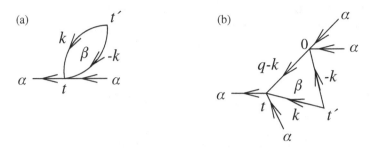

Fig. 4.19 One-loop Feynman diagrams for the vertex functions (a) $\Gamma^{(1,1)}$ and (b) $\Gamma^{(1,3)}$ in the time domain.

as in the first line of Eq. (4.82). Notice that the label t drops out, and we could have set it to zero right away. The 'Hartree' loop in Fig. 4.19(a) thus carries neither external momentum nor frequency.

The one-loop integral for the four-point vertex function $\Gamma^{(1,3)}$, depicted in Fig. 4.19(b), similarly becomes

$$\int_k D(q-k)^a \, 2Dk^a \int_{-\infty}^{0} e^{-\Delta(k)(t-2t')} \, dt' \, e^{-\Delta(q-k)t} \, \Theta(t)$$

$$= \int_k \frac{D(q-k)^a}{r+k^2} \, e^{-[\Delta(k)+\Delta(q-k)]t} \, \Theta(t) \,.$$

After performing the Fourier transform (with $\omega \to -\omega$), we find for this first-order contribution

$$\int_k \frac{D(q-k)^a}{r+k^2} \int_0^\infty e^{-[i\omega+\Delta(k)+\Delta(q-k)]t} \, dt = \int_k \frac{1}{r+k^2} \frac{D(q-k)^a}{i\omega+\Delta(k)+\Delta(q-k)} \,,$$

precisely as in Eq. (4.101).

With this excursion into the time domain, we end our inevitably rather technical exposition of dynamic perturbation theory. We had started with the Gaussian relaxational models A and B, and in Section 3.2 first constructed the lowest-order terms in the perturbation expansion by simply iterating the non-linear Langevin equation of motion. Yet a more tractable representation has now been achieved by means of the mapping of this stochastic process onto the response functional, which formally resembles the path-integral representation of a quantum field theory or equilibrium statistical mechanics, albeit with novel auxiliary response fields. Subsequently we gained deeper insights into the structure of the perturbation series by considering the N-point cumulants (connected Feynman diagrams), and finally the vertex functions (1PI graphs). The latter actually contain all the essential information, and can be viewed as renormalized counterparts of the terms in the original 'bare' response functional. Therefore, all physical quantities such as dynamic response or correlation functions can be expressed in terms of the vertex functions, which naturally enter as the fundamental entities for the renormalization program to be discussed in the following Chapter 5.

Furthermore, although we have developed the theory for the almost simplest case, namely the purely relaxational $O(n)$-symmetric models in the high-temperature phase, the generalization of this formalism to other systems of coupled non-linear Langevin equations is straightforward (and will in fact be discussed explicitly for a host of examples in subsequent chapters of this text). The Gaussian part of the response functional generally defines the propagators of the model under consideration, while the non-linearities are encoded in the structure of the vertices. The perturbation series for the vertex functions is then organized in terms of a loop expansion for the ensuing 1PI Feynman diagrams.

4.5 *Appendix:* discretization, Jacobian, and response loops

In Section 4.1, we employed the Langevin equation (4.1) in order to perform a formal variable transformation from the Gaussian noise distribution (4.3) to a probability distribution for the stochastic fields ψ^α, see Eq. (4.4); and then again when we utilized the set of delta function constraints in Eq. (4.7). This non-linear variable transformation is of course accompanied by a functional determinant $\mathcal{D}[\zeta]/\mathcal{D}[\psi]$. In order to compute this Jacobian, we discretize in time $t_l = t_0 + l\,\tau$, $t_M = t_f$ as in Section 2.4.1, set $\psi_l^\alpha = \psi^\alpha(t_l)$, $\zeta_l^\alpha = \zeta^\alpha(t_l)$, whence the equation of motion (4.1) becomes a difference equation. For now let us choose an arbitrary linear splitting for the generalized forces $F_l = F[\psi(t_l)]$, namely with $0 \le \kappa \le 1$:

$$\frac{\psi_{l+1}^\alpha - \psi_l^\alpha}{\tau} = \kappa\, F_{l+1}^\alpha + (1-\kappa)\, F_l^\alpha + \zeta_l^\alpha . \tag{4.104}$$

With this choice, the Jacobian matrix for the transformation from the variables $\{\zeta_l^\alpha\}$ to $\{\psi_{l+1}^\alpha\}$ has entries only in the diagonal, $1/\tau - \kappa\partial F_l^\alpha/\partial\psi_l^\alpha$, and the lower sub-diagonal, $-1/\tau - (1-\kappa)\partial F_l^\alpha/\partial\psi_l^\alpha$. The functional determinant, at each point x and for all components α, is thus given by the product of the diagonal matrix elements,

$$\frac{1}{\tau^M} \prod_l \left(1 - \kappa\tau\,\frac{\partial F_l^\alpha}{\partial\psi_l^\alpha}\right) \approx \frac{1}{\tau^M}\,\exp\left(-\kappa\sum_l \tau\,\frac{\partial F_l^\alpha}{\partial\psi_l^\alpha}\right).$$

The factor $1/\tau^M$ is then readily absorbed in the functional measure. Taking the limit $\tau \to 0$, we arrive at the additional term $\kappa \int d^d x \int dt \sum_\alpha \delta F^\alpha[\psi]/\delta\psi^\alpha(x, t)$ both in the Onsager–Machlup functional (4.5) and in the Janssen–De Dominicis functional (4.10). Notice that the splitting in Eq. (4.104) corresponds to setting $\Theta(0) = \kappa$ for the Heaviside step function. The Itô representation for Langevin equations uses the forward discretization $\kappa = 0$, see Section 2.4, whence the Jacobian contribution to the action vanishes.

Let us now investigate the subset of diagrams stemming from the non-linear contribution in $-\int dt \sum_\alpha \tilde{\psi}^\alpha F^\alpha[\psi]$ of the response functional, which contain closed response loops, see e.g. Fig. 4.2(a) and (b). These must involve a contraction of an internal field ψ^β with the response field $\tilde{\psi}^\alpha$, wherefrom we see that the response loop terms can be formally written as an effective vertex

$$-\int d^d x \int dt \sum_{\alpha,\beta} \langle \tilde{\psi}^\alpha \psi^\beta \rangle \frac{\delta F^\alpha[\psi]}{\delta \psi^\beta(x,t)} = -\int d^d x \int dt \sum_\alpha \Theta(0) \frac{\delta F^\alpha[\psi]}{\delta \psi^\alpha(x,t)} ,$$

where we have used Eq. (3.29). Consequently the terms from those Feynman diagrams that contain closed response loops are exactly canceled by the contributions from the functional determinant. As we have just established, this holds true for any choice of the discretization splitting, provided $\Theta(0) = \kappa$ is set consistently. This fact also ensures causality, for one could easily construct acausal Feynman graphs if closed response loops were allowed. At last, we note that technically any application of the differential calculus in the dynamic functionals requires in fact a symmetric discretization, i.e., $\kappa = 1/2$.[6] Nevertheless, no terms from closed response loops will eventually emerge, and we shall therefore adopt the Itô representation with $\kappa = 0$ throughout this text, and omit closed response loops entirely from the start.

Problems

4.1 *Path-integral normalization factor in the Gaussian approximation*
   Calculate the normalization constant appearing in Eq. (4.9) for the Gaussian relaxational models, namely a stochastic process of duration $t_f - t_0$ in a volume V:

$$C_0 = \exp\left[-\frac{n}{2} V(t_f - t_0) \int \frac{d^d q}{(2\pi)^d} Dq^a(r + q^2) \right] .$$

   Hint: utilize the Gaussian integrals in Section 4.1, and recall the free energy calculation in Section 1.2. A more rigorous computation utilizes a *symmetric* time discretization (i.e., $\kappa = 1/2$ in the notation of Section 4.5).

4.2 *Dynamic susceptibility for models with reversible mode couplings*
   Show that the dynamic susceptibilities for models with reversible mode-coupling terms, see Eq. (3.59), are related to the auxiliary fields in the response

---

[6] An adequate discussion of the ensuing subtleties can be found in Janssen (1992); see also Honkonen (2011).

functional through

$$\chi^{\alpha\beta}(x - x', t - t') = \langle \psi^{\alpha}(x, t) D^{\beta}(i\nabla)^{\alpha\beta} \widetilde{\psi}^{\beta}(x', t') \rangle$$

$$- \left\langle \psi^{\alpha}(x, t) \int d^d \bar{x} \sum_{\gamma} \left\{ \psi^{\beta}(x', t'), \psi^{\gamma}(\bar{x}, t') \right\} \widetilde{\psi}^{\gamma}(\bar{x}, t') \right\rangle.$$

4.3 *Four-point correlation functions to first order in u*
For the four-point function $\langle S^{\alpha} S^{\beta} \widetilde{S}^{\gamma} \widetilde{S}^{\delta} \rangle$, confirm Eqs. (4.55) and (4.57) by explicit evaluation of the allowed contractions, and compare with Fig. 4.6. Draw the Feynman graphs for the zeroth- and first-order contributions to $\langle S^{\alpha} S^{\beta} S^{\gamma} \widetilde{S}^{\delta} \rangle$ and $\langle S^{\alpha} S^{\beta} S^{\gamma} S^{\delta} \rangle$, and write down the corresponding analytic expressions to $O(u)$.

4.4 *Four-point cumulants*
Confirm Eqs. (4.60), (4.56), and (4.61) directly using the definition of cumulants (4.58). Derive the more general expressions in the ordered phase as well, where $\langle S^{\alpha} \rangle \neq 0$.

4.5 *Connected four-point functions to $O(u^2)$: one-loop diagrams*
Draw all possible Feynman diagrams to $O(u^2)$ for the connected four-point functions $\langle S^{\alpha} S^{\beta} \widetilde{S}^{\gamma} \widetilde{S}^{\delta} \rangle_c$, $\langle S^{\alpha} S^{\beta} S^{\gamma} \widetilde{S}^{\delta} \rangle_c$, and $\langle S^{\alpha} S^{\beta} S^{\gamma} S^{\delta} \rangle_c$. Convince yourself that the second-order terms contain a single closed loop, and that the one-particle irreducible graphs to this order are given by the one-loop diagrams for the vertex functions $\Gamma^{(1,1)}$, $\Gamma^{(1,3)}$, and $\Gamma^{(2,2)}$.

4.6 *Two-point vertex functions to two-loop order*
Evaluate the two-loop diagrams for the vertex functions $\Gamma^{(1,1)}(q, \omega)$ and $\Gamma^{(2,0)}(q, \omega)$ depicted in Fig. 4.13 and Fig. 4.15, respectively. Carry out the internal frequency integrals, confirm Eqs. (4.82) and (4.91), and check the fluctuation-dissipation theorem (4.96) to this order. Also, draw the corresponding Feynman graphs using correlation propagators instead of the two-point noise vertices.

4.7 *Four-point vertex functions and cumulants*
Derive the relations (4.97) and (4.98) expressing the four-point cumulants $\langle S^{\alpha} \widetilde{S}^{\beta} \widetilde{S}^{\gamma} \widetilde{S}^{\delta} \rangle_c$ and $\langle S^{\alpha} S^{\beta} \widetilde{S}^{\gamma} \widetilde{S}^{\delta} \rangle_c$ in terms of vertex functions and (full) propagators, see Figs. (4.16) and (4.17). Construct the corresponding diagrams for the cumulants $\langle S^{\alpha} S^{\beta} S^{\gamma} \widetilde{S}^{\delta} \rangle_c$ and $\langle S^{\alpha} S^{\beta} S^{\gamma} S^{\delta} \rangle_c$.

4.8 *One-loop diagram for $\Gamma^{(2,2)}$ in the time domain*
Compute the one-loop integral corresponding to Fig. 4.18 for the vertex function $\Gamma^{(2,2)}$ in the time domain. Therefrom, Eq. (4.103) can be obtained via Fourier transformation.

References

Amit, D. J., 1984, *Field Theory, the Renormalization Group, and Critical Phenomena*, Singapore: World Scientific, chapters 4, 5.

Bausch, R., H. K. Janssen, and H. Wagner, 1976, Renormalized field theory of critical dynamics, *Z. Phys. B* **24**, 113–127.

Cardy, J., 1996, *Scaling and Renormalization in Statistical Physics*, Cambridge: Cambridge University Press, chapters 5, 10.

De Dominicis, C., 1976, Techniques de renormalisation de la théorie des champs et dynamique des phénomènes critiques, *J. Phys. (France)* Colloq. C1, C247–C253.

Honkonen, J., 2011, Ito and Stratonovich calculuses in stochastic field theory, preprint arXiv: 1102.1581.

Itzykson, C. and J. M. Drouffe, 1989, *Statistical Field Theory*, Vol. I, Cambridge: Cambridge University Press, chapter 5.

Janssen, H. K., 1976, On a Lagrangean for classical field dynamics and renormalization group calculations of dynamical critical properties, *Z. Phys. B* **23**, 377–380.

Janssen, H. K., 1979, Field-theoretic methods applied to critical dynamics, in: *Dynamical Critical Phenomena and Related Topics*, ed. C.P. Enz, Lecture Notes in Physics, Vol. **104**, Heidelberg: Springer, 26–47.

Janssen, H. K., 1992, On the renormalized field theory of nonlinear critical relaxation, in: *From Phase Transitions to Chaos*, eds. G. Györgyi, I. Kondor, L. Sasvári, and T. Tél, Singapore: World Scientific, 68–91.

Kamenev, A., 2011, *Field Theory of Non-equilibrium Systems*, Cambridge: Cambridge University Press, chapters 4, 8.

Le Bellac, M., 1991, *Quantum and Statistical Field Theory*, Oxford: Oxford University Press.

Ma, S.-k., 1976, *Modern Theory of Critical Phenomena*, Reading: Benjamin–Cummings, chapters 9, 14.

Martin, P. C., E. D. Siggia, and H. A. Rose, 1973, Statistical dynamics of classical systems, *Phys. Rev. A* **8**, 423–437.

McComb, W. D., 2004, *Renormalization Methods: a Guide for Beginners*, Oxford: Oxford University Press, chapter 10.

Parisi, G., 1988, *Statistical Field Theory*, Redwood City: Addison–Wesley, chapters 5, 6.

Ramond, P., 1981, *Field Theory – a Modern Primer*, Reading: Benjamin–Cummings, chapter 4.

Vasil'ev, A. N., 2004, *The Field Theoretic Renormalization Group in Critical Behavior Theory and Stochastic Dynamics*, Boca Raton: Chapman & Hall/CRC, chapters 1–5.

Zinn-Justin, J., 1993, *Quantum Field Theory and Critical Phenomena*, Oxford: Clarendon Press, chapter 6.

Further reading

De Dominicis, C., E. Brézin, and J. Zinn-Justin, 1975, Field-theoretic techniques and critical dynamics. I. Ginzburg-Landau stochastic models without energy conservation, *Phys. Rev. B* **12**, 4945–4953.

Folk, R. and G. Moser, 2006, Critical dynamics: a field theoretical approach, *J. Phys. A: Math. Gen.* **39**, R207–R313.

Halperin, B. I., P. C. Hohenberg, and S.-k. Ma, 1972, Calculation of dynamic critical properties using Wilson's expansion methods, *Phys. Rev. Lett.* **29**, 1548–1551.

Hohenberg, P. C. and B. I. Halperin, 1977, Theory of dynamic critical phenomena, *Rev. Mod. Phys.* **49**, 435–479.

Täuber, U. C., 2007, Field theory approaches to nonequilibrium dynamics, in: *Ageing and the Glass Transition*, eds. M. Henkel, M. Pleimling, and R. Sanctuary, Lecture Notes in Physics **716**, Berlin: Springer, chapter 7, 295–348.

5

Dynamic renormalization group

In this central and essential chapter, we develop the dynamic renormalization group approach to time-dependent critical phenomena. Again, we base our exposition on the simple $O(n)$-symmetric relaxational models A and B; the generalization to other dynamical systems is straightforward. We begin with an analysis of the infrared and ultraviolet singularities appearing in the dynamic perturbation expansion. Although we are ultimately interested in the infrared critical region, we first take care of the ultraviolet divergences. Below and at the critical dimension $d_c = 4$, only a finite set of Feynman diagrams carries ultraviolet singularities, which we evaluate by means of the dimensional regularization prescription, and then eliminate via multiplicative as well as additive renormalization (the latter takes into account the fluctuation-induced shift of the critical temperature). The renormalization group equation then permits us to explore the ensuing scaling behavior of the correlation and vertex functions of the renormalized theory upon varying the arbitrary renormalization scale. Of fundamental importance is the identification of an infrared-stable renormalization group fixed point, which describes scale invariance and hence allows the derivation of the critical power laws in the infrared limit from the renormalization constants determined in the ultraviolet regime. This program is explicitly carried through for the relaxational models A and B. For model B, we derive a scaling relation connecting the dynamic exponent to the Fisher exponent η. The critical exponents ν, η, and z are computed to first non-trivial order in an ϵ expansion around the upper critical dimension $d_c = 4$. In four dimensions, logarithmic corrections to the mean-field exponents, valid for $d > 4$, emerge. In addition, we discuss the scaling behavior of model A in the low-temperature phase. A spontaneous breaking of the continuous $O(n)$-symmetry ($n \geq 2$) leads to the appearance of $n - 1$ massless Goldstone modes which cause power law singularities in the entire ordered phase.

5.1 Primitive divergences and scaling dimensions

In Chapter 4, we have developed a systematic perturbational approach to dynamic phenomena described by non-linear Langevin equations. However, near a critical point, this perturbation expansion is not in general directly applicable, as it is plagued by severe *infrared* (IR) singularities. Above the critical dimension, in addition *ultraviolet* (UV) divergences emerge. As we shall ultimately employ these UV poles and specifically their residues for the determination of the asymptotic scaling behavior in the critical region, we first need to characterize and classify these various divergences systematically.[1]

5.1.1 IR and UV divergences, critical dimension

For example, consider the four-point vertex function $\Gamma^{(1,3)}$ for the relaxational models A and B at vanishing external wavevector and frequency, which we employ to define the 'renormalized' non-linear coupling u, with fluctuation contributions taken into account. For $r \to 0$ the one-loop diagram, Eq. (4.102), is proportional to $\int_k k^{-4} = S_d \int_0^\infty k^{d-5} dk$, which diverges in $d \leq 4$ dimensions. Any finite external wavenumber q or 'mass' r of course removes this infrared singularity, which means that outside the critical regime simple perturbation theory can be applied.

More precisely, for $q = 0$ and $\omega = 0$ the integral in (4.102) reads

$$u \int \frac{d^d k}{(2\pi)^d} \frac{1}{(r + k^2)^2} = \frac{1}{2^{d-1} \pi^{d/2} \Gamma(d/2)} u \int_0^\infty \frac{k^{d-1}}{(r + k^2)^2} dk$$

$$\propto u r^{-2+d/2} \int_0^\infty \frac{x^{d-1}}{(1 + x^2)^2} dx \qquad (d < 4). \qquad (5.1)$$

Therefore, provided the effective coupling $ur^{(d-4)/2}$ is small, the perturbation series should lead to meaningful physical results. While this prefactor is well-behaved above four dimensions, and in fact vanishes as $r \to 0$, we see that for $d < 4$ the fluctuation corrections render the effective non-linear coupling divergent in the critical regime, and the perturbation expansion breaks down. In this situation, one has to expect that the fluctuations affect the physical properties in a *non-analytical* manner, i.e., modify the mean-field power laws near the critical point. Already by means of this very simple consideration, we see that $d_c = 4$ plays a special role. Above this *upper critical dimension*, thermal fluctuations should not alter the simple power laws obtained from mean-field theory (yet certainly fluctuation corrections will be present for the critical amplitudes and scaling functions). In low dimensions $d < d_c$, the critical fluctuations are sufficiently strong to quantitatively

---

[1] For more extended systematic treatments of divergences in field theory, we refer to the standard literature, e.g., Amit (1984), Itzykson and Drouffe (1989), and Zinn-Justin (1993).

affect the power laws near the critical point, and sometimes even qualitatively change the character of the phase diagram. In the perturbation expansion, this is reflected through the appearance of severe infrared divergences.

On the other hand, the integral (5.1) is proportional to Euler's gamma function $\Gamma(2 - d/2)$, and thus develops a singularity at four dimensions. The origin of this is clearly to be sought in the ultraviolet behavior of the fluctuation integral. To this end, we formally introduce an upper wavenumber cutoff Λ. Notice that in condensed matter systems, such a cutoff is usually provided physically by the Brillouin zone boundary originating in the underlying lattice periodicity, i.e. $\Lambda \sim 2\pi/a_0$, where a_0 is the lattice constant. In and above four dimensions, respectively, as a function of Λ the integral (5.1) behaves as

$$\int_0^\Lambda \frac{k^{d-1}}{(r+k^2)^2} \, dk \sim \begin{cases} \ln(\Lambda^2/r), & d = 4, \\ \Lambda^{d-4}, & d > 4. \end{cases} \tag{5.2}$$

For $d \geq 4$ dimensions, the integral diverges as $\Lambda \to \infty$, and we find that the simple pole of the associated gamma function in $d = 4$ stems from the logarithmic divergence $\sim \ln(\Lambda^2/r)$. The dimension $d_c = 4$ thus plays a special role also for the ultraviolet divergences: while the loop integral is IR-convergent and UV-divergent for $d > 4$, and vice versa for $d < 4$, both types of singularity emerge coincidentally in four dimensions, appearing as a logarithmic dependence on the ratio of the mass r and the cutoff Λ^2. We shall see later that this is in fact the origin of *logarithmic corrections* to the mean-field power laws which appear precisely at d_c.

Instead of explicitly computing the one-loop corrections, the upper critical dimension may be inferred more easily through a direct scaling analysis (see also Section 1.4). To this end, we consider the response functional for the relaxational models (4.14) and (4.15). Let us then measure lengths in units of inverse wavenumbers, and, for simplicity, time in such units that the relaxation constant D is dimensionless. In terms of an arbitrary momentum scale μ, we write this symbolically as

$$[x] = \mu^{-1}, \quad [q] = \mu, \quad [D] = \mu^0,$$
$$[t] = \mu^{-2-a}, \quad [\omega] = \mu^{2+a}, \quad [A] = \mu^0, \tag{5.3}$$

since the dynamic functional itself enters in the exponential of the statistical weight and therefore has to be dimensionless. Consequently, the *scaling dimensions* of the fields become

$$[\widetilde{S}^\alpha(x, t)] = \mu^{1+d/2}, \quad [S^\alpha(x, t)] = \mu^{-1+d/2},$$
$$[\widetilde{S}^\alpha(q, \omega)] = \mu^{-1-a-d/2}, \quad [S^\alpha(q, \omega)] = \mu^{-3-a-d/2}. \tag{5.4}$$

For the distance r from the critical point and the non-linear coupling u one obtains

$$[r] = \mu^2 , \quad [u] = \mu^{4-d} , \tag{5.5}$$

and we notice again that the dimensionless effective coupling to be generated from these quantities is $ur^{(d-4)/2}$. The non-linearity u thus becomes dimensionless (*marginal*) at the upper critical dimension; it is said to be *relevant* for $d < d_c$, since the corresponding effective coupling grows as $r \to 0$. Another clearly relevant coupling is r, with the positive scaling dimension 2, as is the external field, $[h^\alpha(x, t)] = \mu^{1+d/2}$. Had we allowed for a term $\propto v[S^\alpha(x)^2]^3$ in the Hamiltonian (4.12), we would have found $[v] = \mu^{6-2d}$, which becomes marginal in three dimensions. Near $d_c = 4$, the scaling dimension of v is negative, and the effective dimensionless coupling vr^{d-3} vanishes as $r \to 0$. The higher-order non-linearity v is thus said to be *irrelevant* for the critical power laws of our model (as is u above d_c, albeit dangerously), and can be discarded for the study of asymptotic critical properties. This is also true for higher powers in the gradient expansion $\propto (\nabla^k S^\alpha)^2$ with $k > 1$, which have scaling dimension $-2(k - 1)$; see also the discussion in Section 1.4.

Notice, however, that in two dimensions $[r] = [u] = [v] = \mu^2$ are all equally relevant, as are in fact all higher-order even couplings, for according to Eq. (5.4) the order parameter fields $S^\alpha(x, t)$ become dimensionless at $d = 2$. For the $O(n)$-symmetric Hamiltonian with $n \geq 2$ and purely short-range interactions, this is related to the fact that, as we shall discuss in Section 5.4, long-range order (with a homogeneous order parameter) is not possible at finite temperatures in less than two dimensions (for the XY model with $n = 2$, there exists however a Berezinskii–Kosterlitz–Thouless vortex binding/unbinding transition). One therefore refers to $d_{lc} = 2$ as the *lower critical dimension* for a continuous phase transition in systems with Heisenberg rotational symmetry. Near two dimensions, a power expansion of the free energy density in terms of $S^\alpha(x, t)^2$ is then inappropriate, and the theory needs to be based on a different effective Hamiltonian (e.g., the non-linear sigma model, see Section 7.4).

For the study of critical phenomena, we are of course ultimately interested in the infrared behavior of our theory, and not in the (usually unphysical) ultraviolet divergences. Yet the correct infrared power laws can hardly be naively extracted from a divergent perturbation expansion. On the other hand, in the vicinity of a critical point we expect *scale invariance*, and we should in principle be able to obtain useful results in the IR regime if we properly understand the scaling behavior of the theory in the ultraviolet, i.e., as $\Lambda \to \infty$. In fact, this becomes manifest in four dimensions, where according to Eq. (5.5) the coupling u is dimensionless, and the theory therefore naturally scale-invariant. Indeed, at the critical dimension, both the UV- and IR-divergences appear combined as logarithmic singularities $\sim \ln(\Lambda^2/r)$, compare with Eq. (5.2), and can be extracted *either* by taking the limit $r \to 0$ *or* by letting the ultraviolet cutoff $\Lambda \to \infty$.

Indeed, our strategy will be to absorb the UV singularities into appropriately defined *renormalized* parameters of our model (Section 5.2), and study their dependence under scale transformations. Provided scale invariance holds, we can then therefrom infer the correct infrared behavior by means of the renormalization group equation (Section 5.3), which also provides a link between the critical regime and the non-critical region, where the perturbation expansion is applicable. A formally small expansion parameter for the perturbation series is provided by the deviation $\epsilon = d_c - d$ from the upper critical dimension. Thus, we now proceed to more systematically analyze the UV singularities in the perturbation expansion of the relaxational models A and B.

5.1.2 Power counting and renormalizability

We define the *primitive* or *superficial degree of divergence* $\delta^l_{\tilde{N},N}$ for a given Feynman diagram in Fourier space to order l for the vertex function $\Gamma^{(\tilde{N},N)}$ simply by counting the powers of the momenta in the numerator and denominator, respectively. For example, for the integral (5.1) we have $\delta^1_{1,3} = d - 4$, where the d stems from the factor $k^{d-1}dk$, and -4 comes from the denominator $(r + k^2)^2$. A positive value of $\delta^l_{\tilde{N},N}$ implies a superficial ultraviolet divergence $\sim \Lambda^{\delta^l_{\tilde{N},N}}$; $\delta^l_{\tilde{N},N} = 0$ means that the corresponding integral is logarithmically divergent as $\Lambda \to \infty$, whereas UV-convergent integrals are characterized by negative primitive degrees of divergence. Notice that the actual degree of divergence may well differ from the primitive one, since, e.g., a superficially convergent diagram may actually contain a divergent subgraph. However, as these UV singularities will be taken care of later by the renormalization of the lower loop-order diagrams, we need only consider the primitively divergent integrals.

Let us first investigate the two-point vertex function $\Gamma^{(1,1)}(q, \omega)$. According to Eq. (4.82), we find the values $\delta^1_{1,1} = d - 2$ and $\delta^2_{1,1} = 2d - 6$ for the one- and two-loop contributions to $\partial \Gamma^{(1,1)}(q, 0)/\partial q^a|_{q=0}$, respectively. In four dimensions, both the first- and second-order terms are thus quadratically divergent; yet, as we shall see below, this singularity can be absorbed into a fluctuation-induced (downward) shift of the transition temperature. The modifications through fluctuations of, say, the relaxation constant D etc. are rather given by the derivatives $\partial \Gamma^{(1,1)}(q, 0)/\partial q^{2+a}|_{q=0}$, $\partial [\partial \Gamma^{(1,1)}(q, 0)/\partial q^a]_{q=0}/\partial r$, and $\partial [\partial \Gamma^{(1,1)}(q, \omega)/\partial q^a]_{q=0}/\partial \omega|_{\omega=0}$. This further reduces the degree of divergence to $l(d - 4)$ at order l, which implies logarithmic UV singularities at $d_c = 4$, just as for $\Gamma^{(1,3)}$ which renormalizes the non-linear coupling u. For $d > d_c = 4$, the superficial degree of divergence grows with increasing loop order, while it becomes reduced for $d < d_c$.

Next consider $\partial \Gamma^{(2,0)}(q,0)/\partial q^a$; there is no one-loop term for this vertex function, and according to Eq. (4.91) the primitive degree of divergence of the second-order contribution is $\delta_{2,0}^2 = 2(d - 4) - a$. Yet for the case of conserved order parameter dynamics (model B, $a = 2$), the prefactor vanishes as $q \to 0$, see Eq. (4.92), and the entire vertex function is actually regular to all orders in the perturbation expansion. For $a = 0$ (model A) we again find that $d_c = 4$ separates the UV-convergent and divergent regimes. In addition, we notice that the superficial degree of divergence is reduced both by increasing $\widetilde{N} + N$, and in the transformation $(\widetilde{N}, N) \to (\widetilde{N} + 1, N - 1)$ (with $N \geq 1$). For example, for the four-point vertex function $\Gamma^{(2,2)}$ we find from Eq. (4.103) $\delta_{22}^1 = d - 6 - a$, as compared to $\delta_{13}^1 = d - 4$.

In fact, one may readily derive (Problem 5.1) the general expression

$$\delta_{\widetilde{N},N}^l = d_{\widetilde{N},N} - a\,\widetilde{N} - l\,(4 - d) \tag{5.6}$$

for the primitive degree of divergence for the lth order contribution to the reduced vertex function $\Gamma^{(\widetilde{N},N)}$ (with the overall delta functions stemming from translational invariance in space and time separated off), which has the scaling dimension

$$d_{\widetilde{N},N} = -\frac{d}{2}\left(\widetilde{N} + N - 2\right) - \widetilde{N} + N + 2 + a \,. \tag{5.7}$$

Thus, indeed $\delta_{\widetilde{N}+1,N-1}^l - \delta_{\widetilde{N},N}^l = -(2 + a)$, and in any fixed dimension only a finite number of vertex functions can be UV-divergent. Furthermore, $\delta_{\widetilde{N},N}^{l+1} - \delta_{\widetilde{N},N}^l = -(4 - d)$, which means that for $d < d_c = 4$ in addition only a limited number of loop diagrams carry UV divergences. These singularities can therefore be absorbed into a *finite* number of appropriately redefined parameters, and the field theory is said to be *(super-)renormalizable*. This is still true in four dimensions, where the primitive degree of divergence is independent of the loop order. However, for $d > d_c$ the loop integrals become increasingly singular with growing l, and additional renormalized parameters would be required at each subsequent level in the perturbation expansion. Hence the renormalization program cannot be accomplished with a finite number of renormalizations to all orders in perturbation theory, and the model is *non-renormalizable*. Table 5.1 summarizes the different roles of the upper and lower critical dimensions, both mathematically in the perturbation theory formalism and regarding their physical implications, here specifically for the $O(n)$-symmetric relaxational models A and B.

For the perturbative treatment of the renormalizable situation, it suffices to study the worst case at the upper critical dimension. For the relaxational models, the two-point vertex function $\partial \Gamma^{(1,1)}(q,0)/\partial q^a|_{q=0}$ is quadratically divergent near $d_c = 4$,

Table 5.1 *Mathematical and physical distinctions of the dimensional regimes $d < d_c$, $d = d_c$, and $d > d_c$, for the $O(n)$-symmetric purely relaxational models A and B (or static Φ^4 field theory).*

dimension interval	perturbation series	model A/B or Φ^4 field theory	critical behavior
$d \leq d_{lc} = 2$	IR-singular UV-convergent	ill-defined u relevant	no long-range order ($n \geq 2$)
$2 < d < 4$	IR-singular UV-convergent	super-renormalizable u relevant	non-classical exponents
$d = d_c = 4$	logarithmic IR-/ UV-divergence	renormalizable u marginal	logarithmic corrections
$d > 4$	IR-regular UV-divergent	non-renormalizable u irrelevant	mean-field exponents

whereas its higher derivatives $\partial \Gamma^{(1,1)}(q, 0)/\partial q^{2+a}|_{q=0}$, $\partial [\partial \Gamma^{(1,1)}(q, 0)/\partial q^{a}]_{q=0}/\partial r$, and $\partial [\partial \Gamma^{(1,1)}(q, \omega)/\partial q^{a}]_{q=0}/\partial \omega|_{\omega=0}$, as well as the four-point vertex function $\partial \Gamma^{(1,3)}(-3q/2, 0; q/2, 0; q/2, 0; q/2, 0)/\partial q^{a}|_{q=0}$, and, for $a = 0$, $\Gamma^{(2,0)}(0, 0)$ all diverge logarithmically as $\Lambda \to \infty$. All higher vertex functions and further derivatives are already regular. In addition to the T_c shift, which takes care of the quadratic divergence, we therefore require five (four in the case $a = 2$) renormalized quantities in all, which we may choose as renormalized counterparts of the fields $\tilde{S}^{\alpha}$ and S^{α}, and the parameters D, r, and u.

5.2 Renormalization via dimensional regularization

We now proceed to perform this renormalization program explicitly for the relaxational models. First, dimensional regularization is introduced as a useful means to define the values for loop integrals in non-integer real dimensions $d \leq d_c$. Next, we absorb quadratic ultraviolet divergences at d_c into a fluctuation-induced shift of the critical temperature (additive renormalization). Subsequently, the remaining logarithmic singularities are multiplicatively renormalized. For models A and B, we determine the renormalization constants to lowest non-trivial order in the non-linear coupling u.

5.2.1 Dimensionally regularized loop integrals

As we have seen in the preceding section, it is useful to consider the results from field theory as functions of the (effective) dimension d in which the fluctuations take place. Mathematically, a convenient way to formally even consider non-integer values of d is analytic continuation by means of *dimensional regularization*. As a characteristic example, let us have a look at the integral

$$I_d^{(\sigma,s)}(\tau) = \int \frac{d^d k}{(2\pi)^d} \frac{k^{2\sigma}}{(\tau + k^2)^s} \,, \tag{5.8}$$

which is of the form that typically occurs in the momentum loops for the relaxational models. As the integrand here is isotropic, the angular integrals simply yield the area of the d-dimensional unit sphere $K_d = 2\pi^{d/2}/\Gamma(d/2)$, whence

$$I_d^{(\sigma,s)}(\tau) = \frac{1}{2^{d-1}\pi^{d/2}\,\Gamma(d/2)} \int_0^\infty \frac{k^{d-1+2\sigma}}{(\tau + k^2)^s} \, dk \,. \tag{5.9}$$

In this expression, we may obviously treat the spatial dimension d as an arbitrary real parameter. In fact, the integral is readily evaluated (Problem 5.2):

$$I_d^{(\sigma,s)}(\tau) = \frac{\Gamma(\sigma + d/2)\,\Gamma(s - \sigma - d/2)}{2^d \pi^{d/2}\,\Gamma(d/2)\,\Gamma(s)} \, \tau^{\sigma-s+d/2} \,, \tag{5.10}$$

and Euler's gamma functions provide the desired interpolation between integer values of d. Moreover, in dimensional regularization we even assign the value on the right-hand side of Eq. (5.10) to the integral (5.8) in parameter regimes where it is actually divergent. This corresponds to successive integrations by part until the integral becomes finite, and omitting infinite surface contributions. Notice that the UV singularity at $d = 2(s - \sigma)$ then appears as a *pole* in the gamma function $\Gamma(s - \sigma - d/2)$. Thereby we have constructed a practical tool both to identify the UV singularities of our theory (carefully keeping $\tau \neq 0$) and to isolate the associated residua which we shall require later for the computation of the scaling exponents.

At this point we caution that dimensional regularization may become a somewhat dangerous procedure when applied to a massless field theory. For $\sigma = 0$ and in the limit $\tau \to 0$ our above formula reads

$$I_d^{(0,s)}(0) = \int \frac{d^d k}{(2\pi)^d} \frac{1}{k^{2s}} = 0 \tag{5.11}$$

for $d > 2s$, where the integral is UV-divergent! This strange result may be interpreted as a formal cancelation with the infrared divergence emerging for $d < 2s$. In order to avoid such ambiguities from the mixing of UV and IR singularities, in a massless field theory one should compute the loop integrals at a fixed *non-zero* external wavevector (or frequency), which then provides a 'mass' term that ensures finite infrared behavior. As we shall be primarily interested in the relaxational models near the upper critical dimension $d_c = 4$, i.e., we shall typically encounter integrals such as $I_d^{(0,2)}(\tau)$, let us introduce an abbreviation A_d for the ensuing residuum (geometric factor) stemming from the angular integrals,

$$I_d^{(0,2)}(\tau) = \int \frac{d^d k}{(2\pi)^d} \frac{1}{(\tau + k^2)^2} = \frac{\Gamma(2 - d/2)}{2^d \pi^{d/2}} \tau^{-2+d/2}$$

$$= \frac{A_d}{4 - d} \tau^{-(4-d)/2}, \quad A_d = \frac{\Gamma(3 - d/2)}{2^{d-1} \pi^{d/2}}. \tag{5.12}$$

5.2.2 Additive renormalization: critical temperature shift

Our goal is to define *renormalized* parameters of the field theory for, here, the relaxational models A and B, by means of the appropriate vertex functions. As a first step, let us look at the static susceptibility $\chi(q, 0) = Dq^a G(q, 0) = Dq^a \Gamma^{(1,1)}(-q, 0)^{-1}$ in the limit $q \to 0$. According to Eq. (4.82), we find to two-loop order

$$\chi(0, 0)^{-1} = r + \frac{n + 2}{6} u \int_k \frac{1}{r + k^2} - \left(\frac{n + 2}{6} u\right)^2 \int_k \frac{1}{r + k^2} \int_{k'} \frac{1}{(r + k'^2)^2}$$

$$- \frac{n + 2}{18} u^2 \int_k \frac{1}{r + k^2} \int_{k'} \frac{1}{r + k'^2} \frac{1}{r + (k + k')^2} + O(u^3). \tag{5.13}$$

The location of the critical point is defined by that value of $r = r_c$, for which the static susceptibility diverges. As $\chi(0, 0)^{-1} > 0$ at the mean-field critical point $r = 0$, we see that fluctuations lead to a *downward shift* of the critical temperature. The criticality condition $\chi(0, 0)^{-1} = 0$ at $r = r_c$ then yields the implicit equation

$$r_c = -\frac{n + 2}{6} u \int_k \frac{1}{r_c + k^2} + \left(\frac{n + 2}{6} u\right)^2 \int_k \frac{1}{r_c + k^2} \int_{k'} \frac{1}{(r_c + k'^2)^2}$$

$$+ \frac{n + 2}{18} u^2 \int_k \frac{1}{r_c + k^2} \int_{k'} \frac{1}{r_c + k'^2} \frac{1}{r_c + (k + k')^2} + O(u^3) \tag{5.14}$$

for this T_c shift. To first order in u, we simply find

$$r_c = -\frac{n + 2}{6} u \int_k \frac{1}{k^2} + O(u^2) = -\frac{n + 2}{6} \frac{u K_d}{(2\pi)^d} \frac{\Lambda^{d-2}}{d - 2} + O(u^2) \tag{5.15}$$

in $d > 2$ dimensions, if we introduce the UV cutoff Λ again. Notice that the integral on the right-hand side of (5.15) diverges for $d \leq 2$; this indicates the breakdown of a power expansion of the effective functional with respect to the fields S^α, which become dimensionless in two dimensions.

The fluctuation-induced shift of the critical temperature constitutes a *non-universal* feature of the model, depending on the microscopic details of the inter-actions and lattice structure etc., as indicated by the explicit appearance of the cutoff Λ. It is also instructive to solve the implicit equation (5.14) for r_c in the one-loop approximation. Applying dimensional regularization, the integral becomes

$$I_d^{(0,1)}(r_c) = \frac{\Gamma(1 - d/2)}{2^d \pi^{d/2}} |r_c|^{-1 + d/2} = -\frac{2A_d}{(d - 2)(4 - d)} |r_c|^{(d-2)/2}, \tag{5.16}$$

and solving for r_c gives

$$|r_c| = \left[\frac{2A_d}{(d - 2)(4 - d)} \frac{n + 2}{6} u \right]^{2/(4-d)}. \tag{5.17}$$

This expression displays an essential singularity as $d \uparrow 4$. In a pure ϵ expansion about $d_c = 4$, one consequently finds $r_c = 0$ in dimensional regularization. It should be remembered, though, that of course the physical T_c shift is non-zero, irrespective of the applied regularization procedure.

The fluctuation-induced T_c shift exemplifies an *additive renormalization*, whereupon a new *counterterm*, here r_c, that is not present on the tree diagram level, is introduced into the theory. Notice that r_c has to be determined consistently to each loop order. It is now convenient to reparametrize our result for the dynamic response function in terms of the deviation $\tau = r - r_c$ from the true critical point. For example, to first order in u Eqs. (4.82) and (5.14) lead to

$$\chi(q, \omega)^{-1} = -\frac{i\omega}{Dq^a} + q^2 + \tau \left[1 - \frac{n + 2}{6} u \int_k \frac{1}{k^2(\tau + k^2)} \right] + O(u^2). \tag{5.18}$$

Obviously, by introducing the T_c shift, the quadratic (near four dimensions) UV divergence in the susceptibility or the two-point vertex function $\Gamma^{(1,1)}(q, \omega)$ has been removed.

5.2.3 Multiplicative renormalization, minimal subtraction

Near the upper critical dimension $d_c = 4$, this leaves us with logarithmically UV-divergent integrals in the perturbation expansion. As a second step in the renormalization program, we define new and UV-finite counterparts of our original fields,

$$S_R^\alpha = Z_S^{1/2} S^\alpha, \quad \tilde{S}_R^\alpha = Z_{\tilde{S}}^{1/2} \tilde{S}^\alpha, \tag{5.19}$$

where the logarithmic divergences are to be absorbed into the factors Z_S and $Z_{\tilde{S}}$. This implies that the renormalized cumulants become

$$G_R^{(N,\tilde{N})} = Z_S^{N/2} Z_{\tilde{S}}^{\tilde{N}/2} G^{(N,\tilde{N})}. \tag{5.20}$$

Equation (4.81) then shows that

$$\Gamma_R^{(1,1)} = Z_{\tilde{S}}^{-1/2} Z_S^{-1/2} \Gamma^{(1,1)}, \tag{5.21}$$

and as a consequence of relations such as (4.90), (4.97), etc. we generally deduce for the renormalized vertex functions

$$\Gamma_R^{(\tilde{N},N)} = Z_{\tilde{s}}^{-\tilde{N}/2} Z_S^{-N/2} \Gamma^{(\tilde{N},N)} . \qquad (5.22)$$

Furthermore we choose to define renormalized parameters according to

$$D_R = Z_D D , \quad \tau_R = Z_\tau \tau \, \mu^{-2} , \quad u_R = Z_u u A_d \, \mu^{d-4} , \qquad (5.23)$$

where the inclusion of the as yet arbitrary scale factors μ renders the novel parameters dimensionless.

Upon invoking Eq. (4.18) which relates the dynamic susceptibility χ to $D \, G^{(1,1)}$, we find for the renormalized response function $\chi_R = Z_D Z_S^{1/2} Z_{\tilde{s}}^{1/2} \chi$. Moreover, the fluctuation-dissipation theorem (4.25), which must of course hold both in the original ('bare') and renormalized versions of our dynamic theory, implies the non-trivial identity $Z_D Z_S^{1/2} Z_{\tilde{s}}^{1/2} = Z_S$, or

$$Z_D = \left(Z_S / Z_{\tilde{s}} \right)^{1/2} , \quad \chi_R = Z_S \, \chi . \qquad (5.24)$$

Of course, these identities also follow from Eq. (4.96), the fluctuation-dissipation theorem written in terms of the two-point vertex functions. Altogether, we therefore need to determine *three* independent static renormalization constants, namely Z_S, Z_τ, and Z_u, and one Z factor stemming from the dynamics (either $Z_{\tilde{s}}$ or Z_D). For the relaxational model B with conserved order parameter, we found that $\Gamma^{(1,1)}(q = 0, \omega)$ is actually finite to all orders in perturbation theory, see Eq. (4.84), or

$$a = 2 : \quad Z_{\tilde{s}} Z_S = 1 , \quad Z_D = Z_S , \qquad (5.25)$$

where we have inserted (5.24). The same identity follows from Eq. (4.92). Hence, for conserved order parameter dynamics, the dynamic renormalization may be expressed in terms of entirely static quantities. Consequently, as we shall see, the dynamic critical exponent z is related to the static Fisher exponent η through a scaling relation.

In all the above definitions of renormalized quantities in terms of the bare ones, the prescription is that the *renormalization constants (Z factors)* be chosen in such a manner that the new, renormalized theory becomes UV-finite, i.e., the ultraviolet divergences are to be absorbed into Z_S, $Z_{\tilde{s}}$, Z_D, Z_τ, and Z_u (four of which are independent). In terms of the renormalized quantities, the response functional for the relaxational models reads

$$
\mathcal{A}[\widetilde{S}, S] = \int d^d x \int dt \sum_{\alpha} \left\{ \frac{\widetilde{S}_R^{\alpha}}{(Z_{\widetilde{S}} Z_S)^{1/2}} \left[\frac{\partial}{\partial t} + \frac{D_R}{Z_D} (i\nabla)^a \left(\frac{\mu^2 \tau_R}{Z_\tau} + r_c - \nabla^2 \right) \right] S_R^{\alpha} \right.
$$

$$
\left. - \frac{D_R}{Z_{\widetilde{S}} Z_D} \widetilde{S}_R^{\alpha} (i\nabla)^a \widetilde{S}_R^{\alpha} + \frac{D_R \mu^{4-d} u_R}{6 A_d Z_{\widetilde{S}}^{1/2} Z_S^{3/2} Z_D Z_u} \widetilde{S}_R^{\alpha} (i\nabla)^a S_R^{\alpha} \sum_{\beta} S_R^{\beta} S_R^{\beta} \right\}
$$

$$
= \int d^d x \int dt \sum_{\alpha} \left\{ \widetilde{S}_R^{\alpha} \left[\frac{Z_D}{Z_S} \frac{\partial}{\partial t} + \frac{D_R}{Z_S} (i\nabla)^a \left(\frac{\mu^2 \tau_R}{Z_\tau} + r_c - \nabla^2 \right) \right] S_R^{\alpha} \right.
$$

$$
\left. - \frac{Z_D}{Z_S} D_R \widetilde{S}_R^{\alpha} (i\nabla)^a \widetilde{S}_R^{\alpha} + \frac{D_R \mu^{4-d} u_R}{6 A_d Z_S^2 Z_u} \widetilde{S}_R^{\alpha} (i\nabla)^a S_R^{\alpha} \sum_{\beta} S_R^{\beta} S_R^{\beta} \right\}, \quad (5.26)
$$

where Eq. (5.24) was employed to eliminate $Z_{\widetilde{S}}$.

Now recall from Section 4.4 that the coupling coefficients in this action are represented by the renormalized vertex functions. Thus, Z_S may be determined by rendering $\partial \Gamma^{(1,1)}(q, \omega)/\partial q^{2+a}$ finite, while the combination $Z_S Z_\tau$ should absorb the UV divergences in $\partial \Gamma^{(1,1)}(q, \omega)/\partial q^a$. As for these Z factors we may take the static limit $\omega \to 0$, we could also consider the static susceptibility, and then $Z_S Z_\tau$ and Z_S follow from $\chi(q, 0)^{-1}$ and $\partial \chi(q, 0)^{-1}/\partial q^2$, respectively. The dynamic renormalization Z_D/Z_S can then either be inferred from $\partial \Gamma^{(1,1)}(q, \omega)/\partial \omega$, or the renormalized noise vertex $\partial \Gamma^{(2,0)}(q, \omega)/\partial q^a$. For model B, we have $Z_D = Z_S$, and we see explicitly that merely the static terms in the functional require renormalization. Finally, $Z_{\widetilde{S}}^2 Z_u$ is fixed by $\partial \Gamma^{(1,3)}(-\frac{3q}{2}, -\frac{3\omega}{2}; \frac{q}{2}, \frac{\omega}{2}; \frac{q}{2}, \frac{\omega}{2}; \frac{q}{2}, \frac{\omega}{2})/\partial q^a$. Since all other vertex functions are UV-convergent near $d_c = 4$, this concludes our *multiplicative* renormalization program. In Section 5.3, we shall see how these renormalization constants determine the scaling behavior for the (renormalized) fields and model parameters of our theory. Near a renormalization group fixed point, where scale invariance holds, they will allow for the explicit computation of critical exponents. Moreover, general identities such as Eqs. (5.24) and (5.25) imply scaling relations that reduce the number of independent exponents.

It should be noted that there is still ample freedom of choice for the renormalization constants, of which we have as yet only demanded that they remove all ultraviolet divergences. First of all, we need to evaluate the vertex or correlation functions outside the critical regime, in order to avoid mixing of infrared and ultraviolet singularities. But we may still decide whether to calculate at fixed external momentum $q = \mu$, frequency $i\omega = D_R \mu^2$, and/or non-zero distance τ_R (i.e., $\tau \propto Z_\tau^{-1} \mu^2$) from the critical point. The final result for the scaling exponents should not depend on this choice of a *normalization point* (NP). Yet it is important to realize that the renormalized field theory will explicitly depend on the renormalization scale factor μ, as anticipated in the definitions (5.23) of dimensionless renormalized couplings. In the original model with finite UV cutoff Λ, the lattice spacing defines such an intrinsic length scale as well (see Section 1.4). In the field theory formulation, the role of Λ is taken over by μ. Second, besides the ultraviolet singularities, the multiplicative renormalization factors may contain any UV-finite parts of the associated vertex or correlation functions. A convenient unambiguous prescription, once the normalization point has been fixed, is to demand that *only* the UV-divergent contributions be included in the Z factors. This defines the *minimal subtraction* procedure, which leaves as much information in the (now UV-regular) scaling functions as possible. In terms of dimensional regularization, the minimal subtraction prescription means that the renormalization constants consist solely of a Laurent series of poles $1/\epsilon^k$ and their residua, where $\epsilon = d_c - d$.

5.2.4 Renormalization constants to lowest order

We shall next carry through this renormalization program explicitly to lowest non-trivial order for the $O(n)$-symmetric relaxational models A and B.[2] Our goal is to formally transfer the UV singularities into renormalization constants via isolating the $1/\epsilon$ poles in the correlation or vertex functions, while carefully keeping track of their residues. Starting with the static order parameter response function, Eqs. (5.18), (5.24), and (5.23) tell us that to one-loop order

$$\chi_R(0,0)^{-1} = \frac{\mu^2 \tau_R}{Z_S Z_\tau} \left[1 - \frac{n+2}{6} u \int_k \frac{1}{k^2(\tau + k^2)} + O(u^2) \right], \qquad (5.27)$$

$$\left. \frac{\partial}{\partial q^2} \chi_R(q,0)^{-1} \right|_{q=0} = Z_S^{-1}\left[1 + O(u^2) \right]. \qquad (5.28)$$

This immediately demonstrates the absence of field renormalization to this order,

$$Z_S = 1 + O(u^2) = 1 + O\left(u_R^2\right). \qquad (5.29)$$

On the other hand, in order to determine Z_τ we need to evaluate the integral

$$\int_k \frac{1}{k^2(\tau + k^2)} = \frac{1}{\tau}\left(\int_k \frac{1}{k^2} - \int_k \frac{1}{\tau + k^2} \right) = \frac{2 A_d}{(d-2)(4-d)} \tau^{-(4-d)/2}$$

at the normalization point $\tau_R = 1$, or $\tau = Z_\tau^{-1}\mu^2 = \mu^2[1 + O(u)]$, and then employ the minimal subtraction procedure, whereupon $2/(d-2)$ becomes replaced with

---

[2] Halperin, Hohenberg, and Ma (1972); De Dominicis, Brézin, and Zinn-Justin (1975); Bausch, Janssen, and Wagner (1976); Hohenberg and Halperin (1977); Janssen (1979, 1992); Vasil'ev (2004); Folk and Moser (2006); Kamenev (2011).

1 (the residuum at $\epsilon = 0$ or $d = 4$). Thus

$$Z_\tau = 1 - \frac{n+2}{6} \frac{u A_d \mu^{-\epsilon}}{\epsilon} + O(u^2) = 1 - \frac{n+2}{6} \frac{u_R}{\epsilon} + O\left(u_R^2\right), \quad (5.30)$$

with $\epsilon = 4 - d$; here Eq. (5.23) was used to replace the bare coupling u with its renormalized dimensionless counterpart u_R.

The remaining static Z factor follows from Eq. (4.102) for the four-point vertex function

$$\frac{\partial}{\partial q^a} \Gamma_R^{(1,3)}\left(-\frac{3q}{2}, 0; \frac{q}{2}, 0; \frac{q}{2}, 0; \frac{q}{2}, 0\right)\bigg|_{q=0}$$

$$= \left(\frac{3}{2}\right)^a \frac{\mu^\epsilon D_R u_R}{A_d Z_S^2 Z_u} \left[1 - \frac{n+8}{6} u \int_k \frac{1}{(\tau + k^2)^2}\right], \quad (5.31)$$

where we have replaced r to this order by τ. With the integral (5.12),

$$Z_u = 1 - \frac{n+8}{6} \frac{u A_d \mu^{-\epsilon}}{\epsilon} + O(u^2) = 1 - \frac{n+8}{6} \frac{u_R}{\epsilon} + O\left(u_R^2\right). \quad (5.32)$$

Finally, for model A we have to additionally consider the renormalization of the relaxation constant D. Yet, in Section 4.4, we found no fluctuation correction to the two-point noise vertex to one-loop order; consequently

$$Z_D = 1 + O(u^2) = 1 + O\left(u_R^2\right). \quad (5.33)$$

For a conserved order parameter, this follows already from the general identity (5.25) and Eq. (5.29).

Thus, in order to arrive at a non-trivial result for Z_S and Z_D, we must proceed to second order in u. With the aid of Eq. (4.82) we find for the static susceptibility to two-loop order

$$\frac{\partial}{\partial q^2} \chi_R(q,0)^{-1}\bigg|_{q=0} = Z_S^{-1}\left[1 - \frac{n+2}{18} u^2 \frac{\partial D(q)}{\partial q^2}\bigg|_{q=0} + O(u^3)\right], \quad (5.34)$$

$$D(q) = \int_k \frac{1}{\tau + k^2} \int_{k'} \frac{1}{\tau + k'^2} \frac{1}{\tau + (q - k - k')^2}. \quad (5.35)$$

As shown in the Appendix (Section 5.5), the UV-singular contribution near $d_c = 4$, i.e., the residuum of the associated $1/\epsilon$ pole, is $\partial D(q)/\partial q^2|_{q=0}^{\text{sing.}} = -A_d^2 \tau^{-\epsilon}/8\epsilon$, which at the normalization point $\tau = \mu^2$ leads to the field renormalization constant

$$Z_S = 1 + \frac{n+2}{144} \frac{(u A_d \mu^{-\epsilon})^2}{\epsilon} + O(u^3) = 1 + \frac{n+2}{144} \frac{u_R^2}{\epsilon} + O\left(u_R^3\right). \quad (5.36)$$

In order to find Z_D for the non-conserved case (model A with $a = 0$), we may either consider

$$\Gamma_R^{(2,0)}(0, 0) = -2D_R \frac{Z_D}{Z_S} \left[1 + \frac{n+2}{18} u^2 \int_k \frac{1}{\tau + k^2} \right.$$

$$\left. \times \int_{k'} \frac{1}{\tau + k'^2} \frac{1}{\tau + (k + k')^2} \frac{1}{3\tau + k^2 + k'^2 + (k + k')^2} \right], \quad (5.37)$$

where we have used Eqs. (5.24) and (4.91); or, with Eq. (4.82),

$$\frac{\partial}{\partial(i\omega)} \Gamma_R^{(1,1)}(0, \omega) \bigg|_{\omega=0} = \frac{Z_D}{Z_S} \left[1 + \frac{n+2}{18} u^2 \int_k \frac{1}{\tau + k^2} \int_{k'} \frac{1}{\tau + k'^2} \right.$$

$$\left. \times \frac{1}{\tau + (k + k')^2} \frac{1}{3\tau + k^2 + k'^2 + (k + k')^2} \right]. \quad (5.38)$$

A slightly tedious calculation (Problem 5.3) eventually results in

$$Z_D = 1 - \frac{n+2}{144} \left(6 \ln \frac{4}{3} - 1 \right) \frac{(u A_d \mu^{-\epsilon})^2}{\epsilon} + O(u^3)$$

$$= 1 - \frac{n+2}{144} \left(6 \ln \frac{4}{3} - 1 \right) \frac{u_R^2}{\epsilon} + O(u_R^3). \quad (5.39)$$

5.3 Renormalization group and dimensional expansion

We are now in a position to determine the scaling behavior of our theory. To this end, we exploit our knowledge of the dependence of the renormalized quantities on the momentum (inverse length) scale defined by the choice of the normalization point. The renormalization group equation then serves to relate the parameters of the theory at different scales, thus leading to running couplings. At a renormalization group fixed point, we encounter scale-invariant behavior. Thus our formal treatment of the ultraviolet poles near $d_c = 4$ dimensions allows us to systematically access the physically relevant infrared singularities. Linearizing the RG flow equations for the scale-dependent parameters near an infrared-stable fixed point finally yields the critical exponents.

5.3.1 Renormalization group equation, running couplings

Our starting point is the seemingly innocuous statement that the *unrenormalized* correlation and vertex functions do of course not depend on the arbitrary scale μ given by the normalization point (NP). Hence, if we hold the bare parameters D,

τ, and u fixed, we must have

$$0 = \mu \left. \frac{\mathrm{d}}{\mathrm{d}\mu} \right|_{D,\tau,u} \Gamma^{(\tilde{N},N)} = \mu \frac{\mathrm{d}}{\mathrm{d}\mu} \left[Z_{\tilde{S}}^{\tilde{N}/2} Z_S^{N/2} \Gamma_R^{(\tilde{N},N)}(\mu, D_R, \tau_R, u_R) \right]. \quad (5.40)$$

In the last step, we have used Eq. (5.22) to replace the bare vertex function $\Gamma^{(\tilde{N},N)}(D, \tau, u)$ with the renormalized $\Gamma_R^{(\tilde{N},N)}(\mu, D_R, \tau_R, u_R)$. By means of (5.23), we can translate the total derivative with respect to μ into partial derivatives, noting that the Z factors do depend on the renormalization scale:

$$0 = Z_{\tilde{S}}^{\tilde{N}/2} Z_S^{N/2} \mu \left[\frac{\partial}{\partial \mu} + \frac{\tilde{N}}{2} \frac{\partial \ln Z_{\tilde{S}}}{\partial \mu} + \frac{N}{2} \frac{\partial \ln Z_S}{\partial \mu} + \frac{\partial D_R}{\partial \mu} \frac{\partial}{\partial D_R} + \frac{\partial \tau_R}{\partial \mu} \frac{\partial}{\partial \tau_R} \right.$$
$$\left. + \frac{\partial u_R}{\partial \mu} \frac{\partial}{\partial u_R} \right] \Gamma_R^{(\tilde{N},N)}(\mu, D_R, \tau_R, u_R) ,$$

where the first partial derivative with respect to μ refers to the explicit dependence of the vertex functions on the momentum scale. This leads to the *renormalization group (RG) equation*, which tells us how the *renormalized* vertex functions depend on the scale μ and the renormalized couplings,

$$\left[\mu \frac{\partial}{\partial \mu} + \frac{\tilde{N}}{2} \gamma_{\tilde{S}} + \frac{N}{2} \gamma_S + \gamma_D D_R \frac{\partial}{\partial D_R} + \gamma_\tau \tau_R \frac{\partial}{\partial \tau_R} + \beta_u \frac{\partial}{\partial u_R} \right]$$
$$\times \Gamma_R^{(\tilde{N},N)}(\mu, D_R, \tau_R, u_R) = 0 . \quad (5.41)$$

Here we have introduced *Wilson's flow functions*

$$\gamma_{\tilde{S}} = \mu \left. \frac{\partial}{\partial \mu} \right|_0 \ln Z_{\tilde{S}} , \quad \gamma_S = \mu \left. \frac{\partial}{\partial \mu} \right|_0 \ln Z_S , \quad (5.42)$$

$$\gamma_D = \mu \left. \frac{\partial}{\partial \mu} \right|_0 \ln \frac{D_R}{D} = \mu \left. \frac{\partial}{\partial \mu} \right|_0 \ln Z_D , \quad (5.43)$$

$$\gamma_\tau = \mu \left. \frac{\partial}{\partial \mu} \right|_0 \ln \frac{\tau_R}{\tau} = -2 + \mu \left. \frac{\partial}{\partial \mu} \right|_0 \ln Z_\tau , \quad (5.44)$$

and the particularly important *RG beta function* for the renormalized non-linear coupling,

$$\beta_u = \mu \left. \frac{\partial}{\partial \mu} \right|_0 u_R = u_R \left(d - 4 + \mu \left. \frac{\partial}{\partial \mu} \right|_0 \ln Z_u \right) . \quad (5.45)$$

The subscript '0' is meant as a reminder that the derivatives with respect to the momentum scale μ are to be taken at fixed bare parameter values.

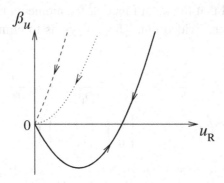

Fig. 5.1 One-loop renormalization group beta function for the relaxational models (Φ^4 field theory) for $\epsilon < 0$ (dashed), $\epsilon = 0$ (dotted), and $\epsilon > 0$ (full).

All these flow functions follow directly from the renormalization constants. Hence, the fluctuation-dissipation theorem via (5.24) implies the exact result

$$\gamma_D = \frac{1}{2}(\gamma_S - \gamma_{\tilde{S}}) . \tag{5.46}$$

For model B with conserved order parameter, Eq. (5.25) yields in addition

$$a = 2 : \quad \gamma_D = \gamma_S = -\gamma_{\tilde{S}} . \tag{5.47}$$

To lowest non-trivial order, we find for the static parameters

$$\gamma_S = -\frac{n+2}{72} u_R^2 + O\left(u_R^3\right) , \tag{5.48}$$

$$\gamma_\tau = -2 + \frac{n+2}{6} u_R + O\left(u_R^2\right) , \tag{5.49}$$

$$\beta_u = u_R \left[-\epsilon + \frac{n+8}{6} u_R + O\left(u_R^2\right) \right] , \tag{5.50}$$

with $\epsilon = 4 - d$. Figure 5.1 depicts this beta function in various dimensions. For non-conserved order parameter dynamics (model A), furthermore

$$a = 0 : \quad \gamma_D = \frac{n+2}{72} \left(6 \ln \frac{4}{3} - 1 \right) u_R^2 + O\left(u_R^3\right) . \tag{5.51}$$

As a first step towards a solution of the partial differential RG equation, it is useful to split off its scaling dimension $d_{\tilde{N},N}$ in momentum space, see Eq. (5.7), and thus introduce the reduced (renormalized) *dimensionless* vertex functions $\hat{\Gamma}_R^{(\tilde{N},N)}$,

$$\Gamma_R^{(\tilde{N},N)}(\mu, D_R, \tau_R, u_R, \{q_i\}, \{\omega_i\})$$

$$= \mu^{d_{\tilde{N},N}} \, \hat{\Gamma}_R^{(\tilde{N},N)} \left(D_R, \tau_R, u_R, \left\{ \frac{q_i}{\mu} \right\}, \left\{ \frac{\omega_i}{\mu^{2+a}} \right\} \right) , \tag{5.52}$$

where it is understood that the delta function for momentum and energy conservation has already been divided out. This allows us to eliminate the explicit μ derivative,

$$
\left[d_{\tilde{N},N} + \frac{\tilde{N}}{2}\gamma_{\tilde{s}} + \frac{N}{2}\gamma_{s} + \gamma_{D}\, D_{R}\, \frac{\partial}{\partial D_{R}} + \gamma_{\tau}\, \tau_{R}\, \frac{\partial}{\partial \tau_{R}} + \beta_{u}\, \frac{\partial}{\partial u_{R}} \right]
$$

$$
\times \hat{\Gamma}_{R}^{(\tilde{N},N)}\left(D_{R}, \tau_{R}, u_{R}, \left\{ \frac{q_{i}}{\mu} \right\}, \left\{ \frac{\omega_{i}}{\mu^{2+a}} \right\} \right) = 0 . \tag{5.53}
$$

We now employ the *method of characteristics*, a standard tool to tackle partial differential equations (utilized in Section 2.3.2 for the generating function of the stochastic population dynamics example), and define $\tilde{\mu}(\ell) = \mu\ell$, where ℓ is a real continuous parameter. Next, flowing dimensionless parameters are introduced via the ordinary differential equations

$$
\ell\, \frac{d\tilde{D}(\ell)}{d\ell} = \tilde{D}(\ell)\,\gamma_{D}(\ell) , \quad \tilde{D}(1) = D_{R} , \tag{5.54}
$$

$$
\ell\, \frac{d\tilde{\tau}(\ell)}{d\ell} = \tilde{\tau}(\ell)\,\gamma_{\tau}(\ell) , \quad \tilde{\tau}(1) = \tau_{R} , \tag{5.55}
$$

or equivalently,

$$
\tilde{D}(\ell) = D_{R}\, \exp\left(\int_{1}^{\ell} \gamma_{D}(\ell')\, \frac{d\ell'}{\ell'} \right) , \quad \tilde{\tau}(\ell) = \tau_{R}\, \exp\left(\int_{1}^{\ell} \gamma_{\tau}(\ell')\, \frac{d\ell'}{\ell'} \right) . \tag{5.56}
$$

Similarly the *running coupling* $\tilde{u}(\ell)$ is defined according to

$$
\ell\, \frac{d\tilde{u}(\ell)}{d\ell} = \beta_{u}(\ell) , \quad \tilde{u}(1) = u_{R} , \tag{5.57}
$$

where finally $\gamma_{s}(\ell) = \gamma_{s}(\tilde{u}(\ell))$, etc., and $\beta_{u}(\ell) = \beta_{u}(\tilde{u}(\ell))$.[3] If we now consider the vertex functions with all renormalized parameters replaced by their running counterparts, $\hat{\Gamma}^{(\tilde{N},N)}(\ell) = \hat{\Gamma}_{R}^{(\tilde{N},N)}(\tilde{D}(\ell), \tilde{\tau}(\ell), \tilde{u}(\ell))$, we notice that its corresponding RG equation becomes a simple ordinary differential equation as well, namely

$$
\left[d_{\tilde{N},N} + \frac{\tilde{N}}{2}\gamma_{\tilde{s}}(\ell) + \frac{N}{2}\gamma_{s}(\ell) + \ell\, \frac{d}{d\ell} \right] \hat{\Gamma}^{(\tilde{N},N)}(\ell) = 0 . \tag{5.58}
$$

---

[3] Occasionally it is useful to choose other parametrizations, e.g., $l = -\ln\ell$, which maps the interval $[0, 1]$ onto $[\infty, 0]$. The flow equations then read

$$
\frac{d\tilde{D}(l)}{dl} = -\tilde{D}(l)\,\gamma_{D}(l) , \quad \frac{d\tilde{\tau}(l)}{dl} = -\tilde{\tau}(l)\,\gamma_{\tau}(l) , \quad \frac{d\tilde{u}(l)}{dl} = -\beta_{u}(l) ,
$$

with initial conditions $\tilde{D}(0) = D_{R}$, $\tilde{\tau}(0) = \tau_{R}$, and $\tilde{u}(0) = u_{R}$; cf. Section 1.4 and Problem 5.4.

Its solution reads

$$\hat{\Gamma}^{(\tilde{N},N)}(\ell) = \hat{\Gamma}^{(\tilde{N},N)}(1)\,\ell^{-d_{\tilde{N},N}}\,\exp\left(-\int_1^\ell \left[\tilde{N}\gamma_{\tilde{S}}(\ell') + N\gamma_S(\ell')\right]\frac{d\ell'}{2\ell'}\right). \quad (5.59)$$

Collecting everything, the original RG equation (5.41) is solved by

$$\Gamma_R^{(\tilde{N},N)}(\mu, D_R, \tau_R, u_R, \{q_i\}, \{\omega_i\})$$

$$= \exp\left(\int_1^\ell \left[\tilde{N}\gamma_{\tilde{S}}(\ell') + N\gamma_S(\ell')\right]\frac{d\ell'}{2\ell'}\right)(\mu\ell)^{d_{\tilde{N},N}}$$

$$\times \hat{\Gamma}_R^{(\tilde{N},N)}\left(\tilde{D}(\ell), \tilde{\tau}(\ell), \tilde{u}(\ell), \left\{\frac{q_i}{\mu\ell}\right\}, \left\{\frac{\omega_i}{(\mu\ell)^{2+a}}\right\}\right). \quad (5.60)$$

Physically, through varying the flow parameter ℓ, we change the scale at which our model is explored. The critical infrared regime, in which we are ultimately interested, is reached as $\ell \to 0$. The flowing parameters $\tilde{D}(\ell)$, $\tilde{\tau}(\ell)$, and $\tilde{u}(\ell)$ can be interpreted as the effective values of the relaxation constant, temperature, and non-linear coupling on different length scales. They evolve under scale transformations $\mu \to \mu\ell$ according to the flow equations (5.54)–(5.57); notice that these are determined entirely by the Z factors from the *ultraviolet* renormalization of the theory. Equation (5.60) at last relates the vertex functions and thus also the correlation functions at different scales. One of the most important merits of the RG equation is that the right-hand side may be evaluated outside the critical region, where simple perturbation theory is applicable. This can be achieved by fixing one of its arguments at a non-zero value, say $\tilde{\tau}(\ell) = 1$, or $|q| = \mu\ell$. Subsequently taking the limit $\ell \to 0$ then tells us how the renormalized vertex functions on the left-hand side of Eq. (5.60) behave in the *infrared* regime.

5.3.2 Fixed points and critical exponents

Our simple relaxational models are characterized by the strength of the non-linear coupling u. Under scale transformations, its dimensionless renormalized counterpart changes as prescribed by the differential RG flow equation (5.57). Clearly, a *scale-invariant* regime is described by a *fixed point* $\tilde{u}(\ell) \to u^* = $ const., and the fixed point condition is

$$\beta_u(u^*) = 0. \quad (5.61)$$

Expanding in the vicinity of the fixed point $\delta\tilde{u}(\ell) = \tilde{u}(\ell) - u^*$, $|\delta\tilde{u}(\ell)| \ll 1$, we are led to $\ell\,d\delta\tilde{u}(\ell)/d\ell \approx (d\beta_u/du)|_{u^*}\,\delta\tilde{u}(\ell)$, and thus $\delta\tilde{u}(\ell) \sim \ell^{\beta'_u(u^*)}$. Consequently, the fixed point u^* is *infrared-stable*, i.e., $\tilde{u}(\ell) \to u^*$ as $\ell \to 0$, provided

$$\beta'_u(u^*) > 0. \quad (5.62)$$

Let us now assume we have found such a stable fixed point, with $\tilde{u}(\ell) \approx u^*$ for $\ell \le \ell^* < 1$. With $\gamma_{D/\tau}(u^*) = \gamma^*_{D/\tau}$ and $\int_{\ell^*}^{\ell} \gamma^*_{D/\tau} d\ell'/\ell' = \gamma^*_{D/\tau} \ln(\ell/\ell^*)$, the solutions in Eq. (5.56) become

$$\tilde{D}(\ell) \approx D_R \exp\left[\int_1^{\ell^*} \gamma_D(\ell') \frac{d\ell'}{\ell'}\right] \left(\frac{\ell}{\ell^*}\right)^{\gamma^*_D} = \overline{D}\, \ell^{\gamma^*_D}\,, \tag{5.63}$$

$$\tilde{\tau}(\ell) \approx \tau_R \exp\left[\int_1^{\ell^*} \gamma_\tau(\ell') \frac{d\ell'}{\ell'}\right] \left(\frac{\ell}{\ell^*}\right)^{\gamma^*_\tau} = \overline{\tau}\, \ell^{\gamma^*_\tau}\,. \tag{5.64}$$

The flowing parameters thus obey *power laws* governed by the *universal anomalous dimensions* γ^*_D and γ^*_τ, the fixed point values of Wilson's flow functions. The *amplitudes* $\overline{D}$ and $\overline{\tau}$, however, are *non-universal*, as they depend on the initial conditions and the RG trajectories in the entire interval between $\ell = 1$ and ℓ^*.

Specializing Eq. (5.60) to the two-point vertex function $\Gamma_R^{(1,1)}(q, \omega)$, with $d_{1,1} = 2 + a$, we find in the vicinity of an IR-stable fixed point

$$\Gamma_R^{(1,1)}(\mu, D_R, \tau_R, u_R, q, \omega) = C\,\mu^{2+a}\, \ell^{2+a+(\gamma^*_S+\gamma^*_S)/2}$$

$$\times \hat{\Gamma}_R^{(1,1)}\left(\overline{D}\, \ell^{\gamma^*_D}, \overline{\tau}\, \ell^{\gamma^*_\tau}, u^*, \frac{q}{\mu\ell}, \frac{\omega}{(\mu\ell)^{2+a}}\right), \tag{5.65}$$

with a non-universal amplitude C. From our explicit result (4.82) we furthermore know that

$$\Gamma_R^{(1,1)}(\mu, D_R, \tau_R, u_R, q, \omega) = D_R\,\mu^{2+a}\, \tilde{\Gamma}_R^{(1,1)}\left(\tau_R, u_R, \frac{q}{\mu}, \frac{\omega}{D_R\,\mu^{2+a}}\right).$$

Hence we may pull out an additional factor $\overline{D}\, \ell^{\gamma^*_D}$; using the relation (5.46), and applying the *matching condition* $\ell = |q|/\mu$, which fixes the third argument on the right-hand side to 1, we arrive at the *scaling form*

$$\Gamma_R^{(1,1)}(\mu, D_R, \tau_R, u_R, q, \omega) = C\,\overline{D}\,\mu^{-\gamma^*_S}\, |q|^{2+a+\gamma^*_S}$$

$$\times \tilde{\Gamma}_R^{(1,1)}\left(\overline{\tau}\,\mu^{-\gamma^*_\tau}\,|q|^{\gamma^*_\tau}, u^*, 1, \frac{\omega\,\mu^{\gamma^*_D}}{\overline{D}\,|q|^{2+a+\gamma^*_D}}\right). \tag{5.66}$$

Thus, by means of the renormalization group equation we have just *derived* the dynamic scaling law in the vicinity of an infrared-stable RG fixed point.

Finally, we can also identify the two independent static and the dynamic exponents for the relaxational models A and B. Equations (4.94) and (5.66) yield the scaling form for the dynamic susceptibility, which we write with simplified arguments as

$$\chi_R(\tau_R, q, \omega)^{-1} = |q|^{2+\gamma^*_S}\, \hat{\chi}_R\left(\tau_R\,|q|^{\gamma^*_\tau}, \frac{\omega}{D_R\,|q|^{2+a+\gamma^*_D}}\right)^{-1}. \tag{5.67}$$

Comparison with Eq. (3.27) establishes the identities

$$\eta = -\gamma_S^* , \quad \nu = -1/\gamma_\tau^* , \quad z = 2 + a + \gamma_D^* . \tag{5.68}$$

For model A with non-conserved order parameter, we thus encounter *three* independent scaling exponents, namely z in addition to the static critical exponents η and ν. For conserved order parameter dynamics, however, z is not an independent exponent, since the identity $\gamma_D = \gamma_S$ implies the *model B scaling relation*

$$z = 4 - \eta . \tag{5.69}$$

5.3.3 Dimensional expansion; logarithmic corrections

The one-loop RG beta function (5.50) displays two zeros, namely the Gaussian fixed point $u_0^* = 0$, and the non-trivial *Heisenberg fixed point*

$$u_{\mathrm{H}}^* = \frac{6\epsilon}{n + 8} + O(\epsilon^2) , \tag{5.70}$$

which is positive only for $d < d_c = 4$ dimensions (see Fig. 5.1). As $\beta_u'(u^*) = -\epsilon + \frac{n+8}{3} u^*$, we see that the Gaussian fixed point $u_0^* = 0$ is stable for $\epsilon < 0$ or $d > 4$. Then $\gamma_S^* = 0 = \gamma_D^*$ and $\gamma_\tau^* = -2$, leading to the mean-field exponents $\eta = 0$, $\nu = 1/2$, and $z = 2 + a$. As expected, above the critical dimension $d_c = 4$ classical scaling applies. It should be noted, though, that there are of course fluctuation corrections to the scaling functions; yet these are not sufficiently strong as to modify the leading singularities. Despite the fact that the running coupling $\tilde{u}(\ell)$ flows to zero under scale transformations, the non-linearity u may not safely be omitted. In fact, the non-linear term drives the phase transition. Above four dimensions, u is therefore said to be a *dangerously irrelevant* variable.

For $d < d_c = 4$, the Gaussian fixed point becomes unstable, whereas $\beta_u'(u_{\mathrm{H}}^*) = \epsilon > 0$. In the vicinity of the non-trivial Heisenberg fixed point, non-classical critical exponents emerge, which can be computed perturbatively. Yet the expansion parameter, essentially u_{H}^*, is not generally minute at all. In order to render the perturbation expansion meaningful, we may consider the deviation from the upper critical dimension ϵ as small. Equation (5.70) is then correct up to terms of order ϵ^2, and likewise the expressions for the anomalous scaling dimensions and universal critical exponents should be understood as (asymptotic) series expansions in $\epsilon = d_c - d$. Inserting the fixed point (5.70) into Eqs. (5.48) and (5.49) yields the static critical exponents to lowest non-trivial order in ϵ, respectively,

$$\eta = \frac{n + 2}{2(n + 8)^2} \epsilon^2 + O(\epsilon^3) , \tag{5.71}$$

$$\frac{1}{\nu} = 2 - \frac{n + 2}{n + 8} \epsilon + O(\epsilon^2) . \tag{5.72}$$

Notice that we do not need the two-loop beta function for the $O(\epsilon^2)$ contribution to η, because γ_S vanishes to one-loop order. For model A with non-conserved order parameter, Eq. (5.51) leads to

$$z = 2 + c\,\eta \,, \quad c = 6\ln\frac{4}{3} - 1 + O(\epsilon)\,. \tag{5.73}$$

Precisely at the borderline dimension $d_c = 4$, the RG flow equation for the non-linear running coupling becomes

$$\ell\,\frac{d\tilde{u}(\ell)}{d\ell} = \frac{n+8}{6}\tilde{u}(\ell)^2\,, \quad \tilde{u}(1) = u_R\,; \tag{5.74}$$

compare with Fig. 5.1 for $\epsilon = 0$. Its solution reads

$$\tilde{u}(\ell) = \frac{u_R}{1 - \frac{n+8}{6}u_R\,\ln\ell}\,, \tag{5.75}$$

i.e., $\tilde{u}(\ell)$ vanishes logarithmically as $\ell \to 0$, in contrast with the much faster power law approach to the fixed point for $\epsilon \neq 0$. At d_c, this gives rise to *logarithmic corrections* to those mean-field power laws that acquire $O(\epsilon)$ corrections in $d < d_c$. Upon inserting the one-loop result (5.49) and $\tilde{u}(\ell) \approx 6/(n+8)|\ln\ell|$ into the flow equation (5.55), we find for $\ell \ll 1$

$$\ell\,\frac{d\tilde{\tau}(\ell)}{d\ell} \approx \tilde{\tau}(\ell)\left(-2 + \frac{n+2}{n+8}\frac{1}{|\ln\ell|}\right)\,, \tag{5.76}$$

with the approximate solution

$$\tilde{\tau}(\ell) \sim \tau_R\,\ell^{-2}\,|\ln\ell|^{-(n+2)/(n+8)}\,. \tag{5.77}$$

Matching $\tilde{\tau}(\ell) = 1$ at $\ell \sim 1/\xi\mu$, one obtains $\tau_R \sim \xi^{-2}\,(\ln\xi)^{(n+2)/(n+8)}$. Upon solving iteratively for the correlation length, therefore

$$\xi \sim \tau_R^{-1/2}\,|\ln\tau_R|^{(n+2)/2(n+8)}\,. \tag{5.78}$$

We end this section with a remark on irrelevant couplings such as the strength v of an additional sixth-order term in the Hamiltonian (4.12). We had already seen that its scaling dimension at the Gaussian fixed point is $[v] = \mu^{6-2d}$. At the Heisenberg fixed point, we expect an $O(\epsilon)$ correction, and hence predict for the corresponding dimensionless running coupling that asymptotically

$$\tilde{v}(\ell) \sim \ell^{2-O(\epsilon)} \to 0\,. \tag{5.79}$$

We conclude that the critical regime will indeed not be affected by this irrelevant parameter. Yet its anomalous scaling dimension enters the *corrections to scaling*, i.e., those power laws that describe the deviations from the leading critical singularities.

5.4 Broken rotational symmetry and Goldstone modes

We now turn our attention to the scaling behavior in the ordered phase of systems with a spontaneously broken *continuous* symmetry. The ensuing massless Goldstone modes induce infrared singularities in certain physical quantities in the entire low-temperature phase, not just at the critical point. These *coexistence anomalies* are amenable to a renormalization group treatment as well, and are governed by a zero-temperature RG coexistence fixed point.[4]

5.4.1 The O(n)-symmetric model A in the ordered phase

For simplicity, we restrict ourselves to the $O(n)$-symmetric model A with purely relaxational dynamics for a non-conserved order parameter. In the thermodynamic phase with homogeneous long-range order, we need to account for the spontaneous symmetry breaking in order parameter space. Let us choose the nth component to have a non-zero expectation value, cf. $\langle S^n \rangle = (6|r|/u)^{1/2}$ in mean-field theory. In general, it is convenient to parametrize $\langle S^n \rangle = (3/u)^{1/2} m$. We may then introduce new fields $\pi^\alpha(x,t) = S^\alpha(x,t)$ for $\alpha = 1, \ldots, n-1$ and $\sigma(x,t) = S^n(x,t) - (3/u)^{1/2} m$, and just rename the associated Martin–Siggia–Rose auxiliary fields $\tilde{\pi}^\alpha(x,t) = \tilde{S}^\alpha(x,t)$, $\tilde{\sigma}(x.t) = \tilde{S}^n(x,t)$. The novel fields all have vanishing expectation values.

Upon inserting this parametrization into the dynamic response functional (with $a = 0$), the new harmonic part acquires contributions from the original interaction term (4.15),

$$\mathcal{A}_0[\tilde{\pi}, \pi, \tilde{\sigma}, \sigma] = \int d^d x \int dt \left(\tilde{\sigma}(x,t) \left[\frac{\partial}{\partial t} + D\left(r + \frac{3m^2}{2} - \nabla^2 \right) \right] \sigma(x,t) \right.$$

$$+ \sum_{\alpha=1}^{n-1} \tilde{\pi}^\alpha(x,t) \left[\frac{\partial}{\partial t} + D\left(r + \frac{m^2}{2} - \nabla^2 \right) \right] \pi^\alpha(x,t)$$

$$- D\tilde{\sigma}(x.t) \left[\tilde{\sigma}(x,t) + h^n(x,t) \right]$$

$$\left. - D\sum_{\alpha=1}^{n-1} \tilde{\pi}^\alpha(x,t) \left[\tilde{\pi}^\alpha(x,t) + h^\alpha(x,t) \right] \right). \tag{5.80}$$

[4] See Brézin and Wallace (1973), Nelson (1976) for Goldstone singularities in static quantities; and Mazenko (1976), Schäfer (1978) for dynamical coexistence anomalies; our renormalization group analysis closely follows Lawrie (1981), Täuber and Schwabl (1992).

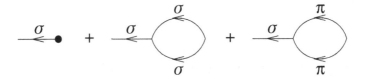

Fig. 5.2 One-loop Feynman diagrams for the computation of $\langle\sigma\rangle$ or $\Gamma_\sigma^{(1,0)}$.

As a consequence of the symmetry breaking, the anharmonic contribution contains both four- and three-point vertices,

$$\mathcal{A}_{\text{int}}[\tilde\pi, \pi, \tilde\sigma, \sigma]$$

$$= D\frac{u}{6}\int d^dx \int dt\, \left(\tilde\sigma\sigma\sigma\sigma + \sum_\alpha (\tilde\sigma\sigma\pi^\alpha\pi^\alpha + \tilde\pi^\alpha\pi^\alpha\sigma\sigma)\right.$$

$$\left. + \sum_{\alpha,\beta} \tilde\pi^\alpha\pi^\alpha\pi^\beta\pi^\beta + \sqrt{\frac{3}{u}}\,m\left[3\tilde\sigma\sigma\sigma + \sum_\alpha (\tilde\sigma\pi^\alpha\pi^\alpha + 2\tilde\pi^\alpha\pi^\alpha\sigma)\right]\right). \quad (5.81)$$

In addition, a *source* term appears,

$$\mathcal{A}_{\text{sr}}[\tilde\sigma] = D\sqrt{\frac{3}{u}}\,m\left(r + \frac{m^2}{2}\right)\int d^dx \int dt\, \tilde\sigma(x,t)\,. \quad (5.82)$$

We now impose our requirement that $\langle\sigma\rangle = 0$, or equivalently $\Gamma_\sigma^{(1,0)} = 0$. To one-loop order, the corresponding Feynman diagrams are depicted in Fig. 5.2, where we have represented the source with the symbol •. Thus we obtain the *equation of state* relating the parameters r and m,

$$0 = r + \frac{m^2}{2} + \frac{u}{2}\int_k \frac{1}{r + 3m^2/2 + k^2} + \frac{n-1}{6}\,u\int_k \frac{1}{r + m^2/2 + k^2} + O(u^2)\,.$$

$$(5.83)$$

Notice that for $r = r_c$, we must have $m = 0$; and indeed, in this limit Eq. (5.83) reduces to the one-loop result for the fluctuation-induced T_c shift, see Eq. (5.14). For $T < T_c$, i.e. non-vanishing m, the critical exponent β describing the onset of the order parameter may be determined from this equation of state, see Problem 5.5.

Equation (5.83) can now be employed to eliminate the temperature parameter r in each successive order in perturbation theory. To first order in u one has

$$r + \frac{m^2}{2} = -\frac{u}{2}\int_k \frac{1}{m^2 + k^2} - \frac{n-1}{6}\,u\int_k \frac{1}{k^2} + O(u^2) = C\,, \quad (5.84)$$

which defines $C = O(u)$. With this substitution, the Gaussian dynamic Janssen–De Dominicis functional becomes finally

$$\mathcal{A}_0[\tilde{\pi}, \pi, \tilde{\sigma}, \sigma] = \int d^d x \int dt \left(\tilde{\sigma}(x, t) \left[\frac{\partial}{\partial t} + D \left(m^2 - \nabla^2 \right) \right] \sigma(x, t) \right.$$

$$+ \sum_{\alpha=1}^{n-1} \tilde{\pi}^\alpha(x, t) \left(\frac{\partial}{\partial t} - D\nabla^2 \right) \pi^\alpha(x, t)$$

$$- D \tilde{\sigma}(x, t) \left[\tilde{\sigma}(x, t) + h^n(x, t) \right]$$

$$\left. - D \sum_{\alpha=1}^{n-1} \tilde{\pi}^\alpha(x, t) \left[\tilde{\pi}^\alpha(x, t) + h^\alpha(x, t) \right] \right), \quad (5.85)$$

while the source (5.82) disappears in favor of the *counterterms*

$$\mathcal{A}_{ct}[\tilde{\pi}, \pi, \tilde{\sigma}, \sigma] = D C \int d^d x \int dt \left[\tilde{\sigma}(x, t) \left(\sigma(x, t) + \sqrt{\frac{3}{u}} m \right) \right.$$

$$\left. + \sum_{\alpha=1}^{n-1} \tilde{\pi}^\alpha(x, t) \pi^\alpha(x, t) \right]. \quad (5.86)$$

We could now construct the perturbation expansion in the ordered phase, taking the counterterms into account to the appropriate orders in u. In the vicinity of the critical point, however, where $m \to 0$ and the symmetry between the transverse fluctuations with respect to the spontaneous order parameter π^α and the longitudinal mode σ is restored, we do not expect any novel critical exponents as compared to approaching the transition from above. In fact the only new parameter in the theory is the (bare) mass m^2 of the longitudinal fluctuations; we define its renormalized counterpart via

$$m_R^2 = Z_m m^2 \mu^{-2} . \quad (5.87)$$

But m only appeared in the reparametrization of the order parameter. As $S_R^n = Z_S^{1/2} S^n$, and $u_R \propto Z_u u$, we find the identity

$$Z_m = Z_S Z_u . \quad (5.88)$$

Consequently, no novel renormalizations appear in the ordered phase, and all its critical exponents are related to those in the high-temperature phase through scaling relations.

5.4.2 Goldstone modes and the coexistence limit

Furthermore, we notice that the transverse propagator in the action (5.85) is massless. This is actually an exact statement, and follows from the rotational invariance of the problem. If we apply an infinitesimal rotation in order parameter space, say $\delta\Phi^\alpha(x,t) = -\varepsilon\Phi^n(x,t)$, $\delta\Phi^n(x,t) = \varepsilon\Phi^\alpha(x,t)$ (and accordingly for the response fields), the corresponding variation of the generating functional of the vertex functions must vanish,

$$
0 = \delta\Gamma[\tilde\Phi, \Phi] = -\varepsilon\int d^d x \int dt \left[\frac{\delta\Gamma}{\delta\tilde\Phi^\alpha(x,t)}\tilde\Phi^n(x,t) - \frac{\delta\Gamma}{\delta\tilde\Phi^n(x,t)}\tilde\Phi^\alpha(x,t)\right.
$$
$$
\left. + \frac{\delta\Gamma}{\delta\Phi^\alpha(x,t)}\Phi^n(x,t) - \frac{\delta\Gamma}{\delta\Phi^n(x,t)}\Phi^\alpha(x,t)\right]. \tag{5.89}
$$

This relation between effectively different vertex functions is called the *Ward–Takahashi identity* associated with the continuous $O(n)$ symmetry. Its most important consequence for the properties of the ordered phase with spontaneously broken rotational invariance, i.e., $\Phi^\alpha = \langle S^\alpha\rangle = 0$, and $\Phi^n = \langle S^n\rangle = \text{const.} \neq 0$, is readily obtained by taking a functional derivative with respect to $\tilde\Phi^\alpha(x',t')$:

$$
0 = -\frac{\delta\Gamma}{\delta\tilde\Phi^n(x',t')} + \langle S^n\rangle\int d^d x\int dt\, \frac{\delta^2\Gamma}{\delta\tilde\Phi^\alpha(x',t')\,\delta\Phi^\alpha(x,t)}. \tag{5.90}
$$

For vanishing source fields $\tilde\jmath = 0 = j$ this implies *Goldstone's theorem*

$$
\langle S^n\rangle\,\Gamma^{(1,1)}_{\alpha;\alpha}(q=0, \omega=0) = 0\,, \tag{5.91}
$$

i.e., for non-vanishing order parameter $\langle S^n\rangle \neq 0$ the transverse fluctuations must be massless. The π fluctuations thus represent the *Goldstone modes* associated with the broken continuous symmetry, and their vanishing mass reflects the fact that physically an infinitesimal global rotation of the order parameter cannot cost any free energy.

    In the low-temperature phase, we thus have to distinguish between the longitudinal and transverse order parameter response function. The longitudinal thermodynamic susceptibility is naturally given by $\chi^{nn} = \delta\langle S^n\rangle/\delta h^n$. On the other hand, upon noticing that for model A the formal source can be identified with the physical external field according to $\tilde\jmath^n = Dh^n$, Eq. (5.90) yields a remarkable expression for the transverse susceptibility:

$$
\chi^{\alpha\alpha} = \delta\langle S^\alpha\rangle/\delta h^\alpha = \langle S^n\rangle/h^n\,, \tag{5.92}
$$

which reflects rotational invariance. Again, we see that for vanishing field $h^n \to 0$ the transverse response function diverges. The vanishing mass of the $n-1$ transverse fluctuations or Goldstone modes indicates that certain physical properties will be governed by *power laws* not merely in the vicinity of the critical point, but throughout the entire ordered phase. This algebraic behavior in systems with spontaneously broken continuous symmetry is referred to as emergence of *coexistence anomalies*.

In the long-wavelength, low-frequency limit we note that the longitudinal fluctuations are suppressed as a consequence of their mean-field mass m as compared with the massless transverse Goldstone modes. In order to infer the asymptotic properties of the coexistence limit, let us absorb m into new fields $\tilde{s} = m\tilde{\sigma}$ and $s = m\sigma$, and then take the limit $m \to \infty$ in the dynamic response functional. Then, as expected, the harmonic contribution contains fluctuations only for the Goldstone modes

$$\mathcal{A}_0 \to \int d^d x \int dt \left(D\tilde{s}\,s + \sum_{\alpha=1}^{n-1} \left[\tilde{\pi}^\alpha \left(\frac{\partial}{\partial t} - D\nabla^2 \right) \pi^\alpha - D\tilde{\pi}^\alpha (\tilde{\pi}^\alpha + h^\alpha) \right] \right),$$

(5.93)

while the counterterm reads

$$\mathcal{A}_{ct} \to DC \int d^d x \int dt \left(\sqrt{\frac{3}{u}}\,\tilde{s} + \sum_{\alpha=1}^{n-1} \tilde{\pi}^\alpha \pi^\alpha \right),$$

(5.94)

and the non-linearities become

$$\mathcal{A}_{int} \to D\frac{u}{6} \int d^d x \int dt \left[\sum_{\alpha,\beta} \tilde{\pi}^\alpha \pi^\alpha \tilde{\pi}^\beta \pi^\beta + \sqrt{\frac{3}{u}} \sum_\alpha (\tilde{s}\pi^\alpha \pi^\alpha + 2\tilde{\pi}^\alpha \pi^\alpha s) \right].$$

(5.95)

Applying the non-linear transformation (with Jacobian one)

$$\tilde{\varphi} = \tilde{s} + \sqrt{\frac{u}{3}} \sum_\alpha \tilde{\pi}^\alpha \pi^\alpha, \quad \varphi = s + \frac{1}{2}\sqrt{\frac{u}{3}} \sum_\alpha \pi^\alpha \pi^\alpha + \sqrt{\frac{3}{u}}\,C,$$

(5.96)

we see that the effective action in the coexistence limit, expressed in the new fields, becomes Gaussian:

$$\mathcal{A}_{eff}[\tilde{\pi}, \pi, \tilde{\varphi}, \varphi] = \int d^d x \int dt \left(D\tilde{\varphi}(x, t)\,\varphi(x, t) \right.$$

$$+ \sum_{\alpha=1}^{n-1} \left[\tilde{\pi}^\alpha(x, t) \left(\frac{\partial}{\partial t} - D\nabla^2 \right) \pi^\alpha(x, t) \right.$$

$$\left. \left. - D\tilde{\pi}^\alpha(x, t) \left[\tilde{\pi}^\alpha(x, t) + h^\alpha(x, t) \right] \right] \right).$$

(5.97)

Consequently, we find the asymptotically exact result

$$\chi^{\alpha\alpha}(q, \omega) = \frac{D}{\Gamma_{\alpha;\alpha}^{(1,1)}(-q, -\omega)} = \frac{1}{-i\omega/D + q^2}$$

(5.98)

for the transverse dynamical susceptibility. The Goldstone modes in the ordered phase of model A with non-conserved order parameter are purely diffusive. There is

(a) (b)

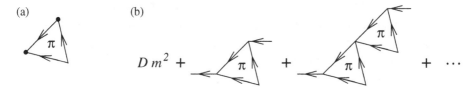

$$D\,m^2 + \qquad + \qquad + \cdots$$

Fig. 5.3 (a) Transverse loop for the composite operator $\langle [\pi^\alpha\,\pi^\alpha]\,[\tilde{\pi}^\alpha\,\pi^\alpha]\rangle$, and (b) diagrammatic representation of the geometric series for $\Gamma_{n;n}^{(1,1)}(q,\omega)$ in terms of a string of transverse loops.

no field renormalization in the coexistence limit, and the dynamic scaling exponent is $z = 2$. Through combining Eqs. (5.98) and Eq. (5.92), we may include the external field h^n along the direction of the order parameter, which then serves as a mass term:

$$\chi^{\alpha\alpha}(q,\omega,h^n) = \frac{1}{-i\omega/D + q^2 + h^n/\langle S^n\rangle}\,. \qquad (5.99)$$

The longitudinal response function is readily obtained from Eq. (5.96),

$$m^2\,\langle\sigma\tilde{\sigma}\rangle = \langle\varphi\tilde{\varphi}\rangle + \frac{u}{6}\sum_{\alpha,\beta}\left\langle\left[\pi^\alpha\,\pi^\alpha\right]\left[\tilde{\pi}^\beta\,\pi^\beta\right]\right\rangle, \qquad (5.100)$$

using that the averages $\langle\tilde{\varphi}\rangle$, $\langle[\tilde{\pi}^\alpha\,\pi^\alpha]\rangle$, $\langle\varphi\,[\tilde{\pi}^\alpha\,\pi^\alpha]\rangle$, and $\langle[\pi^\alpha\,\pi^\alpha]\,\tilde{\varphi}\rangle$ all vanish. The square brackets here indicate *composite operators*, which are comprised of fields at coinciding space-time points; consequently, their Fourier transforms are to be taken as

$$[\tilde{\pi}^\alpha\,\pi^\alpha](q,\omega) = \int d^d x \int dt\,\tilde{\pi}^\alpha(x,t)\,\pi^\alpha(x,t)\,e^{-i(q\cdot x-\omega t)}, \qquad (5.101)$$

and similarly for $[\pi^\alpha\,\pi^\alpha](q,\omega)$. We may then read off $\langle\varphi\tilde{\varphi}\rangle = 1/D$ from the action (5.97), and explicitly compute the composite operator average $\langle[\pi^\alpha\,\pi^\alpha]\,[\tilde{\pi}^\alpha\,\pi^\alpha]\rangle$, which in this harmonic theory, via factorization according to Wick's theorem, simply reduces to the one-loop Feynman diagram depicted in Fig. 5.3(a). The corresponding analytic expression for the asymptotic longitudinal dynamic susceptibility reads

$$\chi^{nn}(q,\omega) = \frac{1}{m^2}\left(1 + \frac{n-1}{3}u\int\frac{d^d k}{(2\pi)^d}\frac{1}{k^2}\frac{1}{-i\omega/D + (q-k)^2 + k^2}\right). \qquad (5.102)$$

Rendering the integral dimensionless, we see that it is proportional to q^{d-4} or to $\omega^{(d-4)/2}$. For $d > 4$ dimensions, we therefore recover the mean-field result that the longitudinal modes simply become massive away from the critical point, and their fluctuations are suppressed. However, for $d < d_c = 4$ dimensions, the transverse fluctuation loop diverges in the infrared, and our exact asymptotic expression

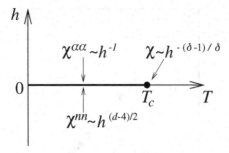

Fig. 5.4 Coexistence and critical singularities for isotropic systems as function of the external field $h = h^n$ in the h-T phase diagram (in dimensions $d < 4$).

(5.102) tells us that in fact the leading long-wavelength and low-frequency behavior is given by

$$\chi^{nn}(q, 0) \sim 1/q^{4-d} , \quad \chi^{nn}(0, \omega) \sim 1/\omega^{(4-d)/2} . \qquad (5.103)$$

In contradiction to mean-field theory, the longitudinal susceptibility diverges also in the entire low-temperature phase! In this sense, the longitudinal fluctuations are not massive at all for $2 < d < d_c = 4$ and $n > 1$, but are rendered massless through their coupling to the Goldstone modes. As function of h^n, one similarly finds for the longitudinal thermodynamic susceptibility

$$\chi^{nn}(h^n) \sim \left(\langle S^n \rangle / h^n \right)^{(4-d)/2} . \qquad (5.104)$$

Notice that at the critical point, we have instead $\chi(h) \sim h^{-\gamma/\beta\delta} = h^{-(\delta-1)/\delta}$, where we have used the scaling relations (1.72). The overall situation is summarized schematically in Fig. 5.4. Actual experimental results for the longitudinal susceptibility as a function of the internal magnetic field, obtained from mutual inductance measurements, are shown in Fig. 5.5 for the Heisenberg ferromagnet EuS at various temperatures. The data nicely follow the predicted power law (5.104) in three dimensions; the deviations at very low fields for the temperature nearest to the critical point might indicate the onset of the critical scaling law.

For $d \to 2$, both χ^{nn} and $\chi^{\alpha\alpha}$ are governed by the same power laws, indicating that there is no symmetry breaking at finite temperatures in a two-dimensional Heisenberg system (*Mermin–Wagner–Hohenberg theorem*).[5] The above result (5.102) implies that in the coexistence limit, the longitudinal two-point vertex function $\Gamma^{(1,1)}_{n;n}(q, \omega)$ can be written as a geometric series of transverse loops, as shown in Fig. 5.3(b). This corresponds to summing the so-called *one-vertex irreducible* Feynman diagrams, which are also precisely those terms that survive in the *spherical model* limit $n \to \infty$ (see Section 8.2.2). The model A vertex functions

---

[5] Mermin and Wagner (1966), Wagner (1966), Hohenberg (1967).

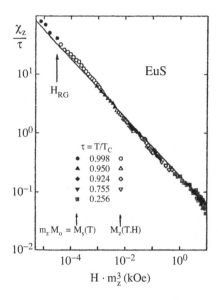

Fig. 5.5 Measured longitudinal susceptibility for the Heisenberg ferromagnet EuS vs. internal magnetic field H at various temperatures. The full line represents the $H^{-1/2}$ singularity, Eq. (5.104) in $d = 3$. The data with filled symbols were obtained by scaling to the reduced spontaneous magnetization as indicated. At $T/T_c \approx 0.998$, the crossover from Goldstone singular to critical behavior is expected at H_{RG}. [Data and figure reproduced with permission from: J. Kötzler, D. Görlitz, R. Dombrowski, and M. Pieper, *Z. Phys. B Cond. Matt.* **94**, 9 (1994); DOI: 10.1007/BF01307647; copyright (1994) by Springer.]

in the ordered phase are hence asymptotically given by the leading contributions to the $1/n$ expansion.

5.4.3 RG description: coexistence fixed point

It is instructive to see how the above asymptotic power laws are obtained by means of the renormalization group approach. We merely need to determine the renormalization constant Z_m, as defined in Eq. (5.87), by evaluating $\Gamma_{n;n}^{(1,1)}(q, \omega)$ at an appropriate normalization point outside the infrared-singular regime. As we are dealing with a massless field theory now, we have to choose either a finite momentum or a finite frequency, e.g., $i\omega/2D = \mu^2$. From Eq. (5.102) we then infer

$$Z_m = \left(1 + \frac{n-1}{6} \frac{2A_d \, \mu^{d-4}}{(d-2)(4-d)} u\right)^{-1} = Z_u \,, \qquad (5.105)$$

since $Z_S = 1$ in the coexistence limit, which is essentially a one-loop theory. As we have already summed the perturbation series to all orders, there is no need for a small parameter and we may evaluate Z_m at *fixed* dimension. A brief calculation yields the associated Wilson flow function

$$\gamma_m = \mu \left.\frac{\partial}{\partial\mu}\right|_0 \ln\frac{m_R^2}{m^2} = -2 + \frac{n-1}{6}\frac{2A_d\,\mu^{d-4}}{d-2}Z_m\,u = -2 + \frac{n-1}{3(d-2)}u_R\,,$$

$$(5.106)$$

and similarly the RG beta function

$$\beta_u = \mu\left.\frac{\partial}{\partial\mu}\right|_0 u_R = u_R\left(d-4+\frac{n-1}{3(d-2)}u_R\right).$$

$$(5.107)$$

Thus, for $d > d_c = 4$ dimensions, the Gaussian fixed point $u^* = 0$ is stable, and $\gamma_m^* = -2$, which is the naive scaling dimension of the longitudinal mass m. This leads to ordinary mean-field behavior, as expected above the upper critical dimension. For $2 < d < 4$ on the other hand, a novel infrared-stable non-trivial coexistence fixed point emerges,

$$u_C^* = \frac{3(d-2)(4-d)}{n-1}\,,$$

$$(5.108)$$

with anomalous scaling dimension $\gamma_m^* = 2 - d$. Upon solving the renormalization group equation for the longitudinal dynamic susceptibility, and employing its mean-field form, we find

$$\chi_R^{nn}(\mu, D, m_R, q, \omega)^{-1} = \mu^2\ell^2\left(\frac{-i\omega}{D\,\mu^2\ell^2} + \frac{q^2}{\mu^2\ell^2} + \tilde{m}(\ell)^2\right)$$

$$= -\frac{i\omega}{D} + q^2 + \mu^2\,\overline{m}^2\,\ell^{2+\gamma_m^*}\,.$$

$$(5.109)$$

The infrared limit is taken care of via the matching condition

$$\mu^2\ell^2 = \left|-\frac{i\omega}{D} + q^2\right| = \left|\chi^{\alpha\alpha}(q,\omega)^{-1}\right|.$$

$$(5.110)$$

Inserting this into Eq. (5.109), the longitudinal response function is governed by the leading algebraic singularity

$$\chi_R^{nn}(q,\omega) \propto \left|-\frac{i\omega}{D} + q^2\right|^{(d-4)/2}\,,$$

$$(5.111)$$

which reduces to the previous results (5.103) for either $\omega = 0$ or $q = 0$.

The coexistence fixed point (5.108) vanishes as $d \to 2$. Below the lower critical dimension $d_{lc} = 2$, there is no stable positive fixed point, which is related to the fact that a system with continuous order parameter symmetry cannot display long-range order in less than two dimensions. In our RG formulation, the Mermin–Wagner–Hohenberg theorem is encoded in the anomalous scaling dimension of the parameter m. For $d < 2$, we find that $\tilde{m}(\ell) \to 0$, which contradicts our initial assumption of the existence of a non-zero uniform order parameter. Finally, u_C^* diverges as $n \to 1$. For a single order parameter component, i.e., the Ising model, the effective non-linear coupling in the low-temperature phase is $u_{\text{eff}} = u\, m^{d-4}$, compare with Eq. (5.1), and the corresponding running coupling vanishes for $d < 4$, as $\tilde{m}(\ell) \to \infty$ under the renormalization group flow. Far below the critical point, the Ising model is therefore simply described by mean-field theory. This is in stark contrast to the $O(n)$-symmetric Heisenberg models with continuous rotational invariance ($n \geq 2$), where the $n - 1$ transverse Goldstone modes cause strong fluctuations and infrared singularities in the entire ordered phase.

5.5 *Appendix:* integrals in dimensional regularization

In this Appendix, we first list a number of useful formulas for dimensionally regularized momentum integrals:

$$\int_k \frac{1}{(m^2 + 2q \cdot k + k^2)^s} = \frac{\Gamma(s - d/2)}{(4\pi)^{d/2}\Gamma(s)} \frac{1}{(m^2 - q^2)^{s-d/2}} \,,$$

$$\int_k \frac{k_i}{(m^2 + 2q \cdot k + k^2)^s} = -\frac{\Gamma(s - d/2)}{(4\pi)^{d/2}\Gamma(s)} \frac{q_i}{(m^2 - q^2)^{s-d/2}} \,,$$

$$\int_k \frac{k_i\, k_j}{(m^2 + 2q \cdot k + k^2)^s} = \frac{1}{(4\pi)^{d/2}\Gamma(s)} \left[q_i\, q_j \frac{\Gamma(s - d/2)}{(m^2 - q^2)^{s-d/2}} \right.$$
$$\left. + \frac{1}{2} \delta_{ij} \frac{\Gamma(s - 1 - d/2)}{(m^2 - q^2)^{s-1-d/2}} \right] \,,$$

$$\int_k \frac{k_i\, k_j\, k_k}{(m^2 + 2q \cdot k + k^2)^s} = -\frac{1}{(4\pi)^{d/2}\Gamma(s)} \left[q_i\, q_j\, q_k \frac{\Gamma(s - d/2)}{(m^2 - q^2)^{s-d/2}} \right.$$
$$\left. + \frac{1}{2} \left(\delta_{ij}\, q_k + \delta_{jk}\, q_i + \delta_{ik}\, q_j\right) \frac{\Gamma(s - 1 - d/2)}{(m^2 - q^2)^{s-1-d/2}} \right] \,,$$

$$\int_k \frac{k_i\, k_j\, k_k\, k_l}{(m^2 + 2q \cdot k + k^2)^s} = \frac{1}{(4\pi)^{d/2}\Gamma(s)} \left[q_i\, q_j\, q_k\, q_l \frac{\Gamma(s - d/2)}{(m^2 - q^2)^{s-d/2}} \right.$$
$$+ \frac{1}{2} \left(\delta_{ij} q_k q_l + \delta_{ik} q_j q_l + \delta_{il} q_j q_k + \delta_{jk} q_i q_l \right.$$
$$\left. + \delta_{jl} q_i q_k + \delta_{kl} q_i q_j\right) \frac{\Gamma(s - 1 - d/2)}{(m^2 - q^2)^{s-1-d/2}}$$
$$\left. + \frac{1}{4} \left(\delta_{ij}\, \delta_{kl} + \delta_{ik}\, \delta_{jl} + \delta_{il}\, \delta_{jk}\right) \frac{\Gamma(s - 2 - d/2)}{(m^2 - q^2)^{s-2-d/2}} \right] \,,$$

etc. Each equation follows from the previous one via taking the derivative with respect to q_i.

In order to evaluate integrals which contain products of different terms in the denominator, Feynman's parametrization is useful:

$$\frac{1}{A^r \, B^s} = \frac{\Gamma(r+s)}{\Gamma(r)\Gamma(s)} \int_0^1 \frac{x^{r-1} (1-x)^{s-1}}{[x \, A + (1-x) \, B]^{r+s}} \, dx \,. \tag{5.112}$$

For the price of an additional parameter integration, one thereby achieves momentum integrals of the above form, albeit with a more complex denominator. Yet a decomposition into partial fractions, if possible, may be preferable.

As an application, let us extract the UV-singular part of the two-loop integral $D(q)$ introduced in Eq. (5.35). Feynman's parametrization (5.112) and the k' integration yield

$$D(q) = \int_0^1 dx \int_k \frac{1}{\tau + k^2} \int_{k'} \frac{1}{[\tau + x(q-k)^2 - 2x(q-k)k' + k'^2]^2}$$

$$= \frac{\Gamma(2 - d/2)}{(4\pi)^{d/2}} \int_0^1 dx \int_k \frac{1}{\tau + k^2} \frac{1}{[\tau + x(1-x)(q-k)^2]^{2-d/2}} \,.$$

Using Feynman's parametrization once more, the k integral can be performed as well:

$$D(q) = \frac{\Gamma(3 - d/2)}{(4\pi)^{d/2}} \int_0^1 \frac{dx}{[x(1-x)]^{2-d/2}}$$

$$\times \int_k \int_0^1 \frac{y^{1-d/2} \, dy}{\left[\tau \left(\frac{y}{x(1-x)} + 1 - y\right) + y \, q^2 - 2y(qk) + k^2\right]^{3-d/2}}$$

$$= \frac{\Gamma(3 - d)}{(4\pi)^d} \int_0^1 \frac{dx}{[x(1-x)]^{2-d/2}} \int_0^1 \frac{y^{1-d/2} \, dy}{\left[\tau \left(\frac{y}{x(1-x)} + 1 - y\right) + y(1-y)q^2\right]^{3-d}} \,.$$

Consequently,

$$\frac{\partial D(q)}{\partial q^2}\bigg|_{q=0} = -\frac{\Gamma(4-d)}{(4\pi)^d} \int_0^1 \frac{dx}{[x(1-x)]^{2-d/2}} \int_0^1 \frac{y^{2-d/2}(1-y) \, dy}{\left[\tau \left(\frac{y}{x(1-x)} + 1 - y\right)\right]^{4-d}}$$

$$= -\frac{\Gamma(1+\epsilon)}{\Gamma(1+\epsilon/2)^2} \frac{A_d^2 \, \tau^{-\epsilon}}{4\epsilon} \int_0^1 dx \, [x(1-x)]^{-\epsilon/2}$$

$$\times \int_0^1 dy \, y^{-\epsilon/2}(1-y) \left[\frac{y}{x(1-x)} + 1 - y\right]^{-\epsilon} \,,$$

where we have inserted $\epsilon = 4 - d$. In order to obtain the renormalization constant Z_S with minimal subtraction, we merely need the residuum at the ultraviolet pole $1/\epsilon$. Noting that $\Gamma(1+\epsilon)/\Gamma(1+\epsilon/2)^2 = 1 + O(\epsilon^2)$, and that the parameter integrals are regular in the limit $\epsilon \to 0$, we are left with

$$\left.\frac{\partial D(q)}{\partial q^2}\right|_{q=0}^{\text{sing.}} = -\frac{A_d^2\,\tau^{-\epsilon}}{4\,\epsilon}\int_0^1 (1-y)\,dy = -\frac{A_d^2\,\tau^{-\epsilon}}{8\,\epsilon}.$$

Problems

5.1 *Scaling dimension and primitive degree of divergence of* $\Gamma^{(\tilde{N},N)}$

Convince yourself that for the relaxational models A and B the scaling dimension of the reduced vertex function $\Gamma^{(\tilde{N},N)}$ (with the delta functions from translational invariance separated off) is given by Eq. (5.7), and then deduce the primitive degree of divergence for the lth order contribution (5.6).

5.2 *Dimensionally regularized integrals*

Using

$$\int_0^\infty x^{n-1}\,e^{-ax}\,dx = \frac{\Gamma(n)}{a^n}, \qquad \int_0^\infty x^{n-1}\,e^{-ax^2}\,dx = \frac{\Gamma(n/2)}{2\,a^{n/2}},$$

derive Eq. (5.10). For which values of d, σ, and s is the integral UV-divergent? Thus, confirm the validity of the formulas in the Appendix (Section 5.5).

5.3 *Relaxation constant renormalization for model A*

Evaluate the singular part of the integral in Eq. (5.37) or (5.38), and thus show that at $\tau = \mu^2$

$$\frac{Z_S}{Z_D} = 1 + \frac{n+2}{24}\,\ln\frac{4}{3}\,\frac{(u\,A_d\,\mu^{-\epsilon})^2}{\epsilon} + O(u^3),$$

which leads to the two-loop result (5.39) for the renormalization constant Z_D.

5.4 *Matching and identification of the scaling exponents*

Use the differential RG flow equations in the form given in the footnote on p. 188 and the matching condition $\tilde{\tau}(l) = 1$ to derive the relations (5.68) and (5.69) between the anomalous scaling dimensions and critical exponents for the relaxational models A and B.

5.5 *Order parameter critical exponent*

Show explicitly that to one-loop order, the ultraviolet poles in the equation of state (5.83) are taken care of by the Z factors of the theory in the high-temperature phase. Assuming this holds to all orders in perturbation theory, derive the scaling relation $\beta = \nu\,(d - 2 + \eta)/2$ for the order parameter critical exponent, $\langle S^n\rangle \sim (-\tau)^\beta$. Compute β to first order in $\epsilon = 4 - d$, and determine the logarithmic corrections to $\beta = 1/2$ in $d_c = 4$ dimensions.

References

Amit, D. J., 1984, *Field Theory, the Renormalization Group, and Critical Phenomena*, Singapore: World Scientific, chapters 6–9.

Bausch, R., H. K. Janssen, and H. Wagner, 1976, Renormalized field theory of critical dynamics, *Z. Phys. B Cond. Matt.* **24**, 113–127.

Brézin, E. and D. J. Wallace, 1973, Critical behavior of a classical Heisenberg ferromagnet with many degrees of freedom, *Phys. Rev. B* **7**, 1967–1974.

De Dominicis, C., E. Brézin, and J. Zinn-Justin, 1975. Field-theoretic techniques and critical dynamics. I. Ginzburg-Landau stochastic models without energy conservation, *Phys. Rev. B* **12**, 4945–4953.

Folk, R. and G. Moser, 2006, Critical dynamics: a field theoretical approach, *J. Phys. A: Math. Gen.* **39**, R207–R313.

Halperin, B. I., P. C. Hohenberg, and S.-k. Ma, 1972, Calculation of dynamic critical properties using Wilson's expansion methods, *Phys. Rev. Lett.* **29**, 1548–1551.

Hohenberg, P. C., 1967, Existence of long-range order in one and two dimensions, *Phys. Rev.* **158**, 383–386.

Hohenberg, P. C. and B. I. Halperin, 1977, Theory of dynamic critical phenomena, *Rev. Mod. Phys.* **49**, 435–479.

Itzykson, C. and J. M. Drouffe, 1989, *Statistical Field Theory*, Vol. I, Cambridge: Cambridge University Press, chapter 5.

Janssen, H. K., 1979, Field-theoretic methods applied to critical dynamics, in: *Dynamical Critical Phenomena and Related Topics*, ed. C. P. Enz, Lecture Notes in Physics, Vol. **104**, Heidelberg: Springer-Verlag, 26–47.

Janssen, H. K., 1992, On the renormalized field theory of nonlinear critical relaxation, in: *From Phase Transitions to Chaos*, eds. G. Györgyi, I. Kondor, L. Sasvári, and T. Tél, Singapore: World Scientific, 68–91.

Kamenev, A., 2011, *Field Theory of Non-equilibrium Systems*, Cambridge: Cambridge University Press, chapter 8.

Kötzler, J., D. Görlitz, R. Dombrowski, and M. Pieper, 1994, Goldstone-mode induced susceptibility-singularity extending to T_c of the Heisenberg ferromagnet EuS, *Z. Phys. B Cond. Matt.* **94**, 9–12.

Lawrie, I. D., 1981, Goldstone modes and coexistence in isotropic N-vector models, *J. Phys. A: Math. Gen.* **14**, 2489–2502.

Mazenko, G. F., 1976, Effect of Nambu–Goldstone modes on wave-number- and frequency-dependent longitudinal correlation functions, *Phys. Rev. B* **14**, 3933–3936.

Mermin, N. D. and H. Wagner, 1966, Absence of ferromagnetism or antiferromagnetism in one- or two-dimensional isotropic Heisenberg models, *Phys. Rev. Lett.* **17**, 1133–1136.

Nelson, D. R., 1976, Coexistence-curve singularities in isotropic ferromagnets, *Phys. Rev. B* **13**, 2222–2230.

Schäfer, L., 1978, Static and dynamic correlation functions of a Landau–Ginzburg model near the magnetization curve, *Z. Phys. B Cond. Matt.* **31**, 289–300.

Täuber, U. C. and F. Schwabl, 1992, Critical dynamics of the $O(n)$-symmetric relaxational models below the transition temperature, *Phys. Rev. B* **46**, 3337–3361.

Vasil'ev, A. N., 2004, *The Field Theoretic Renormalization Group in Critical Behavior Theory and Stochastic Dynamics*, Boca Raton: Chapman & Hall / CRC, chapter 5.

Wagner, H., 1966, Long-wavelength excitations and the Goldstone theorem in many-particle systems with 'broken symmetries', *Z. Phys.* **195**, 273–299.

Zinn-Justin, J., 1993, *Quantum Field Theory and Critical Phenomena*, Oxford: Clarendon Press, chapters 7, 9, 10, 22–27, 34.

Further reading

Adzhemyan, L. T., S. V. Novikov, and L. Sladkoff, 2008, Calculation of the dynamical critical exponent in the model A of critical dynamics to order ϵ^4, preprint arXiv:0808.1347, 1–5.

Cardy, J., 1996, *Scaling and Renormalization in Statistical Physics*, Cambridge: Cambridge University Press, chapters 5, 10.

Dengler, R., H. Iro, and F. Schwabl, 1985, Dynamical scaling functions for relaxational critical dynamics, *Phys. Lett. A* **111**, 121–124.

Dohm, V., 2013, Crossover from Goldstone to critical fluctuations: Casimir forces in confined $O(n)$-symmetric systems, *Phys. Rev. Lett.* **110**, 107207-1–5.

Dupuis, N., 2011, Infrared behavior in systems with a broken continuous symmetry: classical $O(N)$ model versus interacting bosons, *Phys. Rev. E* **83**, 031120-1–17.

Lawrie, I. D., 1985, Goldstone mode singularities in specific heats and non-ordering susceptibilities of isotropic systems, *J. Phys. A: Math. Gen.* **18**, 1141–1152.

Ma, S.-k., 1976, *Modern Theory of Critical Phenomena*, Reading: Benjamin–Cummings.

Mazenko, G. F., 2003, *Fluctuations, Order, and Defects*, Hoboken: Wiley–Interscience, chapter 5.

McComb, W. D., 2004, *Renormalization Methods: a Guide for Beginners*, Oxford: Oxford University Press.

Parisi, G., 1988, *Statistical Field Theory*, Redwood City: Addison–Wesley.

Pawlak, A. and R. Erdem, 2011, Dynamic response function in Ising systems below T_c, *Phys. Rev. B* **83**, 094415-1–8.

Pawlak, A. and R. Erdem, 2013, Effect of magnet fields on dynamic response function in Ising systems, *Phys. Lett. A* **377**, 2487–2493.

Schorgg, A. M. and F. Schwabl, 1994, Theory of ultrasound attenuation at incommensurate phase transitions, *Phys. Rev. B* **49**, 11 682–11 703.

Täuber, U. C., 2007, Field theory approaches to nonequilibrium dynamics, in: *Ageing and the Glass Transition*, eds. M. Henkel, M. Pleimling, and R. Sanctuary, Lecture Notes in Physics **716**, Berlin: Springer, chapter 7, 295–348.

Täuber, U. C. and F. Schwabl, 1993, Influence of cubic and dipolar anisotropies on the static and dynamic coexistence anomalies of the time-dependent Ginzburg–Landau models, *Phys. Rev. B* **48**, 186–209.

6

Hydrodynamic modes and reversible mode couplings

Equipped with the field theory representation of non-linear Langevin equations, the tools of dynamic perturbation theory, and the dynamic renormalization group introduced in Chapters 4 and 5, we are now in the position to revisit models for dynamic critical behavior that entail reversible mode couplings and other conserved hydrodynamic modes. We have already encountered some of these in Section 3.3. In models C and D, respectively, a non-conserved or conserved n-component order parameter is coupled to a conserved scalar field, the energy density. Through a systematic renormalization group analysis, we may critically assess the earlier predictions from scaling theory, and discuss the stability of fixed points characterized by strong dynamic scaling, wherein the order parameter and conserved non-critical mode fluctuate with equal rates, and weak dynamic scaling regimes, where these characteristic time scales differ. Next we investigate isotropic ferromagnets (model J), with the conserved spin density subject to reversible precession in addition to diffusive relaxation. Exploiting rotational invariance, we can now firmly establish the scaling relation $z = (d + 2 - \eta)/2$. Similar symmetry arguments yield a scaling relation for the dynamic exponents associated with the order parameter and the non-critical fields in the $O(n)$-symmetric SSS model that encompasses model E for planar ferromagnets and superfluid helium 4 (for $n = 2$), and model G for isotropic antiferromagnets ($n = 3$). There exist competing strong- and weak-scaling fixed points, with the former stable to one-loop order, and characterized by $z = d/2$ for all slow modes. Lastly, we study model H for the critical dynamics of binary liquids; its decisive symmetry is invariance under Galilean transformations. For the models with reversible non-linear mode couplings, a self-consistent one-loop approximation for the line widths recovers the results of mode-coupling theory, which has been remarkably successful in describing the dynamic scaling functions measured in experiments.

6.1 Coupling to a conserved scalar field

As discussed in Chapter 3, in order to appropriately capture the long-time and long-wavelength properties of a dynamical system, in addition to the order parameter fluctuations all conserved modes need to be accounted for carefully. For example, in Section 3.3 we have argued on the basis of scaling theory that the *static* coupling to the conserved energy density modifies the dynamic critical exponent of the relaxational model A with non-conserved order parameter to $z = 2 + \alpha/\nu$, provided $\alpha > 0$, i.e., the energy fluctuations diverge at the critical point. Armed with the renormalization group as a powerful tool, we are now in a position to check the predictions of dynamic scaling theory, and may systematically study the effect of additional conserved fields on dynamic critical behavior.[1]

6.1.1 Perturbation expansion and renormalization

Let us consider models C and D with non-conserved and conserved n-component vector order parameters, respectively. Based on a coupling Hamiltonian of the form (3.48) with S^2 replaced by $\vec{S}^2$ as appropriate for $O(n)$ order parameter symmetry, a purely relaxational ansatz for the ensuing dynamics yields coupled Langevin equations that generalize Eqs. (3.50) and (3.51) in a straightforward manner. Upon absorbing the specific heat C into the energy density field, $\rho \to \sqrt{C}\rho$, and likewise $g \to \sqrt{C}g$, we find for the order parameter dynamics of models C ($a = 0$) and D ($a = 2$)

$$\frac{\partial S^\alpha(x, t)}{\partial t} = -D\,(i\nabla)^a \left[(r - \nabla^2)S^\alpha(x, t) + \frac{u}{6}\,S^\alpha(x, t)\sum_\beta S^\beta(x, t)^2 \right.$$

$$\left. + g\,\rho(x, t)\,S^\alpha(x, t) \right] + \zeta^\alpha(x, t)\,, \tag{6.1}$$

and for the relaxation of the conserved energy density,

$$\frac{\partial \rho(x, t)}{\partial t} = \lambda\,\nabla^2 \left[\rho(x, t) + \frac{g}{2}\sum_\alpha S^\alpha(x, t)^2 \right] + \eta(x, t)\,. \tag{6.2}$$

The stochastic forces are assumed to have vanishing averages, and setting $k_B T = 1$ their correlations obey the Einstein relations

$$\langle \zeta^\alpha(x, t)\zeta^\beta(x', t')\rangle = 2D\,(i\nabla_x)^a\,\delta(x - x')\delta(t - t')\delta^{\alpha\beta}\,, \tag{6.3}$$

$$\langle \eta(x, t)\eta(x', t')\rangle = -2\lambda\,\nabla_x^2\,\delta(x - x')\delta(t - t')\,. \tag{6.4}$$

---

[1] Halperin, Hohenberg, and Ma (1974); Brézin and De Dominicis (1975); Folk and Moser (2003, 2004); Vasil'ev (2004).

Cast into the Janssen–De Dominicis response functional, we obtain, in addition to the model A/B action (4.14), (4.15), the new Gaussian part

$$A_0[\widetilde{\rho}, \rho] = \int d^d x \int dt \left[\widetilde{\rho}(x, t) \left(\frac{\partial}{\partial t} - \lambda \nabla^2 \right) \rho(x, t) + \lambda \, \widetilde{\rho}(x, t) \, \nabla^2 \widetilde{\rho}(x, t) \right],$$
(6.5)

which yields the diffusive energy density propagator

$$\langle \rho(q, \omega) \widetilde{\rho}(q', \omega') \rangle = \frac{1}{-i\omega + \lambda q^2} (2\pi)^{d+1} \delta(q + q') \delta(\omega + \omega')$$
(6.6)

along with the conserved noise vertex, and the two three-point vertices

$$A_{C/D}[\widetilde{S}, S, \widetilde{\rho}, \rho] = D g \int d^d x \int dt \sum_\alpha \left[\widetilde{S}^\alpha(x, t) (i\nabla)^a S^\alpha(x, t) \rho(x, t) \right.$$

$$\left. - \frac{1}{2w} \, \widetilde{\rho}(x, t) \, \nabla^2 S^\alpha(x, t)^2 \right]$$
(6.7)

that couple the vectorial order parameter to the scalar energy density. Here, we have introduced the ratio

$$w = \frac{D}{\lambda}$$
(6.8)

of the order parameter relaxation constant D and thermal conductivity λ.

In terms of an arbitrary momentum scale μ, the scaling dimensions of the conserved fields and additional parameters, following from the form of the action (6.5), (6.7) and Eqs. (5.3), (5.4), are

$$[\widetilde{\rho}(x, t)] = \mu^{d/2} = [\rho(x, t)],$$
(6.9)

$$[\lambda] = \mu^a, \quad [w] = \mu^{-a}, \quad [g] = \mu^{2-d/2}.$$
(6.10)

Just as the non-linearity u, the coupling between order parameter and conserved energy density g thus becomes marginal in $d_c = 4$ dimensions. Moreover, whereas the time scale ratio w is marginal for model C with non-conserved order parameter, it constitutes an *irrelevant* parameter for model D, $[w] = \mu^{-2}$. Under renormalization, we therefore expect $w \to 0$, which means that near the phase transition the conserved model D order parameter will relax much more slowly than the non-critical mode.

In the diagrammatic representation of the perturbation expansion, we choose to depict the propagator (6.6) as a dashed line, see Fig. 6.1(a). The noise vertex for the conserved field as well as the anharmonic coupling vertices $\propto g$ stemming from the coupled Langevin equations for the order parameter and energy densities are shown in Figs. 6.1(b), (c), and (d), respectively. Henceforth, we need only be concerned with the fluctuation corrections to the purely *dynamical* quantities D

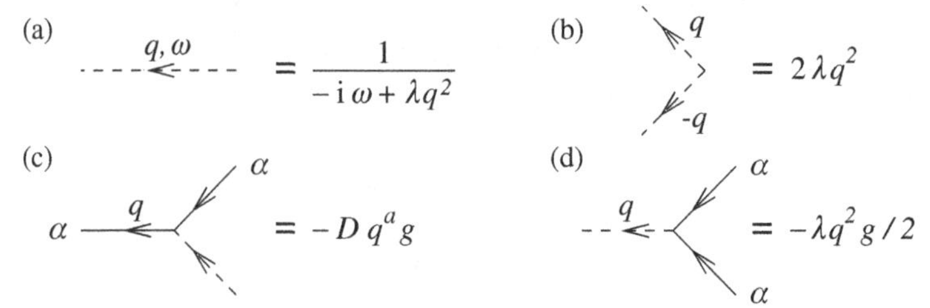

Fig. 6.1 Elements of dynamic perturbation theory for the $O(n)$-symmetric models C and D, supplementing those depicted in Fig. 4.1: (a) energy density propagator; (b) two-point noise vertex for the conserved field; (c) and (d) anharmonic three-point vertices from the Langevin equations for S^α and ρ.

and λ, in addition to the coupling g, since the Einstein relations guarantee that the static properties of all the relaxational models are those of the equilibrium $O(n)$-symmetric Φ^4 theory, but with shifted non-linear coupling $u \to u' = u - 3\,g^2$ (Problem 6.1).

In accord with Eqs. (5.19) and (5.23), we introduce the renormalized fields

$$\rho_R = Z_\rho^{1/2}\rho\,, \quad \widetilde{\rho}_R = Z_{\widetilde{\rho}}^{1/2}\widetilde{\rho} \tag{6.11}$$

and dimensionless renormalized parameters

$$\lambda_R = Z_\lambda \lambda\,\mu^{-a}\,, \quad g_R^2 = Z_g g^2 A_d\,\mu^{-\epsilon}\,, \tag{6.12}$$

with $\epsilon = 4 - d$. We must now label the cumulants and vertex functions with four superscripts, two each for the incoming and outgoing order parameter and conserved fields. The renormalized vertex functions then become

$$\Gamma_R^{(\widetilde{N},N;\widetilde{M},M)} = Z_{\widetilde{S}}^{-\widetilde{N}/2}Z_S^{-N/2}Z_{\widetilde{\rho}}^{-\widetilde{M}/2}Z_\rho^{-M/2}\,\Gamma^{(\widetilde{N},N;\widetilde{M},M)}\,. \tag{6.13}$$

But the new renormalization constants are not all independent. As for the order parameter itself, the fluctuation-dissipation theorem for the dynamic response and correlation functions of the energy density implies $Z_\lambda = (Z_\rho/Z_{\widetilde{\rho}})^{1/2}$ in analogy with Eq. (5.24). Moreover, precisely as for model B, owing to the momentum dependence of the three-point vertex with an outgoing $\widetilde{\rho}$ leg, see Fig. 6.1(d), one finds to *all* orders in the perturbation expansion for the two-point vertex functions of the conserved scalar density:

$$\Gamma^{(0,0;1,1)}(q = 0, \omega) = i\omega\,, \tag{6.14}$$

$$\left.\frac{\partial}{\partial q^2}\,\Gamma^{(0,0;2,0)}(q, \omega)\right|_{q=0} = -2\lambda\,, \tag{6.15}$$

whence we infer the exact relations

$$Z_\rho = Z_{\widetilde{\rho}}^{-1} = Z_\lambda\,. \tag{6.16}$$

For model D, in addition Eq. (5.25) holds, i.e., $Z_S = Z_{\widetilde{S}}^{-1} = Z_D$.

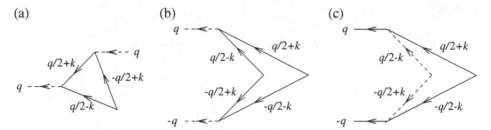

Fig. 6.2 Feynman diagrams for the model C/D two-point vertex functions (a) $\Gamma^{(0,0;1,1)}(q, \omega)$, (b) $\Gamma^{(0,0;2,0)}(q, \omega)$, and (c) $\Gamma^{(2,0;0,0)}(q, \omega)$ to one-loop order.

Figure 6.2 depicts the one-loop Feynman graphs for the two-point vertex functions $\Gamma^{(0,0;1,1)}(q, \omega)$, $\Gamma^{(0,0;2,0)}(q, \omega)$, and $\Gamma^{(2,0;0,0)}(q, \omega)$. A straightforward calculation yields for the diagram in Fig. 6.2(a):

$$\Gamma^{(0,0;1,1)}(q, \omega) = i\omega + \lambda q^2 \left[1 - \frac{n}{2} g^2 \int_k \frac{1}{r + \left(\frac{q}{2} + k\right)^2} \frac{1}{r + \left(\frac{q}{2} - k\right)^2} \right.$$

$$\left. \times \left(1 - \frac{i\omega}{i\omega + \Delta\left(\frac{q}{2} + k\right) + \Delta\left(\frac{q}{2} - k\right)} \right) \right], \quad (6.17)$$

where we have employed the abbreviation (4.83). This explicit one-loop result of course confirms Eq. (6.14). Upon replacing r with τ,

$$\left. \frac{\partial}{\partial q^2} \Gamma_R^{(0,0;1,1)}(q, 0) \right|_{q=0} = \frac{\lambda_R}{Z_\lambda} \left[1 - \frac{n}{2} g^2 \int_k \frac{1}{(\tau + k^2)^2} \right], \quad (6.18)$$

whence at the normalization point $\tau = \mu^2$ with the aid of Eq. (5.12)

$$Z_\lambda = 1 - \frac{n}{2} \frac{g^2 A_d \mu^{-\epsilon}}{\epsilon}. \quad (6.19)$$

Next we turn to the renormalization of the noise vertices, Fig. 6.2(b) and (c). For the conserved non-critical field, one arrives at

$$\Gamma^{(0,0;2,0)}(q, \omega) = -2\lambda q^2 \left[1 + \lambda q^2 \frac{n}{2} g^2 \int_k \frac{1}{r + \left(\frac{q}{2} + k\right)^2} \frac{1}{r + \left(\frac{q}{2} - k\right)^2} \right.$$

$$\left. \times \operatorname{Re} \frac{1}{i\omega + \Delta\left(\frac{q}{2} + k\right) + \Delta\left(\frac{q}{2} - k\right)} \right]. \quad (6.20)$$

Since two of the vertices in Fig. 6.1(d) enter in Fig. 6.2(b), the one-loop correction vanishes $\propto q^4$, in agreement with (6.15). However, we still need the renormalization of the order parameter relaxation constant D, which we may gather from Fig. 6.2(c), whence

$$\Gamma^{(2,0;0,0)}(q, \omega) = -2Dq^a \left[1 + Dq^a g^2 \int_k \frac{1}{r + \left(\frac{q}{2} + k\right)^2} \right.$$

$$\left. \times \operatorname{Re} \frac{1}{i\omega + \Delta\left(\frac{q}{2} + k\right) + \lambda\left(\frac{q}{2} - k\right)^2} \right]. \quad (6.21)$$

For model C ($a = 0$), this becomes

$$\Gamma^{(2,0;0,0)}(0, 0) = -\frac{2D}{Z_{\tilde{S}}Z_D} \left[1 + g^2 \int_k \frac{1}{\tau + k^2} \frac{1}{\tau + (1 + 1/w)k^2} \right], \quad (6.22)$$

which yields in minimal subtraction

$$Z_{\tilde{S}}Z_D = \frac{Z_S}{Z_D} = 1 + \frac{g^2}{1 + 1/w} \frac{A_d \, \mu^{-\epsilon}}{\epsilon}.$$

Hence, since $Z_S = 1 + O(u'^2)$, we find to first order in the non-linear couplings u' and g^2,

$$Z_D = 1 - \frac{g^2}{1 + 1/w} \frac{A_d \, \mu^{-\epsilon}}{\epsilon}. \quad (6.23)$$

For conserved order parameter (model D, $a = 2$), on the other hand, the integral in (6.21) is UV-finite, and therefore we arrive again at the relations (5.25). The same result is obtained by formally taking the limit $w \to 0$, as anticipated above. Identical conclusions follow from the analysis of the two-point vertex function $\Gamma^{(1,1;0,0)}(q, \omega)$, see Problem 6.1(a).

Finally, we need to determine the renormalization of the coupling g^2. The one-loop Feynman diagrams contributing to the three-point vertex function $\Gamma^{(0,2;1,0)}(-q, 0; \frac{q}{2}, 0; \frac{q}{2}, 0)$ are shown in Fig. 6.3. The corresponding analytic expression reads

$$\Gamma^{(0,2;1,0)}\left(-q, 0; \frac{q}{2}, 0; \frac{q}{2}, 0\right)$$

$$= \lambda q^2 g \left[1 - \frac{n+2}{6} u \int_k \frac{1}{r + \left(\frac{q}{2} + k\right)^2} \frac{1}{r + \left(\frac{q}{2} - k\right)^2} \right.$$

$$+ D^2 g^2 \int_k \frac{\left(\frac{q}{2} + k\right)^a \left(\frac{q}{2} - k\right)^a}{\Delta\left(\frac{q}{2} + k\right) + \Delta\left(\frac{q}{2} - k\right)} \left(\frac{1}{\Delta\left(\frac{q}{2} + k\right) + \lambda k^2} + \frac{1}{\Delta\left(\frac{q}{2} - k\right) + \lambda k^2} \right)$$

$$\left. + 2D^2 g^2 w \int_k \frac{\left(\frac{q}{2} + k\right)^2}{r + k^2} \frac{(q+k)^a}{\Delta(q + k) + \Delta(k)} \frac{1}{\Delta(k) + \lambda\left(\frac{q}{2} + k\right)^2} \right], \quad (6.24)$$

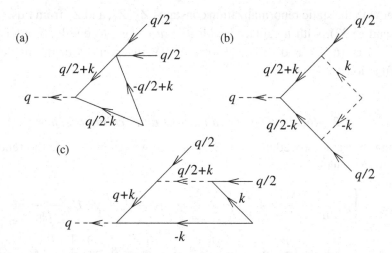

Fig. 6.3 One-loop Feynman diagrams for the model C/D three-point vertex function $\Gamma^{(0,2;1,0)}(-q, 0; \frac{q}{2}, 0; \frac{q}{2}, 0)$.

where we have symmetrized with respect to the internal momentum k in the first term, i.e., the contribution from Fig. 6.3(a). Thus,

$$
\frac{\partial}{\partial q^2} \Gamma_R^{(0,2;1,0)} \left(-q, 0; \frac{q}{2}, 0; \frac{q}{2}, 0\right)\bigg|_{q=0} = \frac{\lambda g}{Z_S (Z_{\tilde{\rho}} Z_g)^{1/2} Z_\lambda}
$$

$$
\times \left[1 - \frac{n+2}{6} u \int_k \frac{1}{(\tau + k^2)^2} + g^2 w \int_k \frac{1}{\tau + k^2} \frac{1}{w k^a (\tau + k^2) + k^2} \right.
$$

$$
\left. + g^2 \int_k \frac{k^2}{(\tau + k^2)^2} \frac{1}{w k^a (\tau + k^2) + k^2} \right], \tag{6.25}
$$

whence we obtain for model C, using Eq. (6.16), in dimensional regularization with minimal subtraction at $\tau = \mu^2$

$$
Z_S (Z_\lambda Z_g)^{1/2} = 1 - \frac{n+2}{6} \frac{u A_d \mu^{-\epsilon}}{\epsilon} + \frac{g^2 A_d \mu^{-\epsilon}}{\epsilon},
$$

since the two terms $\propto g^2$ combine readily to the integral (5.12).

Remarkably, the same final result follows for model D, where only the last integral in (6.25) contributes in the limit $w \to 0$. In both cases, therefore, upon inserting Eq. (6.19) we find to one-loop order

$$
Z_g = 1 - \frac{n+2}{3} \frac{u A_d \mu^{-\epsilon}}{\epsilon} + \frac{n+4}{2} \frac{g^2 A_d \mu^{-\epsilon}}{\epsilon}
$$

$$
= 1 - \frac{n+2}{3} \frac{u' A_d \mu^{-\epsilon}}{\epsilon} - \frac{n}{2} \frac{g^2 A_d \mu^{-\epsilon}}{\epsilon}. \tag{6.26}
$$

Together with the static renormalization constants Z_S, Z_τ, and Z_u from Eqs. (5.29), (5.30), and (5.32), with u replaced with $u' = u - 3\,g^2$, to one-loop order the Z factors (6.19), (6.23), and (6.26) absorb all UV divergences in the model C/D vertex functions (6.13).

6.1.2 Model C/D fixed points, dynamic scaling regimes

Following the same procedure as in Section 5.3, we readily derive the renormalization group equation

$$
\left[\mu \frac{\partial}{\partial \mu} + \frac{\widetilde{N}}{2} \gamma_{\widetilde{S}} + \frac{N}{2} \gamma_S + \frac{\widetilde{M}}{2} \gamma_{\widetilde{\rho}} + \frac{M}{2} \gamma_\rho + \gamma_D\, D_R \frac{\partial}{\partial D_R} \right.
$$

$$
\left. + \gamma_\lambda\, \lambda_R \frac{\partial}{\partial \lambda_R} + \gamma_\tau\, \tau_R \frac{\partial}{\partial \tau_R} + \beta_{u'} \frac{\partial}{\partial u'_R} + \beta_g \frac{\partial}{\partial g^2_R} \right]
$$

$$
\times \Gamma_R^{(\widetilde{N},N;\widetilde{M},M)}(\mu, D_R, \lambda_R, \tau_R, u'_R, g^2_R) = 0 \,, \tag{6.27}
$$

where we have defined

$$
\gamma_{\widetilde{\rho}} = \mu \frac{\partial}{\partial \mu}\bigg|_0 \ln Z_{\widetilde{\rho}} \,, \qquad \gamma_\rho = \mu \frac{\partial}{\partial \mu}\bigg|_0 \ln Z_\rho \,, \tag{6.28}
$$

$$
\gamma_\lambda = \mu \frac{\partial}{\partial \mu}\bigg|_0 \ln \frac{\lambda_R}{\lambda} = -a + \mu \frac{\partial}{\partial \mu}\bigg|_0 \ln Z_\lambda \,, \tag{6.29}
$$

and the RG beta functions

$$
\beta_{u'} = \mu \frac{\partial}{\partial \mu}\bigg|_0 u'_R = u'_R \left(d - 4 + \mu \frac{\partial}{\partial \mu}\bigg|_0 \ln Z'_u \right) \,, \tag{6.30}
$$

$$
\beta_g = \mu \frac{\partial}{\partial \mu}\bigg|_0 g^2_R = g^2_R \left(d - 4 + \mu \frac{\partial}{\partial \mu}\bigg|_0 \ln Z_g \right) \,, \tag{6.31}
$$

$$
\beta_w = \mu \frac{\partial}{\partial \mu}\bigg|_0 w_R = w_R\,(\gamma_D - \gamma_\lambda) \tag{6.32}
$$

in addition to Wilson's flow functions (5.42)–(5.44). By means of the beta functions we may introduce running couplings $\widetilde{u}'(\ell)$, $\widetilde{g}^2(\ell)$, and $\widetilde{w}(\ell)$ as in Eq. (5.57).

Within the one-loop approximation, and for model C ($a = 0$), the new flow functions read explicitly

$$
\gamma_\lambda = \gamma_\rho = -\gamma_{\widetilde{\rho}} = \frac{n}{2}\, g^2_R \,, \tag{6.33}
$$

$$
\gamma_S = 0 \,, \qquad \gamma_D = -\frac{\gamma_{\widetilde{S}}}{2} = \frac{g^2_R}{1 + 1/w_R} \,, \tag{6.34}
$$

whereas $\gamma_D = 0$ for model D. Consequently, we arrive at

$$\beta_w = w_R \, g_R^2 \left(\frac{1}{1 + 1/w_R} - \frac{n}{2} \right). \tag{6.35}$$

For model C, the RG beta function for the relaxation rate ratio w yields three fixed points, provided $g_R^2 > 0$, namely $w_0^* = \infty$, $w_C^* = n/(2 - n)$, which is positive for $0 < n < 2$, and $w_D^* = 0$. In order to establish the regimes of stability for these fixed points, we calculate

$$\frac{\partial \beta_w}{\partial w_R} = g_R^2 \left[\frac{w_R \, (2 + w_R)}{(1 + w_R)^2} - \frac{n}{2} \right], \tag{6.36}$$

and recall that this derivative must be positive at an IR-stable fixed point $w^* < \infty$. Therefore, $w_C^* = 1$ is stable for model C with scalar order parameter ($n = 1$), whereas for $n \geq 2$ the renormalization group flow drives $\tilde{w}(\ell)$ towards $w_0^* = \infty$. In contrast, a conserved order parameter yields the sole fixed point $w_D^* = 0$. For model C with non-conserved order parameter, this model D fixed point remains unstable for all values of n.

Notice that the dynamic parameter w_R does not affect the RG flow for the static couplings u_R' and g_R^2, since

$$\beta_{u'} = u_R' \left(-\epsilon + \frac{n + 8}{6} u_R' \right), \tag{6.37}$$

$$\beta_g = g_R^2 \left(-\epsilon + \frac{n + 2}{3} u_R' + \frac{n}{2} g_R^2 \right). \tag{6.38}$$

The associated stability matrix reads

$$\Lambda = \begin{pmatrix} \partial \beta_{u'} / \partial u_R' & \partial \beta_{u'} / \partial g_R^2 \\ \partial \beta_g / \partial u_R' & \partial \beta_g / \partial g_R^2 \end{pmatrix} \tag{6.39}$$

$$= \begin{pmatrix} -\epsilon + (n + 8) u_R'/3 & 0 \\ (n + 2)g_R^2/3 & -\epsilon + (n + 2)u_R'/3 + n \, g_R^2 \end{pmatrix}.$$

In $d > d_c = 4$ dimensions ($\epsilon < 0$), the Gaussian fixed point $u_0^* = 0$ becomes stable. From Eq. (6.38) we see that then necessarily $g_0^* = 0$: both non-linearities become irrelevant above the upper critical dimension, implying mean-field dynamic critical exponents $z_S = 2 + a$ and $z_\rho = 2$. For $\epsilon > 0$, inserting the Heisenberg fixed point $u_H^* = 6\epsilon/(n + 8)$ into Eq. (6.38) yields a non-trivial fixed point $g_C^{*2} = 2(4 - n)\epsilon/n(n + 8) > 0$ provided $n < 4$. Indeed, one finds that this model C fixed point is IR-stable in this regime, whereas $\tilde{g}(\ell) \to g_0^* = 0$ for $n \geq 4$. The stability of the fixed points in the (u_R', g_R^2) plane thus depends on both the spatial dimension d *and* the number of order parameter components n.

Since by means of the hyperscaling relation (1.77) the critical exponent of the specific heat becomes

$$\alpha = 2 - dv = \frac{4 - n}{2(n + 8)} \epsilon + O(\epsilon^2) \,, \tag{6.40}$$

we may rewrite to this order in ϵ

$$g_C^{*2} = \frac{4}{n} \alpha = \frac{2}{n} \frac{\alpha}{v} \,. \tag{6.41}$$

A non-zero fixed-point coupling between the order parameter and the energy densities hence requires that the critical exponent α be positive. To one-loop order, this is the case if $n < 4$. The dynamical critical exponents z_S and z_ρ that describe the divergences of the characteristic relaxation times for the order parameter and the conserved density, respectively, follow from the solutions of the renormalization group equations for the corresponding two-point correlation functions near an infrared-stable RG fixed point,

$$z_S = 2 + a + \gamma_D^* \,, \quad z_\rho = 2 + a + \gamma_\lambda^* \,. \tag{6.42}$$

For model C ($a = 0$) with non-conserved order parameter, we find three dynamic scaling regimes to one-loop order. (i) For a scalar order parameter with Ising symmetry ($n = 1$), the stable critical one-loop fixed point is

$$u_I^* = \frac{2\epsilon}{3} \,, \quad g_C^{*2} = \frac{2\epsilon}{3} \,, \quad w_C^* = 1 \,. \tag{6.43}$$

This describes a *strong scaling regime* with equal dynamic exponents for the order parameter and the conserved field,

$$z_S = z_\rho = 2 + \frac{\alpha}{v} = 2 + \frac{\epsilon}{3} + O(\epsilon^2) \,, \tag{6.44}$$

as anticipated in Section 3.3. (ii) For an $O(n)$-symmetric order parameter with $2 \leq n < 4$, the stable critical fixed point becomes

$$u_H^* = \frac{6\epsilon}{n + 8} \,, \quad g_C^{*2} = \frac{2(4 - n)\epsilon}{n(n + 8)} \,, \quad w_0^* = \infty \,, \tag{6.45}$$

leading to *weak dynamic scaling* with

$$z_S = 2 + \frac{2(4 - n)}{n(n + 8)} \epsilon + O(\epsilon^2) = 2 + \frac{2}{n} \frac{\alpha}{v} \leq z_\rho = 2 + \frac{\alpha}{v} \,. \tag{6.46}$$

For $n > 2$, the order parameter and scalar conserved density fluctuate on different time scales. For $n = 2$, on the other hand, to order ϵ we recover $z_S = z_\rho$. (iii) Finally, for $n \geq 4$, one has

$$u_H^* = \frac{6\epsilon}{n + 8} \,, \quad g_0^{*2} = 0 \,, \quad w_0^* = \infty \,, \tag{6.47}$$

whence to one-loop order $z_S = z_\rho = 2$.

More generally, for $\alpha < 0$ the order parameter and conserved energy density dynamics become *decoupled* in the vicinity of the critical point. For the order parameter, purely model A dynamics ensues, while the non-critical energy density relaxes diffusively,

$$z_S = 2 + c\,\eta \,, \quad z_\rho = 2 \,, \tag{6.48}$$

with $c = 6\ln\frac{4}{3} - 1 + O(\epsilon)$, see Eq. (5.73). For the $O(n)$-symmetric Landau–Ginzburg–Wilson Hamiltonian in $d > 2$ dimensions, $\alpha < 0$ ensues for $n \geq 2$. In effect, as has now been established through higher-order loop calculations,[2] this eliminates the weak dynamic scaling regime (ii), leaving us with the strong scaling exponents (6.44) for $n = 1$, and model A order parameter scaling (6.48) for $n \geq 2$.

For model D, the conserved order parameter ($a = 2$) always relaxes much slower than the also conserved, but non-critical energy density near the phase transition, whence $\widetilde{w}(\ell) \to w_D^* = 0$ asymptotically. Notice that the identity (5.25) implies $\gamma_D = \gamma_S$, and hence the model B scaling relation (5.69). Here, we have only *two* different dynamic scaling regimes: (i) for $\alpha > 0$ ($n < 2$), the conserved field is influenced by the critical order parameter variable,

$$u_H^* = \frac{6\epsilon}{n+8} \,, \quad g_C^{*2} = \frac{2(4-n)\epsilon}{n(n+8)} \,, \quad w_D^* = 0 \,, \tag{6.49}$$

whence the dynamic critical exponents become

$$z_S = 4 - \eta \,, \quad z_\rho = 2 + \frac{\alpha}{\nu} \,, \tag{6.50}$$

whereas (ii) for $\alpha \leq 0$ ($n \geq 2$)

$$u_H^* = \frac{6\epsilon}{n+8} \,, \quad g_0^{*2} = 0 \,, \quad w_D^* = 0 \,, \tag{6.51}$$

with the decoupled model B dynamics described by the scaling exponents

$$z_S = 4 - \eta \,, \quad z_\rho = 2 \,. \tag{6.52}$$

6.2 Reversible mode couplings in isotropic ferromagnets

We next apply our field theory tools and the dynamic renormalization group to dynamical models that incorporate reversible non-linear mode couplings, and thus establish a solid basis for our earlier scaling theory results for the dynamic critical exponents z (Section 3.3). We start with model J for isotropic ferromagnets,[3] subsequently study the $O(n)$-symmetric Sasvári–Schwabl–Szépfalusy (SSS) model

---

[2] Folk and Moser (2003, 2004).
[3] Ma and Mazenko (1974, 1975); Gunton and Kawasaki (1975); Bausch, Janssen, and Wagner (1976); Hohenberg and Halperin (1977); Janssen (1979); Zinn-Justin (1993); Frey and Schwabl (1994); Vasil'ev (2004); Folk and Moser (2006).

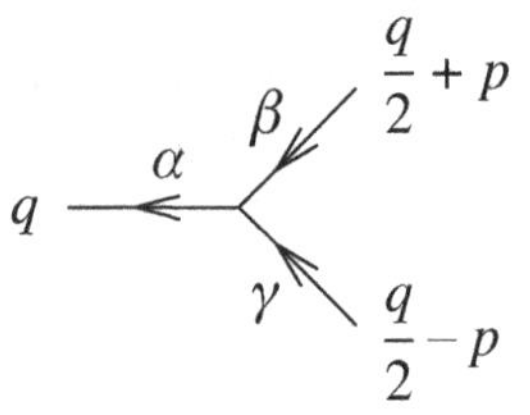

Fig. 6.4 Mode-coupling three-point vertex for model J.

that contains, as special cases for $n = 2$ and $n = 3$ respectively, model E for planar ferromagnets (i.e., the dynamic XY model) and model G for isotropic antiferromagnets, and finally investigate model H for the critical dynamics at the gas–liquid transition, or more generally, the phase separation in binary fluids.

6.2.1 Model J dynamic field theory and renormalization

In Section 3.3, we established the Langevin equations (3.67) governing the three conserved order parameter components in isotropic ferromagnets (model J). These consist of reversible precession of the magnetization density in the effective field in addition to diffusive relaxation,

$$\frac{\partial S^\alpha(x, t)}{\partial t} = g \sum_{\beta\gamma} \epsilon^{\alpha\beta\gamma} S^\beta(x, t) \left[\nabla^2 S^\gamma(x, t) + h^\gamma(x, t)\right]$$

$$+ D\nabla^2 \left[\left(r - \nabla^2 + \frac{u}{6}\vec{S}(x, t)^2\right) S^\alpha(x, t) - h^\alpha(x, t)\right] + \zeta^\alpha(x, t),$$

$$(6.53)$$

with $\langle \zeta^\alpha \rangle = 0$ and the noise correlations $(k_B T = 1)$

$$\langle \zeta^\alpha(x, t)\, \zeta^\beta(x', t') \rangle = -2D\,\nabla_x^2\, \delta(x - x')\delta(t - t')\,\delta^{\alpha\beta}\,. \qquad (6.54)$$

The corresponding response functional contains, in addition to the model B contributions (4.14) and (4.15), with $a = 2$, the mode-coupling vertex

$$\mathcal{A}_J[\widetilde{S}, S] = -g \int d^d x \int dt \sum_{\alpha,\beta,\gamma} \epsilon^{\alpha\beta\gamma} \widetilde{S}^\alpha(x, t) S^\beta(x, t)[\nabla^2 S^\gamma(x, t) + h^\gamma(x, t)].$$

$$(6.55)$$

Its graphical representation (for $h = 0$) is depicted in Fig. 6.4; notice that we have symmetrized with respect to the incoming order parameter fields, $-\frac{g}{2}(q_\gamma^2 - q_\beta^2) = g\,(q \cdot p)$ with the wavevectors assigned here. This three-point vertex supplements the elements of the diagrammatic perturbation expansion of model B as shown in Fig. 4.1.

Power counting by means of Eqs. (5.3) and (5.4) yields the scaling dimension for the reversible mode-coupling strength,

$$[g] = \mu^{3-d/2} , \tag{6.56}$$

wherefrom we infer that its upper critical dimension, where $[g] = \mu^0$, is $d'_c = 6$. Critical statics and dynamics for the isotropic ferromagnet are therefore characterized by *different* critical dimensions, and cannot be treated within a single ϵ expansion framework. Fortunately, though, in thermal *equilibrium*, time-dependent phenomena can be treated as fully decoupled from the statics whose critical properties are governed by the scaling exponents of the three-component $(n = 3)$ Landau–Ginzburg–Wilson Hamiltonian, i.e., the two independent critical exponents $\eta = 5\epsilon^2/242 + O(\epsilon^3)$ and $\nu = 1/2 + 5\epsilon/44 + O(\epsilon^2)$, see Eqs. (5.71) and (5.72).

Moreover, since the mode-coupling vertex is proportional to the outgoing momentum, we still have to all orders in perturbation theory (see Fig. 6.5(a) for the corresponding one-loop Feynman graph)

$$\Gamma^{(1,1)}(q = 0, \omega) = i\omega , \tag{6.57}$$

as for models B and D. Consequently, with the definitions (5.19) for the renormalized fields we obtain again

$$Z_{\tilde{s}} Z_S = 1 . \tag{6.58}$$

The model B result (4.92), however, does not apply, since both the tree and the one-loop contribution to the noise vertex are proportional to q^2, see Fig. 6.5(b). Thus, for model J we still need to determine two dynamical renormalization constants as defined via

$$D_R = Z_D D , \quad g_R^2 = Z_g g^2 B_d \mu^{d-6} , \tag{6.59}$$

with the convenient choice for the geometric factor

$$B_d = \frac{\Gamma(4 - d/2)}{2^d \, d \, \pi^{d/2}} . \tag{6.60}$$

Yet the very fact that the spin components S^α constitute the generators of the rotation group yields specific Ward identities that relate Z_g with static renormalizations. To this end, consider switching on a spatially uniform magnetic field $h(t)$ at time $t = 0$. According to the equation of motion (6.53), this induces an additional magnetization given by the self-consistent relation

$$\langle S^\alpha(x, t)\rangle_h = g \int_0^t dt' \sum_{\beta,\gamma} \epsilon^{\alpha\beta\gamma} \langle S^\beta(x, t')\rangle_h h^\gamma(t) . \tag{6.61}$$

Upon introducing the *non-linear susceptibility*, see Eq. (2.16),

$$R^{\alpha;\beta\gamma}(x, t; x-x', t-t') = \frac{\delta^2 \langle S^\alpha(x, t) \rangle}{\delta h^\beta(0, 0)\, \delta h^\gamma(x', t')}\bigg|_{h=0}, \qquad (6.62)$$

one readily obtains by means of (6.61) the non-trivial identity

$$\int d^d x'\, R^{\alpha;\beta\gamma}(x, t; x-x', t-t') = g\, \epsilon^{\alpha\beta\gamma}\, \chi^{\beta\beta}(x, t)\, \Theta(t)\, \Theta(t-t'). \qquad (6.63)$$

The same relation must hold for the renormalized quantities, whence we infer $Z_S^{3/2} = Z_g^{1/2} Z_S$ or

$$Z_g = Z_S. \qquad (6.64)$$

The remaining renormalization constant Z_D could now be determined by rendering the two-point vertex funtion $\Gamma^{(2,0)}(q, \omega)$ finite, see Problem 6.2(b). We shall follow the alternative route and consider the dynamic response function $\chi(q, \omega)$. Yet because the external magnetic field h appears also in the mode-coupling action (6.55), the simple relation (4.18) valid for the relaxational models does not hold, but acquires an additional contribution,

$$\chi^{\alpha\beta}(x-x', t-t') = -D\, \langle S^\alpha(x, t)\, \nabla^2\, \widetilde{S}^\beta(x', t') \rangle$$
$$+ g \sum_{\gamma,\delta} \epsilon^{\beta\gamma\delta} \langle S^\alpha(x, t)[\widetilde{S}^\gamma(x', t')\, S^\delta(x', t')]\rangle, \qquad (6.65)$$

see Problem 4.2. In the high-temperature phase, $\chi^{\alpha\beta}(q, \omega) = \chi(q, \omega)\delta^{\alpha\beta}$. Defining in the usual manner the vertex function associated with the Martin–Siggia–Rose auxiliary field $\widetilde{S}^\alpha$ and the *composite operator* $X^{\beta\gamma} = [\widetilde{S}^\beta S^\gamma]$, see Eq. (5.101), as the negative of the one-particle irreducible contributions to the corresponding cumulant $\langle S^\alpha(q, \omega)[\widetilde{S}^\beta S^\gamma](q', \omega')\rangle_c = G^{(1;X)}_{\alpha;[\beta\gamma]}(q, \omega)\, (2\pi)^{d+1}\, \delta(q+q')\delta(\omega+\omega')$, i.e., with the external propagator divided out,

$$\Gamma^{(1;X)}_{\alpha;[\beta\gamma]}(q, \omega) = -G(q, \omega)^{-1}\, G^{(1;X)}_{\alpha;[\beta\gamma]}(q, \omega), \qquad (6.66)$$

we obtain in Fourier space for $T > T_c$ the equivalent expression

$$\chi(q, \omega) = \Gamma^{(1,1)}(-q, -\omega)^{-1}\left[Dq^2 - g \sum_{\beta,\gamma} \epsilon^{\alpha\beta\gamma}\, \Gamma^{(1;X)}_{\alpha;[\beta\gamma]}(q, \omega)\right], \qquad (6.67)$$

which amends Eq. (4.94) for the conserved relaxational model B.

The new model J one-loop contributions to the two-point vertex functions $\Gamma^{(1,1)}(q, \omega)$, $\Gamma^{(1,X)}_{\alpha;[\beta\gamma]}(q, \omega)$, and $\Gamma^{(2,0)}(q, \omega)$ are depicted in Fig. 6.5. Combining with the three-component model B one-loop contribution, compare with Eq. (4.70), one

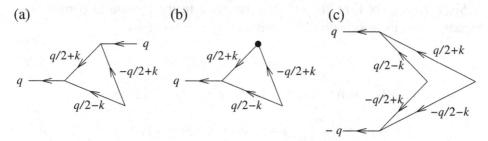

(a)   (b)   (c)

Fig. 6.5 One-loop Feynman diagrams for the model J two-point vertex functions (a) $\Gamma^{(1,1)}(q,\omega)$, (b) $\Gamma^{(1;X)}_{\alpha;[\beta\gamma]}(q,\omega)$ for the composite operator required for the computation of the dynamic susceptibility $\chi(q,\omega)$, and (c) the noise vertex $\Gamma^{(2,0)}(q,\omega)$.

finds explicitly (see Problem 6.2)

$$\Gamma^{(1,1)}(q,\omega) = i\omega + Dq^2\left[r + q^2 + \frac{5}{6}u\int_k \frac{1}{r+k^2}\right]$$

$$+ 2g^2\,(r+q^2)\,I_r(-q,-\omega)\,, \qquad (6.68)$$

$$\Gamma^{(1;X)}_{\alpha;[\beta\gamma]}(q,\omega) = -g\,I_r(q,\omega)\,, \qquad (6.69)$$

where, with the abbreviation $\Delta(q) = Dq^2(r+q^2)$,

$$I_r(q,\omega) = \int_k \frac{(q\cdot k)^2}{\left[r + \left(\frac{q}{2}+k\right)^2\right]\left[r + \left(\frac{q}{2}-k\right)^2\right]}\,\frac{1}{-i\omega + \Delta\left(\frac{q}{2}+k\right) + \Delta\left(\frac{q}{2}-k\right)}\,.$$

$$(6.70)$$

By means of Eq. (6.67) and a little algebra we see that to first order in u and g^2 the dynamic response function can be written in the form

$$\chi(q,\omega) = \frac{\chi(q)\,\Gamma_\chi(q,\omega)}{-i\omega + \Gamma_\chi(q,\omega)}\,, \qquad (6.71)$$

with the relaxation coefficient or line width

$$\Gamma_\chi(q,\omega) = \chi(q)^{-1}\left[Dq^2 + 2g^2\,I_\tau(q,\omega)\right]. \qquad (6.72)$$

Precisely as for the relaxational models, Eq. (5.18), the inverse static susceptibility to this order reads

$$\chi(q)^{-1} = r + q^2 + \frac{5}{6}u\int_k \frac{1}{r+k^2} = q^2 + \tau\left[1 - \frac{5}{6}u\int_k \frac{1}{k^2(\tau + k^2)}\right], \qquad (6.73)$$

after accounting for the fluctuation-induced T_c shift (5.14).

Since $I_\tau(q, \omega) \propto 1/D$, the effective coupling in the dynamical perturbation expansion and its renormalized counterpart are, with Eq. (6.64),

$$f = \frac{g^2}{D^2}, \quad f_R = Z_S Z_D^{-2} f B_d \mu^{d-6}. \tag{6.74}$$

The associated RG beta function then reads, using Eqs. (5.42) and (5.43),

$$\beta_f = \mu \left.\frac{\partial}{\partial \mu}\right|_0 f_R = f_R(d - 6 + \gamma_S - 2\gamma_D). \tag{6.75}$$

The subsequent analysis of the renormalization group equation proceeds as in Section 5.3, taking into account the additional coupling f_R. The dynamic critical exponent is again identified with the anomalous scaling dimension of the relaxation rate D, $z = 4 + \gamma_D^*$, see Eq. (5.68). However, since rotational invariance has already fixed the renormalization constant Z_g for the mode-coupling vertex, z must be given in terms of the static critical exponents. Indeed, *provided* there exists a *finite* infrared-stable fixed point $0 < f^* < \infty$, Eq. (6.75) implies $2\gamma_D^* - \gamma_S^* = d - 6$, or, since $\eta = -\gamma_S^*$,

$$z = \frac{d + 2 - \eta}{2}, \tag{6.76}$$

which is the exact scaling relation anticipated in Section 3.3.3.

Indeed, we may utilize Eq. (6.72) at the normalization point $q = 0$, $\omega = 0$, $\tau_R = 1$ to define the renormalized relaxation rate D_R through absorbing the UV divergence in the integral $I_{\tau=\mu^2}$ into Z_D in minimal subtraction. With

$$\frac{\partial}{\partial q^2} \int_k \frac{(q \cdot k)^2}{k^2 (\tau + k^2)^3} = \frac{1}{d} \int_k \frac{1}{(\tau + k^2)^3} = \frac{B_d}{6 - d} \tau^{-(6-d)/2},$$

we obtain to one-loop order

$$Z_D = 1 + \frac{f B_d \mu^{-\varepsilon}}{\varepsilon} + O(u^2, f^2), \tag{6.77}$$

where we have introduced $\varepsilon = d_c' - d = 6 - d$. Consequently, we compute

$$\gamma_D = -f_R + O(u_R^2, f_R^2), \tag{6.78}$$

and with $\gamma_S = 0 + O(u_R^2)$ find

$$\beta_f = f_R\left[-\varepsilon + 2f_R + O(f_R^2)\right], \tag{6.79}$$

which is of the form plotted in Fig. 5.1. For $\varepsilon \le 0$, there is only the Gaussian fixed point $f_0^* = 0$, and the reversible mode coupling is irrelevant, whence $z = 4$. For $\varepsilon > 0$, the non-trivial model J fixed point

$$f_J^* = \frac{\varepsilon}{2} + O(\varepsilon^2) \tag{6.80}$$

becomes stable, with $\gamma_D^* = -\varepsilon/2$ and $z = 4 - \varepsilon/2 = 1 + d/2$, in agreement with the exact result for $4 < d < 6$. For $d \leq 4$, we need to reinstate γ_S into (6.79), which leads to a corrected fixed point value $f_J^* = (\eta + \varepsilon)/2$, and thereby recovers the scaling relation (6.76).

6.2.2 Mode-coupling theory

Whereas the dynamic critical exponent (6.76) is fully determined by rotational symmetry, the different upper critical dimensions for the static and dynamic properties of isotropic ferromagnets impede accurate perturbational calculations of the associated scaling functions. Yet we can employ Eq. (6.71) to make contact with *mode-coupling* theory, which provides a quite successful tool to numerically evaluate dynamic scaling functions, even when the presence of long-range dipolar interactions produces interesting and non-trivial crossover features.

Originally, mode-coupling theory is based on the Mori–Zwanzig projector method and formulated in terms of the Kubo relaxation function $\Phi(q, \omega) = \chi(q)/[-i\omega + \Gamma_\chi(q, \omega)]$, or its self-energy (6.72), with the static susceptibility as the crucial input function. It turns out, though, that the ensuing mode-coupling equations are equivalent to a self-consistent one-loop theory for the line width. To this end, we shall henceforth employ the *Lorentzian* approximation, where the frequency dependence of $\Gamma_\chi(q, \omega) \approx \Gamma_\chi(q)$ is neglected, and replace the 'bare' quantity $\Delta(q)$ in the denominator by its fully renormalized counterpart $\Gamma_\chi(q)$. Moreover, instead of the factors $(r + q^2)^{-1}$ we insert the full static response function $\chi(q)$, and recall that the vertex factor $(q \cdot k)^2$ in Eq. (6.70) that originates from the reversible mode coupling (3.66) actually stems from a difference in inverse susceptibilities,

$$v(q, k) = \frac{1}{4}\left[\chi\left(\frac{q}{2} + k\right)^{-1} - \chi\left(\frac{q}{2} - k\right)^{-1}\right]^2. \qquad (6.81)$$

For $T > T_c$, we are thereby led to the following *self-consistent* equation for the line width

$$\Gamma_\chi(q) \approx \chi(q)^{-1}\left[Dq^2 + 2g^2 \int \frac{d^d k}{(2\pi)^d} v(q, k) \frac{\chi\left(\frac{q}{2} + k\right)\chi\left(\frac{q}{2} - k\right)}{\Gamma_\chi\left(\frac{q}{2} + k\right) + \Gamma_\chi\left(\frac{q}{2} - k\right)}\right]. \qquad (6.82)$$

If we insert the critical scaling $\chi(q) \sim |q|^{-2+\eta}$ and $\Gamma_\chi(q) \sim |q|^z$, we obtain the scaling relation (6.76) from the fluctuation contribution, describing a *slower* critical

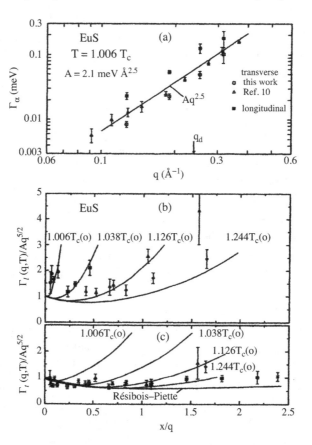

Fig. 6.6 Measured line widths for the ferromagnet EuS from inelastic polarized neutron scattering near a Bragg reflection with $q \neq 0$. The top and bottom panels show comparisons of the longitudinal and transverse (with respect to the wavevector q) line widths from Lorentzian fits, normalized to $A |q|^{5/2}$ with $A = 2.1$ meV $\text{Å}^{5/2}$, with the dipolar dynamic scaling functions obtained from self-consistent mode-coupling theory (full lines). The *Résibois–Piette* scaling function describes the isotropic model J limit. [Data and figure reproduced with permission from: P. Böni, D. Görlitz, J. Kötzler, and J. L. Martínez, *Phys. Rev. B* **43**, 8755 (1991), DOI: 10.1103/PhysRevB.43.8755; copyright (1991) by the American Physical Society.]

decay than the model B prediction $z = 4 - \eta$ which follows for $g = 0$, provided $d < 6$ (since $\eta = 0$ near $d'_c = 6$). In the ferromagnetic phase, the spin precession terms produce propagating modes with the quadratic dispersion (3.77), see Section 3.3.3. As sketched in Fig. 3.8, upon approaching T_c, these spin waves become overdamped, and eventually their line width displays the standard scaling laws associated with critical slowing-down.

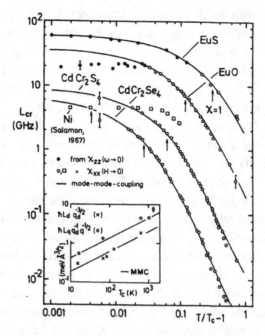

Fig. 6.7 Critical part of the Onsager kinetic coefficient above T_c for the ferromagnets EuS, EuO, CdCr$_2$S$_4$, CdCr$_2$Se$_4$, and Ni, compared with the corresponding scaling function obtained from mode-coupling theory (full line); the inset provides the values of non-universal amplitudes. [Data and figure reproduced with permission from: J. Kötzler, *Phys. Rev. B* **38**, 12027 (1988), DOI: 10.1103/PhysRevB.38.12027; copyright (1988) by the American Physical Society.]

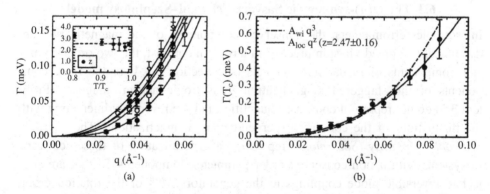

Fig. 6.8 Inelastic neutron scattering data showing dynamic scaling for the itinerant ferromagnet Ni$_3$Al. (a) Measured line widths for the spin wave peak at various temperatures; from top to bottom: $T/T_c = 0.98$ (open diamonds ◇), 0.96 (filled diamonds), 0.94 (filled squares), 0.90 (open circles ○), and 0.80 (filled circles ●); the data match power laws $\Gamma_\chi(q) \sim |q|^z$ with $z \approx 2.5$ (inset). (b) Quasielastic line width for the paramagnetic fluctuations at T_c; the data fit yields $z = 2.47 \pm 0.16$ in good agreement with the prediction (6.76). [Data and figures reproduced with permission from: P. Semadeni, B. Roessli, P. Böni, P. Vorderwisch, and T. Chatterji, *Phys. Rev. B* **62**, 1083–1088 (2000), DOI: 10.1103/PhysRevB.62.1083; copyright (2000) by the American Physical Society.]

In ferromagnets with localized magnetic moments, close to T_c and at sufficiently long wavelengths the long-range dipolar interaction between the spins becomes significant. Dipolar forces are anisotropic with respect to the external wavevector q, and upon approaching the critical point the longitudinal static susceptibility $\chi^\parallel(q) = \sum_{i,j=1}^{d} \chi_{ij}(q)\, P_{ij}^\parallel(q)$, where $P_{ij}^\parallel(q) = q_i q_j / q^2$ denotes the longitudinal projector onto the components parallel to q, remains finite. Thus rotational invariance is broken, and the isotropic model J critical dynamics becomes drastically modified at a characteristic dipolar crossover scale q_d. This is seen in Fig. 6.6, which depicts the scaled longitudinal (b) and transverse (c) line widths. Specifically, the neutron scattering data on EuS for $\chi^\parallel(q)$ show marked deviations from the isotropic scaling $\Gamma_\chi(q) \sim |q|^{5/2}$ in three dimensions. Yet by means of an appropriate generalization of the mode-coupling equation (6.82) to properly include dipolar interactions a quantitative description of the experiments can be achieved. Figure 6.7 demonstrates that the scaling functions obtained from mode-coupling theory fit experimental results for the ferromagnets EuS, EuO, $CdCr_2S_4$, $CdCr_2Se_4$, and Ni remarkably well. Figure 6.8 finally illustrates that the isotropic model J dynamic scaling holds also for *itinerant* ferromagnets, here the compound Ni_3Al, wherein the magnetic moments are not localized. Although the Heisenberg model (3.62) thus does not apply here, these systems are still rotationally invariant in spin space, whence the dynamic critical exponent (6.76) follows from symmetry considerations.

6.3 The $O(n)$-symmetric Sasvári–Schwabl–Szépfalusy model

In isotropic ferromagnets, the conserved components of the magnetization are dynamically coupled via spin precession terms. In contrast, the two order parameter components of planar ferromagnets (see Section 3.3.4) or the three components of the staggered magnetization in isotropic antiferromagnets (Problem 3.5) do not represent conserved quantities, but are reversibly interacting with the fluctuations of the conserved (perpendicular) magnetization.[4] The *Sasvári–Schwabl–Szépfalusy (SSS) model* represents a generalization of these scenarios to systems with n non-conserved order parameter components S^α that are subject to reversible mode couplings to the generators $M^{\alpha\beta}$ of the rotation group $O(n)$.[5]

---

[4] Halperin, Hohenberg, and Siggia (1976); Gunton and Kawasaki (1976); Freedman and Mazenko (1976); Hohenberg and Halperin (1977); De Dominicis and Peliti (1977, 1978).
[5] Sasvári, Schwabl, and Szépfalusy (1975); Sasvári and Szépfalusy (1977); Janssen (1977); De Dominicis and Peliti (1978); Folk and Moser (2006).

6.3.1 SSS model description

Originally motivated by structural phase transitions where the displacement field serves as the order parameter, and $M^{\alpha\beta} = -M^{\beta\alpha}$ may be identified with the set of $n(n-1)/2$ conserved generalized angular momenta, the SSS model is defined through the Poisson brackets

$$\{S^{\alpha}(x), S^{\beta}(x')\} = 0 \,, \tag{6.83}$$

$$\{S^{\alpha}(x), M^{\beta\gamma}(x')\} = g[\delta^{\alpha\beta} S^{\gamma}(x) - \delta^{\alpha\gamma} S^{\beta}(x)]\delta(x - x') \,, \tag{6.84}$$

$$\{M^{\alpha\beta}(x), M^{\gamma\delta}(x')\} = -g\Big[\delta^{\alpha\gamma} M^{\beta\delta}(x) + \delta^{\beta\delta} M^{\alpha\gamma}(x) - \delta^{\alpha\delta} M^{\beta\gamma}(x)$$

$$- \delta^{\beta\gamma} M^{\alpha\delta}(x)\Big]\delta(x - x') \,. \tag{6.85}$$

Since the conserved generalized angular momenta $M^{\alpha\beta}$ represent non-critical degrees of freedom, we may write the effective Hamiltonian as

$$\mathcal{H}[S, M] = \mathcal{H}[S] + \int d^d x \sum_{\alpha < \beta} \left[\frac{1}{2} M^{\alpha\beta}(x)^2 - H^{\alpha\beta}(x) M^{\alpha\beta}(x)\right], \tag{6.86}$$

with the standard $O(n)$-symmetric Landau–Ginzburg–Wilson Hamiltonian $\mathcal{H}[S]$, Eq. (4.12), while $H^{\alpha\beta}$ denotes the thermodynamically conjugate field to the variable $M^{\alpha\beta}$.

By means of the Poisson brackets (6.83)–(6.85) and the general prescription in Section 3.3.2 we obtain the reversible force terms in the Langevin equations for the fluctuations of the fields S^{α} and $M^{\alpha\beta}$, respectively:

$$F_{\text{rev}}^{\alpha}[S, M](x) = -g \sum_{\beta} S^{\beta}(x) \frac{\delta\mathcal{H}[S, M]}{\delta M^{\alpha\beta}(x)} \,, \tag{6.87}$$

$$F_{\text{rev}}^{\alpha\beta}[S, M](x) = -g \left[S^{\alpha}(x) \frac{\delta\mathcal{H}[S, M]}{\delta S^{\beta}(x)} - S^{\beta}(x) \frac{\delta\mathcal{H}[S, M]}{\delta S^{\alpha}(x)}\right]$$

$$- g \sum_{\gamma} \left[M^{\alpha\gamma}(x) \frac{\delta\mathcal{H}[S, M]}{\delta M^{\beta\gamma}(x)} - M^{\beta\gamma}(x) \frac{\delta\mathcal{H}[S, M]}{\delta M^{\alpha\gamma}(x)}\right]. \tag{6.88}$$

Notice that the finite-temperature terms in Eq. (3.59) all vanish. Upon performing the functional derivatives with the Hamiltonian (6.86) we arrive at the explicit stochastic equations of motion and noise correlations for the n components of the order parameter:

$$\frac{\partial S^{\alpha}(x, t)}{\partial t} = F^{\alpha}_{\text{rev}}[S, M](x, t) - D \frac{\delta \mathcal{H}[S, M]}{\delta S^{\alpha}(x, t)} + \zeta^{\alpha}(x, t)$$

$$= -g \sum_{\beta} S^{\beta}(x, t)[M^{\alpha\beta}(x, t) - H^{\alpha\beta}(x, t)] \tag{6.89}$$

$$- D\left[\left(r - \nabla^2 + \frac{u}{6} \sum_{\beta} S^{\beta}(x, t)^2\right) S^{\alpha}(x, t) - h^{\alpha}(x, t)\right]$$

$$+ \zeta^{\alpha}(x, t),$$

$$\langle \zeta^{\alpha}(x, t) \zeta^{\beta}(x', t')\rangle = 2D \, \delta(x - x')\delta(t - t')\delta^{\alpha\beta}, \tag{6.90}$$

and the conserved generators ($\alpha < \beta$):

$$\frac{\partial M^{\alpha\beta}(x, t)}{\partial t} = F^{\alpha\beta}_{\text{rev}}[S, M](x, t) + \lambda\nabla^2 \frac{\delta \mathcal{H}[S, M]}{\delta M^{\alpha\beta}(x, t)} + \eta^{\alpha\beta}(x, t)$$

$$= g \, S^{\alpha}(x, t)[\nabla^2 S^{\beta}(x, t) + h^{\beta}(x, t)]$$

$$- g \, S^{\beta}(x, t)[\nabla^2 S^{\alpha}(x, t) + h^{\alpha}(x, t)]$$

$$+ g \sum_{\gamma}[M^{\alpha\gamma}(x, t)H^{\beta\gamma}(x, t) - M^{\beta\gamma}(x, t)H^{\alpha\gamma}(x, t)]$$

$$+ \lambda\nabla^2[M^{\alpha\beta}(x, t) - H^{\alpha\beta}(x, t)] + \eta^{\alpha\beta}(x, t), \tag{6.91}$$

with the stochastic forces obeying $\langle\eta^{\alpha\beta}(x, t)\rangle = 0$ and

$$\langle \eta^{\alpha\beta}(x, t)\,\eta^{\gamma\delta}(x', t')\rangle = -2\lambda\,\nabla^2_x \,\delta(x - x')\delta(t - t')(\delta^{\alpha\gamma}\delta^{\beta\delta} - \delta^{\alpha\delta}\delta^{\beta\gamma}). \tag{6.92}$$

Now recall that for planar ferro- and isotropic antiferromagnets the Poisson brackets $\propto \tilde{g}$ between the order parameter components do *not* vanish. Yet in either case $\tilde{g}$ represents an irrelevant coupling, and can hence be omitted as far as asymptotic critical properties are concerned. Thus, for the purpose of studying critical dynamics, the SSS model as defined through Eqs. (6.89)–(6.92) precisely reduces to model E for planar ferromagnets for $n = 2$, with the identification $M^{12} = M$, whereas for $n = 3$ we recover model G for isotropic antiferromagnets, see Problem 3.5. Consequently, as in these models, we expect the dynamic critical exponents $z = d/2$ in dimensions $d \leq d'_c = 4$, provided strong dynamic scaling ($z_S = z_M$) holds.

In our field theory representation, we may decompose the response functional corresponding to the above coupled Langevin equations into four terms, $\mathcal{A}[\tilde{S}, S, \tilde{M}, M] = \mathcal{A}_0[\tilde{S}, S] + \mathcal{A}_{\text{int}}[\tilde{S}, S] + \mathcal{A}_0[\tilde{M}, M] + \mathcal{A}_{\text{SSS}}[\tilde{S}, S, \tilde{M}, M]$. Here, the first two terms comprise the usual $O(n)$-symmetric model A action (4.14)

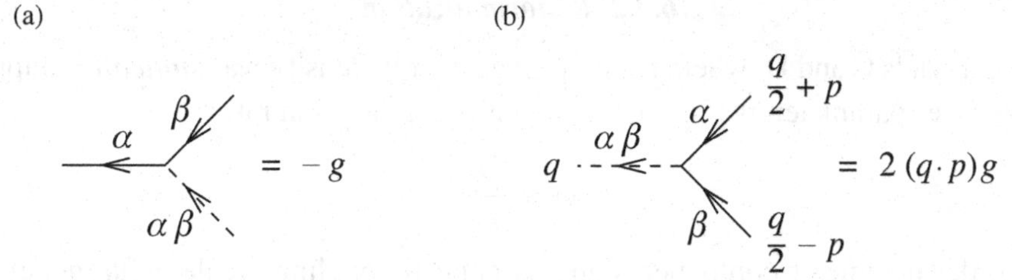

(a)   (b)

Fig. 6.9 Mode-coupling vertices for the $O(n)$-symmetric SSS model.

and (4.15) with $a = 0$, whose pictorial representation is shown in Fig. 4.1. The third contribution is the action for the conserved fields,

$$\mathcal{A}_0[\tilde{M}, M] = \int d^d x \int dt \sum_{\alpha < \beta} \left[\tilde{M}^{\alpha\beta}(x, t) \left(\frac{\partial}{\partial t} - \lambda \nabla^2 \right) M^{\alpha\beta}(x, t) \right.$$

$$\left. + \lambda \, \tilde{M}^{\alpha\beta}(x, t) \, \nabla^2 (\tilde{M}^{\alpha\beta}(x, t) + H^{\alpha\beta}(x, t)) \right],$$

$$(6.93)$$

which yields the generalized angular momentum propagator

$$\langle M^{\alpha\beta}(q, \omega) \, \tilde{M}^{\gamma\delta}(q', \omega') \rangle$$

$$= \frac{1}{-i\omega + \lambda q^2} (2\pi)^{d+1} \delta(q + q') \delta(\omega + \omega') (\delta^{\alpha\gamma} \delta^{\beta\delta} - \delta^{\alpha\delta} \delta^{\beta\gamma}) \quad (6.94)$$

that as in Section 6.1 will be depicted by a dashed line, see Fig. 6.1(a), and at last we have the non-linear mode-coupling terms

$$\mathcal{A}_{SSS}[\tilde{S}, S, \tilde{M}, M]$$

$$= g \int d^d x \int dt \sum_{\alpha, \beta} \left[\tilde{S}^\alpha(x, t) S^\beta(x, t) [M^{\alpha\beta}(x, t) - H^{\alpha\beta}(x, t)] - \tilde{M}^{\alpha\beta}(x, t) \right.$$

$$\times \frac{1}{2} \left(S^\alpha(x, t) [\nabla^2 S^\beta(x, t) + h^\beta(x, t)] - S^\beta(x, t) \right.$$

$$\times [\nabla^2 S^\alpha(x, t) + h^\alpha(x, t)] + \sum_\gamma [M^{\alpha\gamma}(x, t)$$

$$\left. \left. \times H^{\beta\gamma}(x, t) - M^{\beta\gamma}(x, t) H^{\alpha\gamma}(x, t)] \right) \right]. \quad (6.95)$$

The corresponding vertices (for $h = 0 = H$) are displayed in Fig. 6.9. Note the respective similarities of Fig. 6.9(a) to Fig. 6.1(c) for model C and of the symmetrized vertex in Fig. 6.9(b) to Fig. 6.4 for model J.

6.3.2 Renormalization

As for models C and D, where the conserved energy density was *statically* coupled
to the order parameter, we introduce the ratio of relaxation rates

$$w = \frac{D}{\lambda} \,. \tag{6.96}$$

The order parameter components and associated couplings scale as in model A,
Eqs. (5.3)–(5.5) with $a = 0$, while the scaling dimensions of the conserved angular
momenta follow in analogy with model C, see Eqs. (6.9) and (6.10),

$$[\tilde{M}^{\alpha\beta}(x, t)] = \mu^{d/2} = [M^{\alpha\beta}(x, t)] \,, \tag{6.97}$$

$$[\lambda] = \mu^0 = [w] \,, \quad [g] = \mu^{2-d/2} \,. \tag{6.98}$$

Thus, the dynamic upper critical dimension $d'_c = 4$ for the reversible mode-coupling
vertices coincides with the static critical dimension d_c. Accordingly, supplementing Eqs. (5.19) and (5.23), we follow the scheme of model C to introduce the
multiplicative renormalization constants

$$M_R^{\alpha\beta} = Z_M^{1/2} M^{\alpha\beta} \,, \quad \tilde{M}_R^{\alpha\beta} = Z_{\tilde{M}}^{1/2} \tilde{M}^{\alpha\beta} \,, \tag{6.99}$$

$$\lambda_R = Z_\lambda \lambda \,, \quad g_R^2 = Z_g g^2 A_d \mu^{-\epsilon} \,, \tag{6.100}$$

where $\epsilon = 4 - d$. As for models C and D, the cumulants and vertex functions carry
four superscript labels, two each for the incoming and outgoing order parameter
and conserved fields, respectively, and the renormalized vertex functions read

$$\Gamma_R^{(\tilde{N},N;\tilde{M},M)} = Z_{\tilde{S}}^{-\tilde{N}/2} Z_S^{-N/2} Z_{\tilde{M}}^{-\tilde{M}/2} Z_M^{-M/2} \Gamma^{(\tilde{N},N;\tilde{M},M)} \,. \tag{6.101}$$

The subsequent analysis proceeds along the route explained in detail in the
previous two sections. First, we note that the external wavevector dependence of
the mode-coupling vertex in Fig. 6.9(b) implies

$$\Gamma^{(0,0;1,1)}(q = 0, \omega) = i\omega \,, \tag{6.102}$$

whence $Z_{\tilde{M}} Z_M = 1$ to all orders in perturbation theory. Next, following the procedure in Section 6.2, we define the non-linear (mixed) susceptibility

$$R^{\alpha;\beta}(x, t; x - x', t - t') = \frac{\delta^2 \langle S^\alpha(x, t) \rangle}{\delta h^\beta(0, 0) \, \delta H^{\alpha\beta}(x', t')}\bigg|_{h=0=H} \,, \tag{6.103}$$

and exploit the fact that the generalized angular momenta $M^{\alpha\beta}$ generate rotations
in the space of the order parameter fields S^α, which yields the Ward identity
(Problem 6.3)

$$\int d^d x' \, R^{\alpha;\beta}(x, t; x - x', t - t') = g \, \chi^{\alpha\beta}(x, t) \, \Theta(t) \, \Theta(t - t') \,. \tag{6.104}$$

(a)   (b)

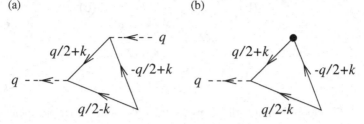

Fig. 6.10 One-loop Feynman diagrams for the SSS model conserved field two-point vertex functions (a) $\Gamma^{(0,0;1,1)}(q, \omega)$, and (b) $\Gamma^{(0;1;X)}(q, \omega)$.

Since this relation must hold for the corresponding renormalized quantities as well, we infer $Z_S Z_M^{1/2} = Z_g^{1/2} Z_S$, i.e., $Z_g = Z_M$. Yet the non-critical nature of the conserved angular momentum fluctuations as encoded in the Hamiltonian (6.86) implies that the static limit of the corresponding linear dynamic susceptibility

$$\Xi^{\alpha\beta,\gamma\delta}(x - x', t - t') = \frac{\delta \langle M^{\alpha\beta}(x, t) \rangle}{\delta H^{\gamma\delta}(x', t')}\bigg|_{h=0=H}, \tag{6.105}$$

$$\Xi^{\alpha\beta,\gamma\delta}(q, \omega) = \Xi(q, \omega)(\delta^{\alpha\gamma}\delta^{\beta\delta} - \delta^{\alpha\delta}\delta^{\beta\gamma}), \tag{6.106}$$

simply becomes $\Xi(q, 0) = 1$, and hence, since $\Xi_R = Z_M \Xi$, see Eq. (5.24),

$$Z_g = Z_M = Z_{\tilde{M}} = 1 \tag{6.107}$$

to all orders in the perturbation expansion.

Of course, the renormalization constants Z_τ, Z_u, and Z_S are just those of the $O(n)$-symmetric Landau–Ginzburg–Wilson theory, see Eqs. (5.30), (5.32), and (5.36). The exact result (6.107) thus leaves us with the task of determining the remaining dynamical Z factors $Z_{\tilde{S}}$, Z_D, and Z_λ. To this end, we again consider the renormalization of the linear susceptibilities of the order parameter and the conserved fields; similarly to model J, these involve cumulants with the composite operators $X^{\alpha\beta} = [\tilde{S}^\alpha S^\beta]$ and $Y^{\alpha\beta} = [\tilde{M}^{\alpha\beta} S^\beta]$, see Problem 6.3, namely $\langle S^\alpha(q, \omega)[\tilde{M}^{\beta\gamma} S^\gamma](q', \omega') \rangle_c = G^{(1;0;Y)}_{\alpha;[\alpha\gamma]}(q, \omega)(2\pi)^{d+1}\delta(q+q')\delta(\omega+\omega')\delta^{\alpha\beta}$ and $\langle M^{\alpha\beta}(q, \omega)[\tilde{S}^\gamma S^\delta](q', \omega') \rangle_c = G^{(0;1;X)}(q, \omega)(2\pi)^{d+1}\delta(q+q')\delta(\omega+\omega')(\delta^{\alpha\gamma}\delta^{\beta\delta} - \delta^{\alpha\delta}\delta^{\beta\gamma})$. Defining the vertex functions associated with cumulants containing these composite fields in analogy with Eq. (6.66), this leads to

$$\chi(q, \omega) = \Gamma^{(1,1;0,0)}(-q, -\omega)^{-1}\left[D + g \sum_\beta \Gamma^{(1;0;Y)}_{\alpha;[\alpha\beta]}(q, \omega)\right], \tag{6.108}$$

$$\Xi(q, \omega) = \Gamma^{(0,0;1,1)}(-q, -\omega)^{-1}\left[\lambda q^2 - 2g\, \Gamma^{(0;1;X)}(q, \omega)\right]. \tag{6.109}$$

Let us begin with the susceptibility for the generalized angular momenta. Figure 6.10 shows the one-loop diagrams for the two-point vertex functions

(a)   (b)   (c)

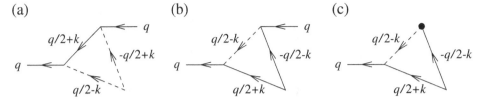

Fig. 6.11 One-loop Feynman diagrams for the SSS model order parameter two-point vertex functions (a), (b) $\Gamma^{(1,1;0,0)}(q,\omega)$, and (c) $\Gamma^{(1;0;Y)}_{\alpha;[\alpha\beta]}(q,\omega)$.

contributing to (6.109); notice that they are all comprised of order parameter fluctuation loops. The corresponding analytic expressions read (Problem 6.4)

$$\Gamma^{(0,0;1,1)}(q,\omega) = i\omega + \lambda q^2 + 4g^2\, I_{\Xi,r}(-q,-\omega)\,, \tag{6.110}$$

$$\Gamma^{(0;1;X)}(q,\omega) = -2g\, I_{\Xi,r}(q,\omega)\,, \tag{6.111}$$

with the wavevector integral, cf. Eq. (6.70),

$$I_{\Xi,r}(q,\omega) = \int_k \frac{(q\cdot k)^2}{\left[r+\left(\frac{q}{2}+k\right)^2\right]\left[r+\left(\frac{q}{2}-k\right)^2\right]}\,\frac{1}{-i\omega+2D\left(r+\frac{q^2}{4}+k^2\right)}\,, \tag{6.112}$$

whence we may write the dynamic response function as

$$\Xi(q,\omega) = \frac{\Gamma_\Xi(q,\omega)}{-i\omega+\Gamma_\Xi(q,\omega)}\,, \tag{6.113}$$

which confirms $\Xi(q,0)=1$. The relaxation coefficient reads to one-loop order

$$\Gamma_\Xi(q,\omega) = \lambda\, q^2 + 4g^2\, I_{\Xi,\tau}(q,\omega)\,. \tag{6.114}$$

At the normalization point $q=0$, $\omega=0$, $\tau_R=1$, we then obtain

$$Z_\lambda = 1 + \frac{f A_d\,\mu^{-\epsilon}}{2\epsilon} + O(u^2,\,f^2)\,, \tag{6.115}$$

with the effective dynamical coupling

$$f = \frac{g^2}{D\lambda}\,,\qquad f_R = (Z_D Z_\lambda)^{-1} f A_d\,\mu^{-\epsilon}\,, \tag{6.116}$$

where we have employed Eq. (6.107), and the geometric factor A_d is given in Eq. (5.12). The result (6.115) also follows from the renormalization of the noise strength, see Fig. 6.2(b) and Problem 6.4.

The one-loop Feynman diagrams required for the order parameter response function (6.108), in addition to the model A Hartree loop, are depicted in Fig. 6.11. Graphs (a) and (b) for $\Gamma^{(1,1;0,0)}(q,\omega)$ both contain highly UV-singular terms that, however, cancel each other as a consequence of detailed balance, and the static susceptibility $\chi(q)$ becomes identical with Eq. (5.18) for $\omega = 0$. We may then write to first order in u,

$$\Gamma^{(1,1;0,0)}(q, \omega) = i\omega + \chi(q)^{-1}[D + (n-1)g^2 I_{\chi,r}(-q, -\omega)], \qquad (6.117)$$

$$\sum_\beta \Gamma^{(1;0;Y)}_{\alpha;[\alpha\beta]}(q, \omega) = (n-1)g\, I_{\chi,r}(q, \omega), \qquad (6.118)$$

where (Problem 6.4)

$$I_{\chi,r}(q, \omega) = \int_k \frac{1}{r + \left(\frac{q}{2} + k\right)^2} \frac{1}{-i\omega + D\left[r + \left(\frac{q}{2} + k\right)^2\right] + \lambda\left(\frac{q}{2} - k\right)^2}, \qquad (6.119)$$

and the dynamic order parameter susceptibility assumes the form (6.71) with the relaxation coefficient

$$\Gamma_\chi(q, \omega) = \chi(q)^{-1}[D + (n-1)g^2 I_{\chi,\tau}(q, \omega)]. \qquad (6.120)$$

Since $\partial\Gamma^{(1,1;0,0)}(q, \omega)/\partial\omega$ is UV-finite, we obtain $Z_{\tilde{s}}Z_S = 1$, and therefore to one-loop order

$$Z_{\tilde{s}} = 1 + O(u^2, f^2). \qquad (6.121)$$

Finally we extract the ϵ pole of the integral $I_{\chi,\mu^2}(0, 0)$ and find in minimal subtraction

$$Z_D = 1 + \frac{n-1}{1+w} \frac{fA_d\,\mu^{-\epsilon}}{\epsilon} + O(u^2, f^2), \qquad (6.122)$$

which we could also have obtained directly from renormalizing the order parameter noise vertex, see Fig. 6.2(c) and Problem 6.4.

6.3.3 Strong and weak dynamic scaling regimes

In order to identify possible dynamic scaling regimes, we look for RG fixed points f^* for the renormalized dynamic mode coupling (6.116). Defining Wilson's flow functions as usual, the associated RG beta function becomes

$$\beta_f = \mu \left.\frac{\partial}{\partial\mu}\right|_0 f_R = f_R(d - 4 - \gamma_D - \gamma_\lambda). \qquad (6.123)$$

Therefore, provided $0 < f^* < \infty$, necessarily $\gamma_D^* + \gamma_\lambda^* = d - 4$. Since the dynamic critical exponents are given by $z_S = 2 + \gamma_D^*$ and $z_M = 2 + \gamma_\lambda^*$, compare with Eq. (6.42), this results in the SSS model scaling relation

$$z_S + z_M = d. \qquad (6.124)$$

Equation (6.124) also follows from mode-coupling (or self-consistent one-loop) theory, if we apply the Lorentzian approximation, see Section 6.2.2 and Problem 6.5. Assuming that the characteristic fluctuation time scales for the order parameter and conserved fields are identical, we are then immediately led to

$$z_S = z_M = \frac{d}{2} \,. \tag{6.125}$$

So let us now investigate the scaling behavior within the one-loop approximation. Explicitly, Eqs. (6.122) and (6.115) yield

$$\gamma_D = -\frac{n-1}{1+w_R} f_R \,, \quad \gamma_\lambda = -\frac{1}{2} f_R \,, \tag{6.126}$$

whereupon the RG beta functions become

$$\beta_w = \mu \left. \frac{\partial}{\partial \mu} \right|_0 w_R = w_R(\gamma_D - \gamma_\lambda) = w_R f_R \left(\frac{1}{2} - \frac{n-1}{1+w_R} \right) \,, \tag{6.127}$$

$$\beta_f = f_R \left[-\epsilon + f_R \left(\frac{1}{2} + \frac{n-1}{1+w_R} \right) \right] \,. \tag{6.128}$$

Equation (6.127) allows for the non-trivial fixed point $w_s^* = 2n - 3$ in addition to $w_w^* = 0$ and $w_{w'}^* = \infty$. In order to assess the stability of the various RG fixed points, we compute the stability matrix

$$
\begin{aligned}
\Lambda &= \begin{pmatrix} \partial \beta_f / \partial f_R & \partial \beta_f / \partial w_R \\ \partial \beta_w / \partial f_R & \partial \beta_w / \partial w_R \end{pmatrix} \\
&= \begin{pmatrix} -\epsilon + f_R[1 + 2(n-1)/(1+w_R)] & -(n-1)f_R^2/(1+w_R)^2 \\ w_R[1/2 - (n-1)/(1+w_R)] & f_R[1/2 - (n-1)/(1+w_R)^2] \end{pmatrix} \,.
\end{aligned}
\tag{6.129}
$$

In $d > d_c = 4$ dimensions ($\epsilon < 0$), the model A fixed point $f_0^* = 0$ is infrared-stable, the conserved fields decouple from the order parameter and the dynamic exponents acquire their mean-field values $z_S = 2 = z_M$.

For $d < 4$ ($\epsilon > 0$), $f^* > 0$ becomes stable. As illustrated in Fig. 6.12, there are three competing fixed points. (i) At the *strong dynamic scaling* fixed point

$$f_s^* = \epsilon \,, \quad w_s^* = 2n - 3 \,, \tag{6.130}$$

we find $\gamma_D^* = -\epsilon/2 = \gamma_\lambda^*$, and therefore indeed Eq. (6.125). To one-loop order, the strong scaling fixed point is stable, since the corresponding eigenvalues of the stability matrix, ϵ and $(2n - 3)\epsilon/4(n - 1)$, are both positive (for $n \geq 2$). We had anticipated this result in Section 3.3.4 by means of various scaling arguments. (ii) Next, there is a *weak dynamic scaling* fixed point

$$f_w^* = \frac{2\epsilon}{2n - 1} \,, \quad w_w^* = 0 \tag{6.131}$$

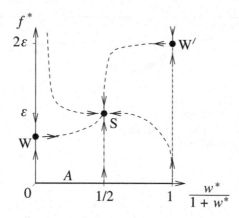

Fig. 6.12 The strong scaling (S) and two weak dynamic scaling (W, W') fixed points and RG flow for the SSS model (plotted here for $n = 2$).

that is unstable in the w_R direction, and for which

$$z_S = 2 - \frac{2(n-1)\epsilon}{2n-1} < z_M = 2 - \frac{\epsilon}{2n-1} . \qquad (6.132)$$

Of course, these exponents satisfy the scaling relation (6.124). (iii) Finally, a second unstable weak dynamic scaling regime is governed by

$$f_{w'}^* = 2\epsilon , \quad w_{w'}^* = \infty , \qquad (6.133)$$

with dynamic scaling exponents

$$z_S = 2 , \quad z_M = d - 2 . \qquad (6.134)$$

Owing to the fact that all loop diagrams for the order parameter fluctuations produce factors of $1 + w$ in the denominators, and hence the conserved fields do not couple into the order parameter dynamics, this result should hold beyond the one-loop approximation, up to corrections of order η.

The RG flow in the parameter space spanned by the mode coupling f_R and the variable $w_R/(1 + w_R)$ (which maps the domain for the time scale ratio w_R to the interval $[0, 1]$) is shown in Fig. 6.12. The model A fixed line ($f_0^* = 0$, w_R arbitrary) as well as the two weak scaling fixed points are unstable, and the RG flow runs toward the strong scaling fixed point (6.130). Notice, however, that the second eigenvalue of the stability matrix Λ at (f_s^*, w_s^*) is only $\epsilon/4$ for $n = 2$, and turns out to be even smaller when evaluated to two-loop order. Thus, the strong scaling fixed point is in fact located quite close to its stability boundary. For planar ferromagnets, one therefore expects slow transients in approaching the universal critical regime.

These are indeed found to be quite prominent in the critical dynamics of super-fluid helium 4, and in order to aptly capture experimental data, one needs to solve

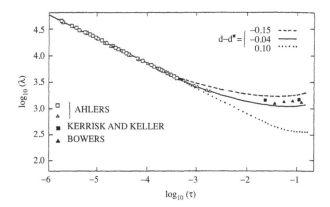

Fig. 6.13 The thermal conductivity λ as function of the relative temperature variable $\tau = (T - T_c)/T_c$ in superfluid helium 4. Measured data are compared with predictions from solving the non-linear RG flow equations of model E (SSS model with $n = 2$) in three dimensions, assuming different values for the borderline dimension d^* for stability of the strong-scaling fixed point. [Data and figure reproduced with permission from: V. Dohm and R. Folk, *Z. Phys. B Cond. Matt.* **40**, 79 (1980), DOI: 10.1007/BF01295074; copyright (1980) by Springer.]

the full non-linear RG flow equations.[6] As an example, Fig. 6.13 depicts the temperature dependence of the diffusion coefficient λ, which in helium is just the thermal conductivity, as predicted from the model E ($n = 2$) RG flow equations in $d = 3$ dimensions. It is apparent that the theory fits the experimental data very well. In order to appropriately characterize other dynamical properties of superfluid helium, one must, however, resort to the so-called model F that incorporates a second diffusive mode. While the strong scaling fixed point is again infrared-stable to one-loop order for model F, it actually becomes unstable in higher loop orders, and Borel-resummed perturbation theory suggests that in fact a weak dynamic scaling fixed point akin to (6.131) governs the asymptotic critical dynamical properties of the normal- to superfluid phase transition.[7]

For isotropic antiferromagnets for which the three-component staggered magnetization serves as order parameter (model G), the critical scaling regime is not as badly affected by slow crossovers as for $n = 2$. In order to make contact with experiment, we conclude with a brief discussion of the $O(n)$-symmetric SSS model in the ordered phase. Below T_c, we note that the spontaneous symmetry breaking in the order parameter sector, say with the component $S^n = \langle S^n \rangle + \sigma$ assuming a non-zero average, see Section 5.4, also implies different dynamic behavior of the generalized angular momenta $M^{\alpha N}$ that couple to the longitudinal fluctuations σ, and the fields $M^{\alpha\beta}$, which interact exclusively with the transverse order parameter

---

[6] Dohm and Folk (1980).   [7] Dohm (2006).

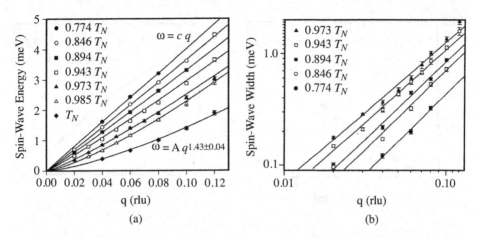

Fig. 6.14 (a) Spin wave dispersion in the Heisenberg antiferromagnet RbMnF$_3$, obtained from neutron scattering data. (b) The Lorentzian halfwidth of the spin wave peaks as function of wavevector at different temperatures, yielding $\Gamma(q) \sim |q|^z$ with dynamic critical exponent $z = 1.41 \pm 0.05$. [Data and figures reproduced with permission from: R. Coldea, R. A. Cowley, T. G. Perring, D. F. McMorrow, and B. Roessli, *Phys. Rev. B* **57**, 5281 (1998), DOI: 10.1103/PhysRevB.57.5281; copyright (1998) by the American Physical Society.]

fields π^α (with $\alpha < \beta < n$). Whereas the $M^{\alpha\beta}$ remain diffusive modes, the fields $M^{\alpha N}$ combine with the π^α to produce propagating excitations with the essentially linear dispersion relation (3.95), see Section 3.3.4 and Problem 3.5. Upon approaching the critical point, these Goldstone modes (spin waves or magnons for magnetic systems, second sound for helium 4) become overdamped and display standard dynamic critical scaling behavior.

Experimentally, RbMnF$_3$ (with critical Néel temperature $T_N \approx 83$ K) is considered the best realization of an isotropic three-dimensional Heisenberg antiferromagnet. As is evident in Fig. 6.14(a), neutron scattering data clearly show the crossover from propagating spin waves with linear dispersion $\omega(q) \approx c|q|$ to overdamped critical modes with $\Gamma(q) \sim |q|^z$, where $z = 1.43 \pm 0.04$. As is shown in Fig. 6.14(b), fitting the wavenumber dependence of the line widths at different temperatures to power laws yields an average $z = 1.41 \pm 0.05$, which within the experimental error bars is in fair agreement with the theoretical prediction $z = 1.5$. This is also confirmed in dynamic Monte Carlo simulations of the antiferromagnetic Heisenberg model (3.62) with negative nearest-neighbor exchange couplings $J < 0$ on a three-dimensional cubic lattice. In this system one finds $T_c \approx 1.44 |J|$, and direct fitting of the spin wave peak locations yields $z = 1.45 \pm 0.07$, see Fig. 6.15(a). Using a dynamic finite-size scaling method, this numerical estimate

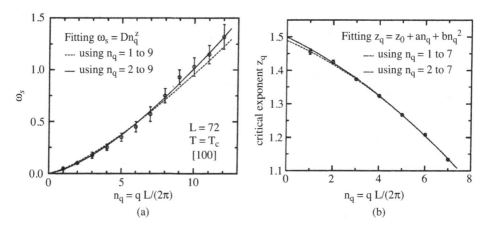

Fig. 6.15 Monte Carlo simulations for the critical dynamics of a Heisenberg antiferromagnet on a three-dimensional cubic lattice: (a) spin wave peaks as function
of wavenumber; (b) finite-size scaling analysis for the dynamic critical exponent z.
[Data and figures reproduced with permission from: S.-H. Tsai and D. P. Landau,
Phys. Rev. B **67**, 104411 (2003), DOI: 10.1103/PhysRevB.67.104411; copyright
by the American Physical Society.]

can be further refined to $z = 1.50 \pm 0.02$ if the lowest wavenumber is discarded,
in excellent agreement with the theoretical prediction.

6.4 Critical dynamics of binary fluids

As our last example for the critical dynamics of systems with relevant reversible
mode couplings, we study *model H* wherein a conserved vector field J interacts with
an also conserved scalar order parameter S.[8] Since in the Gaussian approximation
$z_J = 2$ and $z_S = 4$, we can hardly expect a strong dynamic scaling regime. Rather
we anticipate that the order parameter will fluctuate on much longer characteristic
time scales than the conserved non-critical modes, as in model D, Eqs. (6.50)
and (6.52), or at the SSS model weak scaling fixed point (6.133) with (6.134).
Yet, whereas in these cases the order parameter dynamics follows that of the
relaxational models B and A, respectively, and the mode coupling is manifest only
in the dynamics of the conserved fields, here the critical scalar density S will be
markedly affected by the reversible coupling to the vector fields J.

6.4.1 Model H and Galilean invariance

Model H describes the dynamics near the critical point in binary fluids, which
encompasses gas–liquid transitions. Correspondingly, the appropriate choice for

---

[8] Siggia, Halperin, and Hohenberg (1976); Gunton and Kawasaki (1976); Hohenberg and Halperin (1977);
De Dominicis and Peliti (1978); Vasil'ev (2004); Folk and Moser (2006).

the order parameter is basically the deviation of the grand-canonical potential density from its critical value, namely the linear combination $S(x) = [e(x) - e_c] - (T\bar{s} + \bar{\mu})[\rho(x) - \rho_c]$ of the energy and mass densities $e(x)$ and $\rho(x)$, where $\bar{s}$ and $\bar{\mu}$ represent the equilibrium entropy (excitations propagate adiabatically in a fluid) and chemical potential per unit mass. Energy and mass conservation imply continuity equations of the form (3.13), which in Fourier space reads $i q \cdot J(q, t) = -\partial S(q, t)/\partial t$. Next we define the longitudinal and transverse components of the current density J with respect to the wavevector q $(i, j = 1, \ldots, d)$:

$$J_i^{\|}(q) = \sum_j P_{ij}^{\|}(q) J_j(q), \quad P_{ij}^{\|}(q) = \frac{q_i q_j}{q^2}, \tag{6.135}$$

$$J_i^{\perp}(q) = \sum_j P_{ij}^{\perp}(q) J_j(q), \quad P_{ij}^{\perp}(q) = \delta_{ij} - \frac{q_i q_j}{q^2}, \tag{6.136}$$

with the projectors satisfying $\sum_j P_{ij}^{\|/\perp}(q) P_{jk}^{\|/\perp}(q) = P_{ik}^{\|/\perp}(q)$, or in matrix notation $P^{\|/\perp}(q)^2 = P^{\|/\perp}(q)$. Hence the longitudinal current component can be expressed through the fluctuations of the conserved density via

$$J_i^{\|}(q, t) = -i \frac{q_i}{q^2} \frac{\partial S(q, t)}{\partial t}. \tag{6.137}$$

This leaves only the transverse current components $J_i^{\perp}$ as independent hydrodynamic variables.

In a fluid, the particle current is proportional to the momentum density. Recall that the momentum operator generates spatial translations, $[p_i, f(x)] = \frac{\hbar}{i} \nabla_i f(x)$. After coarse-graining, this commutator suggests the following Poisson brackets involving the current densities:

$$\{J_i(x), S(x')\} = g \nabla_i S(x) \delta(x - x'), \tag{6.138}$$

$$\{J_i(x), J_j(x')\} = g \nabla_i J_j(x) \delta(x - x'). \tag{6.139}$$

From our discussion of the weakly interacting van-der-Waals gas in Chapter 1 we know that the critical behavior of the liquid–gas phase transition is governed by the Ising universality class, described by the Hamiltonian (1.49). Since the current density is a non-critical variable, we again simply add a Gaussian term to arrive at

$$\mathcal{H}[S, J] = \mathcal{H}[S] + \int d^d x \left[\frac{1}{2} J(x)^2 - A(x) J(x) \right], \tag{6.140}$$

where we have scaled the static current response function to unity, and $A(x)$ is the external vector potential thermodynamically conjugate to the current density $J(x)$. Thus we arrive at the reversible forces in the coupled Langevin equations for

$S(x, t)$ and $J(x, t)$, respectively:

$$\frac{\partial S(x, t)}{\partial t} = F_{\text{rev}}[S, J](x, t) + D \nabla^2 \frac{\delta \mathcal{H}[S, J]}{\delta S(x, t)} + \zeta(x, t), \qquad (6.141)$$

$$F_{\text{rev}}[S, J](x, t) = g \nabla S(x, t) \cdot \frac{\delta \mathcal{H}[S, J]}{\delta J(x, t)}$$

$$= g \nabla S(x, t) \cdot [J(x, t) - A(x, t)], \qquad (6.142)$$

$$\frac{\partial J^\perp(x, t)}{\partial t} = F_{\text{rev}}^\perp[S, J](x, t) + \lambda \nabla^2 \frac{\delta \mathcal{H}[S, J]}{\delta J^\perp(x, t)} + \eta(x, t), \qquad (6.143)$$

$$F_{\text{rev}}^\perp[S, J](x, t) = -g \, P^\perp \left[\frac{\delta \mathcal{H}[S, J]}{\delta S(x, t)} \nabla S(x, t) - \left(\frac{\delta \mathcal{H}[S, J]}{\delta J(x, t)} \cdot \nabla \right) J(x, t) \right]$$

$$= g \, P^\perp \Big([\nabla^2 S(x, t) + h(x, t)] \nabla S(x, t)$$

$$+ [J(x, t) - A(x, t)] \cdot \nabla J(x, t) \Big), \qquad (6.144)$$

where $P^\perp$ projects onto the transverse components. Note that after symmetrizing, the terms $\propto r, u$ from the functional derivative $\delta \mathcal{H}/\delta S$ vanish under the operation of the transverse projector. Of course, $\langle \zeta \rangle = 0 = \langle \eta \rangle$, and the stochastic noise correlations read

$$\langle \zeta(x, t) \zeta(x', t') \rangle = -2D \nabla_x^2 \delta(x - x')\delta(t - t'), \qquad (6.145)$$

$$\langle \eta_i(x, t) \eta_j(x', t') \rangle = -2\lambda \big(\delta_{ij} \nabla_x^2 - \nabla_i \nabla_j \big)\delta(x - x')\delta(t - t'). \qquad (6.146)$$

Linearizing these Langevin equations of motion (with $g = 0$), we find two diffusive solutions, namely the *Landau–Placzek* mode with $\omega_S(q) = -iDrq^2$, whence we may interpret $\lambda_T = Dr$ as essentially the thermal conductivity, and $d - 1$ transverse overdamped *shear* modes with $\omega_J(q) = -i\lambda q^2$, and λ proportional to the fluid viscosity. In addition, the longitudinal particle current coupled with local pressure variations supports propagating sound waves with linear dispersion $\omega_P(q) = \pm c_s|q|$. These 'faster' excitations should, however, not affect the leading dynamic critical behavior of the fluid, and will therefore not be taken into account here. Notice furthermore that for $g = -1$, the second term in Eq. (6.144) and the reversible force for the order parameter (6.142) just represent the convective terms that ensure Galilean invariance of the fluid motion. Indeed, it is a straightforward exercise (Problem 6.6) to show that the coupled Langevin equations (6.141)–(6.144) remain invariant under the *generalized Galilean coordinate transformations*

$$x \to x' = x - g \, v \, t, \quad t \to t' = t \qquad (6.147)$$

with constant vector v, whereupon

$$S(x, t) \to S'(x', t') = S(x - g v t, t), \qquad (6.148)$$

$$J(x, t) \to J'(x', t') = J(x - g v t, t) - v. \qquad (6.149)$$

The scaling dimensions of the conserved order parameter field and associated parameters are given by Eqs. (5.3)–(5.5) with $a = 2$. The conserved currents in turn scale in analogy with Eqs. (6.9) and (6.97), whence

$$[\lambda] = \mu^2, \quad [g] = \mu^{3-d/2}, \qquad (6.150)$$

see also Eq. (6.10), in order to render the scaling dimensions of all terms in the Langevin equations (6.141) and (6.143) identical. From our experience with the SSS model, we infer that the *effective* coupling stemming from the vertex in Eq. (6.141) and the first non-linearity in Eq. (6.143) is

$$f = \frac{g^2}{D\lambda}, \quad [f] = \mu^{4-d}, \qquad (6.151)$$

with critical dimension $d_c' = 4 = d_c$. The second term in (6.144) for $J_i^\perp$ generates the vertex $g P^\perp \tilde{J}_i J \cdot \nabla J_i$, and the associated current fluctuation loops lead to the effective coupling

$$\tilde{f} = \frac{g^2}{\lambda^2}, \quad [\tilde{f}] = \mu^{2-d}, \qquad (6.152)$$

which is *irrelevant* in the renormalization group sense as compared with the static non-linearity u and f: $[\tilde{f}/f] = \mu^{-2}$ has a negative scaling dimension. The leading power laws should thus not be affected by the coupling $\tilde{f}$, and as long as we are interested only in the universal critical properties of model H we may thus safely disregard the second term in Eq. (6.144) for our subsequent perturbative analysis.

After omitting the irrelevant vertex, the Janssen–De Dominicis functional $A[\tilde{S}, S, \tilde{J}, J] = A_0[\tilde{S}, S] + A_{\text{int}}[\tilde{S}, S] + A_0[\tilde{J}, J] + A_H[\tilde{S}, S, \tilde{J}, J]$ consists of the scalar ($\alpha = \beta = 1$) model B ($a = 2$) contributions (4.14) and (4.15), see Fig. 3.5 ($k_B T = 1$), the Gaussian action for the transverse currents

$$A_0[\tilde{J}, J] = \int d^d x \int dt \sum_{i=1}^{d} \left[\tilde{J}_i^\perp(x, t) \left(\frac{\partial}{\partial t} - \lambda \nabla^2 \right) J_i^\perp(x, t) \right.$$

$$\left. + \lambda \tilde{J}_i^\perp(x, t) \nabla^2 \left(\tilde{J}_i^\perp(x, t) + A_i^\perp(x, t) \right) \right], \qquad (6.153)$$

which gives the additional propagator

$$\langle J_i(q, \omega) \tilde{J}_j(q', \omega') \rangle = \frac{P_{ij}^\perp(q)}{-i\omega + \lambda q^2} (2\pi)^{d+1} \delta(q + q') \delta(\omega + \omega'), \qquad (6.154)$$

(a)   (b)

Fig. 6.16 Mode-coupling vertices for model H.

to be graphically represented by a dashed line again, and the non-linear mode-coupling terms

$$\mathcal{A}_{\mathrm{H}}[\tilde{S}, S, \tilde{J}, J] = -g \int d^d x \int dt \sum_{i=1}^{d} \left(\tilde{S}(x, t) [\nabla_i S(x, t)][J_i(x, t) - A_i(x, t)] \right.$$

$$\left. + \sum_{j=1}^{d} \tilde{J}_i^{\perp}(x, t) P_{ij}^{\perp}[\nabla^2 S(x, t) + h(x, t)][\nabla_j S(x, t)] \right). \quad (6.155)$$

The ensuing vertices (for $h = 0 = A$) are shown in Fig. 6.16; the symmetrized wavevector-dependent vertex factor here reads

$$v_i(q, p) = \frac{1}{2} \sum_{j} P_{ij}^{\perp}(q + p)(q^2 p_j + p^2 q_j). \quad (6.156)$$

6.4.2 Renormalization and dynamic critical exponents

Following our earlier prescriptions, we introduce the renormalized transverse currents, viscosity, and effective dynamical coupling according to

$$J_{iR}^{\perp} = Z_J^{1/2} J_i^{\perp}, \quad \tilde{J}_{iR}^{\perp} = Z_{\tilde{J}}^{1/2} \tilde{J}_i^{\perp}, \quad (6.157)$$

$$\lambda_{\mathrm{R}} = Z_{\lambda} \lambda \mu^{-2}, \quad f_{\mathrm{R}} = Z_g (Z_D Z_{\lambda})^{-1} f A_d \mu^{-\epsilon}, \quad (6.158)$$

where once more $\epsilon = 4 - d$, in addition to Eqs. (5.19) and (5.23) for the order parameter fluctuations and associated parameters. The vertex functions involving the scalar order parameter fields S and currents are defined as in Eq. (6.13) with ρ replaced by J. We shall also need the composite operators $X_i = [\tilde{S} \nabla_i S]$ and $Y = [\tilde{J} P^{\perp} \nabla S]$, as well as their cumulants and associated vertex functions, see Eq. (6.66). All vertex functions involving the currents are proportional to the transverse projector (6.136), i.e., $\Gamma_{i;j}^{(0,0;1,1)}(q, \omega) = \Gamma^{(0,0;1,1)}(q, \omega) P_{ij}^{\perp}(q)$, and similarly for $\Gamma^{(0,0;2,0)}(q, \omega)$ and $\Gamma^{(0;1;X)}(q, \omega)$, and therefore also for the dynamic susceptibility

$$\Xi_{ij}^{\perp}(x - x', t - t') = \frac{\delta \langle J_i^{\perp}(x, t) \rangle}{\delta A_j(x', t')}\bigg|_{h=0=A}, \tag{6.159}$$

$$\Xi_{ij}^{\perp}(q, \omega) = \Xi(q, \omega) P_{ij}^{\perp}(q). \tag{6.160}$$

The mode-coupling terms (6.155) in the action again lead to additional contributions, as compared with the simple relaxational models, to the dynamic response functions that include composite operator vertex functions, namely

$$\chi(q, \omega) = \Gamma^{(1,1;0,0)}(-q, -\omega)^{-1}\left[Dq^2 - g\,\Gamma^{(1;0;Y)}(q, \omega)\right], \tag{6.161}$$

$$\Xi(q, \omega) = \Gamma^{(0,0;1,1)}(-q, -\omega)^{-1}\left[\lambda q^2 + g\,\Gamma^{(0;1;X)}(q, \omega)\right]. \tag{6.162}$$

Inspection of Figs. 6.16 and 4.1(c) (with $a = 2$) shows that all vertices carry the wavevectors of the outgoing fields. Consequently, we find to all orders in the perturbation expansion

$$\Gamma^{(1,1;0,0)}(q = 0, \omega) = i\omega, \quad \Gamma^{(0,0;1,1)}(q = 0, \omega) = i\omega, \tag{6.163}$$

whence we infer

$$Z_{\tilde{S}}Z_S = 1 \tag{6.164}$$

and $Z_{\tilde{J}}Z_J = 1$. But since the non-critical conserved currents only appear quadratically in the Hamiltonian (6.140), the static response function is exactly $\Xi(q, 0) = 1$, so $Z_J = 1 = Z_{\tilde{J}}$.

Furthermore, in order to preserve Galilean invariance, Eq. (6.149) must hold in the renormalized theory as well. Consequently the mode-coupling strength must scale according to $g^{-1}\mu^3 \sim v \sim J$ and the product gJ cannot acquire any anomalous scaling dimension after renormalization. Formally, we can prove that via constructing a *Ward identity* associated with the symmetry under the transformations (6.148) and (6.149), see Section 5.4.2. The generating functional Γ for the vertex functions must accordingly remain invariant with respect to the combined variations $\delta S(x, t) = S(x', t') - S(x, t) = -gt\,v \cdot \nabla S(x, t)$, $\delta J_i(x, t) = J_i(x', t') - J_i(x, t) = -gt\,v \cdot \nabla J_i(x, t) - v_i$, and $\delta \tilde{S}(x, t) = -gt\,v \cdot \nabla \tilde{S}(x, t)$, $\delta \tilde{J}_i(x, t) = -gt\,v \cdot \nabla \tilde{J}(x, t)$. Since the Galilean boost vector v is arbitrary, we have

$$0 = \int d^d x \int dt \left[g\,t \left(\frac{\delta \Gamma}{\delta \tilde{S}(x, t)} \nabla_i \tilde{S}(x, t) + \frac{\delta \Gamma}{\delta S(x, t)} \nabla_i S(x, t) \right.\right.$$

$$\left.\left. + \frac{\delta \Gamma}{\delta \tilde{J}(x, t)} \cdot \nabla_i \tilde{J}(x, t) + \frac{\delta \Gamma}{\delta J(x, t)} \cdot \nabla_i J(x, t) \right) + \frac{\delta \Gamma}{\delta J_i(x, t)} \right].$$

$$\tag{6.165}$$

Taking functional derivatives with respect to $\widetilde{S}(x', t')$ and $S(x'', t'')$, and observing the definition (4.80) yields

$$g(t'' - t') \frac{\partial \Gamma^{(1,1;0,0)}(x'' - x', t'' - t')}{\partial(x_i'' - x_i')} = \int d^d x \int dt \, \Gamma_i^{(1,1;0,1)}(x', t'; x'', t''; x, t) \,,$$

or after Fourier transform

$$- i q_i \, g \, \frac{\partial \Gamma^{(1,1;0,0)}(q, \omega)}{\partial(i\omega)} = \Gamma_i^{(1,1;0,1)}(-q, -\omega; q, \omega; 0, 0) \,. \tag{6.166}$$

An explicit evaluation of the three-point vertex function confirms this statement, see Problem 6.7. The identity (6.166) must hold for the renormalized vertex functions as well, whence $Z_g = Z_J^{-1}$.

In summary, exploiting the conservation laws, the symmetry, and static properties all together tells us that

$$Z_g = Z_J^{-1} = Z_{\tilde{J}} = 1 \,. \tag{6.167}$$

With Eq. (6.158), this result immediately leads to the following RG beta function for the effective mode-coupling strength f:

$$\beta_f = \mu \left. \frac{\partial}{\partial \mu} \right|_0 f_R = f_R(d - 6 - \gamma_D - \gamma_\lambda) \,. \tag{6.168}$$

Therefore, at any non-trivial finite RG fixed point $0 < f^* < \infty$, we have $\gamma_D^* + \gamma_\lambda^* = d - 6$. Since the dynamic critical exponents are given by $z_S = 4 + \gamma_D^*$ and $z_J = 4 + \gamma_\lambda^*$, compare with Eq. (6.42), this results in the scaling relation

$$z_S + z_J = d + 2 \,, \tag{6.169}$$

in analogy with Eq. (6.124) for the SSS model. The same relation again follows from a self-consistent one-loop or mode-coupling approach using the Lorentzian line approximation, see Problem 6.5.

Explicitly, the model H Feynman diagrams for the two-point vertex functions to one-loop order look identical to Figs. 6.10 and 6.11 for the SSS model. The corresponding analytical expressions for the current vertex functions become (Problem 6.7)

$$\Gamma^{(0,0;1,1)}(q, \omega) = i\omega + \lambda q^2 + \frac{2g^2}{d - 1} \, I_{\Xi,r}(-q, -\omega) \,, \tag{6.170}$$

$$\Gamma^{(0;1;X)}(q, \omega) = \frac{2g}{d - 1} \, I_{\Xi,r}(q, \omega) \,, \tag{6.171}$$

where the wavevector integral reads with $\Delta(q) = Dq^2(r + q^2)$:

$$I_{\Xi,r}(q, \omega) = \int_k \frac{(q \cdot k)^2}{\left[r + \left(\frac{q}{2} + k\right)^2\right]\left[r + \left(\frac{q}{2} - k\right)^2\right]} \frac{k^2 - (q \cdot k)^2/q^2}{-i\omega + \Delta\left(\frac{q}{2} + k\right) + \Delta\left(\frac{q}{2} - k\right)},$$

(6.172)

see also Eqs. (6.70) and (6.112). The dynamic susceptibility for the conserved currents then takes the form (6.113), which confirms Eq. (6.163) and $\Xi(q, 0) = 1$, and where the relaxation coefficient is to one-loop order

$$\Gamma_\Xi(q, \omega) = \lambda q^2 + \frac{2g^2}{d - 1} I_{\Xi,\tau}(q, \omega).$$

(6.173)

The order parameter vertex functions become to first order in u

$$\Gamma^{(1,1;0,0)}(q, \omega) = i\omega + \chi(q)^{-1}\left[Dq^2 + g^2 I_{\chi,r}(-q, -\omega)\right],$$ (6.174)

$$\Gamma^{(1;0;Y)}(q, \omega) = -g I_{\chi,r}(q, \omega),$$ (6.175)

where (Problem 6.7)

$$I_{\chi,r}(q, \omega) = \int_k \frac{1}{r + \left(\frac{q}{2} + k\right)^2} \frac{q^2 - \left[\frac{q^2}{2} - (q \cdot k)\right]^2 / \left(\frac{q}{2} - k\right)^2}{-i\omega + \Delta\left(\frac{q}{2} + k\right) + \lambda\left(\frac{q}{2} - k\right)^2}.$$

(6.176)

This is again in accord with Eq. (6.163), and the dynamic order parameter suscep-tibility assumes the form (6.71) with the relaxation coefficient

$$\Gamma_\chi(q, \omega) = \chi(q)^{-1}\left[Dq^2 + g^2 I_{\chi,\tau}(q, \omega)\right].$$

(6.177)

Keeping in mind that $w = D/\lambda \to 0$ asymptotically, we determine the renormalization constants for the relaxation rates from $\partial I_{\chi,\mu^2}(q, 0)/\partial q^2$ and $\partial I_{\Xi,\mu^2}(q, 0)/\partial q^2$, Eqs. (6.177) and (6.173), as follows:

$$Z_D = 1 + f \int_k \frac{k^2 - (q \cdot k)^2/q^2}{k^4(\mu^2 + k^2)} = 1 + f\left(1 - \frac{1}{d}\right)\int_k \frac{1}{k^2(\mu^2 + k^2)}$$

$$= 1 + \frac{2(d - 1)}{d(d - 2)} \frac{f A_d \mu^{-\epsilon}}{\epsilon} \quad \to \quad 1 + \frac{3}{4} \frac{f A_d \mu^{-\epsilon}}{\epsilon},$$ (6.178)

$$Z_\lambda = 1 + \frac{f}{d - 1} \frac{\partial}{\partial q^2} \int_k \frac{(q \cdot k)^2[k^2 - (q \cdot k)^2/q^2]}{k^2(\mu^2 + k^2)^3}$$

$$= 1 + \frac{f}{d(d - 1)}\left(1 - \frac{3}{d + 2}\right)\int_k \frac{k^2}{(\mu^2 + k^2)^3}$$

$$= 1 + \frac{1}{4(d + 2)} \frac{f A_d \mu^{-\epsilon}}{\epsilon} \quad \to \quad 1 + \frac{1}{24} \frac{f A_d \mu^{-\epsilon}}{\epsilon}.$$ (6.179)

(for the dimensionally regularized integrals, see Eq. (5.10) and Section 5.5). In the last step, we have applied the minimal subtraction procedure, and have replaced d with 4 in the residua of the ϵ poles. Once again, we could have also obtained these results from renormalizing the noise vertices for the order parameter and conserved transverse currents.

The Z factors (6.178) and (6.179) yield Wilson's flow functions

$$\gamma_D = -\frac{3}{4} f_R , \quad \gamma_\lambda = -2 - \frac{1}{24} f_R , \tag{6.180}$$

and upon inserting into the RG beta function (6.168), we find the non-trivial fixed point

$$f_H^* = \frac{24}{19} \epsilon + O(\epsilon^2) , \tag{6.181}$$

which is infrared-stable for $d < d_c = 4$. To one-loop order, we thus arrive at last at the dynamic critical exponents for the scalar conserved order parameter and the conserved transverse currents, respectively,

$$z_S = 4 + \gamma_D^* = 4 - \frac{18}{19} \epsilon + O(\epsilon^2) , \tag{6.182}$$

$$z_J = 4 + \gamma_\lambda^* = 2 - \frac{1}{19} \epsilon + O(\epsilon^2) . \tag{6.183}$$

The scaling relation (6.169) is of course satisfied. Notice that the deviation of z_J from the mean-field value 2 is quite small, which makes an experimental verification of the anomalous dynamics for the conserved currents difficult. Yet the order parameter kinetics is rendered considerably faster as compared to model B which does not entail reversible couplings to other hydrodynamic modes. Interestingly, the critical dynamics of an *isothermal* non-ideal fluid becomes effectively reduced to model A in the asymptotic limit.[9]

Problems

6.1 *Model C/D vertex functions*

(a) For the relaxational models C and D, compute the two-point order parameter vertex function $\Gamma^{(1,1;0,0)}(q, \omega)$ in the one-loop approximation. Check that the renormalization constant Z_τ comes out as for model A/B, when the result is expressed in terms of $u' = u - 3 g^2$, and confirm Eq. (6.23) for model C to this order.

[9] Gross and Varnik (2012).

Hint: fluctuations render the mean energy density $\langle\rho\rangle$ non-zero, which must be accounted for through an appropriate counterterm (Section 5.4).

(b) Draw and evaluate the one-loop Feynman diagrams for the model C/D three-point vertex function $\Gamma^{(1,1;0,1)}(-q, 0; \frac{q}{2}, 0; \frac{q}{2}, 0)$. Therefrom deduce the renormalization constant Z_g.

6.2 *Model J vertex functions*

(a) For model J, confirm the one-loop results (6.68) for $\Gamma^{(1,1)}(q, \omega)$ and (6.69) for $\Gamma^{(1,X)}_{\alpha;[\beta\gamma]}(q, \omega)$ and convince yourself that the dynamic response and Kubo relaxation functions can indeed be written in the form (6.71), with the relaxation coefficient (6.72).

(b) Evaluate the Feynman diagram in Fig. 6.5(c) for the two-point vertex function $\Gamma^{(2,0)}(q, \omega)$, and confirm the one-loop result (6.77) for the renormalization constant Z_D.

6.3 *Response functions for the Sasvári–Schwabl–Szépfalusy model*

(a) Show that the dynamic susceptibilities for the order parameter and generalized angular momenta of the SSS model are given by

$$\chi^{\alpha\beta}(x-x', t-t') = D\langle S^\alpha(x, t)\, \widetilde{S}^\beta(x', t')\rangle$$
$$- g\sum_\gamma \langle S^\alpha(x, t)[\widetilde{M}^{\beta\gamma}(x', t')\, S^\gamma(x', t')]\rangle,$$

$$\Xi^{\alpha\beta,\gamma\delta}(x-x', t-t') = -\lambda\langle M^{\alpha\beta}(x, t)\, \nabla^2\, \widetilde{M}^{\gamma\delta}(x', t')\rangle$$
$$+ 2g\langle M^{\alpha\beta}(x, t)[\widetilde{S}^\gamma(x', t')\, S^\delta(x', t')]\rangle$$
$$+ 2g\sum_\lambda \langle M^{\alpha\beta}(x, t)[\widetilde{M}^{\gamma\lambda}(x', t')\, M^{\delta\lambda}(x', t')]\rangle,$$

and use these results to derive the relations (6.108) and (6.109).

(b) Consider switching on a spatially uniform field $H(t)$ at time $t = 0$, and employ the Langevin equation (6.89) to derive the SSS model Ward identity (6.104), and thence $Z_g = Z_M$.

6.4 *SSS model vertex functions and renormalization*

(a) Confirm the one-loop results (6.110)–(6.115) and (6.117)–(6.122) for the vertex and linear response functions of the SSS model.

(b) Derive Z_λ and Z_D directly through renormalization of the noise vertices $\Gamma^{(0,0;2,0)}(q, \omega)$ and $\Gamma^{(2,0;0,0)}(q, \omega)$, see Figs. 6.2(b) and (c).

6.5 *Mode-coupling theory for the SSS model and model H*

(a) Follow the procedure in Section 6.2.2 and generalize the one-loop results (6.120) and (6.114) for the SSS model line widths to obtain the self-consistent mode-coupling equations in the Lorentzian approximation

$$\chi(q)\,\Gamma_\chi(q) \approx D + (n-1)\,g^2 \int \frac{d^d k}{(2\pi)^d}\, \frac{\chi\left(\frac{q}{2}+k\right)\Xi\left(\frac{q}{2}-k\right)}{\Gamma_\chi\left(\frac{q}{2}+k\right)+\Gamma_\Xi\left(\frac{q}{2}-k\right)},$$

$$\Xi(q)\,\Gamma_\Xi(q) \approx \lambda\,q^2 + 4g^2 \int \frac{d^d k}{(2\pi)^d}\, v(q,k)\, \frac{\chi\left(\frac{q}{2}+k\right)\chi\left(\frac{q}{2}-k\right)}{\Gamma_\chi\left(\frac{q}{2}+k\right)+\Gamma_\chi\left(\frac{q}{2}-k\right)},$$

with the vertex factor

$$v(q,k) = \frac{1}{4}\left[\chi\left(\frac{q}{2}+k\right)^{-1} - \chi\left(\frac{q}{2}-k\right)^{-1}\right]^2.$$

(b) Discuss the ensuing scaling regimes that follow from inserting the critical power laws $\Gamma_\chi(q) \sim |q|^{z_S}$ and $\Gamma_\Xi(q) \sim |q|^{z_M}$.

(c) In a similar manner, employ Eqs. (6.177) and (6.173) to derive the analogous mode-coupling equations for model H,

$$\chi(q)\,\Gamma_\chi(q) \approx D\,q^2 + g^2 \int \frac{d^d k}{(2\pi)^d}\left(q^2 - \frac{\left[\frac{q^2}{2}-(q\cdot k)\right]^2}{\left(\frac{q}{2}-k\right)^2}\right)$$

$$\times \frac{\chi\left(\frac{q}{2}+k\right)\Xi\left(\frac{q}{2}-k\right)}{\Gamma_\chi\left(\frac{q}{2}+k\right)+\Gamma_\Xi\left(\frac{q}{2}-k\right)},$$

$$\Xi(q)\,\Gamma_\Xi(q) \approx \lambda\,q^2 + \frac{2g^2}{d-1} \int \frac{d^d k}{(2\pi)^d}\left(k^2 - \frac{(q\cdot k)^2}{q^2}\right)$$

$$\times \frac{v(q,k)\,\chi\left(\frac{q}{2}+k\right)\chi\left(\frac{q}{2}-k\right)}{\Gamma_\chi\left(\frac{q}{2}+k\right)+\Gamma_\chi\left(\frac{q}{2}-k\right)},$$

and therefrom deduce the scaling relation (6.169).

6.6 *Generalized Galilean invariance for model H*
Confirm that the coupled Langevin equations (6.141)–(6.144) for binary fluids (model H) remain invariant under the generalized Galilean coordinate transformations (6.147)– (6.149).

6.7 *Model H vertex functions*
(a) Confirm the explicit one-loop results (6.170)–(6.177) for the two-point vertex and response functions of model H.
(b) By evaluating the one-loop Feynman diagrams for the three-point vertex function $\Gamma_i^{(1,1;0,1)}(-q,-\omega;q,\omega;0,0)$, check the Ward identity (6.166) to this order.

References

Bausch, R., H. K. Janssen, and H. Wagner, 1976, Renormalized field theory of critical dynamics, *Z. Phys. B Cond. Matt.* **24**, 113–127.

Böni, P., D. Görlitz, J. Kötzler, and J. L. Martínez, 1991, Dynamics of longitudinal and transverse fluctuations above T_c in EuS, *Phys. Rev. B* **43**, 8755–8758.

Brézin, E. and C. De Dominicis, 1975, Field-theoretic techniques and critical dynamics. II. Ginzburg–Landau stochastic models with energy conservation, *Phys. Rev. B* **12**, 4954–4962.

Coldea, R., R. A. Cowley, T. G. Perring, D. F. McMorrow, and B. Roessli, 1998, Critical behavior of the three-dimensional Heisenberg antiferromagnet $RbMnF_3$, *Phys. Rev. B* **57**, 5281–5290.

De Dominicis, C. and L. Peliti, 1977, Deviations from dynamic scaling in helium and antiferromagnets, *Phys. Rev. Lett.* **38**, 505–508.

De Dominicis, C. and L. Peliti, 1978, Field-theory renormalization and critical dynamics above T_c: helium, antiferromagnets, and liquid-gas systems, *Phys. Rev. B* **18**, 353–376.

Dohm, V., 2006, Model F in two-loop order and the thermal conductivity near the superfluid transition of ^{4}He, *Phys. Rev. B* **73**, 092503-1–4.

Dohm, V. and R. Folk, 1980, Nonasymptotic critical dynamics near the superfluid transition in ^{4}He, *Z. Phys. B Cond. Matt.* **40**, 79–93.

Folk, R. and G. Moser, 2003, Critical dynamics of model C resolved, *Phys. Rev. Lett.* **91**, 030601-1–4.

Folk, R. and G. Moser, 2004, Critical dynamics of stochastic models with energy conservation (model C), *Phys. Rev. E* **69**, 036101-1–18.

Folk, R. and G. Moser, 2006, Critical dynamics: a field theoretical approach, *J. Phys. A: Math. Gen.* **39**, R207–R313.

Freedman, R. and G. F. Mazenko, 1976, Critical dynamics of isotropic antiferromagnets using renormalization-group methods: $T \geq T_N$, *Phys. Rev. B* **13**, 4967–4983.

Frey, E. and F. Schwabl, 1994, Critical dynamics of magnets, *Adv. Phys.* **43**, 577–683.

Gross, M. and F. Varnik, 2012, Critical dynamics of an isothermal compressible nonideal fluid, *Phys. Rev. E* **86**, 061119-1–15.

Gunton, J. D. and K. Kawasaki, 1975, Critical transport anomalies in $4 - \epsilon$ and $6 - \epsilon$ dimensions, *J. Phys. A: Math. Gen.* **8**, L9–L12.

Gunton, J. D. and K. Kawasaki, 1976, Renormalization group equations in critical dynamics. II. Binary liquids, superfluid helium and magnets, *Progr. Theor. Phys.* **56**, 61–76.

Halperin, B. I., P. C. Hohenberg, and S.-k. Ma, 1974, Renormalization-group methods for critical dynamics: I. Recursion relations and effects of energy conservation, *Phys. Rev. B* **10**, 139–153.

Halperin, B. I., P. C. Hohenberg, and S.-k. Ma, 1976, Renormalization-group methods for critical dynamics: II. Detailed analysis of the relaxational models, *Phys. Rev. B* **13**, 4119–4131.

Halperin, B. I., P. C. Hohenberg, and E. D. Siggia, 1974, Renormalization-group calculations of divergent transport coefficients at critical points, *Phys. Rev. Lett.* **32**, 1289–1292.

Halperin, B. I., P. C. Hohenberg, and E. D. Siggia, 1976, Renormalization-group treatment of the critical dynamics of superfluid helium, the isotropic antiferromagnet, and the easy-plane ferromagnet, *Phys. Rev. B* **13**, 1299–1328; err. *Phys. Rev. B* **21**, 2044–2045 (1980).

Hohenberg, P. C. and B. I. Halperin, 1977, Theory of dynamic critical phenomena, *Rev. Mod. Phys.* **49**, 435–479.

Janssen, H. K., 1977, Renormalized field theory for the critical dynamics of $O(n)$-symmetric systems, *Z. Phys. B Cond. Matt.* **26**, 187–189.

Janssen, H. K., 1979, Field-theoretic methods applied to critical dynamics, in: *Dynamical Critical Phenomena and Related Topics*, ed. C. P. Enz, Lecture Notes in Physics, Vol. **104**, Heidelberg: Springer-Verlag, 26–47.

Kötzler, J., 1988, Universality of the dipolar dynamic crossover of cubic ferromagnets above T_c, *Phys. Rev. B* **38**, 12027–12030.

Ma, S.-k. and G. F. Mazenko, 1974, Critical dynamics of ferromagnets in $6 - \epsilon$ dimensions, *Phys. Rev. Lett.* **33**, 1383–1385.

Ma, S.-k. and G. F. Mazenko, 1975, Critical dynamics of ferromagnets in $6 - \epsilon$ dimensions: general discussion and detailed calculation, *Phys. Rev. B* **11**, 4077–4100.

Sasvári, L. and P. Szépfalusy, 1977, Dynamic critical properties of a stochastic n-vector model, *Physica* **87A**, 1–34.

Sasvári, L., F. Schwabl, and P. Szépfalusy, 1975, Hydrodynamics of an n-component phonon system, *Physica* **81A**, 108–128.

Semadeni, P., B. Roessli, P. Böni, P. Vorderwisch, and T. Chatterji, 2000, Critical fluctuations in the weak itinerant ferromagnet Ni_3Al: a comparison between self-consistent renormalization and mode-mode coupling theory, *Phys. Rev. B* **62**, 1083–1088.

Siggia, E. D., B. I. Halperin, and P. C. Hohenberg, 1976, Renormalization-group treatment of the critical dynamics of the binary-fluid and gas-liquid transitions, *Phys. Rev. B* **13**, 2110–2123.

Tsai, S.-H. and D. P. Landau, 2003, Critical dynamics of the simple-cubic Heisenberg antiferromagnet $RbMnF_3$: extrapolation to $q = 0$, *Phys. Rev. B* **67**, 104411-1–6.

Vasil'ev, A. N., 1993, *The Field Theoretic Renormalization Group in Critical Behavior Theory and Stochastic Dynamics*, Boca Raton: Chapman & Hall / CRC, chapter 5.

Zinn-Justin, J., 1993, *Quantum Field Theory and Critical Phenomena*, Oxford: Clarendon Press, chapter 34.

Further reading

Bhattacharjee, J. K., 1996, Critical dynamics of systems with reversible mode coupling terms: spherical limit, *Europhys. Lett.* **34**, 525–530.

Bhattacharjee, J. K., U. Kaatze, and S. Z. Mirzaev, 2010, Sound attenuation near the demixing point of binary liquids: interplay of critical dynamics and noncritical kinetics, *Rep. Prog. Phys.* **73**, 066601, 1–36.

Campellone, M. and J.-P. Bouchaud, 1997, Self-consistent screening approximation for critical dynamics, *J. Phys. A: Math. Gen.* **30**, 3333–3343.

Chaikin, P. M. and T. C. Lubensky, 1995, *Principles of Condensed Matter Physics*, Cambridge: Cambridge University Press, chapter 8.

Chen, A., E. H. Chimowitz, S. De, and Y. Shapir, 2005, Universal dynamic exponent at the liquid–gas transition from molecular dynamics, *Phys. Rev. Lett.* **95**, 255701-1–4.

Das, S. K., M. E. Fisher, J. V. Sengers, J. Horbach, and K. Binder, 2006, Critical dynamics in a binary fluid: simulations and finite-size scaling, *Phys. Rev. Lett.* **97**, 025702-1–4.

Dohm, V., 1987, Renormalization-group theory of critical phenomena near the lambda transition of ^4He, *J. Low Temp. Phys.* **69**, 51–75.

Dohm, V., 1991, Renormalization-group flow equations of model F, *Phys. Rev. B* **44**, 2697–2712.

Dohm, V. and R. Folk, 1982, Nonlinear dynamic renormalization-group analysis above and below the lambda transition in ^4He, *Physica* **109 & 110B**, 1549–1556.

Hohenberg, P. C., 1982, Critical phenomena in ^4He, *Physica* **109 & 110B**, 1436–1446.

Mesterházy, D., J. H. Stockemer, L. F. Palhares, and J. Berges, 2013, Dynamic universality class of Model C from the functional renormalization group, *Phys. Rev. B* **88**, 174301-1–4.

Roy, S. and S. K. Das, 2011, Transport phenomena in fluids: finite-size scaling for critical behavior, *Europhys. Lett. (EPL)* **94**, 36001-1–6.

7

Phase transitions in quantum systems

This chapter addresses phase transitions and dynamic scaling occurring in systems comprised of interacting indistinguishable quantum particles, for which entanglement correlations are crucial. It first describes how the dynamics (in real time) and thermodynamics (in imaginary time) of quantum many-particle Hamiltonians can be mapped onto field theories based on coherent-state path integrals. While bosons are described by complex-valued fields, fermions are represented by anticommuting Grassmann variables. Since quantum-mechanical systems are of inherently dynamical nature, the corresponding field theory action entails $d + 1$ dimensions, with time playing a special role. For Hamiltonians that incorporate only two-particle interactions, we can make contact with the previously studied Langevin equations, yet with effectively multiplicative rather than additive noise. As an illustration, this formalism is applied to deduce fundamental properties of weakly interacting boson superfluids. Whereas Landau–Ginzburg theory already provides a basic hydrodynamic description, the Gaussian approximation allows the computation of density correlations, the Bose condensate fraction, and the normal- and superfluid densities from the particle current correlations. We next establish that quantum fluctuations are typically irrelevant for thermodynamic critical phenomena, provided that $T_c > 0$, and readily extend finite-size scaling theory to the imaginary time axis to arrive at general scaling forms for the free energy. Intriguing novel phenomena emerge in the realm of genuine quantum phase transitions at zero temperature, governed by other control parameters such as particle density, interaction or disorder strengths. The quantum Ising model in a transverse magnetic field and the localization transition in disordered boson superfluids are briefly discussed as instructive examples. Finally, classical vs. quantum critical behavior in quantum antiferromagnets and the dimensional crossover connecting these scaling regimes are treated in more detail.

7.1 Coherent-state path integrals

Quantum particles have the peculiar property to be strictly indistinguishable. This implies subtle correlations even in non-interacting quantum systems of identical particles which are not present at all in the classical realm. Indeed, these entanglement correlations have been the source of many puzzling 'paradoxa' that arise in attempts to capture such genuine quantum phenomena in terms of classical concepts. In the formalism of quantum theory, the indistinguishability of identical quantum particles (with the same mass, spin, charge, and other charge-like quantum numbers) implies that the many-particle Hamiltonian governing the time evolution according to the Schrödinger equation must be invariant under all permutations $P_{ij} = P_{ji}^{-1}$ of particle labels $i, j = 1, \ldots, N$: $P_{ij} H P_{ji} = H$. When acting on proper quantum many-particle states, the permutation operator P_{ij} can thus only produce an overall phase factor $e^{i\varphi_{ij}}$. The spin-statistics theorem valid in local, causal, and Lorentz-covariant quantum field theories makes the remarkable statement that in dimensions $d > 2$ only two phases are possible: either $\varphi_{ij} = 0$, for particles with integer spin $S = 0, 1, 2, \ldots$; or $\varphi_{ij} = \pi$ for particles with half-integer spin $S = \frac{1}{2}, \frac{3}{2}, \ldots$. The only allowed quantum many-particle states are therefore eigenstates of the permutation operator with real eigenvalues ± 1. Particles whose states are not affected at all by label permutation are referred to as bosons, while for fermions, the associated states pick up a phase $e^{i\pi} = -1$ under particle exchange. These (anti-)symmetry properties of bosonic and fermionic states are most economically encoded in the Fock or occupation number representation (also somewhat misleadingly called 'second quantization'). In the following, it will be explained how upon utilizing a different basis, namely that of coherent states, bosonic and (later) fermionic quantum many-particle systems can be represented through $(d + 1)$-dimensional path integrals.

7.1.1 Coherent states for boson systems

Let us begin by constructing the Fock space for a quantum system with N identical bosons.[1] For $N = 0$, we merely have the 'vacuum' state $|0\rangle$. A single-particle state labeled by a certain set of quantum numbers α is then constructed through the action of a creation operator $a_\alpha^\dagger$: $|\alpha\rangle = a_\alpha^\dagger |0\rangle$. Now proceed to a two-particle state $|\alpha\,\beta\rangle = a_\alpha^\dagger a_\beta^\dagger |0\rangle$; since permutation symmetry requires this state to be identical to $|\beta\,\alpha\rangle$, we infer the commutation relation

$$\left[a_\alpha^\dagger, a_\beta^\dagger\right] = 0 = [a_\alpha, a_\beta],$$ (7.1)

[1] See, e.g., Negele and Orland (1988); Schwabl (2008).

where we have introduced the adjoint single-particle annihilation operators a_α. The ladder operator algebra (in immediate analogy with the quantum-mechanical treatment of harmonic oscillators) is completed by further demanding that

$$[a_\alpha, a_\beta^\dagger] = \delta_{\alpha\beta} . \tag{7.2}$$

This allows us to form normalized many-particle states with n_α particles in the same single-particle state α. By means of (7.2) it is easy to see that

$$a_\alpha | \cdots n_\alpha \cdots \rangle = \sqrt{n_\alpha} | \cdots n_\alpha - 1 \cdots \rangle , \tag{7.3}$$

$$a_\alpha^\dagger | \cdots n_\alpha \cdots \rangle = \sqrt{n_\alpha + 1} | \cdots n_\alpha + 1 \cdots \rangle , \tag{7.4}$$

whence $a_\alpha | \cdots n_\alpha = 0 \cdots \rangle = 0$, and the occupation number operator is given by $\hat{n}_\alpha = a_\alpha^\dagger a_\alpha$ with integer eigenvalues $n_\alpha = 0, 1, \ldots$. A general N-particle Fock space state is then uniquely characterized by the set of occupation numbers $\{n_\alpha\}$ with $N = \sum_\alpha n_\alpha$, and is by construction properly symmetrized with respect to the single-particle labels:

$$|\{n_\alpha\}\rangle = \prod_\alpha \frac{1}{\sqrt{n_\alpha!}} \left(a_\alpha^\dagger\right)^{n_\alpha} |0\rangle = \sqrt{\frac{\prod_\alpha n_\alpha!}{N!}} \sum_{P'} P'_{\alpha_1 \ldots \alpha_N} |\alpha_1 \cdots \alpha_N\rangle , \tag{7.5}$$

where $P'_{\alpha_1 \cdots \alpha_N}$ symbolizes the set of all *distinct* permutations of the state labels α_i.

Now consider transforming to a new orthonormal and complete basis of single-particle states $|\alpha'\rangle$: $\langle \alpha'|\beta'\rangle = \delta_{\alpha'\beta'}$ and $\sum_{\alpha'} |\alpha'\rangle\langle\alpha'| = 1$. Since $|\alpha'\rangle = \sum_\alpha \langle\alpha|\alpha'\rangle |\alpha\rangle$, we see that the associated creation operators follow via the same unitary transformation $a_{\alpha'}^\dagger = \sum_\alpha \langle\alpha|\alpha'\rangle a_\alpha^\dagger$, and consequently the commutators (7.1) and (7.2) are preserved, e.g. $[a_{\alpha'}, a_{\beta'}^\dagger] = \sum_{\alpha\beta} \langle\alpha'|\alpha\rangle \langle\beta|\beta'\rangle [a_\alpha, a_\beta^\dagger] = \delta_{\alpha'\beta'}$. Specifically, employing the position eigenstates $|x\rangle$ involves the wave functions $\psi_\alpha(x) = \langle x|\alpha\rangle$, and produces the field operators $\hat{\psi}(x) = \sum_\alpha \psi_\alpha(x) a_\alpha$ that obey the commutation relation $[\hat{\psi}(x), \hat{\psi}^\dagger(x')] = \delta(x - x')$, the continuous analog of Eq. (7.2).

Consider a single-particle operator $T = \sum_{i=1}^N T_i^{(1)}$ in its eigenstate basis, $T^{(1)}|\alpha\rangle = t_\alpha |\alpha\rangle$. In the associated Fock space representation, this operator becomes $T = \sum_\alpha t_\alpha a_\alpha^\dagger a_\alpha$, with eigenvalues $\sum_\alpha t_\alpha n_\alpha$. In an arbitrary basis, one has instead

$$T = \sum_{\alpha\beta} \langle\alpha|T^{(1)}|\beta\rangle a_\alpha^\dagger a_\beta . \tag{7.6}$$

For example, the kinetic energy reads in the position and momentum representation, respectively,

$$T = -\frac{\hbar^2}{2m} \int d^d x \, \hat{\psi}^\dagger(x) \nabla^2 \hat{\psi}(x) = \frac{1}{2m} \int \frac{d^d p}{(2\pi\hbar)^d} \, p^2 \, \hat{\psi}^\dagger(p) \hat{\psi}(p) . \tag{7.7}$$

In a similar manner, two-particle operators $V = \frac{1}{2} \sum_{i \neq j} V_{ij}^{(2)}$ assume the following Fock space form in an arbitrary basis:

$$V = \frac{1}{2} \sum_{\alpha\beta\gamma\delta} \langle \alpha\beta | V^{(2)} | \gamma\delta \rangle \, a_\alpha^\dagger a_\beta^\dagger a_\delta a_\gamma \,, \tag{7.8}$$

exemplified by pair interactions in position space,

$$V = \frac{1}{2} \int d^d x \int d^d x' \, V(x - x') \, \hat{\psi}^\dagger(x) \hat{\psi}^\dagger(x') \hat{\psi}(x') \hat{\psi}(x) \,. \tag{7.9}$$

The properly symmetrized Fock states (7.5) form a basis of the many-particle Hilbert space for bosons. We shall find an alternative representation in terms of *coherent states* convenient. These are defined as right eigenstates of the annihilation operators, with *complex* eigenvalues ϕ_α:

$$a_\alpha | \phi \rangle = \phi_\alpha | \phi \rangle \,. \tag{7.10}$$

Let us now expand $|\phi\rangle = \sum_{\{n_\alpha\}} \phi_{\{n_\alpha\}} |\{n_\alpha\}\rangle$, choosing the expansion coefficient for the vacuum state to be $\phi_{\{n_\alpha=0\}} = \langle 0 | \phi \rangle = 1$. By means of Eq. (7.3) we then immediately obtain the recursion relation $\phi_{\cdots n_\alpha \cdots} = (\phi_\alpha / \sqrt{n_\alpha}) \phi_{\cdots n_\alpha - 1 \cdots}$, which implies $\phi_{\{n_\alpha\}} = \prod_\alpha \phi_\alpha^{n_\alpha} / \sqrt{n_\alpha !}$. This leads to an explicit formula for the coherent states,

$$|\phi\rangle = \sum_{\{n_\alpha\}} \prod_\alpha \frac{\phi_\alpha^{n_\alpha}}{n_\alpha !} (a_\alpha^\dagger)^{n_\alpha} |0\rangle = e^{\sum_\alpha \phi_\alpha a_\alpha^\dagger} |0\rangle \,. \tag{7.11}$$

Obviously, the particle number is not fixed in a coherent state; in fact, as a consequence of the exponential form (7.11), in such a state the occupation number probability distribution $P(n_\alpha)$ is a Poissonian (see Problem 7.1). Also, note that Eq. (7.11) allows us to identify

$$a_\alpha^\dagger | \phi \rangle = \frac{\partial}{\partial \phi_\alpha} | \phi \rangle \,. \tag{7.12}$$

For the adjoint states $\langle \phi |$, i.e., *left* eigenstates of the creation operators $a_\alpha^\dagger$, we correspondingly obtain[2]

$$\langle \phi | a_\alpha^\dagger = \phi_\alpha^* \langle \phi | \,, \quad \langle \phi | = \langle 0 | e^{\sum_\alpha \phi_\alpha^* a_\alpha} \,, \quad \langle \phi | a_\alpha = \frac{\partial}{\partial \phi_\alpha^*} \langle \phi | \,. \tag{7.13}$$

It is then a straightforward exercise (Problem 7.2) to compute the overlap of two coherent states,

$$\langle \phi | \phi' \rangle = e^{\sum_\alpha \phi_\alpha^* \phi_\alpha'} \,. \tag{7.14}$$

---

[2] Notice there cannot be any right eigenstates of the creation operators $a_\alpha^\dagger$ (or left eigenstates of a_α), since these increase the particle number in any Fock state by one.

As a consequence, the expectation value of any normal-ordered product of creation and annihilation operator $A(\{a_\alpha^\dagger\}, \{a_\alpha\})$ becomes

$$\langle\phi|A(\{a_\alpha^\dagger\}, \{a_\alpha\})|\phi'\rangle = A(\{\phi_\alpha^*\}, \{\phi_\alpha'\})\, e^{\sum_\alpha \phi_\alpha^* \phi_\alpha'} , \qquad (7.15)$$

whereas

$$\langle\phi|[a_\alpha, a_\beta^\dagger]|\phi'\rangle = \left[\frac{\partial}{\partial\phi_\alpha^*}, \phi_\beta^*\right] e^{\sum_\alpha \phi_\alpha^* \phi_\alpha'} = \delta_{\alpha\beta}\, \langle\phi|\phi'\rangle , \qquad (7.16)$$

consistent with the fundamental boson commutator (7.2).

Finally, the closure relation, i.e., the decomposition of the Fock space unit operator in terms of coherent states, reads (see Problem 7.2)

$$\int \prod_\alpha \frac{d\phi_\alpha^*\, d\phi_\alpha}{2\pi i}\, e^{-\sum_\alpha \phi_\alpha^* \phi_\alpha} |\phi\rangle\langle\phi| = 1 , \qquad (7.17)$$

which demonstrates that the coherent states (7.11) actually form an overcomplete basis of the many-particle Hilbert space. Correspondingly, we find for the operator trace of a normal-ordered product $A(\{a_\alpha^\dagger\}, \{a_\alpha\})$:

$$\text{Tr}\, A = \int \prod_\alpha \frac{d\phi_\alpha^*\, d\phi_\alpha}{2\pi i}\, e^{-\sum_\alpha \phi_\alpha^* \phi_\alpha} \langle\phi|A(\{a_\alpha^\dagger\}, \{a_\alpha\})|\phi\rangle$$

$$= \int \prod_\alpha \frac{d\phi_\alpha^*\, d\phi_\alpha}{2\pi i}\, A(\{\phi_\alpha^*\}, \{\phi_\alpha\}) . \qquad (7.18)$$

7.1.2 Coherent-state path integrals for interacting bosons

We can now utilize the previous results to construct a path integral representation for the *coherent-state propagator* $\langle\phi(t_f)|U(t_f, t_0)|\phi(t_0)\rangle$, i.e. the matrix element of the unitary time evolution operator introduced in Section 2.1.[3] Recall that U satisfies the time-dependent Schrödinger equation (2.1) with initial condition $U(t_0, t_0) = 1$, and thus can be formally written as a time-ordered product $U(t_f, t_0) = \mathcal{T} \exp[-i \int_{t_0}^{t_f} H(t')\, dt'/\hbar]$, see Eqs. (2.29) and (2.30). Once again, we proceed by splitting the time interval $[t_0, t_f]$ into M discrete steps: $\tau = (t_f - t_0)/M$, $t_l = t_0 + l\,\tau$, $l = 0, 1, \ldots, M$, where we identify $t_M = t_f$. In the limit $M \to \infty$, the time intervals τ become infinitesimal, and we obtain $U(t_f, t_0) = \lim_{\tau \to 0} \prod_{l=0}^{M-1} \exp[-iH(t_l)\,\tau/\hbar]$. Since the quantum many-particle Hamiltonian can be taken to be normal-ordered, as in Eqs. (7.6) and (7.8), we have by means of (7.15) to first order in τ:

$$\langle\phi(t_l)|\, e^{-iH\tau/\hbar}\, |\phi(t_{l-1})\rangle = e^{\sum_\alpha \phi_\alpha^*(t_l)\, \phi_\alpha(t_{l-1}) - iH(\{\phi_\alpha^*(t_l)\}, \{\phi_\alpha(t_{l-1})\})\, \tau/\hbar} .$$

---

[3] Here we largely follow Negele and Orland (1988); see also Kamenev (2011).

At each intermediate time step t_l $(l = 1, \ldots, M - 1)$ we then insert the closure relation (7.17) to finally arrive at

$$U(\phi_f^*, t_f; \phi_0, t_0) = \langle \phi(t_f) | \, U(t_f, t_0) \, | \phi(t_0) \rangle$$

$$= \lim_{\tau \to 0} \int \prod_\alpha \prod_{l=1}^{M-1} \frac{\mathrm{d}\phi_\alpha^*(t_l) \, \mathrm{d}\phi_\alpha(t_l)}{2\pi \mathrm{i}} \, \mathrm{e}^{\sum_\alpha \phi_\alpha^*(t_M) \, \phi_\alpha(t_{M-1})}$$

$$\times \mathrm{e}^{\mathrm{i}\left[\sum_\alpha \sum_{l=1}^{M-1} \mathrm{i}\hbar \, \phi_\alpha^*(t_l) \, [\phi_\alpha(t_l) - \phi_\alpha(t_{l-1})]/\tau - \sum_{l=1}^{M} H(\{\phi_\alpha^*(t_l)\}, \{\phi_\alpha(t_{l-1})\})\right]\tau/\hbar} \, .$$

(7.19)

Here, the creation and annihilation operators in the Hamiltonian are replaced by the coherent-state eigenvalues, i.e., complex fields which serve as the integration variables for the functional integral that extends over all paths connecting the prescribed initial and end points.

In the continuous-time notation, we may write this in more compact form:

$$U(\phi_f^*, t_f; \phi_0, t_0) = \int_{\phi(t_0) = \phi_0}^{\phi^*(t_f) = \phi_f^*} \mathcal{D}[\phi^*, \phi] \, \mathrm{e}^{\sum_\alpha \phi_\alpha^*(t_f) \phi_\alpha(t_f)} \, \mathrm{e}^{\frac{\mathrm{i}}{\hbar} \int_{t_0}^{t_f} L(\{\phi_\alpha^*(t)\}, \{\phi_\alpha(t)\}) \, \mathrm{d}t} \, ,$$

(7.20)

where we have introduced the *Lagrangian*

$$L(\{\phi_\alpha^*(t)\}, \{\phi_\alpha(t)\}) = \sum_\alpha \phi_\alpha^*(t) \, \mathrm{i}\hbar \, \frac{\partial \phi_\alpha(t)}{\partial t} - H(\{\phi_\alpha^*(t)\}, \{\phi_\alpha(t)\}) \, ,$$

(7.21)

and defined the functional integration measure as

$$\mathcal{D}[\phi^*, \phi] = \lim_{M \to \infty} \prod_\alpha \prod_{l=1}^{M-1} \frac{\mathrm{d}\phi_\alpha^*(t_l) \, \mathrm{d}\phi_\alpha(t_l)}{2\pi \mathrm{i}} \, .$$

(7.22)

Note that, if needed, Eq. (7.19) provides the proper time discretization to evaluate the path integral.

In a completely analogous manner, we may express the grand-canonical partition function $\mathcal{Z}(\beta, \mu) = \mathrm{Tr} \, \mathrm{e}^{-\beta(H - \mu N)}$ at temperature $T = 1/k_B \beta$ and fixed chemical potential μ in terms of a coherent-state path integral. To this end, we employ the trace formula (7.18), note that $H - \mu N = H(\{a_\alpha^\dagger\}, \{a_\alpha\}) - \mu \sum_\alpha a_\alpha^\dagger a_\alpha$, and discretize the inverse temperature interval $[0, \beta]$ according to $t_l = l\tau$ $(l = 0, \ldots, M)$, where $\tau = \beta\hbar/M$. Note that aside from the shift in the Hamiltonian by the particle number operator, this amounts to transforming the propagator (7.19) to an *imaginary time* interval $t_f - t_0 \to -\mathrm{i}\beta\hbar$, with *periodic boundary conditions*

$$\phi_\alpha(\beta\hbar) = \phi_\alpha(0) \, .$$

(7.23)

Upon inserting the closure relation (7.17) at each intermediate imaginary time step t_l, we then obtain

$$Z(\beta, \mu) = \int \prod_\alpha \frac{d\phi_\alpha^* \, d\phi_\alpha}{2\pi i} \, e^{-\sum_\alpha \phi_\alpha^* \phi_\alpha} \, \langle \phi | e^{-\beta(H - \mu N)} | \phi \rangle$$

$$= \lim_{\tau \to 0} \int \prod_\alpha \prod_{l=1}^{M} \frac{d\phi_\alpha^*(t_l) \, d\phi_\alpha(t_l)}{2\pi i}$$

$$\times e^{-\sum_{l=1}^{M} [\sum_\alpha \phi_\alpha^*(t_l)(\hbar[\phi_\alpha(t_l) - \phi_\alpha(t_{l-1})]/\tau - \mu \, \phi_\alpha(t_{l-1})) + H(\{\phi_\alpha^*(t_l)\}, \{\phi_\alpha(t_{l-1})\})] \tau / \hbar} . \quad (7.24)$$

Employing the functional measure (7.22), but with a product over M terms, this can be written concisely as

$$Z(\beta, \mu) = \int_{\phi(0) = \phi(\beta\hbar)} \mathcal{D}[\phi^*, \phi] \, e^{-S[\phi^*, \phi]/\hbar} , \quad (7.25)$$

with the exponential configurational weight given by the coherent-state path integral action

$$S[\phi^*, \phi] = \int_0^{\beta\hbar} \left[\sum_\alpha \phi_\alpha^*(\tau) \left(\hbar \frac{\partial}{\partial \tau} - \mu \right) \phi_\alpha(\tau) + H(\{\phi_\alpha^*(\tau)\}, \{\phi_\alpha(\tau)\}) \right] d\tau .$$

$$(7.26)$$

In contrast with the fixed end points in Eq. (7.20) for the propagator, in the path integral representation of the grand-canonical partition function we need to integrate over the initial values of the fields as well, with periodic boundary conditions, as indicated in Eq. (7.25).

In order to actually perform functional integrals, it is often best to take recourse to the original discretized form. As an explicit example, we evaluate the grand-canonical partition function (7.24) for non-interacting bosons. In their single-particle energy eigenstate basis, the Hamiltonian reads $H = \sum_\alpha \epsilon_\alpha a_\alpha^\dagger a_\alpha$, with eigenvalues $E(\{n_\alpha\}) = \sum_\alpha \epsilon_\alpha n_\alpha$ and the single-particle energies ϵ_α. The action thus becomes diagonal in the single-particle states $S[\phi^*, \phi] = \sum_\alpha S_\alpha(\phi_\alpha^*, \phi_\alpha)$, and consequently $Z(\beta, \mu) = \prod_\alpha Z_\alpha(\beta, \mu)$ factorizes into a product of single-particle contributions:

$$Z_\alpha(\beta, \mu) = \lim_{\tau \to 0} \int \prod_{l=1}^{M} \frac{d \, \mathrm{Re} \, \phi_\alpha(t_l) \, d \, \mathrm{Im} \, \phi_\alpha(t_l)}{\pi} \, e^{-\sum_{l,l'=1}^{M} \phi_\alpha^*(t_l) S_{\alpha \, ll'} \, \phi_\alpha(t_{l'})} , \quad (7.27)$$

with the $M \times M$ matrix

$$
S_\alpha = \begin{pmatrix}
1 & 0 & 0 & \cdots & -\zeta_\alpha \\
-\zeta_\alpha & 1 & 0 & \cdots & 0 \\
0 & -\zeta_\alpha & 1 & \cdots & 0 \\
\vdots & \vdots & \ddots & \ddots & \vdots \\
0 & 0 & \cdots & -\zeta_\alpha & 1
\end{pmatrix} ,
$$

recall Eq. (7.23), and the abbreviation $\zeta_\alpha = 1 - (\epsilon_\alpha - \mu)\tau/\hbar$. We just need to perform Gaussian integrals; for complex integration variables, we have

$$
\int \prod_l \frac{d\phi_l^* \, d\phi_l}{2\pi i} \, e^{-\sum_{l,l'} \phi_l^* S_{ll'} \phi_{l'} + \sum_l (j_l^* \phi_l + j_l \phi_l^*)} = (\det S)^{-1} \, e^{\sum_{l,l'} j_l^* S_{ll'}^{-1} j_{l'}} , \quad (7.28)
$$

whence $\mathcal{Z}_\alpha(\beta, \mu) = \lim_{\tau \to 0} (\det S_\alpha)^{-1}$. Expanding the determinant with respect to the top row gives $\det S_\alpha = 1 - \zeta_\alpha^M$, and finally taking the limit $\tau = \beta\hbar/M \to 0$, we arrive at the familiar result

$$
\mathcal{Z}_\alpha(\beta, \mu) = \lim_{M \to \infty} \frac{1}{1 - [1 - \beta(\epsilon_\alpha - \mu)/M]^M} = \frac{1}{1 - e^{-\beta(\epsilon_\alpha - \mu)}} . \quad (7.29)
$$

Let us next consider the temporal single-particle Green function, with time-dependent ladder operators in the Heisenberg picture,

$$
\langle T \, a_\alpha(\tau) \, a_{\alpha'}^\dagger(\tau') \rangle = \langle a_\alpha(\tau) \, a_{\alpha'}^\dagger(\tau') \rangle \, \Theta(\tau - \tau') + \langle a_{\alpha'}^\dagger(\tau') \, a_\alpha(\tau) \rangle \, \Theta(\tau' - \tau') ,
$$

$$(7.30)$$

which represents the time-ordered probability amplitude for a particle produced in state α' at time τ' to be destroyed in state α at time τ. In the coherent-state representation, this simply becomes

$$
\langle T \, a_\alpha(\tau) \, a_{\alpha'}^\dagger(\tau') \rangle = \lim_{\eta \downarrow 0} \langle \phi_\alpha(\tau) \, \phi_{\alpha'}^*(\tau' + \eta) \rangle \quad (7.31)
$$

with positive infinitesimal η, since the fields ϕ_α in the discretized action (7.24) always precede the ϕ_α^* in time. Retaining the small positive parameter η will allow us to bypass the discretization (see Problem 7.3) and perform calculations directly in the temporal continuum limit. For non-interacting bosons in the energy eigenstate basis, the quantity in Eq. (7.31) must already be diagonal in the single-particle state indices,

$$
\langle \phi_\alpha(\tau) \, \phi_{\alpha'}^*(\tau' + \eta) \rangle_0 = G_{0\alpha}(\tau - \tau' - \eta) \, \delta_{\alpha\alpha'} . \quad (7.32)
$$

Keeping in mind the periodic boundary condition (7.23), the free-particle action is then readily diagonalized by means of a discrete Fourier transform

$$\phi_\alpha(\tau) = \frac{1}{\beta\hbar} \sum_n \phi_\alpha(\omega_n) e^{-i\omega_n \tau} , \tag{7.33}$$

with the *Matsubara frequencies*

$$\omega_n = \frac{2\pi n}{\beta\hbar} , \quad n = 0, \pm 1, \pm 2, \ldots . \tag{7.34}$$

We thus obtain

$$S_0[\phi^*, \phi] = \frac{1}{\beta} \sum_\alpha \sum_n \left(-i\omega_n + \frac{\epsilon_\alpha - \mu}{\hbar} \right) |\phi_\alpha(\omega_n)|^2 , \tag{7.35}$$

from which we immediately infer the single-particle propagator

$$G_{0\alpha}(\omega_n) = \frac{1}{-i\omega_n + (\epsilon_\alpha - \mu)/\hbar} , \tag{7.36}$$

and in the time domain consequently

$$G_{0\alpha}(\tau) = \frac{1}{\beta\hbar} \sum_n \frac{e^{-i\omega_n \tau}}{-i\omega_n + (\epsilon_\alpha - \mu)/\hbar} . \tag{7.37}$$

Note that

$$\left(\frac{\partial}{\partial\tau} + \frac{\epsilon_\alpha - \mu}{\hbar} \right) G_{0\alpha}(\tau) = \delta(\tau) , \tag{7.38}$$

whence $G_{0\alpha}(\tau)$ is indeed the free-particle Green function, and also

$$\langle n_\alpha \rangle_0 = \lim_{\eta \downarrow 0} \langle \phi_\alpha^*(\tau + \eta) \phi_\alpha(\tau) \rangle_0 = \lim_{\eta \uparrow 0} G_{0\alpha}(\eta) . \tag{7.39}$$

The Appendix (Section 7.5) demonstrates how the Matsubara frequency sum in Eq. (7.37) is evaluated; Eq. (7.213) then finally yields

$$G_{0\alpha}(\tau) = e^{-(\epsilon_\alpha - \mu)\tau/\hbar} [\langle n_\alpha \rangle_0 \, \Theta(-\tau) + (1 + \langle n_\alpha \rangle_0)\Theta(\tau)] , \tag{7.40}$$

with the familiar *Bose–Einstein distribution*

$$\langle n_\alpha \rangle_0 = \frac{1}{e^{\beta(\epsilon_\alpha - \mu)} - 1} , \tag{7.41}$$

which naturally follows from Eq. (7.29) via $\langle n_\alpha \rangle_0 = \beta^{-1}\partial \ln \mathcal{Z}_\alpha(\beta, \mu)/\partial\mu$ as well. The temporal propagator (7.40) can also be directly computed by means of the discretized action, see Problem 7.3.

For interacting particles, of course neither the propagator (7.20) nor the grand-canonical partition function (7.25) can in general be evaluated in closed form. One may, however, resort to a perturbative expansion with respect to the particle inter-actions, precisely with the same techniques introduced in Chapter 4. For example, two-particle interactions (scattering) described by the operator (7.8) enter the action as

$$S_{int}[\phi^*, \phi] = \int_0^{\beta\hbar} \frac{1}{2} \sum_{\alpha,\beta,\gamma,\delta} \langle \alpha\beta|V^{(2)}|\gamma\delta\rangle \, \phi_\alpha^*(\tau) \, \phi_\beta^*(\tau) \, \phi_\delta(\tau) \, \phi_\gamma(\tau) \, d\tau \,, \qquad (7.42)$$

and thus generate a four-point vertex with two incoming and two outgoing lines. These vertices, in addition to the free-particle propagator G_0, constitute the basic elements of the Feynman diagrams that graphically depict the various terms con-tributing to the perturbation series.

Notice that upon formally identifying the fields $\phi_\alpha^*(t)$ in the Lagrangian (7.21) with the Martin–Siggia–Rose auxiliary fields in the Janssen–De Domini-cis response functional (4.10), the quantum dynamics (in real time) of interacting bosons may be recast in the form of a *linear* stochastic differential equation

$$i\hbar \frac{\partial \phi_\alpha(t)}{\partial t} = \epsilon_\alpha \, \phi_\alpha(t) + \zeta_\alpha(t) \,, \qquad (7.43)$$

where the non-linear particle interactions enter in the form of non-diagonal *multi-plicative* Gaussian noise with zero mean $\langle \zeta_\alpha(t) \rangle = 0$ and second moment

$$\langle \zeta_\alpha(t) \, \zeta_\beta(t') \rangle = \sum_{\gamma,\delta} \langle \alpha\beta|V^{(2)}|\gamma\delta\rangle \, \phi_\delta(t) \, \phi_\gamma(t) \, \delta(t - t') \,. \qquad (7.44)$$

7.1.3 Grassmann variables and fermion coherent states

We now follow the preceding procedures and construct a coherent-state functional integral representation for fermionic quantum many-particle systems.[4] We begin by defining the vacuum state $|0\rangle$ and single-particle states $|\alpha\rangle = c_\alpha^\dagger |0\rangle$, with the cre-ation operator $c_\alpha^\dagger$, adjoint to the annihilation operator c_α. Yet for identical fermions, we require the two-particle states to be antisymmetric under particle exchange, $c_\alpha^\dagger c_\beta^\dagger |0\rangle = |\alpha\,\beta\rangle = -|\beta\,\alpha\rangle = -c_\beta^\dagger c_\alpha^\dagger |0\rangle$, which immediately implies the *anticom-mutation* relation (with $[A, B]_+ = AB + BA$ for operators A, B)

$$\left[c_\alpha^\dagger, c_\beta^\dagger\right]_+ = 0 = [c_\alpha, c_\beta]_+ \,. \qquad (7.45)$$

Thus $(c_\alpha^\dagger)^2 = 0$, which expresses Pauli's exclusion principle: no two identical fermions can occupy the same single-particle state. The fermionic ladder oper-ator algebra is completed by further demanding that

$$\left[c_\alpha, c_\beta^\dagger\right]_+ = \delta_{\alpha\beta} \,. \qquad (7.46)$$

---

[4] For more detailed expositions, see Negele and Orland (1988); Shankar (1994); Kamenev (2011).

Indeed, the occupation number operator $\hat{n}_\alpha = c_\alpha^\dagger c_\alpha$ then satisfies $\hat{n}_\alpha^2 = c_\alpha^\dagger(1 - c_\alpha^\dagger c_\alpha)c_\alpha = \hat{n}_\alpha$. Consequently its only eigenvalues are $n_\alpha = 0$ or 1, and its eigenstates obey $c_\alpha^\dagger|0\rangle = |1\rangle$ and $c_\alpha|1\rangle = |0\rangle$. We may then construct an arbitrary N-particle Fock space state via the Slater determinant

$$|\{n_\alpha = 0, 1\}\rangle = \prod_\alpha (c_\alpha^\dagger)^{n_\alpha}|0\rangle = \frac{1}{\sqrt{N!}}\sum_P (-1)^P \, P_{\alpha_1\cdots\alpha_N}|\alpha_1\cdots\alpha_N\rangle, \quad (7.47)$$

where $(-1)^P$ indicates the sign of the permutation P. Upon ordering the single-particle state labels, we note the identities

$$c_\alpha|\cdots n_\alpha\cdots\rangle = n_\alpha\,(-1)^{\sum_{\beta<\alpha} n_\beta}|\cdots n_\alpha - 1\cdots\rangle, \quad (7.48)$$

$$c_\alpha^\dagger|\cdots n_\alpha\cdots\rangle = (1 - n_\alpha)(-1)^{\sum_{\beta<\alpha} n_\beta}|\cdots n_\alpha + 1\cdots\rangle, \quad (7.49)$$

which are easily checked to be consistent with the anticommutator (7.46). Basis changes are effected in the same manner as for bosons, and upon replacing the bosonic operators a_α with the fermionic c_α, the general single- and two-particle operators assume precisely the forms given in Eqs. (7.6)–(7.9), with the fermion field operators of course satisfying $[\hat{\psi}(x), \hat{\psi}(x')]_+ = 0$ and $[\hat{\psi}(x), \hat{\psi}^\dagger(x')]_+ = \delta(x - x')$.

As a consequence of the anticommutation relations (7.45), the eigenvalues of fermion coherent states, defined in analogy to (7.10) through

$$c_\alpha|\xi\rangle = \xi_\alpha|\xi\rangle, \quad (7.50)$$

cannot be complex numbers, but must be anticommuting *Grassmann variables*, $[\xi_\alpha, \xi_\beta]_+ = 0$, and thus $\xi_\alpha^2 = 0$. Let us now consider a Grassmann algebra generated by ξ and its conjugate ξ^*, for which we impose the rules $(\xi^*)^* = \xi$, $(a\,\xi)^* = a^*\xi^*$, with a complex number a, and $(\xi\,\xi')^* = \xi'^*\xi^*$. Any analytic function of ξ must then be linear, $f(\xi) = a_0 + a_1\,\xi$, while an arbitrary operator in this Grassmann algebra can be constructed via $A(\xi^*, \xi) = a_0 + a_1\,\xi + \bar{a}_1\,\xi^* + a_{12}\,\xi^*\xi$. In line with the standard derivative of complex functions, one defines the differential operation

$$\frac{\partial}{\partial\xi}(\xi\,\xi^*) = \xi^* = -\frac{\partial}{\partial\xi}(\xi^*\xi), \quad (7.51)$$

since $[\xi, \xi^*]_+ = 0$. Thus $\frac{\partial}{\partial\xi}f(\xi) = a_1$, while $\frac{\partial}{\partial\xi}A(\xi^*, \xi) = a_1 - a_{12}\,\xi^*$ and $\frac{\partial}{\partial\xi^*}A(\xi^*, \xi) = \bar{a}_1 + a_{12}\,\xi$, whence $\frac{\partial^2}{\partial\xi^*\partial\xi}A(\xi^*, \xi) = -a_{12} = -\frac{\partial^2}{\partial\xi\,\partial\xi^*}A(\xi^*, \xi)$, which demonstrates that the derivative operators anticommute as well:

$$\left[\frac{\partial}{\partial\xi}, \frac{\partial}{\partial\xi^*}\right]_+ = 0. \quad (7.52)$$

This further motivates the formal definition of a Grassmann variable integral through the linear operations

$$\int d\xi\, 1 = 0 = \int d\xi^*\, 1 \,, \quad \int d\xi\, \xi = 1 = \int d\xi^*\, \xi^* \,. \tag{7.53}$$

Note that Eqs. (7.53) provide a *complete* integration table. In fact, direct application of these rules gives $\int d\xi\, f(\xi) = a_1$ and $\int d\xi^* d\xi\, A(\xi^*, \xi) = -a_{12} = -\int d\xi\, d\xi^*\, A(\xi^*, \xi)$; note that Grassmann differentiation and integration are effectively identical operations.

We can now introduce a *Grassmann-valued delta function* through formal integration over the Grassmann variable η:

$$\delta(\xi - \xi') = \int d\eta\, e^{-\eta(\xi - \xi')} = \int d\eta\, [1 - \eta(\xi - \xi')] = -(\xi - \xi') \,. \tag{7.54}$$

For an arbitrary Grassmann-valued function $f(\xi)$ we then indeed see that $\int d\xi'\, \delta(\xi - \xi')\, f(\xi') = -\int d\xi'\, (\xi - \xi')(a_0 + a_1 \xi') = a_0 + a_1 \xi = f(\xi)$. Next, we define and evaluate the *inner product* of two Grassmann-valued functions $f(\xi) = a_0 + a_1 \xi$ and $g(\xi) = b_0 + b_1 \xi$ via

$$\langle f | g \rangle = \int d\xi^* d\xi\, e^{-\xi^* \xi}\, f^*(\xi)\, g(\xi^*)$$

$$= \int d\xi^* d\xi\, (1 - \xi^* \xi)(a_0^* + a_1^* \xi)(b_0 + b_1 \xi^*) = a_0^* b_0 + a_1^* b_1 \tag{7.55}$$

by straightforward application of Eqs. (7.53). Notice that in the same manner a *Gaussian integral* of Grassmann variables becomes

$$\int d\xi^* d\xi\, e^{-a\xi^* \xi} = a \tag{7.56}$$

(in contrast with π/a for a regular complex integration variable ξ). Correspondingly, for general Gaussian integrals involving multiple Grassmann variables one obtains

$$\int \prod_l d\xi_l^*\, d\xi_l\, e^{-\sum_{l,l'} \xi_l^* S_{ll'} \xi_{l'} + \sum_l (j_l^* \xi_l + j_l \xi_l^*)} = (\det S)\, e^{\sum_{l,l'} j_l^* S_{ll'}^{-1} j_{l'}} \,. \tag{7.57}$$

Constructing fermion coherent states will require us to consider an augmented Fock space, wherein expansion coefficients in terms of the N-particle states $|\{n_\alpha = 0, 1\}\rangle$ are allowed to be Grassmann variables. We thus introduce Grassmann algebra generators ξ_α and ξ_α^* associated with each annihilation and creation operator c_α and $c_\alpha^\dagger$, respectively, and furthermore set up the natural mixed anticommutation rules $[\xi_\alpha, c_\beta]_+ = 0 = [\xi_\alpha, c_\beta^\dagger]_+$ along with $(\xi_\alpha c_\beta)^\dagger = c_\beta^\dagger \xi_\alpha^*$, etc. The fermion coherent states satisfying the eigenvalue equation (7.50) can then be written in the form

$$|\xi\rangle = e^{-\sum_\alpha \xi_\alpha c_\alpha^\dagger}|0\rangle = \prod_\alpha \left(1 - \xi_\alpha c_\alpha^\dagger\right)|0\rangle, \tag{7.58}$$

as is readily confirmed, since $c_\alpha(1 - \xi_\alpha c_\alpha^\dagger)|0\rangle = \xi_\alpha|0\rangle = \xi_\alpha(1 - \xi_\alpha c_\alpha^\dagger)|0\rangle$. Similarly, observing that $c_\alpha^\dagger(1 - \xi_\alpha c_\alpha^\dagger) = c_\alpha^\dagger = -\frac{\partial}{\partial\xi_\alpha}(1 - \xi_\alpha c_\alpha^\dagger)$, we also establish

$$c_\alpha^\dagger|\xi\rangle = -\frac{\partial}{\partial\xi_\alpha}|\xi\rangle. \tag{7.59}$$

The adjoint states then satisfy the corresponding relations

$$\langle\xi|c_\alpha^\dagger = \langle\xi|\xi_\alpha^*, \quad \langle\xi| = \langle 0|e^{\sum_\alpha \xi_\alpha^* c_\alpha}, \quad \langle\xi|c_\alpha = \frac{\partial}{\partial\xi_\alpha^*}\langle\xi|, \tag{7.60}$$

and the fermionic coherent-state inner product becomes

$$\langle\xi|\xi'\rangle = \langle 0|\prod_{\alpha,\beta}(1 + \xi_\alpha^* c_\alpha)(1 - \xi_\beta' c_\beta^\dagger)|0\rangle = e^{\sum_\alpha \xi_\alpha^* \xi_\alpha'}. \tag{7.61}$$

For any normal-ordered product of fermionic creation and annihilation operators we thereby obtain

$$\langle\xi|A(\{c_\alpha^\dagger\}, \{c_\alpha\})|\xi'\rangle = A(\{\xi_\alpha^*\}, \{\xi_\alpha'\})e^{\sum_\alpha \xi_\alpha^* \xi_\alpha'}, \tag{7.62}$$

which takes the very same form as for the bosonic case; furthermore, consistent with Eq. (7.46),

$$\langle\xi|[c_\alpha, c_\beta^\dagger]_+|\xi'\rangle = \left[\frac{\partial}{\partial\xi_\alpha^*}, \xi_\beta^*\right]_+ e^{\sum_\alpha \xi_\alpha^* \xi_\alpha'} = \delta_{\alpha\beta}\langle\xi|\xi'\rangle. \tag{7.63}$$

The associated closure relation now becomes (Problem 7.4)

$$\int \prod_\alpha d\xi_\alpha^* \, d\xi_\alpha \, e^{-\sum_\alpha \xi_\alpha^* \xi_\alpha}|\xi\rangle\langle\xi| = 1, \tag{7.64}$$

again akin to (7.17) for bosonic coherent states, and similarly for the operator trace formula, wherein the anticommuting Grassmann variables however induce a crucial minus sign (see Problem 7.4):

$$\mathrm{Tr}\, A = \int \prod_\alpha d\xi_\alpha^* \, d\xi_\alpha \, e^{-\sum_\alpha \xi_\alpha^* \xi_\alpha} \langle -\xi|A(\{c_\alpha^\dagger\}, \{c_\alpha\})|\xi\rangle$$

$$= \int \prod_\alpha d\xi_\alpha^* \, d\xi_\alpha \, e^{-2\sum_\alpha \xi_\alpha^* \xi_\alpha} A(\{-\xi_\alpha^*\}, \{\xi_\alpha\}). \tag{7.65}$$

As an application, we evaluate the grand-canonical partition function for non-interacting fermions $\mathcal{Z}(\beta, \mu) = \mathrm{Tr}\, e^{-\beta(H-\mu N)} = \prod_\alpha \mathcal{Z}_\alpha(\beta, \mu)$. To this end, we first need to apply normal-ordering, and can then use Eq. (7.65):

$$Z_\alpha(\beta, \mu) = \mathrm{Tr}\, e^{-\beta(\epsilon_\alpha - \mu)c_\alpha^\dagger c_\alpha} = \mathrm{Tr}\big[1 + \big(e^{-\beta(\epsilon_\alpha - \mu)} - 1\big)c_\alpha^\dagger c_\alpha\big]$$

$$= \int d\xi_\alpha^* \, d\xi_\alpha (1 - 2\,\xi_\alpha^* \,\xi_\alpha)\big[1 - \big(e^{-\beta(\epsilon_\alpha - \mu)} - 1\big)\xi_\alpha^* \,\xi_\alpha\big]$$

$$= 1 + e^{-\beta(\epsilon_\alpha - \mu)} \, . \tag{7.66}$$

7.1.4 Coherent-state path integrals for fermion systems

We may now proceed to evaluate the coherent-state propagator for fermions precisely following the procedure employed for the bosonic case, namely through discretizing the time interval $[t_0, t_f]$ via $t_l = t_0 + l\,\tau$, $l = 0, 1, \ldots, M$, with $t_M = t_f$, employing Eq. (7.62) and inserting the closure relation (7.64) at each intermediate time step. This gives

$$U(\xi_f^*, t_f; \xi_0, t_0) = \langle \xi(t_f)| \, U(t_f, t_0) \, |\xi(t_0)\rangle$$

$$= \lim_{\tau \to 0} \int \prod_\alpha \prod_{l=1}^{M-1} d\xi_\alpha^*(t_l) \, d\xi_\alpha(t_l) \, e^{\sum_\alpha \xi_\alpha^*(t_M)\,\xi_\alpha(t_{M-1})}$$

$$\times e^{i\left[\sum_\alpha \sum_{l=1}^{M-1} i\hbar\, \xi_\alpha^*(t_l)\,[\xi_\alpha(t_l) - \xi_\alpha(t_{l-1})]/\tau - \sum_{l=1}^{M} H(\{\xi_\alpha^*(t_l)\}, \{\xi_\alpha(t_{l-1})\})\right]\tau/\hbar} \, ,$$

$$\tag{7.67}$$

or in the short-hand continuous-time notation

$$U(\xi_f^*, t_f; \xi_0, t_0) = \int_{\xi(t_0) = \xi_0}^{\xi^*(t_f) = \xi_f^*} \mathcal{D}[\xi^*, \xi]\, e^{\sum_\alpha \xi_\alpha^*(t_f)\,\xi_\alpha(t_f)}\, e^{\frac{i}{\hbar} \int_{t_0}^{t_f} L(\{\xi_\alpha^*(t)\}, \{\xi_\alpha(t)\})\, dt} \, , \tag{7.68}$$

with the Lagrangian defined as in Eq. (7.21), and the Grassmann variable functional integral measure

$$\mathcal{D}[\xi^*, \xi] = \lim_{M \to \infty} \prod_\alpha \prod_{l=1}^{M-1} d\xi_\alpha^*(t_l)\, d\xi_\alpha(t_l) \, . \tag{7.69}$$

In the same manner, the Grassmann coherent-state path integral representation for the grand-canonical partition can be constructed by transforming to imaginary time and $i(t_f - t_0) \to \beta\hbar$. Yet the Grassmann algebra now enforces *antiperiodic boundary conditions*,

$$\xi_\alpha(\beta\hbar) = -\xi_\alpha(0) \, , \tag{7.70}$$

whereupon

$$\mathcal{Z}(\beta,\mu) = \int \prod_\alpha d\xi_\alpha^* \, d\xi_\alpha \, e^{-\sum_\alpha \xi_\alpha^* \xi_\alpha} \langle -\xi | e^{-\beta(H-\mu N)} | \xi \rangle \qquad (7.71)$$

$$= \lim_{\tau \to 0} \int \prod_\alpha \prod_{l=1}^M d\xi_\alpha^*(t_l) \, d\xi_\alpha(t_l)$$

$$\times e^{-\sum_{l=1}^M [\sum_\alpha \xi_\alpha^*(t_l)(\hbar[\xi_\alpha(t_l)-\xi_\alpha(t_{l-1})]/\tau - \mu \xi_\alpha(t_{l-1})) + H(\{\xi_\alpha^*(t_l)\},\{\xi_\alpha(t_{l-1})\})]\tau/\hbar}$$

$$= \int_{\xi(0)=-\xi(\beta\hbar)} \mathcal{D}[\xi^*,\xi] \, e^{-S[\xi^*,\xi]/\hbar} \,, \qquad (7.72)$$

in compact notation, with the action (7.26). For non-interacting fermions with Hamiltonian $H = \sum_\alpha \epsilon_\alpha c_\alpha^\dagger c_\alpha$, this becomes just a product of Gaussian integrals, and the formula (7.57) yields $\mathcal{Z}_\alpha(\beta,\mu) = \lim_{\tau\to 0} \det S_\alpha$. Yet now the matrix element in the upper right corner of S_α reads $+\zeta_\alpha = 1 - (\epsilon_\alpha - \mu)\tau/\hbar$ (rather than $-\zeta_\alpha$ for bosons), whence $\det S_\alpha = 1 + \zeta_\alpha^M$, and in the limit $\tau = \beta\hbar/M \to 0$ one recovers (7.66).

With the goal to set up a perturbation expansion with respect to particle interactions, we next compute the fermionic non-interacting single-particle Green function

$$\langle T c_\alpha(\tau) c_{\alpha'}^\dagger(\tau') \rangle_0 = \lim_{\eta\downarrow 0} \langle \xi_\alpha(\tau) \xi_{\alpha'}^*(\tau'+\eta) \rangle_0 = \lim_{\eta\downarrow 0} G_{0\alpha}(\tau-\tau'-\eta) \delta_{\alpha\alpha'} \quad (7.73)$$

in the coherent-state representation. As for bosons, we introduce the discrete Fourier transform

$$\xi_\alpha(\tau) = \frac{1}{\beta\hbar} \sum_n \xi_\alpha(\omega_n) e^{-i\omega_n \tau} \,, \qquad (7.74)$$

but the antiperiodic boundary condition (7.70) imposes different *Matsubara frequencies*:

$$\omega_n = \frac{(2n+1)\pi}{\beta\hbar} \,, \quad n = 0,\pm1,\pm2,\dots. \qquad (7.75)$$

The free particle action $S_0[\xi^*,\xi]$ then takes the form (7.35), with $\phi_\alpha(\omega_n)$ replaced by $\xi_\alpha(\omega_n)$, and with the result of Problem 7.5(a) we arrive again at the propagator (7.36). Upon evaluating the fermionic Matsubara frequency sum, see Eq. (7.213) in the Appendix (Section 7.5), one finds at last

$$G_{0\alpha}(\tau) = e^{-(\epsilon_\alpha-\mu)\tau/\hbar}[-\langle n_\alpha\rangle_0 \, \Theta(-\tau) + (1-\langle n_\alpha\rangle_0)\,\Theta(\tau)] \,, \qquad (7.76)$$

where we have inserted the *Fermi–Dirac distribution*

$$\langle n_\alpha \rangle_0 = \lim_{\eta \downarrow 0} \left\langle \xi_\alpha^*(\tau + \eta)\, \xi_\alpha(\tau) \right\rangle_0 = -\lim_{\eta \uparrow 0} G_{0\alpha}(\eta) = \frac{1}{e^{\beta(\epsilon_\alpha - \mu)} + 1}. \tag{7.77}$$

We proceed to generate a perturbation expansion for the grand-canonical partition function in the usual manner by decomposing the action into its free and interacting parts, $S[\xi^*, \xi] = S_0[\xi^*, \xi] + S_{\text{int}}[\xi^*, \xi]$, cf. Eq. (7.42) for two-particle interactions, and noticing that

$$\mathcal{Z}(\beta, \mu) = \mathcal{Z}_0(\beta, \mu) \left\langle e^{-S_{\text{int}}[\xi^*, \xi]/\hbar} \right\rangle_0,$$

$$\langle A(\xi_\alpha, \xi_\alpha^*) \rangle = \frac{\left\langle A(\xi_\alpha, \xi_\alpha^*)\, e^{-S_{\text{int}}[\xi^*, \xi]/\hbar} \right\rangle_0}{\left\langle e^{-S_{\text{int}}[\xi^*, \xi]/\hbar} \right\rangle_0}. \tag{7.78}$$

These latter averages can then be evaluated by invoking the fermionic version of *Wick's theorem*, which states that non-vanishing expectation values in the Gaussian ensemble contain an equal number of Grassmann variables ξ_α^* and ξ_α, and are found by summing over all permutations of complete contractions into single-particle propagators (7.73), with appropriate factors of -1 included for odd index permutations:

$$\left\langle \xi_{\alpha_1} \xi_{\alpha_2} \cdots \xi_{\alpha_{N/2}} \xi_{\alpha_{1+N/2}}^* \cdots \xi_{\alpha_{N-1}}^* \xi_{\alpha_N}^* \right\rangle_0$$

$$= \sum_{\substack{\text{permutations} \\ i_1(1)\ldots i_N(N)}} (-1)^P \underbrace{\xi_{\alpha_{i_1(1)}} \xi_{\alpha_{i_2(1+N/2)}}^*}_{} \cdots \underbrace{\xi_{\alpha_{i_{N-1}(N/2)}} \xi_{\alpha_{i_N(N)}}^*}_{}. \tag{7.79}$$

This is demonstrated explicitly for the four-point correlation function in Problem 7.5(b). As a direct consequence, any *closed fermion loop* in a Feynman graph induces a minus sign. Otherwise, the structure of the perturbation expansion and its graphical representation in terms of Feynman diagrams is quite similar to the classical or bosonic case.

7.2 Boson superfluids

Quantum Hamiltonians for identical particles and the associated bosonic many-particle states are required to be symmetric under permutations of the particle labels. The ensuing purely quantum-mechanical correlations, present even in the absence of any interactions, cause the striking phenomenon of *Bose–Einstein condensation* at sufficiently low temperature or high particle density. Following a brief review of condensation in an ideal Bose gas, we utilize the coherent-state path integral formalism to describe superfluidity in interacting boson systems, first on the level of the saddle point approximation or Landau–Ginzburg theory, which yields the *Gross–Pitaevskii equation* for the collective boson field, and next in the framework of the Gaussian approximation. The latter allows us to compute essential physical properties of boson superfluids such as the quasi-particle spectrum of low-energy excitations, the condensate fraction, density and current correlation functions, and the normal- and superfluid densities.[5] We also demonstrate that quantum fluctuations are asymptotically irrelevant for the critical properties near the normal- to superfluid phase transition (if $T_c > 0$), and that its universal critical exponents are those of the classical XY model.

7.2.1 Bose–Einstein condensation

The thermodynamics of a gas of identical non-interacting bosons is encoded in the grand-canonical partition function, or equivalently the grand-canonical potential $\Phi(T, \mu)$, which via the Gibbs–Duhem relation immediately yields the pressure $P(T, \mu)$:

$$\Phi(T, \mu) = -V\, P(T, \mu) = -k_B T \ln \mathcal{Z}(T, \mu) = k_B T \sum_\alpha \ln\left[1 - e^{-(\epsilon_\alpha - \mu)/k_B T}\right],$$

(7.80)

where we have used $\mathcal{Z}(T, \mu) = \prod_\alpha \mathcal{Z}_\alpha(T, \mu)$ and our earlier explict result (7.29). In accord with Eq. (7.41) this gives the mean particle number

$$N(T, \mu) = \langle N \rangle = -\left(\frac{\partial \Phi}{\partial \mu}\right)_T = \sum_\alpha \frac{1}{e^{(\epsilon_\alpha - \mu)/k_B T} - 1} = \sum_\alpha \langle n_\alpha \rangle_0 ,$$

(7.81)

and similarly the mean (internal) energy $E(T, \mu) = \langle H \rangle = -(\partial \ln \mathcal{Z}/\partial \beta)_z = \sum_\alpha \epsilon_\alpha \langle n_\alpha \rangle_0$, with $\beta = 1/k_B T$ and the *fugacity* $z = e^{\mu/k_B T}$.

In the following, we specifically consider a free boson gas confined to a volume V in d dimensions, characterized by the power law dispersion relation $\epsilon(p) = c\,|p|^s$, independent of the (integer) particle spin $S = 0, 1, 2, \ldots$. Sums over single-particle states $\alpha = (p, m_s)$, with $2S + 1$ spin quantum numbers $m_s = S, S - 1, \ldots, -S + 1, -S$, then become in the continuum limit

$$\sum_\alpha F(\epsilon_\alpha) \to \frac{(2S + 1)\,V}{(2\pi\hbar)^d} \int d^d p\, F(\epsilon(\vec{p})) = \int_0^\infty g(\epsilon)\, F(\epsilon)\, d\epsilon ,$$

(7.82)

where we have introduced the *density of states*

$$g(\epsilon) = \frac{(2S + 1)\,V}{(2\pi\hbar)^d} \int d^d p\, \delta(\epsilon - \epsilon(\vec{p})) = \frac{(2S + 1)\,V}{h^d\, c^{d/s}} \frac{2\pi^{d/2}}{s\,\Gamma(d/2)} \epsilon^{-1+d/s}$$

(7.83)

---

[5] This section largely follows Popov (1987); Täuber and Nelson (1997); see also Fetter and Walecka (1971); Negele and Orland (1988); Kamenev (2011).

for our specific power law dispersion relation. Writing $g(\epsilon) = \bar{g}\,\epsilon^{-1+d/s}$ and integrating by parts, we obtain both the thermal and caloric equations of state

$$V\,P(T,\mu) = -k_{\mathrm{B}}T\,\bar{g}\int_0^\infty \epsilon^{-1+d/s}\,\ln\!\left[1 - e^{-(\epsilon-\mu)/k_{\mathrm{B}}T}\right]\!d\epsilon$$

$$= \frac{s}{d}\,\bar{g}\int_0^\infty \frac{\epsilon^{d/s}}{e^{(\epsilon-\mu)/k_{\mathrm{B}}T} - 1}\,d\epsilon = \frac{s}{d}\,E(T,\mu)\,. \tag{7.84}$$

Further evaluation would require an inversion of the relation $N(T,\mu)$ to arrive at the chemical potential $\mu(T,N) \le 0$ as an explicit function of the particle number N; inserting into (7.84) then yields $P(T,N)$ and $E(T,N)$.

In order to evaluate the mean particle number in the grand-canonical ensemble at fixed temperature and chemical potential, we carefully separate out the ground state with momentum $p = 0$ and $\epsilon(p) = 0$:

$$N(T,\mu) = N_0(T,\mu) + N'(T,\mu)\,, \quad N_0(T,\mu) = \frac{2S+1}{e^{-\mu/k_{\mathrm{B}}T} - 1}\,. \tag{7.85}$$

Recall that for the classical ideal gas, $\mu(T,N) < 0$ and thus $z = e^{\mu/k_{\mathrm{B}}T} < 1$. As $\mu \to 0$ and $z \to 1$ from below, the number N_0 of particles at zero energy becomes of order N, leading to *macroscopic ground state occupation*, i.e., the fraction N_0/N remains finite in the thermodynamic limit $N \to \infty$. As we shall see shortly, this Bose–Einstein condensation into the ground state occurs at sufficiently low temperatures or high particle densities, if the total particle number N cannot be accommodated by the excited states. For the computation of the number $N'(T,\mu)$ of particles occupying states with $p \ne 0$, we may again apply the continuum limit to find

$$N'(T,z) = \int_0^\infty \frac{g(\epsilon)}{z^{-1}\,e^{\epsilon/k_{\mathrm{B}}T} - 1}\,d\epsilon = \frac{(2S+1)\,V}{\lambda_{\mathrm{th}}(T)^d}\,\mathcal{G}_{d/s}(z) \tag{7.86}$$

after rendering the integral dimensionless with $x = \epsilon/k_{\mathrm{B}}T$, and introducing the *thermal de Broglie wavelength*

$$\lambda_{\mathrm{th}}(T) = \left[\frac{\Gamma(1+d/2)}{\Gamma(1+d/s)}\right]^{1/d}\frac{h}{\sqrt{\pi}}\left(\frac{c}{k_{\mathrm{B}}T}\right)^{1/s}, \tag{7.87}$$

where $\Gamma(x)$ denotes Euler's gamma function. In addition, we have defined the Bose functions

$$\mathcal{G}_k(z) = \frac{1}{\Gamma(k)}\int_0^\infty \frac{x^{k-1}}{z^{-1}\,e^x - 1}\,dx = \sum_{n=1}^\infty \frac{z^n}{n^k}\,, \tag{7.88}$$

which converge for $|z| < 1$ provided $k > 1$, since then $|\mathcal{G}_k(z)|$ is bounded by Riemann's zeta function $\mathcal{G}_k(1) = \sum_{n=1}^\infty n^{-k} = \zeta(k)$.

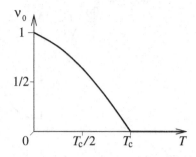

Fig. 7.1 Bose–Einstein condensate fraction $\nu_0(T)$ as function of temperature, here depicted for a non-relativistic ideal boson gas ($s = 2$) in $d = 3$ dimensions.

Thus, $N'(T)\lambda_{\mathrm{th}}(T)^d/(2S+1)V \le \zeta(d/s)$ at the extremal values of $\mu = 0$ and $z = 1$. In dimensions $d \le s$, the zeta function diverges and the inequality is always satisfied. However, for dimensions larger than the *lower critical dimension $d_{\mathrm{lc}} = s$* for Bose–Einstein condensation, the bound is exceeded at fixed temperature once $N(T)/V \ge (N(T)/V)_{\mathrm{c}} = (2S+1)\zeta(d/s)/\lambda_{\mathrm{th}}(T)^d$, or at fixed particle density N/V for $T \le T_{\mathrm{c}}(N/V)$, with the critical temperature

$$T_{\mathrm{c}}(N/V) = \frac{c}{k_B}\left(\frac{h}{\sqrt{\pi}}\right)^s \left[\frac{\Gamma(1+d/2)}{\Gamma(1+d/s)}\frac{N/V}{(2S+1)\zeta(d/s)}\right]^{s/d}. \tag{7.89}$$

For example, for an ideal gas of non-relativistic spinless bosons with $s = 2$ and mass m, i.e., $c = 1/2m$, the lower critical dimension is $d_{\mathrm{lc}} = 2$, and in three dimensions Bose–Einstein condensation occurs at $T_{\mathrm{c}}(N/V) = (h^2/2\pi\,m\,k_B)[N/V\,\zeta(3/2)]^{2/3}$, with $\zeta(3/2) \approx 2.612$. For $T < T_{\mathrm{c}}$, $z = 1 - O(1/N)$, and consequently

$$N'(T) = \frac{(2S+1)V\,\zeta(d/s)}{\lambda_{\mathrm{th}}(T)^d} = N\left[\frac{\lambda_{\mathrm{th}}(T_{\mathrm{c}})}{\lambda_{\mathrm{th}}(T)}\right]^d = N\left(\frac{T}{T_{\mathrm{c}}}\right)^{d/s}, \tag{7.90}$$

whence we find for the Bose–Einstein *condensate fraction*, shown in Fig. 7.1,

$$\nu_0(T) = \lim_{N\to\infty}(N_0/N) = \begin{cases} 0, & T > T_{\mathrm{c}}, \\ 1 - (T/T_{\mathrm{c}})^{d/s}, & T \le T_{\mathrm{c}}. \end{cases} \tag{7.91}$$

This quantity readily serves as the order parameter for Bose–Einstein condensation, characterizing the continuous phase transition at T_{c} that is driven entirely by quantum-mechanical entanglement correlations. Since just below the critical temperature $\nu_0(T) \approx d\,(T_{\mathrm{c}} - T)/s\,T_{\mathrm{c}}$, the associated critical exponent is $\beta = 1$, independent of the dispersion exponent s and even the dimension d.

We may now compute the thermodynamic properties of an ideal Bose gas. First, we observe that the condensate does not contribute to the grand-canonical potential in the thermodynamic limit, since $\Phi_0(T, z)/N = (2S+1)k_B T \ln(1 - z)/N = O(N^{-1}\ln N)$. The condensate therefore does not carry pressure, internal

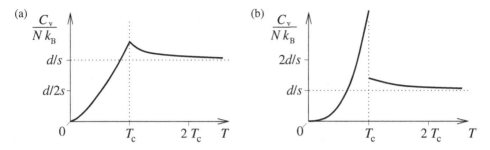

Fig. 7.2 Specific heat $C_v(T)$ of an ideal boson gas as function of temperature, displaying (a) a cusp singularity at the condensation temperature T_c for $s < d \leq 2s$ (plotted for $s = 2, d = 3$); (b) a discontinuity at T_c for $d > 2s$ (here $s = 1, d = 3$).

energy, or entropy. All thermodynamic quantities are fully determined by the excited states, and continuous across the condensation phase transition. In the continuum limit we obtain with the density of states (7.83) for our power law dispersion relation:

$$E(T, V, z) = -\frac{d}{s} \Phi(T, V, z) = \frac{d}{s} V P(T, z) = \frac{d}{s} \frac{(2S+1) V}{\lambda_{\text{th}}(T)^d} k_B T \, \mathcal{G}_{1+d/s}(z) \, .$$

$$(7.92)$$

For $T < T_c$, we may set $\mu = 0$ or $z = 1$, whence

$$E(T, V) = \frac{d}{s} V P(T) = \frac{d}{s} \frac{(2S+1) V}{\lambda_{\text{th}}(T)^d} k_B T \, \zeta(1 + d/s)$$

$$= \frac{d}{s} N'(T) k_B T \frac{\zeta(1 + d/s)}{\zeta(d/s)} = \frac{d}{s} N k_B T \frac{\zeta(1 + d/s)}{\zeta(d/s)} \left(\frac{T}{T_c}\right)^{d/s} .$$

$$(7.93)$$

Straightforward derivatives yield the entropy

$$S(T) = -\left(\frac{\partial \Phi}{\partial T}\right)_{V, \mu = 0} = \left(1 + \frac{d}{s}\right) N k_B \frac{\zeta(1 + d/s)}{\zeta(d/s)} \left(\frac{T}{T_c}\right)^{d/s} \qquad (7.94)$$

and specific heat

$$C_v(T) = \left(\frac{\partial E}{\partial T}\right)_{V, N} = T \left(\frac{\partial S}{\partial T}\right)_{V, N} = \left(1 + \frac{d}{s}\right) \frac{d}{s} N k_B \frac{\zeta(1 + d/s)}{\zeta(d/s)} \left(\frac{T}{T_c}\right)^{d/s} ,$$

$$(7.95)$$

whose maximum value $C_v(T_c) = (1 + d/s) d \, N k_B \, \zeta(1 + d/s)/s \, \zeta(d/s)$ exceeds the classical limit $d \, N k_B/s$, see Fig. 7.2.

In the normal state above T_c, we may set the particle number $N = N'$, and inserting Eq. (7.86) into (7.92) gives

$$E(T, V, \mu) = \frac{d}{s} V P(T, \mu) = \frac{d}{s} N k_B T \frac{\mathcal{G}_{1+d/s}(z)}{\mathcal{G}_{d/s}(z)}. \qquad (7.96)$$

In the limit $z \to 0$, this reduces to the classical ideal gas equations of state $E = \frac{d}{s} PV = \frac{d}{s} N k_B T$. To proceed, we need to be especially careful in taking the partial derivatives, holding the appropriate variables μ or N fixed. Noting that $z \, d\mathcal{G}_k(z)/dz = \mathcal{G}_{k-1}(z)$ and $T (\partial z/\partial T)_\mu = -z \ln z$, we arrive at the entropy as function of fugacity z,

$$S(z) = - \left(\frac{\partial \Phi}{\partial T} \right)_{V,\mu} = N k_B \left[\left(1 + \frac{d}{s} \right) \frac{\mathcal{G}_{1+d/s}(z)}{\mathcal{G}_{d/s}(z)} - \ln z \right]. \qquad (7.97)$$

Furthermore, from Eq. (7.86) we infer $T (\partial \mathcal{G}_{d/s}(z)/\partial T)_{V,N} = -\frac{d}{s} \mathcal{G}_{d/s}(z)$ and consequently $T (\partial z/\partial T)_{V,N} = -\frac{d}{s} z \mathcal{G}_{d/s}(z)/\mathcal{G}_{-1+d/s}(z)$. These relations now allow us to compute the specific heat for $T > T_c$,

$$C_V(z) = \frac{d}{s} N k_B \left[\left(1 + \frac{d}{s} \right) \frac{\mathcal{G}_{1+d/s}(z)}{\mathcal{G}_{d/s}(z)} - \frac{d}{s} \frac{\mathcal{G}_{d/s}(z)}{\mathcal{G}_{d/s-1}(z)} \right]. \qquad (7.98)$$

If $s < d \leq 2s$, $\zeta(\frac{d}{s} - 1)$ diverges, and the second term in (7.98) vanishes as $z \to 1$. The specific heat then displays a cusp singularity at T_c, illustrated in Fig. 7.2(a). On the other hand, for $d > 2s$, there emerges a discontinuity $\Delta C_V(T_c) = \frac{d^2}{s^2} N k_B \zeta(\frac{d}{s})/\zeta(\frac{d}{s} - 1)$, see Fig. 7.2(b); in either case, $\alpha = 0$. The standard scaling relations (1.72), (1.75), and (1.77) then formally yield the other critical exponents $\gamma = 0$, $\delta = 1$, $\eta = 0$, and $\nu = 2/d$. In addition, we infer the dynamic critical exponent $z = s$ from de Broglie's relations $\epsilon = \hbar \omega$ and $p = \hbar k$, and the dispersion $\omega(k) = c |k|^s$.

7.2.2 Landau–Ginzburg theory for interacting bosons

Real boson systems are naturally subject to particle interactions. Even short-range contact interactions turn out to be relevant, generating a distinct universality class for the condensation phase transition. In superfluid helium 4, for example, the repulsive forces between the atoms are quite strong, reducing the condensate fraction to a mere 10 % even at zero temperature, in stark contrast to an ideal Bose gas (see Fig. 7.1). The remarkable feat of achieving Bose–Einstein condensation in trapped cooled atomic gases partly relies on the possibility of tuning the interatomic forces. The resulting condensate is characterized by weak repulsive interactions, and thus well described by the theoretical approaches developed in this subsection.

Since a Bose condensate occupies a single quantum state and thus has a definite phase, the coherent-state path integral formalism is expecially adept for its mathematical description. We employ the imaginary-time representation for the grand-canonical partition function (7.25); when needed, real-time quantum dynamics is recovered by setting $\tau = i t$ (and $\mu = 0$). We use the position eigenstate basis, and consider a Hamiltonian with three contributions, namely kinetic energy (7.7), pair interactions (7.9), and an external potential $U = \int d^d x \, \hat{\psi}^\dagger(x) \, U(x) \, \hat{\psi}(x)$, e.g., modeling the (time-averaged) atomic trap ($U < 0$). Replacing the field operators $\hat{\psi}(x)$ by the corresponding coherent state eigenvalues $\phi(x)$, the action (7.26) reads explicitly

$$
S[\phi^*, \phi] = \int_0^{\beta \hbar} d\tau \int d^d x \left(\phi^*(x, \tau) \left[\hbar \frac{\partial}{\partial \tau} - \frac{\hbar^2}{2m} \nabla^2 - \mu + U(x) \right] \phi(x, \tau) \right.
$$
$$
\left. + \frac{1}{2} |\phi(x, \tau)|^2 \int d^d x' \, V(x - x') \, |\phi(x', \tau)|^2 \right) . \tag{7.99}
$$

The external potential $U(x)$ can thus be viewed as a local modification of the chemical potential μ. In the spirit of Landau–Ginzburg theory, we seek the most likely configuration by minimizing the action

$$
0 = \frac{\delta S[\phi^*, \phi]}{\delta \phi^*(x, \tau)}
$$
$$
= \left[\hbar \frac{\partial}{\partial \tau} - \frac{\hbar^2}{2m} \nabla^2 - \mu + U(x) + \int d^d x' \, V(x - x') \, |\phi(x', \tau)|^2 \right] \phi(x, \tau) .
$$
$$
\tag{7.100}
$$

For short-range repulsive contact interactions, $V(x - x') = V_0 \, \delta(x - x')$ with $V_0 > 0$, this non-local integro-differential equation simplifies to a local non-linear Schrödinger equation in imaginary time,

$$
-\hbar \frac{\partial \phi(x, \tau)}{\partial \tau} = -\frac{\hbar^2}{2m} \nabla^2 \phi(x, \tau) - \left[\mu - U(x) - V_0 \, |\phi(x, \tau)|^2 \right] \phi(x, \tau) , \tag{7.101}
$$

whose real-time counterpart is known as the *Gross–Pitaevskii equation*.

We henceforth consider a vanishing external potential $U(x) = 0$. For stationary, homogeneous solutions, Eq. (7.101) reduces to $(\mu - V_0 |\phi|^2)\phi = 0$, corresponding to the results from Landau theory, see Section 1.2. The order parameter $\phi = 0$ vanishes for chemical potential $\mu < 0$, whereas for $\mu > 0$ one obtains the condensate density $n_0 = |\phi|^2 = \mu/V_0$. The critical point for Bose–Einstein condensation thus remains located at $\mu = 0$, and the grand-canonical potential per volume in the Landau approximation is $\Phi/V \approx k_B T S[n_0]/\hbar V = -\mu^2/2V_0$. This gives the total particle density $n = N/V = -V^{-1}(\partial \Phi/\partial \mu)_{T,v} \approx \mu/V_0 = n_0$. Mean-field theory predicts that virtually all particles agglomerate in the condensate. For the isothermal compressibility one finds $\kappa_T = (V/N^2)(\partial N/\partial\mu)_{T,V} \approx 1/n_0^2 V_0$, from which we infer the sound speed in the condensate, $c_s = 1/\sqrt{n\,m\,\kappa_s} \approx \sqrt{n_0\,V_0/m}$, assuming $\kappa_s \approx \kappa_T$.

Note that we may write the complex order parameter as $\phi = \sqrt{n_0}\,e^{i\Theta}$ with arbitrary but fixed phase Θ. Bose–Einstein condensation therefore entails a spontaneous symmetry breaking of continuous $U(1)$ gauge invariance, isomorphic to a rotational symmetry in the complex ϕ plane. In mean-field theory, the field operator expectation value $\langle \hat{\psi}(x,t) \rangle = \langle \phi(x,t) \rangle$ in the coherent-state representation is simply replaced by the most likely value ϕ. According to the general coherent state formula (7.11), and with $\hat{\psi}(x) = \sum_p e^{ipx/\hbar} a_p/V$ in a finite volume V, the corresponding quantum-mechanical many-particle boson state takes the form

$$\left| \sqrt{n_0}\,e^{i\Theta} \right\rangle = \exp\left[\sqrt{n_0}\,e^{i\Theta} \int \hat{\psi}^\dagger(x)\,d^d x \right] |0\rangle = \exp\left[\sqrt{n_0}\,e^{i\Theta} a^\dagger_{p=0} \right] |0\rangle$$

$$= \sum_{n=0}^{\infty} \sqrt{\frac{n_0^n}{n!}}\,e^{in\Theta}\,|n_{p=0}\rangle . \qquad (7.102)$$

The condensate wave function is thus a coherent superposition of all possible n-particle Fock states occupying the ground state with $p = 0$, characterized by a Poissonian probability distribution $P(n_{p=0}) = n_0^n\,e^{-n_0}/n!$, with expectation value and mean-square fluctuation $\langle n_{p=0} \rangle = n_0 = (\Delta n_{p=0})^2$, see Problem 7.1(b). The well-defined phase Θ comes at the price of maximal particle number uncertainty.

Generalizing to a spatially inhomogeneous condensate with (real) time dependence, we may write

$$\phi(x,t) = \sqrt{n_0(x,t)}\,e^{i\Theta(x,t)} , \qquad (7.103)$$

with *condensate density* $n_0(x,t) = |\phi(x,t)|^2$ and associated quantum-mechanical *current density*

$$j_0(x,t) = \frac{\hbar}{m}\,\mathrm{Im}[\phi^*(x,t)\,\nabla\phi(x,t)] = \frac{\hbar}{m}\,n_0(x,t)\,\nabla\Theta(x,t) , \qquad (7.104)$$

which takes the form $j_0(x,t) = n_0(x,t)\,v_s(x,t)$ if we identify

$$v_s(x,t) = \frac{\hbar}{m}\,\nabla\Theta(x,t) \qquad (7.105)$$

as the *superfluid velocity*. Except for singularities in the phase field $\Theta(x,t)$, this implies irrotational flow $\nabla \times v_s = 0$, and for an incompressible fluid with $\nabla \cdot v_s = 0$, furthermore, $\nabla^2 v_s = 0$, which entails the absence of any viscous damping, hence the name superfluid.

Inserting Eq. (7.103) into the real-time Gross–Pitaevskii equation gives

$$i\hbar \frac{\partial n_0}{\partial t} - 2\hbar n_0 \frac{\partial \Theta}{\partial t} = -\frac{\hbar^2}{m} \sqrt{n_0}\, e^{-i\Theta}\, \nabla \cdot \left[\left(\frac{\nabla n_0}{2\sqrt{n_0}} + i\sqrt{n_0}\, \nabla\Theta \right) e^{i\Theta} \right] + 2n_0^2 V_0 \,,$$

and separating out the imaginary part, one arrives at the *continuity equation*

$$\frac{\partial n_0(x, t)}{\partial t} = -\frac{\hbar}{m} \left[\nabla n_0(x, t) \cdot \nabla\Theta(x, t) + n_0(x, t)\, \nabla^2\Theta(x, t) \right] = -\nabla \cdot j_0(x, t) \,.$$

$$(7.106)$$

Inspecting the real part, one identifies the characteristic *coherence length*

$$\xi = \frac{\hbar}{\sqrt{2m\, n_0 V_0}} = \frac{\hbar}{\sqrt{2}\, m\, c_s} \,, \qquad (7.107)$$

which governs spatial variations of the condensate wave function; e.g., if we impose the presence of a normal state at $x \leq 0$ and a stationary state for $x > 0$ where $v_s = 0$, and set $\mu = n_0 V_0$ in Eq. (7.101), we obtain for this semi-infinite geometry $\phi(x) = \sqrt{n_0} \tanh(x/\sqrt{2}\,\xi)$, see Problem 1.4. Assuming spatially slow variation on the scale of ξ, $\xi|\nabla n_0|/n_0 \ll 1$, gradients of the condensate density can be neglected, and the real part indeed reduces to

$$\hbar \frac{\partial\Theta(x, t)}{\partial t} \approx -\frac{\hbar^2}{2m} [\nabla\Theta(x, t)]^2 - V_0\, n_0(x, t) \,, \qquad (7.108)$$

or equivalently,

$$m \frac{\partial v_s(x, t)}{\partial t} + \nabla \left[\frac{m}{2} v_s(x, t)^2 \right] \approx -V_0\, \nabla n_0(x, t) = -\nabla\mu(x, t) \,, \qquad (7.109)$$

which can be interpreted as a quantum version of Bernoulli's equation, here describing superfluid flow driven by a chemical potential gradient. Indeed, by means of the identity $\frac{1}{2}\nabla v^2 = (v \cdot \nabla)v + v \times (\nabla \times v)$ and $\nabla \times v_s = 0$, this can be written in the form of Euler's hydrodynamic equation for non-viscous flow (i.e., with vanishing kinematic viscosity)

$$m \frac{dv_s(x, t)}{dt} = m \left[\frac{\partial}{\partial t} + v_s(x, t) \cdot \nabla \right] v_s(x, t) \approx -\nabla\mu(x, t) \,. \qquad (7.110)$$

Finally, we may explore the properties of *topological defects* in the complex order parameter field $\phi(x, t)$, namely *vortices*, see also Section 8.2.3. Naturally, we demand that the condensate wave function be unique under phase shift $\Theta \to \Theta + 2\pi k$ with integer k. Consequently, one finds that the circulation of the superfluid velocity (7.105) in a closed loop that winds k times about the origin is quantized:

$$C = \oint v_s \cdot ds = \frac{\hbar}{m} \oint \nabla\Theta \cdot ds = k \frac{h}{m} \,. \qquad (7.111)$$

Assuming cylindrical symmetry about the z axis, the *quantization of circulation* within a circle of radius ρ, i.e., $x = (\rho, z)$, gives $C = v_s 2\pi\rho = k\,h/m$, and thus the superfluid flow field

$$v_s = k\,\frac{\hbar}{m\,\rho}\,e_\Theta\,, \qquad \nabla \times v_s = k\,\frac{h}{m}\,\delta(\rho)\,e_z\,, \tag{7.112}$$

which is irrotational except at the origin, where a straight k-vortex resides (a line defect in three dimensions, a point singularity in $d = 2$, cf. Fig. 8.3). Note also that $v_s(\rho = k\sqrt{2}\,\xi) = c_s$. The integer winding number k plays the role of a topological quantum number, identical to the quantum number associated with the angular momentum operator L_z. The cylindrical symmetry indeed motivates the factorization ansatz in terms of angular momentum eigenfunctions, $\phi_k(\rho, \Theta) = \sqrt{n_0}\,e^{ik\Theta}\,f_k(\rho)$, whereupon the stationary Gross–Pitaevskii equation becomes

$$\xi^2 \left(\frac{d^2}{d\rho^2} + \frac{1}{\rho}\frac{d}{d\rho} - \frac{k^2}{\rho^2} \right) f_k(\rho) = -f_k(\rho) + f_k(\rho)^3\,, \tag{7.113}$$

with the boundary condition $f_k(\rho \to \infty) = 1$. At small distances $\rho \ll \xi$ inside the vortex core $f_k(\rho) \sim (\rho/\xi)^{|k|}$: the order parameter field vanishes at the origin. As expected, the total angular momentum for a single vortex in a disk of radius L is

$$L_z = \frac{1}{\pi L^2} \int_0^L m\,v_s\,\rho\,2\pi\rho\,d\rho = \frac{2k\hbar}{L^2} \int_0^L \rho\,d\rho = k\hbar\,. \tag{7.114}$$

The kinetic energy per unit length provides the *vortex line tension*; here we use the coherence length and the system extension to set the required ultraviolet and infrared cutoffs:

$$\epsilon = \frac{1}{\pi L^2} \int_\xi^L \frac{m}{2}\,v_s^2\,2\pi\rho\,d\rho = \frac{k^2\hbar^2}{m\,L^2}\,\ln\frac{L}{\xi}\,. \tag{7.115}$$

In summary, applying the Landau–Ginzburg approximation to the coherent-state path integral for repulsive bosons largely recovers the phenomenology of super-fluids. Much additional information can be obtained by considering Gaussian fluctuations.

7.2.3 Gaussian correlations and quasi-particle excitations

We first explore the single-particle Green function for bosons in the normal phase with chemical potential $\mu < 0$. The Gaussian approximation then simply omits the interaction contributions (7.42); using momentum eigenstates $\alpha \to q$ and the dispersion relation $\epsilon_0(q) = \hbar^2 q^2/2m$ as appropriate for free non-relativistic particles with mass m, this leaves us with

$$G_0(q, \tau) = e^{-[\epsilon_0(q)-\mu]\tau/\hbar}[\langle n(q)\rangle_0 \, \Theta(-\tau) + (1 + \langle n(q)\rangle_0)\Theta(\tau)] , \qquad (7.116)$$

see Eq. (7.40). Thus we obtain the spatial equal-time Green function

$$G_0(x) = G_0(x, \tau \uparrow 0) = \int \frac{d^d q}{(2\pi)^d} \frac{e^{iq\cdot x}}{e^{\beta[\epsilon_0(q)-\mu]} - 1} \qquad (7.117)$$

in the continuum limit in d dimensions. In the classical regime where the fugacity $z = e^{\beta\mu} \ll 1$, this simplifies to

$$G_0(x) \approx z \int \frac{d^d q}{(2\pi)^d} \, e^{iq\cdot x - \hbar^2 q^2/2mk_B T} = z \, \lambda_{\text{th}}(T)^{-d} \, e^{-\pi x^2/\lambda_{\text{th}}(T)^2} , \qquad (7.118)$$

where $\lambda_{\text{th}}(T) = h/\sqrt{2\pi m k_B T}$ denotes the *thermal de Broglie wavelength*, see Eq. (7.87) with $s = 2$ and $c = 1/2m$, which here characterizes the spatial extension of the (Gaussian) particle correlations for $T > T_c$.

In the vicinity of the normal- to superfluid phase transition where $\mu \uparrow 0$, the correlations are dominated by the large-wavelength modes $q \to 0$, and as long as $T_c > 0$, one may instead approximate

$$G_0(x) \approx z \int \frac{d^d q}{(2\pi)^d} \, e^{iq\cdot x} \frac{k_B T_c}{|\mu| + \hbar^2 q^2/2m} = \frac{e^{-|x|/\xi}}{\lambda_{\text{th}}(T_c)^2 \, |x|} \qquad (7.119)$$

in three dimensions. $G_0(x)$ thus takes the form of an *Ornstein–Zernicke correlation function* with correlation length $\xi = \hbar/\sqrt{2m|\mu|}$, which coincides with the coherence length (7.107); in $d > 2$ dimensions, $G_0(x) \sim 1/|x|^{d-2}$ as $|\mu| \to 0$, see Section 1.1.3. In fact, the expression (7.116) becomes essentially independent of imaginary time $0 \le \tau \le \beta\hbar$, provided $\beta(|\mu| + \hbar^2 q^2/2m) \ll 1$, which is satisfied in the *entire* critical regime $|\mu| \to 0$, $q \to 0$. Consequently, the imaginary-time integral in the action (7.99) is reduced to the integrand value at a local time slice, multiplied by $\beta\hbar$. Near its critical point at $T_c > 0$, quantum fluctuations become irrelevant, and the original $(d + 1)$-dimensional quantum field theory action undergoes a dimensional crossover to a d-dimensional classical effective Hamiltonian, $S[\phi^*, \phi]/\hbar \to \mathcal{H}_{\text{eff}}[\phi^*, \phi]/k_B T$. As the correlation length $\xi \to \infty$, we may also replace any finite-range non-local interaction in Eq. (7.99) by a mere contact repulsion with strength V_0. The effective critical action hence becomes

$$\frac{\mathcal{H}_{\text{eff}}[\phi^*, \phi]}{k_B T} = \int d^d x \left(\frac{\hbar^2}{2m} |\nabla\phi(x)|^2 - \mu \, |\phi(x)|^2 + \frac{V_0}{2} |\phi(x)|^4 \right) , \qquad (7.120)$$

which we recognize as a classical Φ^4 complex field theory with global $U(1)$ gauge symmetry, equivalent to a real $O(2)$-symmetric Landau–Ginzburg–Wilson Hamiltonian (4.12). The critical properties of the continuous normal- to superfluid phase transition are therefore described by the universality class of the two-component Heisenberg or XY model.

The Bose-condensed phase is characterized by anomalous expectation values $\langle \phi \rangle = \sqrt{n_0}\, e^{i\Theta} \neq 0$, also termed *off-diagonal long-range order*. Here, the thermal de Broglie wavelength exceeds the typical particle separation, inducing phase coherence volumes that encompass several indistinguishable bosons. In Feynman's visualization, this implies entangled particle world lines over a range of $\lambda_{th}(T)$, or macroscopic ring exchange processes. Also note that $G(x - x', 0) = \langle \phi(x, \tau)\phi^*(x', \tau + \eta) \rangle \rightarrow \langle \phi \rangle \langle \phi^* \rangle = n_0$ as $|x - x'| \rightarrow \infty$. We shall expand about a homogeneous condensate density n_0, i.e., $n(x, \tau) = n_0 + \pi(x, \tau)$ with $\langle \pi(x, \tau) \rangle = 0$, choosing the mean phase $\langle \Theta(x, \tau) \rangle = 0$. This gives the convenient field parametrization

$$\phi(x, \tau) = \sqrt{n_0 + \pi(x, \tau)}\, e^{i\Theta(x,\tau)} . \tag{7.121}$$

Since the functional determinant from the variables (ϕ^*, ϕ) to (π, Θ) is i (or the one from $(\text{Re}\,\phi, \text{Im}\,\phi)$ to (π, Θ) is $1/2$), $d\phi^* d\phi / 2\pi i$ in the functional measure becomes replaced by $d\pi\, d\Theta / 2\pi$, and we may write the grand-canonical partition function (7.25) as

$$\mathcal{Z}(\beta, \mu) = \int_{\substack{\pi(0)=\pi(\beta\hbar) \\ \Theta(0)=\Theta(\beta\hbar)}} \mathcal{D}[\pi, \Theta]\, e^{-S[\pi,\Theta]/\hbar} . \tag{7.122}$$

Inserting the parametrization (7.121) into the action (7.99) with $U(x) = 0$, chemical potential $\mu = n_0 V_0$, and observing the periodic boundary conditions in imaginary time for the fluctuating fields π and Θ, straightforward manipulations (Problem 7.6) lead to $S[\pi, \Theta] = -n_0^2 V_0 \beta\hbar V / 2 + S_0[\pi, \Theta] + S_{\text{int}}[\pi, \Theta]$. The first term here is just the Landau contribution for spatially uniform fields, whereas the fluctuation corrections in the action have been split into the Gaussian part

$$S_0[\pi, \Theta] = \int_0^{\beta\hbar} d\tau \int d^d x \left[i\hbar\, \pi(x, \tau)\frac{\partial \Theta(x, \tau)}{\partial \tau} + \frac{\hbar^2 n_0}{2m} [\nabla\Theta(x, \tau)]^2 \right.$$
$$\left. + \frac{\hbar^2}{8m\, n_0} [\nabla\pi(x, \tau)]^2 + \frac{1}{2} \int d^d x'\, V(x - x')\, \pi(x, \tau)\pi(x', \tau) \right], \tag{7.123}$$

and the non-linear contributions

$$S_{\text{int}}[\pi, \Theta] = \frac{\hbar^2}{2m} \int_0^{\beta\hbar} d\tau \int d^d x \left(\pi(x, \tau)[\nabla\Theta(x, \tau)]^2 - \frac{\pi(x, \tau)\,[\nabla\pi(x, \tau)]^2}{4n_0\,[n_0 + \pi(x, \tau)]} \right) . \tag{7.124}$$

Note that upon expanding the second term for small density fluctuations π, $S_{int}[\pi, \Theta]$ successively contains n-point π vertices with $n \geq 3$ in addition to the three-point vertex $\Gamma_{\pi\Theta\Theta}$.

The Gaussian action $S_0[\pi, \Theta]$, which includes spatially non-local pair interactions $V(x - x')$, is readily diagonalized through Fourier transformation with the Matsubara frequencies (7.34),

$$\pi(x, \tau) = \int \frac{d^d q}{(2\pi)^d} \frac{1}{\beta\hbar} \sum_n \pi(q, \omega_n) \, e^{i(q \cdot x - \omega_n \tau)} \tag{7.125}$$

(and similarly for the phase fluctuations Θ), which gives

$$S_0[\pi, \Theta] = \frac{1}{2\beta\hbar} \int_q \sum_n \left(\pi(-q, -\omega_n) \; \Theta(-q, -\omega_n) \right) \mathbf{A}(q, \omega_n) \begin{pmatrix} \pi(q, \omega_n) \\ \Theta(q, \omega_n) \end{pmatrix}, \tag{7.126}$$

with the harmonic coupling matrix

$$\mathbf{A}(q, \omega_n) = \begin{pmatrix} V(q) + \hbar^2 q^2 / 4mn_0 & \hbar\,\omega_n \\ -\hbar\,\omega_n & \hbar^2 q^2 n_0 / m \end{pmatrix}. \tag{7.127}$$

The ensuing density and phase correlations can now be immediately read off its inverse (see Section 4.2.2),

$$\begin{pmatrix} \langle \pi(q, \omega_n) \pi(q', \omega_{n'}) \rangle_0 & \langle \pi(q, \omega_n) \Theta(q', \omega_{n'}) \rangle_0 \\ \langle \Theta(q, \omega_n) \pi(q', \omega_{n'}) \rangle_0 & \langle \Theta(q, \omega_n) \Theta(q', \omega_{n'}) \rangle_0 \end{pmatrix}$$

$$= \hbar \, \mathbf{A}^{-1}(q, \omega_n) \, (2\pi)^d \, \delta(q + q') \, \beta\hbar \, \delta_{n, -n'}, \tag{7.128}$$

with

$$\hbar \, \mathbf{A}^{-1}(q, \omega_n) = \frac{1}{\omega_n^2 + \epsilon_B(q)^2/\hbar^2} \begin{pmatrix} \hbar q^2 n_0 / m & -\omega_n \\ \omega_n & m \, \epsilon_B(q)^2 / n_0 \hbar^3 q^2 \end{pmatrix}, \tag{7.129}$$

and where

$$\epsilon_B(q) = \sqrt{n_0 \hbar^2 q^2 V(q)/m + (\hbar^2 q^2 / 2m)^2} \tag{7.130}$$

denotes the *Bogoliubov quasi-particle spectrum*. At long wavelengths ($q \to 0$) this yields a linear acoustic phonon dispersion with the condensate sound speed c_s, $\epsilon_B(q) \approx \hbar\sqrt{n_0 V_0/m} \, |q| = \hbar c_s |q|$, whereas of course the free-particle dispersion $\epsilon_0(q) = \hbar^2 q^2 / 2m$ is recovered for $V(q) = 0$.

For the density correlations (in imaginary time)

$$S(x - x', \tau - \tau') = \langle n(x, \tau) n(x', \tau') \rangle - \langle n \rangle^2 = \langle \pi(x, \tau) \pi(x', \tau') \rangle, \tag{7.131}$$

Equations (7.128) and (7.129) give in the Gaussian approximation

$$S_0(q, \tau) = \frac{q^2 n_0}{\beta m} \sum_n \frac{e^{-i\omega_n \tau}}{\omega_n^2 + \epsilon_B(q)^2/\hbar^2} = \frac{\hbar^2 q^2 n_0}{2m \, \epsilon_B(q)} \frac{e^{(\beta - |\tau|/\hbar)\epsilon_B(q)} + e^{\epsilon_B(q)|\tau|/\hbar}}{e^{\beta \, \epsilon_B(q)} - 1}$$

(7.132)

by means of Eq. (7.214). Thus we find the *static structure factor*

$$S_0(q) = S_0(q, \tau = 0) = \frac{\hbar^2 q^2 n_0}{2m \, \epsilon_B(q)} \coth \frac{\beta \, \epsilon_B(q)}{2},$$

(7.133)

which at low temperatures approximately becomes $S_0(q) \approx n_0 \epsilon_0(q)/\epsilon_B(q) \approx n_0 \hbar |q|/2mc_s$ as $q \to 0$. Transforming to real time $\tau = it > 0$, one obtains

$$S_0(q, \omega) = \int S_0(q, t) e^{i\omega t} \, dt = \frac{\pi \hbar^2 q^2 n_0}{m \, \epsilon_B(q)} \frac{1}{e^{\beta \, \epsilon_B(q)} - 1}$$

$$\times \left[e^{\beta \, \epsilon_B(q)} \delta(\omega - \epsilon_B(q)/\hbar) + \delta(\omega + \epsilon_B(q)/\hbar) \right],$$

(7.134)

which indicates that the elementary excitations detectable in scattering experiments will indeed be governed by the Bogoliubov dispersion (7.130). Notice that the absorption and emission contributions come with the proper Bose–Einstein and detailed balance factors (see Section 2.1.1).

In order to compute the *Green function* and the *anomalous* cumulant,

$$G_c(x - x', \tau - \tau' - \eta) = \langle \phi(x, \tau) \phi^*(x', \tau' + \eta) \rangle - n_0,$$

(7.135)

$$\overline{G}_c(x - x', \tau - \tau' - \eta) = \langle \phi(x, \tau) \phi(x', \tau' + \eta) \rangle - n_0,$$

(7.136)

we expand (7.121) to first order in the density and phase fluctuations, $\phi(x, \tau) \approx \sqrt{n_0} [1 + \pi(x, \tau)/2n_0 + i\Theta(x, \tau)]$, whence in the Gaussian approximation $\langle \phi \, \phi^{(*)} \rangle_0 = n_0 + \langle \pi \, \pi \rangle_0/4n_0 \pm i \langle \pi \, \Theta \rangle_0/2 + i \langle \Theta \, \pi \rangle_0/2 \mp n_0 \langle \Theta \, \Theta \rangle_0$. With the explicit expressions (7.128) and (7.129), this yields

$$G_{c0}(q, \omega_n) = \frac{i\omega_n + n_0 V(q)/\hbar + \hbar q^2/2m}{\omega_n^2 + \epsilon_B(q)^2/\hbar^2},$$

(7.137)

$$\overline{G}_{c0}(q, \omega_n) = \frac{-n_0 V(q)/\hbar}{\omega_n^2 + \epsilon_B(q)^2/\hbar^2};$$

(7.138)

note that $\overline{G}_{c0}(q, \omega_n) = 0$ for $n_0 = 0$, i.e., in the absence of a Bose condensate. Upon defining the Bogoliubov quasi-particle weights

$$u_\pm(q)^2 = \frac{1}{2} \left(\frac{n_0 \, V(q) + \hbar^2 q^2/2m}{\epsilon_B(q)} \pm 1 \right),$$

(7.139)

which satisfy $u_+(q)^2 - u_-(q)^2 = 1$ and $u_+(q)\,u_-(q) = n_0 V(q)/2\epsilon_B(q)$, we arrive at the single-pole decomposition for the Gaussian Green functions

$$G_{c0}(q, \omega_n) = \frac{u_+(q)^2}{-i\omega_n + \epsilon_B(q)/\hbar} + \frac{u_-(q)^2}{i\omega_n + \epsilon_B(q)/\hbar} , \tag{7.140}$$

$$\overline{G}_{c0}(q, \omega_n) = -u_+(q)\,u_-(q) \left(\frac{1}{-i\omega_n + \epsilon_B(q)/\hbar} + \frac{1}{i\omega_n + \epsilon_B(q)/\hbar} \right). \tag{7.141}$$

Performing the Matsubara frequency sum (7.213), this gives in imaginary time $0 \leq \tau \leq \beta\hbar$:

$$G_{c0}(q, \tau) = u_+(q)^2 \frac{e^{-\epsilon_B(q)\tau/\hbar}}{e^{\beta \epsilon_B(q)} - 1} + u_-(q)^2 \frac{e^{\epsilon_B(q)\tau/\hbar}}{1 - e^{-\beta \epsilon_B(q)}} , \tag{7.142}$$

$$\overline{G}_{c0}(q, \tau) = -u_+(q)\,u_-(q) \left(\frac{e^{-\epsilon_B(q)\tau/\hbar}}{e^{\beta \epsilon_B(q)} - 1} + \frac{e^{\epsilon_B(q)\tau/\hbar}}{1 - e^{-\beta \epsilon_B(q)}} \right). \tag{7.143}$$

In the long-wavelength limit $\beta \epsilon_B(q) \approx \beta\hbar c_s|q| \ll 1$, these expressions become independent of τ, whereupon $G_{c0}(q) \approx -\overline{G}_{c0}(q) \approx k_B T m/\hbar^2 q^2$, which is just the transverse correlation function for the effective classical Hamiltonian (7.120) in its low-temperature symmetry-broken phase.

Recalling Eq. (7.39), we can now determine the *condensate depletion*, i.e., the total volume density of particles in excited states $n'(T) = n(T) - n_0 = N'(T)/V = V^{-1} \sum_{q \neq 0} G_c(q, \tau \uparrow 0)$. With Eq. (7.142) and taking the continuum limit, we find in the Gaussian approximation

$$n'(T) \approx \int_q \left(\frac{u_+(q)^2}{e^{\beta \epsilon_B(q)} - 1} + \frac{u_-(q)^2}{1 - e^{-\beta \epsilon_B(q)}} \right) = \int_q \left(u_-(q)^2 + \frac{1 + 2u_-(q)^2}{e^{\beta \epsilon_B(q)} - 1} \right). \tag{7.144}$$

The first term represents the density of particles that are lifted to excited states due to their mutual repulsion, even at zero temperature. Assuming contact interactions, $V(q) \approx V_0$, and inserting Eqs. (7.139) and (7.130), we find for this zero-temperature depletion in d dimensions

$$n'(0) \approx \left(\frac{mn_0 V_0}{2\pi \hbar^2} \right)^{d/2} \frac{1}{\Gamma(d/2)} \int_0^\infty x^{d-2} \left[\sqrt{2 + x^2} - x - \frac{1}{\sqrt{2 + x^2}} \right] dx , \tag{7.145}$$

where substitution with the variable $x = \hbar|q|/\sqrt{2mn_0 V_0}$ has rendered the integral dimensionless. For $d \leq 1$, the right-hand side of Eq. (7.145) diverges at the lower integration bound, which means that even at $T = 0$ essentially all particles occupy excited states, contradicting the implicit assumption of the existence of a macroscopic Bose condensate. In two dimensions, $n'(0) \approx mn_0 V_0/4\pi \hbar^2$ or $n_0 \approx n/(1 + mV_0/4\pi \hbar^2)$, whereas $n'(0) \approx (mn_0 V_0)^{3/2}/3\pi^2 \hbar^3$ for $d = 3$.

In order to evaluate the fraction of particles $\Delta n'(T) = n'(T) - n'(0)$ that are thermally excited into quasi-particle states, we employ the *phonon approximation* $\epsilon_B(q) \approx \hbar c_s |q|$ so that $1 + 2u_-(q)^2 \approx mc_s/\hbar|q|$, whence

$$\Delta n'(T) \approx \frac{mc_s/\hbar}{2^{d-1}\pi^{d/2}\,\Gamma(d/2)} \int_0^\infty \frac{q^{d-2}\,dq}{e^{\beta\hbar c_s q} - 1} = \frac{\Gamma(d-1)\,\zeta(d-1)}{2^{d-1}\pi^{d/2}\,\Gamma(d/2)} \frac{m(k_B T)^{d-1}}{\hbar^d c_s^{d-2}}$$

(7.146)

for $d > 2$; for example, in three dimensions $\Delta n'(T) \approx m(k_B T)^2/12\,\hbar^3 c_s$. In dimensions $d \leq 2$, the integral in Eq. (7.146) is infrared-divergent, and thermal excitations destroy the Bose condensate characterized by spontaneously broken continuous $U(1)$ gauge symmetry and off-diagonal long-range order, $\overline{G}_{c0}(q, \omega_n) \neq 0$. This establishes the lower critical dimension $d_{lc} = 2$ for Bose–Einstein condensation even for interacting bosons, as one would expect for the effective XY model Hamiltonian (7.120) in accord with the general considerations in Sections 1.3 and 5.4.

7.2.4 Current correlations; normal- and superfluid density

Finally, we explore the imaginary-time current cumulants

$$C_{ij}(x - x', \tau - \tau') = \langle j_{0i}(x, \tau)\, j_{0j}(x', \tau')\rangle - \langle j_{0i}\rangle\,\langle j_{0j}\rangle\,. \qquad (7.147)$$

Recalling Eq. (7.104) with the parametrization (7.121), we find $j_0(x, \tau) = \hbar\,[n_0 + \pi(x, \tau)]\,\nabla\Theta(x, \tau)/m$; in Fourier space, the current therefore contains a composite operator,

$$j_i(q, \omega_n) = i\frac{\hbar}{m}\left[n_0 q_i \Theta(q, \omega_n) + \int \frac{d^d k}{(2\pi)^d}\, \frac{1}{\beta\hbar}\sum_m \pi(q-k, \omega_{n-m})\, k_i \Theta(k, \omega_m)\right]\,.$$

(7.148)

Since $\langle\pi\rangle = 0 = \langle\Theta\rangle$, all three-point correlation functions vanish in the Gaussian ensemble, while we may factorize the four-point functions according to Wick's theorem, whereupon

$$C_{0ij}(q, \omega_n)(2\pi)^d \delta(q + q')\beta\hbar\delta_{n,-n'}$$

$$= -\frac{\hbar^2}{m^2}\left(n_0^2 q_i q_j' \langle\Theta(q, \omega_n)\Theta(q', \omega_{n'})\rangle_0 \right.$$

$$+ \int_k\!\int_{k'} \frac{k_i\,k_j'}{(\beta\hbar)^2}\sum_{m,m'}\left[\langle\pi(q-k, \omega_{n-m})\pi(q'-k', \omega_{n'-m'})\rangle_0\langle\Theta(k, \omega_m)\Theta(k', \omega_{m'})\rangle_0 \right.$$

$$\left.\left. + \langle\pi(q-k, \omega_{n-m})\Theta(k', \omega_{m'})\rangle_0\,\langle\Theta(k, \omega_m)\pi(q'-k', \omega_{n'-m'})\rangle_0\right]\right)\,. \qquad (7.149)$$

Inserting the Gaussian correlations from Eqs. (7.128) and (7.129), we obtain

$$C_{0ij}(q, \omega_n) = \frac{q_i\, q_j}{q^2}\, \frac{n_0\, \epsilon_B(q)^2/\hbar m}{\omega_n^2 + \epsilon_B(q)^2/\hbar^2}$$
$$+ \frac{\hbar}{m^2\beta} \int_k \sum_m \frac{k_i\left[k_j(q-k)^2\epsilon_B(k)^2/\hbar^2 k^2 - (q_j - k_j)\,\omega_{n-m}\omega_m\right]}{\left[\omega_{n-m}^2 + \epsilon_B(q-k)^2/\hbar^2\right]\left[\omega_m^2 + \epsilon_B(k)^2/\hbar^2\right]}.$$

$$(7.150)$$

The transverse current correlations

$$C^\perp(q, \omega_n) = \frac{1}{d-1} \sum_{i,j} P_{ij}^\perp(q)\, C_{ij}(q, \omega_n), \qquad (7.151)$$

with the transverse projector $P_{ij}^\perp(q) = \delta_{ij} - q_i q_j/q^2$, see Eq. (6.136), are related to the superfluid's response to shear. We thus define the *normalfluid density* as the transport coefficient (with dimensions of mass density) associated with the transverse current correlations through the limit

$$\rho_n = \frac{m^2}{\hbar} \lim_{q\to 0} C^\perp(q, \omega_n = 0), \qquad (7.152)$$

and the *superfluid density* as its complement $\rho_s = n\,m - \rho_n$.[6] Since the first term in (7.150), which orginates from pure phase fluctuations, is longitudinal, and $\sum_{i,j} P_{ij}^\perp(q)\, k_i(q_j - k_j) = -k^2 + (q \cdot k)^2/q^2$,

$$C_0^\perp(q, \omega_n) = \frac{\hbar}{(d-1)m^2\beta} \int_k \sum_m \frac{q^2 k^2 - (q \cdot k)^2}{q^2 k^2}$$
$$\times \frac{\left[(q-k)^2\epsilon_B(k)^2/\hbar^2 + k^2\omega_{n-m}\omega_m\right]}{\left[\omega_{n-m}^2 + \epsilon_B(q-k)^2/\hbar^2\right]\left[\omega_m^2 + \epsilon_B(k)^2/\hbar^2\right]}. \qquad (7.153)$$

A non-vanishing normalfluid density is therefore exclusively generated by thermally excited fluctuations, which are in fact intimately related to the superfluid's intrinsic *vorticity* $\nabla \times j_0(x, \tau)$. For, introducing vorticity correlations via $V_{ij}(x - x', \tau - \tau') = \langle [\nabla \times j_0(x, \tau)]_i\, [\nabla \times j_0(x, \tau)]_j \rangle$, one readily observes that $V^\perp(q, \omega_n) = q^2\, C^\perp(q, \omega_n)$. In the Gaussian approximation, the normalfluid density (7.152) becomes

$$\rho_n(T) \approx \frac{1}{d\,\beta} \int_q \sum_n q^2\, \frac{-\omega_n^2 + \epsilon_B(q)^2/\hbar^2}{\left[\omega_n^2 + \epsilon_B(q)^2/\hbar^2\right]^2} = \frac{\beta\hbar^2}{4d} \int \frac{d^d q}{(2\pi)^d}\, \frac{q^2}{\sinh^2[\beta\,\epsilon_B(q)/2]}$$

$$(7.154)$$

---

[6] Note that the superfluid density, a transport coefficient, is *not* directly related to the scalar condensate fraction. In anisotropic systems, for example, a superfluid subject to extended defects, both normal- and superfluid densities are naturally tensorial quantities.

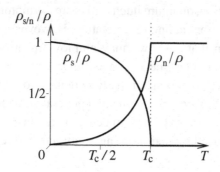

Fig. 7.3 The normal- and superfluid densities $\rho_{n/s}(T)$ as functions of temperature.

by means of Eq. (7.216) in the Appendix (Section 7.5). With

$$\int_0^\infty \frac{x^k}{\sinh^2 x}\,dx = 2^{1-k}\,\Gamma(k+1)\,\zeta(k)\,,$$

the integral for the normalfluid density (7.154) can be evaluated in the phonon approximation $\epsilon_B(q) \approx \hbar c_s|q|$,

$$\rho_n(T) \approx \frac{\Gamma(d+2)\,\zeta(d+1)}{(4\pi)^{d/2}\,\Gamma(1+d/2)}\,\frac{(k_B T)^{d+1}}{\hbar^d\,c_s^{d+2}}\,, \tag{7.155}$$

which gives a good description at low temperatures. For example, with $\zeta(4) = \pi^4/90$, one finds in three dimensions $\rho_n(T) \approx 2\pi^2(k_B T)^4/45\hbar^3 c_s^5$. The superfluid density vanishes at the critical temperature, and $\rho_n(T_c) = nm = \rho$, as sketched in Fig. 7.3.

We conclude that a variety of relevant physical properties of weakly interacting boson superfluids can be addressed with comparatively simple approximations. Indeed, many features of Bose–Einstein condensates produced in cool atomic gases are adequately captured by variants of the Gross–Pitaevskii equation (7.101). However, mere Gaussian correlation functions or perturbative expansions obviously fail to describe strongly interacting Bose liquids such as helium 4. Path integral Monte Carlo simulations based on the action (7.99) have proven a very efficient numerical tool for such systems, and even permit quantitatively accurate computations of key material parameters and response functions.[7]

7.3 Quantum critical phenomena

The coherent-state path integral formalism provides the proper tool to discuss critical phenomena in quantum systems. As we have already seen for the normal- to

---

[7] The quantum Monte Carlo approach to boson superfluids is reviewed in Ceperley (1995).

superfluid phase transition, quantum fluctuations are asymptotically irrelevant compared to thermal fluctuations at finite temperature. However, genuinely novel behavior arises if $T_c = 0$. In that case, we may employ anisotropic scaling theory to describe the critical phenomena near such *quantum phase transitions* that are controlled by some interaction parameter, particle density, or disorder strength rather than temperature.[8] Following a few basic and general considerations, we focus on two instructive examples, namely the Ising chain in a transverse magnetic field and boson localization induced by their pinning to attractive defects.

7.3.1 Phase transitions and quantum fluctuations

In the coherent-state path integral formalism, bosonic/fermionic quantum many-particle Hamiltonians are represented by a field theory action of the form (7.26), with complex/Grassmann fields and periodic/antiperiodic boundary conditions in imaginary time for bosons and fermions, respectively; see also the explicit example (7.99). We may view this coherent-state action as a $(d + 1)$-dimensional field theory, which is, however, bounded by the interval $[0, \beta\hbar]$ in the imaginary-time direction. Let us then consider a critical point at finite temperature $T_c > 0$, with an associated diverging correlation length $\xi \sim |T - T_c|^{-\nu}$. In the imaginary-time domain, there is an associated diverging time scale $t_c \sim \xi^z \sim |T - T_c|^{-z\nu}$, or characteristic frequency $\omega_c \sim t_c^{-1} \sim |T - T_c|^{z\nu}$, see Section 3.1. These power laws define the dynamic critical exponent z; note, however, that in this context z pertains to fully time-reversible quantum dynamics. Critical quantum-mechanical fluctuations are cut off once t_c reaches the extension of the finite temporal interval $\beta\hbar$, in full analogy to the correlation length ξ for static critical fluctuations being limited by the system size L in constrained geometries, compare with Section 1.3.2. Quantum fluctuations are consequently suppressed upon approaching the critical point as $t_c > \beta\hbar \approx \beta_c\hbar = \hbar/k_B T_c$, i.e., for $|T - T_c| < T_c(t_0 k_B T_c/\hbar)^{1/z\nu}$ (where t_0 denotes a microscopic time scale). The asympotic critical power laws are thus governed by purely classical statistical mechanics, namely the d-dimensional field theory that emerges from the coherent-state action at fixed imaginary time.

As an example, consider the finite-size scaling of the singular part of the free energy near a continuous phase transition in a quantum system. With the cutoff scales L in space and $\beta\hbar = \hbar/k_B T$ in imaginary time, we have

$$f_{\text{sing}}(T, h) = |T - T_c|^{2-\alpha} \, \hat{f}_\pm\big(h/|T - T_c|^{\beta\delta}, \xi/L, t_c k_B T/\hbar\big) , \qquad (7.156)$$

[8] For recent overviews, see, e.g., the articles by Continentino (1994); Sondhi, Girvin, Karini, and Shahar (1997); Vojta (2003); Vojta (2008); and the textbooks by Sachdev (2011) and Kamenev (2011).

compare with Eqs. (1.66) and (1.78). In the thermodynamic limit $L \to \infty$, the second argument vanishes; if we subsequently let $T \to T_c$, we may rewrite

$$f_{\text{sing}}(T, h) = |T - T_c|^{dv} \, \hat{f}_{\pm}\big(h/|T - T_c|^{\beta\delta}, 0, t_0 \, k_B T_c^{1+zv}/\hbar \, |T - T_c|^{zv}\big), \quad (7.157)$$

where we have used the hyperscaling relation $2 - \alpha = dv$. If $T_c \neq 0$, the critical singularities in the time domain become rounded off in the same manner as for a finite external field h (conjugate to the order parameter). With the temporal fluctuations frozen out, we recover the standard classical ($\hbar \to 0$) free energy scaling form (1.66).

These arguments become invalid if the critical point is located at $T_c = 0$, since the imaginary-time domain is (half-)infinite in that situation. Quantum fluctuations now play a prominent role, whereas thermal fluctuations are naturally absent at zero temperature. Let the phase transition be controlled by a (dimensionless) external parameter g in the quantum Hamiltonian, e.g., some effective interaction or disorder strength, or scaled particle density. We may then define the correlation length and dynamic critical exponents at such a genuine *quantum critical point* (QCP) via $\xi \sim |g - g_c|^{-\nu}$ and $\omega_c \sim t_c^{-1} \sim |g - g_c|^{z\nu}$. In the coherent-state path integral representation, we now encounter a $(d + 1)$-dimensional field theory. Nevertheless, we may identify this quantum system with an equivalent spatially *anisotropic* classical system, where the imaginary-time 'direction' plays a special role if $z \neq 1$. Specifically, the associated temporal correlation 'length' divergence is governed by the critical exponent $\nu_t = z\nu$, and the dynamic exponent z assumes the role of an anisotropy exponent ratio. With this observation, we may readily generalize the *hyperscaling* relations (1.77) in Section 1.3 by means of the substitution $d\nu \to (d + z)\nu$ to obtain their quantum critical counterparts:

$$\beta = \frac{\nu}{2}(d + z - 2 + \eta), \quad 2 - \alpha = (d + z)\nu; \quad (7.158)$$

in effect, the dimensionality d thus becomes replaced by the (generally non-integer) sum $d + z$. The singular part of the free energy density (7.157) now scales according to

$$f_{\text{sing}}(T, g, h) = |g - g_c|^{(d+z)\nu} \, \hat{f}_{\pm}\big(h/|g - g_c|^{\beta\delta}, 0, t_0 \, k_B T/\hbar \, |g - g_c|^{z\nu}\big). \quad (7.159)$$

Note that $f_{\text{sing}} \sim \xi^{-(d+z)} \sim 1/\xi^d t_c$ at $T = 0$, the inverse spatio-temporal correlation volume, whereas $f_{\text{sing}} \sim \xi^{-d}$ at finite temperatures.

The quantum critical point scenario is illustrated in Fig. 7.4(a), where a QCP at $T = 0$, $g = g_c$, $h = 0$ separates an ordered from a quantum 'disordered' phase. Notice that quantum-mechanical phase coherence persists even for $g > g_c$,

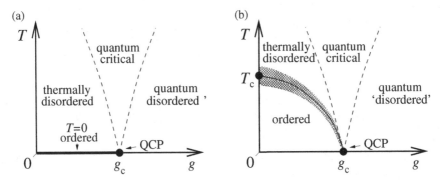

Fig. 7.4 Schematic phase diagram scenarios for quantum systems with continuous phase transitions. (a) Only a quantum critical point (QCP) exists, at temperature $T = 0$ and coupling parameter $g = g_c$; at the dashed crossover line, $k_B T \approx \hbar\omega_c$ (compare also with Fig. 3.8). (b) A critical line connects the quantum critical point with a classical second-order phase transition at $g = 0$, $T_c > 0$; at any temperature $T > 0$, the asymptotic critical behavior within the shaded region is governed by the exponents of the classical critical point. [Figure adapted from Vojta (2008).]

typically inducing long-range correlations in this phase. At finite temperature, the quantum critical regime extends up to $k_B T \approx \hbar\omega_c$ or $|g - g_c| < (t_0\, k_B T/\hbar)^{1/z\nu}$. In some cases, a critical line connects a QCP to a classical critical point at $T = T_c$, $g = 0, h = 0$. This situation is depicted in Fig. 7.4(b): except for $T = 0$, the asymptotic critical behavior along the entire critical line (and inside the shaded critical region) will be governed by the scaling exponents of the classical critical point, but interesting crossover phenomena will affect the system's low-temperature properties. In both situations depicted in Fig. 7.4, the quantum critical point crucially affects a sizeable region of the phase diagram, namely the one emanating from $T = 0$ and $g = g_c$ within the dashed lines. In this regime, physical properties are best captured through analysis of the QCP scaling behavior (and leading corrections thereof), rather than extrapolating from either the thermally or quantum disordered phases.

If the quantum to classical mapping leads to a classical model with upper critical dimension d_c, another immediate consequence of the effective replacement $d \rightarrow d + z$ is that the original quantum system will have a *reduced* critical dimension $d_c - z$. As is characteristic of continuous phase transitions in anisotropic systems, fluctuations are weakened, and one would in general expect dominant quantum fluctuations only in low spatial dimensions. At this point some cautionary remarks are in place. First, the above straightforward finite-size scaling analysis assumed a *single* divergent characteristic length scale ξ. In systems with spontaneously broken continuous symmetries, on the other hand, the presence of massless

Goldstone modes leads to additional infrared divergences aside from those induced by the critically growing correlation length. As we have seen in Section 5.4, these are controlled by a zero-temperature fixed point. Goldstone singularities can therefore considerably complicate QCP scaling features, as can conserved hydrodynamic modes, see Chapter 6. Second, it should be emphasized that the quantum to classical mapping merely pertains to the *thermodynamical* properties near a quantum critical point. Dynamic correlation functions in quantum systems are usually characterized by interference terms that may *not* be adequately captured by computations in imaginary time and subsequent Wick rotation to real time. Specifically, the appearance of topological Berry phases in adiabatically slowly evolving quantum systems generates non-local effects that tend to invalidate Landau–Ginzburg type expansions of effective free energies with respect to local order parameter fields.

7.3.2 Quantum Ising model in a transverse magnetic field

Studying a quantum version of the Ising model helps to illustrate some of the general concepts introduced in the previous section. Let us consider the Hamiltonian

$$H[\{\sigma_i\}] = -\frac{J}{2} \sum_{i,j=1}^{N} \sigma_i^z \sigma_j^z - h_x \sum_{i=1}^{N} \sigma_i^x - h_z \sum_{i=1}^{N} \sigma_i^z , \qquad (7.160)$$

where $J > 0$ denotes a ferromagnetic exchange coupling, h_x and h_z are proportional to the x and z components of an external magnetic field, and σ_i^x and σ_i^z represent spin $S = \frac{1}{2}$ operators (Pauli matrices) at each site i of a regular d-dimensional lattice. If the transverse field vanishes, $h_x = 0$, we may employ an eigenstate basis of the operators σ_i^z, and the Hamiltonian in this basis then assumes the classical Ising model form (1.10). From our considerations in Chapter 1 we know that this system displays spontaneous ferromagnetic order for $h_z = 0$ at low (but finite) temperatures only in dimensions $d > d_{lc} = 1$. In contrast, the one-dimensional Ising chain remains in a paramagnetic state except precisely at zero temperature, see Problem 1.2. For $T = 0$ and $h_z = 0$, switching on the transverse field in the Hamiltonian (7.160) introduces quantum fluctuations that will destroy ferromagnetic order once $|h_x|$ exceeds a critical threshold h_c. At $T = 0$, $h_z = 0$, and $|h_x| = h_c$ we therefore encounter a genuine quantum critical point for the transverse Ising model (7.160) in one dimension.

   In the following, let us consider the case $h_z = 0$; we can then map the canonical partition function for the *quantum* Hamiltonian (7.160) to an equivalent *classical* model in $d + 1$ dimensions. To this end, we basically follow the procedure in

Section 7.1.2 and first discretize imaginary time in the usual manner, $t_l = l\tau = l\beta\hbar/M$ ($l = 0, \ldots, M$). This allows us to rewrite the partition function as

$$Z(\beta) = \mathrm{Tr}\, e^{-\beta H} = \lim_{\tau \to 0} \prod_{l=0}^{M-1} e^{J \sum_{i \neq j} \sigma_i^z(t_l)\sigma_j^z(t_l)\tau/2\hbar} \, e^{h_x \sum_i \sigma_i^x(t_l)\tau/\hbar} , \qquad (7.161)$$

since this factorization is correct up to factors of order τ^2 that stem from the non-vanishing commutators of the exponentials on the right-hand side. Next, we insert the completeness relation $1 = |\chi_+^x\rangle\langle\chi_+^x| + |\chi_-^x\rangle\langle\chi_-^x|$ for the two orthonormal eigenstates $|\chi_\pm^x\rangle$ of σ^x at each time step t_l between the exponentials, but evaluate the trace with the eigenstates $|\chi_\pm^z\rangle$ of σ^z, where $\sigma^{x/z}|\chi_\pm^{x/z}\rangle = \pm|\chi_\pm^{x/z}\rangle$. At t_l, the exchange Hamiltonian then gives the contribution $\exp(K \sum_{i \neq j} \sigma_{i,l}\sigma_{j,l}/2)$ with classical spin variables (namely the z-component eigenvalues) $\sigma_{i,l} = \pm 1$ and dimensionless coupling $K = J\tau/\hbar$. From elementary quantum mechanics, we recall $|\chi_\pm^x\rangle = (|\chi_+^z\rangle \pm |\chi_-^z\rangle)/\sqrt{2}$, and thus find for the transverse field contributions for the terms multiplying $|\chi_\pm^z(t_l)\rangle$: $[\cosh(h_x\tau/\hbar)\langle\chi_\pm^z(t_l)| + \sinh(h_x\tau/\hbar)\langle\chi_\mp^z(t_l)|]|\chi_\pm^z(t_{l-1})\rangle$, which can naturally be cast in the form $\exp[A + K'\sigma_i(t_l)\sigma_i(t_{l-1})]$. Direct comparison of both expressions for the two distinct cases $\sigma_{i,l}\sigma_{i,l-1} = \pm 1$ yields

$$A = \frac{1}{2}\ln\left(\frac{1}{2}\sinh\frac{2h_x\tau}{\hbar}\right), \qquad K' = \frac{1}{2}\ln\coth\frac{h_x\tau}{\hbar} . \qquad (7.162)$$

Upon collecting all contributions, and omitting the constant terms $\sim e^A$, we finally arrive at the effective classical anisotropic Ising model Hamiltonian on a $(d+1)$-dimensional discrete space,

$$\beta H_{\mathrm{eff}}[\{\sigma_{i,l}\}] = -\frac{K}{2}\sum_{i \neq j, l}\sigma_{i,l}\,\sigma_{j,l} - K'\sum_{i,l}\sigma_{i,l}\,\sigma_{i,l-1} . \qquad (7.163)$$

Restricting the exchange interactions in the quantum Hamiltonian (7.160) to nearest neighbors, the spatial and temporal couplings in the equivalent classical Hamiltonian (7.163) take identical forms, and are only distinguished by the different coupling strengths $K \neq K'$. In a coarse-grained continuum description, we may thus readily generalize Eq. (1.49) and write down the anisotropic Landau–Ginzburg–Wilson Hamiltonian in $d + 1$ dimensions

$$\mathcal{H}_{\mathrm{eff}}[S] = \int d^d x \int_0^{\beta\hbar} d\tau \left[\frac{r}{2}S(x,\tau)^2 + \frac{1}{2}[\nabla S(x,\tau)]^2 + \frac{c}{2}\left(\frac{\partial S(x,\tau)}{\partial\tau}\right)^2\right.$$

$$\left. + \frac{u}{4!}S(x,\tau)^4\right] . \qquad (7.164)$$

At non-zero temperature, the imaginary time slab is finite, and upon approaching a classical critical point with $T_c > 0$, the anisotropy along with diverging spatial correlation length $\xi \sim r^{-\nu}$ induces this slab's rescaled width to effectively shrink to zero; the critical theory becomes local in imaginary time and consequently the quantum Hamiltonian (7.164) reduces to the standard classical Landau–Ginzburg–Wilson Hamiltonian (1.49) in d dimensions.

At the zero-temperature quantum critical point of the transverse Ising chain (7.160), however, both spatial and temporal domains are infinite, and the quantum Ising Hamiltonian maps onto a classical Ising model in two dimensions. The anisotropy coefficient c can then be absorbed into a rescaled imaginary time, and consequently the critical exponents for the quantum critical point of the Ising model in a transverse magnetic field are those of the classical Ising universality class for $d = 2$, namely $\nu = 1$ and $\eta = 1/4$, as first computed by Onsager.[9] Since both time and space scale in the same manner in Eq. (7.164), the associated dynamic critical exponent is $z = 1$. We finally remark that, as in the classical case, the elementary excitations in the quantum paramagnetic phase are spin flips, and domain walls in the ferromagnetic phase. An experimental realization for a quasi-one-dimensional Ising spin chain with transverse magnetic field has been established in $CoNb_2O_6$, and both the existence of a continuous quantum phase transition and the nature of the quasi-particle excitations have been unambiguously demonstrated via neutron scattering in this system.[10]

7.3.3 Boson localization

Superfluidity in Bose systems may be suppressed even at zero temperature by introducing pinning centers. As the bosons become localized near attractive defects, long-distance ring exchange processes become less abundant, and consequently macroscopic phase coherence is reduced. Sufficiently strong disorder can then induce a super- to normalfluid continuous quantum phase transition at $T = 0$, with the disorder strength serving as the relevant control parameter, akin to Anderson localization for fermionic charge carriers in disordered (semi-)conductors. In the following, we briefly sketch the scaling theory for this localization transition.[11]

A continuum model for bosons subject to disorder is represented by the action (7.99), where $U(x) < 0$ describes spatially varying attractive pinning forces that locally modify the chemical potential μ. We first consider the superfluid density ρ_s, the transport coefficient related to leading order to longitudinal current correlations, and hence to superfluid phase fluctuations, see Eq. (7.148). In fact, we may view ρ_s

---

[9] Onsager (1944).   [10] Coldea *et al.* (2010).
[11] Fisher and Fisher (1988); Fisher, Weichman, Grinstein, and Fisher (1989); for an introductory overview, see Young (1994).

as the *rigidity* in the superfluid phase, wherein gauge invariance is spontaneously broken, with respect to order parameter phase fluctuations. By dimensional analysis and collecting appropriate prefactors, see also Eq. (7.123), we thus have

$$\delta f_{\nabla\Theta} = \frac{1}{2} \rho_s v_s^2 = \frac{\hbar^2}{2m^2} \rho_s (\nabla\Theta)^2 \qquad (7.165)$$

for the free energy density associated with spatial phase variations. Upon invoking the scaling hypothesis (7.159) for the singular part of the free energy density, one predicts that the superfluid density should vanish as

$$\rho_s \sim |g - g_c|^{(d+z-2)\nu} \qquad (7.166)$$

at the $T = 0$ quantum critical point associated with boson localization, generalizing the classical *Josephson scaling relation* $\rho_s \sim |T - T_c|^{(d-2)\nu}$ for the normal- to superfluid phase transition at finite critical temperature T_c.

In a similar vein, one can argue that the compressibility $\kappa = n^{-2} \partial^2 f / \partial \mu^2$ represents the system's rigidity with respect to temporal phase variations, since these too couple *linearly* to the density fluctuations, see Eq. (7.123). In analogy with Eq. (7.165), this suggests

$$\delta f_{\partial_\tau\Theta} = \frac{\hbar^2}{2} \kappa \left(\frac{\partial\Theta}{\partial\tau} \right)^2 , \qquad (7.167)$$

whence

$$\kappa \sim |g - g_c|^{(d-z)\nu} \qquad (7.168)$$

in the vicinity of the phase transition. Yet for this disorder-driven Anderson localization, one would expect the compressibility to remain finite at the critical point; this is in contrast with an interaction-induced Mott localization transition that would be very sensitive to changes in the particle density. Consequently, we arrive at the non-trivial prediction

$$z = d \qquad (7.169)$$

for the boson localization dynamic critical exponent. However, recent numerical data appear to contradict this simple direct consequence of scaling theory and the assumption of a finite critical compressibility.[12]

7.4 Quantum antiferromagnets

Quantum antiferromagnets represent paradigmatic model systems for quantum critical behavior and the crossover to classical critical scaling at any finite

---

[12] Meier and Wallin (2012).

temperature $T > 0$. This section first introduces the quantum non-linear sigma model as the appropriate continuum representation for the critical dynamics of an $O(n)$-symmetric quantum antiferromagnet. In the classical limit, the fluctuation expansion in its ordered phase corresponds to a low-temperature expansion. The critical exponents may be computed perturbatively in a dimensional expansion about the *lower* critical dimension d_{lc}. The ensuing RG flow equations capture the *dimensional crossover* from the quantum critical point with $d_{lc} = 1$ to the asymptotic classical critical scaling with $d_{lc} = 2$.

7.4.1 The quantum non-linear sigma model

We now proceed to study the following imaginary-time effective action for an n-component bosonic vector field $\hat{\Omega}(x, \tau) = (\vec{\pi}(x, \tau), \sigma(x, \tau))$ whose magnitude is constrained to $|\hat{\Omega}(x, \tau)|^2 = \vec{\pi}(x, \tau)^2 + \sigma(x, \tau)^2 = 1$,[13]

$$S_{\text{eff}}[\hat{\Omega}] = \int_0^{\beta\hbar} d\tau \int d^d x \left[\frac{\chi_\perp}{2} \left| \frac{\partial \hat{\Omega}(x, \tau)}{\partial \tau} \right|^2 + \frac{\rho_s}{2} |\nabla \hat{\Omega}(x, \tau)|^2 - h^n(x, \tau) \sigma(x, \tau) \right],$$

(7.170)

with appropriate periodic boundary conditions in the imaginary time direction $\hat{\Omega}(x, 0) = \hat{\Omega}(x, \beta\hbar)$, and associated partition function

$$Z[h^n] = \int \mathcal{D}[\hat{\Omega}] \, \delta(|\hat{\Omega}(x, \tau)|^2 - 1) \, e^{-S_{\text{eff}}[\hat{\Omega}]/\hbar} .$$

(7.171)

We shall next argue that this action captures the universal large-scale static and long-time dynamic properties of a quantum Heisenberg *antiferromagnet*.

To this end, consider first the classical limit $\beta\hbar \to 0$. The imaginary-time integral and the first term in the action (7.170) then disappear, and the exponent in the partition function assumes a classical Hamiltonian form,

$$\frac{S_{\text{eff,cl}}[\hat{\Omega}]}{\hbar} = \beta \int d^d x \left[\frac{\rho_s}{2} |\nabla \hat{\Omega}(x)|^2 - h^n(x) \sigma(x) \right] = \frac{\mathcal{H}_{\text{eff}}[\hat{\Omega}]}{k_B T} ,$$

(7.172)

with associated partition funtion

$$Z_{\text{cl}}[h^n] = \int \mathcal{D}[\hat{\Omega}] \, \delta(|\hat{\Omega}(x)|^2 - 1) \, e^{-\mathcal{H}_{\text{eff}}[\hat{\Omega}]/k_B T} .$$

(7.173)

Equations (7.172) and (7.173) define the (classical) *non-linear sigma model*; it is easily recognized as describing the continuum limit of the lattice Heisenberg

---

[13] Fisher (1989); Chakravarty, Halperin, and Nelson (1989).

model (3.62) with nearest-neighbor exchange coupling J, and $\rho_s = dJ|\vec{S}|^2/2a_0^{d-2}$ for a hypercubic lattice in d dimensions with lattice constant a_0. For ferromagnetic coupling $J > 0$, one directly identifies $\hat{\Omega}(x) = \vec{S}(x)/|\vec{S}(x)|$ as the magnetization vector normalized to fixed unit length. Yet the identical continuum effective Hamiltonian also describes a Heisenberg antiferromagnet with $J < 0$, where $\hat{\Omega}(x) = \vec{N}(x)/|\vec{S}(x)|$ now represents the *staggered magnetization* unit vector, i.e., the normalized local spin vector multiplied by alternating signs on neighboring lattice sites, see Problem 3.5.

The mean-field approximation of course yields $\langle \vec{\pi} \rangle = 0$ and $\langle \sigma \rangle = 1$, even as the symmetry-breaking field approaches $h^n \downarrow 0$, characteristic of the ordered phase of this $O(n)$-symmetric system at low temperatures, and in dimensions $d > d_{lc} = 2$. Notice, however, that the transverse order parameter susceptibility at finite wavevector q is $\chi(q) = C(q)/k_B T = 1/\rho_s q^2$, signaling that transverse orientation fluctuations constitute the $n - 1$ massless Goldstone modes associated with the spontaneous rotational symmetry breaking; see Eq. (5.98) at zero frequency, Problem 1.5, and also Section 7.2.3. The non-linear sigma model (7.172) provides an alternative description of the second-order phase transition in $O(n)$-symmetric systems by means of an expansion about the low-temperature ordered phase near the *lower* critical dimension $d_{lc} = 2$. This approach is complementary to the $O(n)$-symmetric Landau–Ginzburg–Wilson Hamiltonian (3.69) which may be viewed as the corresponding effective model with a 'softened' constraint on the magnitude of the vector order parameter, and has to be analyzed in the vicinity of the *upper* critical dimension $d_c = 4$.

In the zero-temperature limit $\beta\hbar \to \infty$, the imaginary-time slab becomes infinite, and the quantum non-linear sigma model (7.170) again assumes the form (7.172), but with quantum fluctuations $\propto \hbar$ replacing thermal fluctuations $\propto 1/\beta = k_B T$, and in an effective dimension $d + 1$, since the anisotropy along the imaginary time axis is readily scaled away via $\tau \to \tau' = c_s \tau$, where $c_s = \sqrt{\rho_s/\chi_\perp}$. One thus infers immediately that the lower critical dimension for the existence of spontaneous long-range order in the quantum non-linear sigma model at $T = 0$ is shifted downward to $d'_{lc} = 1$. Moreover, the connection between imaginary time and space directions implies a linear dispersion relation for low-energy excitations in real time, and is consequently suggestive of the zero-temperature dynamics of a quantum *antiferromagnet* with spin wave dispersion $\omega(q) = c_s q$, i.e., $z = 1$ (rather than a ferromagnet with $\omega(q) \sim q^2$, see Sections 3.3, 6.2, and 6.3).

Indeed, stationarity for the action (7.170) in real time yields the equation of motion ($\hat{e}^n$ denotes the unit vector in the n direction)

$$\chi_\perp \frac{\partial^2 \hat{\Omega}(x, t)}{\partial t^2} = \rho_s \nabla^2 \hat{\Omega}(x, t) + h^n(x, t)\hat{e}^n . \qquad (7.174)$$

Now define the generalized 'angular momentum' associated with the order parameter field $\hat{\Omega}(x, t)$, $\vec{M}(x, t) = \chi_\perp \, \hat{\Omega}(x, t) \times \partial \hat{\Omega}(x, t)/\partial t$ or equivalently

$$\frac{\partial \hat{\Omega}(x, t)}{\partial t} = -\chi_\perp^{-1} \, \hat{\Omega}(x, t) \times \vec{M}(x, t) \, ; \qquad (7.175)$$

thus we may decouple the second-order differential equation (7.174) into two first-order differential equations in time, namely Eq. (7.175) and

$$\frac{\partial \vec{M}(x, t)}{\partial t} = \chi_\perp \, \hat{\Omega}(x, t) \times \frac{\partial^2 \hat{\Omega}(x, t)}{\partial t^2} = \hat{\Omega}(x, t) \times \left[\rho_s \nabla^2 \hat{\Omega}(x, t) + h^n(x, t) \, \hat{e}^n \right].$$

$$(7.176)$$

These are just the dissipation-free reversible contributions to the model G dynamics for an isotropic antiferromagnet at $T = 0$, see Section 6.3 and Problem 3.5, with $\vec{N} \propto \sqrt{\rho_s \chi_\perp} \, \hat{\Omega}$ and the mode-coupling strength $g \propto \chi_\perp^{-1}$; recall that the vertex factor in the self-consistent mode-coupling approximation involves the inverse (here transverse) static susceptibility, see Problem 6.5, which explains the notation $\chi_\perp$. For a two-component order parameter, one may equivalently employ a complex field representation, whereupon the second term in the action (7.170) reduces to phase gradients, and the associated stiffness ρ_s can naturally be identified with the superfluid density.

7.4.2 Perturbation expansion and renormalization

In *both* the classical ($\beta\hbar \to 0$) and zero-temperature limits ($\beta\hbar \to \infty$), the renormalization of the quantum non-linear sigma model (7.170) proceeds precisely as for the corresponding classical system (7.172),[14] just respectively in d and $d + 1$ dimensions. At any finite temperature, the extent of the finite slab in imaginary time with the associated Matsubara frequency sums controls the dimensional crossover from effectively $d + 1$ to d fluctuation dimensions. This becomes obvious when we introduce $x_0 = c_s \tau$ and rescaled couplings

$$u = \hbar c_s/\rho_s \, , \quad \tilde{h} = u \, h^n / \hbar c_s \, , \quad g = \beta \hbar \, c_s \qquad (7.177)$$

to rewrite the effective action (7.170) as

$$S_{\text{eff}}[\hat{\Omega}] = \frac{\hbar}{u} \int_0^g dx_0 \int d^d x \left[\frac{1}{2} |\partial \hat{\Omega}(x, x_0)|^2 - \tilde{h}(x, x_0) \sigma(x, x_0) \right], \qquad (7.178)$$

where $\partial = (\partial/\partial x_0, \nabla)$. The integration boundary g naturally controls the dimensional crossover: in the classical limit $g \to 0$ the action (7.178) reduces to the

[14] Amit (1984).

d-dimensional non-linear sigma model Hamiltonian (7.172) or

$$S_{\text{eff, cl}}[\hat{\Omega}] = \frac{\hbar}{t} \int d^d x \left[\frac{1}{2} |\partial \hat{\Omega}(x)|^2 - \tilde{h}(x) \sigma(x) \right],$$

(7.179)

with a new effective coupling that is proportional to the temperature,

$$t = u/g = k_B T / \rho_s .$$

(7.180)

Conversely, in the quantum limit $g \to \infty$ and the action (7.178) assumes the same form as (7.179), namely

$$S_{\text{eff, qu}}[\hat{\Omega}] = \frac{\hbar}{2u} \int d^{d+1} x \left[\frac{1}{2} |\partial \hat{\Omega}(x)|^2 - \tilde{h}(x) \sigma(x) \right],$$

(7.181)

but with a dimensional shift $d \to d + 1$ and coupling $t \to 2u$.

Exploiting the non-linear constraint $\sigma^2 = 1 - \vec{\pi}^2$, we may next integrate out the longitudinal fluctuations from the partition function (7.171):

$$Z[\tilde{h}] = \int \mathcal{D}[\vec{\pi}] \int \mathcal{D}[\sigma] \, \delta\left(\vec{\pi}(x, x_0)^2 + \sigma(x, x_0)^2 - 1 \right) e^{-S_{\text{eff}}[\vec{\pi}, \sigma]/\hbar}$$

$$= \int \mathcal{D}[\vec{\pi}] \, \frac{1}{\sqrt{1 - \vec{\pi}(x, x_0)^2}} \, e^{-S_{\text{eff}}[\vec{\pi}, \sqrt{1 - \vec{\pi}^2}]/\hbar} = \int \mathcal{D}[\vec{\pi}] \, e^{-\tilde{S}_{\text{eff}}[\vec{\pi}]/\hbar} ,$$

(7.182)

where

$$S_{\text{eff}}[\vec{\pi}, \sqrt{1 - \vec{\pi}^2}] = \frac{\hbar}{u} \int_0^g dx_0 \int d^d x \left[\frac{1}{2} |\partial \vec{\pi}(x, x_0)|^2 + \frac{[\vec{\pi}(x, x_0) \cdot \partial \vec{\pi}(x, x_0)]^2}{2[1 - \vec{\pi}(x, x_0)^2]} \right.$$

$$\left. - \tilde{h}(x, x_0) \sqrt{1 - \vec{\pi}(x, x_0)^2} \right],$$

(7.183)

while $\tilde{S}_{\text{eff}}[\vec{\pi}]$ contains additional terms originating in the functional integration measure in (7.182). Upon discretizing in space-time,

$$\prod_{(x, x_0)} \frac{1}{\sqrt{1 - \vec{\pi}(x, x_0)^2}} \to \exp\left(-\frac{1}{2a_0^{d+1}} \int_0^g dx_0 \int d^d x \, \ln\left[1 - \vec{\pi}(x, x_0)^2 \right] \right),$$

(7.184)

as we return to the continuum limit. While the $O(n - 1)$ symmetry in the ordered phase is explicit in the action (7.183), the logarithmic contributions from the Jacobian safeguard the original $O(n)$ rotational invariance. Note that the elementary hypercubic lattice cell volume can be formally recast as a wavevector integral with

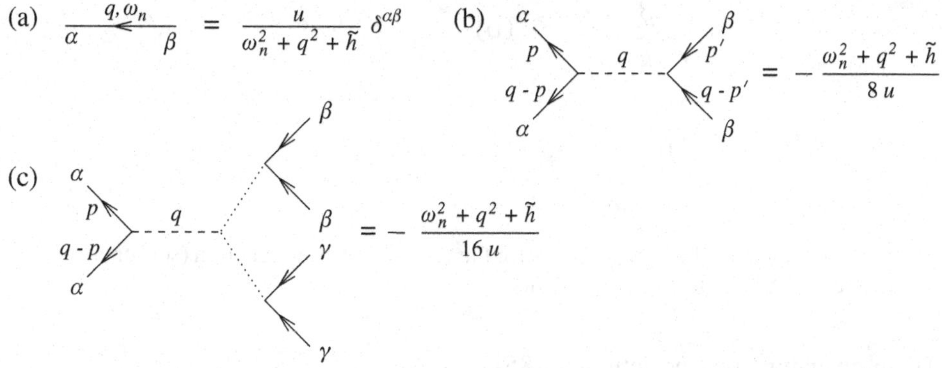

(a) $\dfrac{q,\omega_n}{\alpha \xleftarrow{\hspace{1.5cm}} \beta} = \dfrac{u}{\omega_n^2 + q^2 + \tilde{h}}\,\delta^{\alpha\beta}$ (b)

$$= -\frac{\omega_n^2 + q^2 + \tilde{h}}{8\,u}$$

(c)

$$= -\frac{\omega_n^2 + q^2 + \tilde{h}}{16\,u}$$

Fig. 7.5 Perturbation theory elements for the quantum non-linear sigma model: (a) propagator; (b) four-point vertex; and (c) six-point vertex.

finite ultraviolet cutoff,

$$a_0^{-d} = \frac{\Lambda^d}{(2\pi)^d} = \int_0^\Lambda \frac{\mathrm{d}^d q}{(2\pi)^d}\,.$$

When $\ln[1 - \vec{\pi}^2] = -\vec{\pi}^2 - \frac{1}{2}(\vec{\pi}^2)^2 \cdots$ is expanded in the transverse fluctuations, the Jacobian action (7.184) generates precisely those counterterms that are required to remove the leading severe ultraviolet divergences. Alternatively, we may take recourse to the fact that the wavevector integral of unity vanishes in dimensional regularization, and simply set all those contributions to zero in the perturbational analysis.

With $\sqrt{1 - \vec{\pi}^2} = 1 - \frac{1}{2}\vec{\pi}^2 - \frac{1}{8}(\vec{\pi}^2)^2 - \frac{1}{16}(\vec{\pi}^2)^3 + \cdots$ and $(\vec{\pi} \cdot \partial\vec{\pi}/\sqrt{1 - \vec{\pi}^2})^2$ $= \frac{1}{2}[\partial\vec{\pi}^2 + \frac{1}{8}\partial(\vec{\pi}^2)^2 + \cdots]^2 = \frac{1}{4}(\partial\vec{\pi}^2)^2 + \frac{1}{8}(\partial\vec{\pi}^2)[\partial(\vec{\pi}^2)^2] + \cdots$, the fluctuation expansion for the effective action, quite similar to the one in Section 7.2.2, reads

$$\frac{\tilde{S}_{\mathrm{eff}}[\vec{\pi}]}{\hbar} = \frac{1}{2u} \int_0^g \mathrm{d}x_0 \int \mathrm{d}^d x \left[|\partial\vec{\pi}|^2 + \hbar\,\vec{\pi}^2 + \frac{1}{4}(\partial\vec{\pi}^2)^2 + \frac{\hbar}{4}(\vec{\pi}^2)^2 \right.$$

$$\left. + \frac{1}{8}(\partial\vec{\pi}^2)[\partial(\vec{\pi}^2)^2] + \frac{\hbar}{8}(\vec{\pi}^2)^3 + \cdots \right]. \quad (7.185)$$

As in Eq. (7.125), we define the Fourier-transformed transverse fields

$$\vec{\pi}(x, x_0) = \int \frac{\mathrm{d}^d q}{(2\pi)^d}\,\frac{1}{g}\sum_n \vec{\pi}(q, \omega_n)\,\mathrm{e}^{\mathrm{i}(q\cdot x - \omega_n x_0)}\,, \quad (7.186)$$

with the Matsubara frequencies $\omega_n = 2\pi\,n/g$, n integer. From the Gaussian part of the action (7.185), obviously $\langle\vec{\pi}^2\rangle \propto u$; thus, the expansion (7.185) in successive powers of composite operators $\vec{\pi}^2$ translates to a perturbation series with respect to u. In the classical limit, this becomes a *low-temperature expansion* in powers of t.

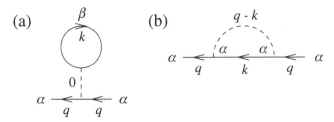

Fig. 7.6 One-loop Feynman graphs for the two-point vertex function (self-energy) of the quantum non-linear sigma model.

The propagator from the action (7.185),

$$G_0(q, \omega_n) = \frac{u}{\omega_n^2 + q^2 + \tilde{h}} \, , \tag{7.187}$$

as well as the four- and six-point vertices are depicted in Fig. 7.5.

To first order in u, only the four-point vertex in Fig. 7.5(b) is required. Figure 7.6 shows the one-loop Feynman graphs for the two-point vertex function or self-energy. The corresponding analytic expressions read (Problem 7.7)

$$u\, G(q, \omega_n)^{-1} = \omega_n^2 + q^2 + \tilde{h} + \frac{n-1}{2} \frac{u\,\tilde{h}}{g} \sum_m \int_k \frac{1}{\omega_m^2 + k^2 + \tilde{h}}$$

$$+ \frac{u}{g} \sum_m \int_k \frac{(\omega_n - \omega_m)^2 + (q - k)^2 + \tilde{h}}{\omega_m^2 + k^2 + \tilde{h}}$$

$$= (\omega_n^2 + q^2)[1 + u\, I_d(g, \tilde{h})] + \tilde{h}\left[1 + \frac{n-1}{2} u\, I_d(g, \tilde{h})\right]. \tag{7.188}$$

The final step uses that terms odd in the internal wavevector k or frequency ω_m vanish, and that contributions proportional to $\int_k 1 = a_0^{-d}$ are either canceled by the logarithmic counterterms from the Jacobian, or set to zero in dimensional regularization.

After performing the Matsubara frequency sum by means of Eq. (7.215) (see Section 7.5), the loop integral becomes

$$I_d(g, \tilde{h}) = \frac{1}{2} \int \frac{d^d k}{(2\pi)^d} \frac{1}{\sqrt{k^2 + \tilde{h}}} \coth\left(\frac{g}{2} \sqrt{k^2 + \tilde{h}}\right). \tag{7.189}$$

In the classical limit,

$$\lim_{g \to 0} g\, I_d(g, \tilde{h}) = \int \frac{d^d k}{(2\pi)^d} \frac{1}{k^2 + \tilde{h}} = \frac{C_d}{2 - d} \tilde{h}^{(d-2)/2}, \quad C_d = \frac{\Gamma(2 - d/2)}{2^{d-1} \pi^{d/2}}$$

$$\tag{7.190}$$

in dimensional regularization, see Eq. (5.10). On the other hand, note that

$$\lim_{g \to \infty} I_d(g, \tilde{h}) = \frac{1}{2} \int \frac{\mathrm{d}^d k}{(2\pi)^d} \frac{1}{\sqrt{k^2 + \tilde{h}}} = \frac{C_{d+1}}{1 - d} \tilde{h}^{(d-1)/2} = \lim_{g \to 0} g\, I_{d+1}(g, \tilde{h}) \,,$$

(7.191)

which once again establishes the dimensional shift $d \to d + 1$ (along with $t = u/g \to u$) that links the classical and zero-temperature quantum non-linear sigma models.

Next we define multiplicative renormalization constants, following our standard conventions as in Eqs. (5.19) and (5.23),

$$\pi_R^\alpha(x, x_0) = Z_\pi^{1/2} \pi^\alpha(x, x_0) \,, \quad h_R^n = Z_h h^n \mu^{-(d+1)} \,, \quad \tilde{h}_R = Z_{\tilde{h}} \tilde{h} \mu^{-2} \,,$$
$$u_R = Z_u u\, C_d\, \mu^{d-1} \,, \quad t_R = Z_t\, t\, C_d\, \mu^{d-2} \,.$$

(7.192)

The cutoff g clearly requires no renormalization, and carries the scaling dimension of length or inverse momentum (since the dynamic critical exponent is $z = 1$); consequently,

$$Z_g = 1 \,, \quad Z_u = Z_t \,.$$

(7.193)

Furthermore, from the scale transformation of the paramagnetic term in the action (7.170), one infers that the anomalous dimension of the external field h^n is just the inverse of that of the order parameter field, i.e.,

$$Z_h = Z_\pi^{-1/2} \,, \quad Z_{\tilde{h}} = Z_u Z_\pi^{-1/2} \,,$$

(7.194)

compare with Eq. (1.91) in Section 1.4. This leaves only two independent renormalization constants that are readily determined to one-loop order from Eq. (7.188), employing $\tilde{h}_R = 1$ ($\tilde{h} = \mu^2$) as normalization point:

$$Z_\pi = 1 + (n - 1)\, u\, I_d(g, \mu^2) + O(u^2) \,,$$

(7.195)

$$Z_u = 1 + (n - 2)\, u\, I_d(g, \mu^2) + O(u^2) \,.$$

(7.196)

7.4.3 RG flow equations and scaling regimes

We may now proceed to investigate the scaling behavior for the quantum non-linear sigma model in its various regimes. To this end, we construct running couplings through our standard procedures; for the renormalized non-linear couplings u_R and

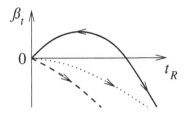

Fig. 7.7 One-loop renormalization group beta function for the non-linear sigma model for $d < d_{\text{lc}}$ (dashed), $d = d_{\text{lc}}$ (dotted), and $d > d_{\text{lc}}$ (full).

t_{R}, one thus obtains the RG beta functions

$$\beta_u = \mu \left.\frac{\partial}{\partial \mu}\right|_0 u_{\text{R}} = u_{\text{R}} \left[d - 1 - (n-2)\, u_{\text{R}} \frac{1-d}{C_d}\, I_d(g, \mu^2) + O\!\left(u_{\text{R}}^2\right) \right], \quad (7.197)$$

$$\beta_t = \mu \left.\frac{\partial}{\partial \mu}\right|_0 t_{\text{R}} = t_{\text{R}} \left[d - 2 - (n-2)\, u_{\text{R}} \frac{2-d}{C_d}\, I_d(g, \mu^2) + O\!\left(u_{\text{R}}^2\right) \right]. \quad (7.198)$$

Notice that Eq. (7.193) implies that $g(\ell) = g(1)\,\ell$. Thus, for any finite initial value $g(1) < \infty$, $g(\ell \to 0) \to 0$, whence asymptotically the classical critical behavior is recovered. Indeed, with (7.190) the RG beta function (7.198) becomes

$$\beta_t(t_{\text{R}}) = t_{\text{R}}\left[d - 2 - (n-2)\, t_{\text{R}} + O\!\left(t_{\text{R}}^2\right) \right], \quad (7.199)$$

which is plotted in Fig. 7.7 for various values of the dimension d.

With respect to its d dependence, the RG beta function for the non-linear sigma model displays the opposite behavior from the one for the Landau–Ginzburg–Wilson Hamiltonian shown in Fig. 5.1: in low dimensions $d < d_{\text{lc}} = 2$ and for $n > 2$, the Gaussian fixed point $t_0^* = 0$ is unstable, indicating that the critical temperature is zero, and the system does not display spontaneous long-range order in accord with the Mermin–Wagner–Hohenberg theorem. For $d > d_{\text{lc}}$, the critical scaling properties are determined by an infrared-unstable non-trivial RG fixed point, namely the critical temperature which continuously emerges from the Gaussian fixed point in the limit $\epsilon = d - d_{\text{lc}} \to 0$,

$$t_{\text{C}}^* = \frac{\epsilon}{n-2} + O(\epsilon^2). \quad (7.200)$$

It is ultraviolet-stable, and thus controls the UV singularities for $d > d_{\text{lc}}$. Note also the formal similarity of the d and n dependence with the coexistence fixed point (5.108) in Section 5.4 that governs the Goldstone singularities in the ordered phase with spontaneously broken rotational symmetry.

Recalling that the inverse two-point correlation function is multiplied by the coupling u, see Eq. (7.188), and with Wilson's flow functions

$$\gamma_\pi = -(n-1)\, t_R + O(t_R^2)\,, \tag{7.201}$$

$$\gamma_u = -(n-2)\, t_R + O(t_R^2)\,, \tag{7.202}$$

we find Fisher's anomalous scaling exponent to first order in ϵ:

$$\eta = \gamma_u^* - \gamma_\pi^* = t_C^* + O(t_C^{*2}) = \frac{\epsilon}{n-2} + O(\epsilon^2)\,. \tag{7.203}$$

Since the RG fixed point (7.200) represents the critical temperature, one may determine the correlation length exponent ν through linearizing the beta function β_t in the vicinity of the critical point, $\beta_t(t_R) \approx \beta_t'^*(t_R - t_C^*)$. This yields the RG flow equation

$$\ell \frac{d\tilde{t}(\ell)}{d\ell} \approx \beta_t'^* [\tilde{t}(\ell) - t_c^*]\,, \tag{7.204}$$

which is solved by $\tilde{t}(\ell) - t_C^* \sim (t_R - t_C^*)\,\ell^{\beta_t'^*}$. The correlation length ξ naturally scales as the inverse flow parameter, $\ell \sim \xi^{-1} \sim |t_R - t_C^*|^\nu$, and integrating Eq. (7.204) until $\tilde{t}(\ell) - t_C^* = O(1)$, we obtain

$$\frac{1}{\nu} = -\beta_t'^* = \epsilon + O(\epsilon^2)\,. \tag{7.205}$$

The non-linear sigma model allows us to compute the two independent classical critical exponents (7.203) and (7.205) in a systematic expansion about the lower critical dimension $d_{lc} = 2$, complementary to the expansion near the upper critical dimension in the Landau–Ginzburg–Wilson field theory.

The correlation length exponent ν formally diverges as $\epsilon \to 0$. Indeed, with Eq. (7.199) and $t_C^* = 0$ we find the RG flow equation for the temperature variable at $d_{lc} = 2$

$$\frac{d\tilde{t}(\ell)}{d\ln \ell} = -(n-2)\, \tilde{t}(\ell)^2\,, \tag{7.206}$$

which becomes exact as $\tilde{t}(\ell) \to 0$. Its solution reads

$$\tilde{t}(\ell) = \frac{t_R}{1 + (n-2)\, t_R \ln \ell}\,, \tag{7.207}$$

whence asympotically

$$\xi(t) \sim e^{1/(n-2)t_R} \sim e^{2\pi/(n-2)t}\,; \tag{7.208}$$

the correlation length diverges exponentially as the temperature $t \to 0$, as is characteristic of the lower critical dimension (cf. Problem 1.2). The case of a two-component order parameter at $d_{lc} = 2$ is clearly special, as (7.206) then suggests an entire line of fixed points, associated with continuously varying critical exponents. Precisely this scenario is played out at the Berezinskii–Kosterlitz–Thouless vortex binding-unbinding transition of the XY model in two dimensions.[15]

Genuine novel, non-classical critical behavior for the quantum non-linear sigma model can emerge only at precisely zero temperature, i.e., $g \to \infty$. With (7.191), the RG beta function (7.197) becomes in this limit

$$\beta_u(u_R) = u_R \left[d - 1 - (n-2) \frac{C_{d+1}}{C_d} u_R + O(u_R^2) \right], \qquad (7.209)$$

which is of the very same structure as its classical counterpart (7.199), yet with reduced lower critical dimension $d'_{lc} = 1$. It yields a new non-trivial quantum critical fixed point

$$u_C^* = \frac{C_d}{C_{d+1}} \frac{\epsilon'}{n-2} + O(\epsilon'^2), \qquad (7.210)$$

which exists and is ultraviolet-stable for $\epsilon' = d - d'_{lc} > 0$. In $d > 1$ dimensions, the critical exponents consequently assume the form (7.203) and (7.205), with $\epsilon \to \epsilon'$. At the lower critical dimension $d'_{lc} = 1$, Eq. (7.208) becomes replaced by

$$\xi(u) \sim e^{1/4\pi(n-2)u_R} \sim e^{1/2(n-2)u}, \qquad (7.211)$$

provided again $n > 2$. For a two-component order parameter field, the quantum version of Berezinskii–Kosterlitz–Thouless physics ensues in $1 + 1$ space-time dimensions.

The dimensional crossover between the genuine quantum critical and classical regimes is governed by the function $I_d(g, \mu^2)$ defined by the wavevector integral (7.189).[16] For initially large values of g, the RG flow first remains close to the quantum critical point before eventually reaching the classical fixed point (7.200) and its associated scaling behavior. In this *renormalized classical regime*, integrating the RG flow along the crossover trajectory results in scaling amplitudes that differ from both the corresponding mean-field and the universal classical critical values. Detailed analysis leads to very satisfactory agreement with experimental data for the two-dimensional Heisenberg antiferromagnet La_2CuO_4.[17]

---

[15] Excellent discussions of the Berezinskii–Kosterlitz–Thouless transition can be found in the textbooks by Chaikin and Lubensky (1995); Nelson (2002); and Van Vliet (2010).

[16] In the framework of Wilson's momentum shell renormalization group approach, a different crossover function ensues, namely essentially the integrand in Eq. (7.189). The one-loop fluctuation correction in the RG flow equation for $\tilde{u}(l)$, where $l = -\ln \ell$, then becomes

$$(n-2) \frac{\tilde{u}(l)^2}{2} \frac{S_d \Lambda^d}{\sqrt{\Lambda^2 + \tilde{h}(l)}} \coth \left(\frac{g(l)}{2} \sqrt{\Lambda^2 + \tilde{h}(l)} \right),$$

which approaches $(n-2) t(l) S_d \Lambda^d$ as $g(l) \to 0$, and $(n-2) u(l) S_d \Lambda^{d-1}/2$ as $g(l) \to \infty$.

[17] Chakravarty, Halperin, and Nelson (1989); for an up-to-date discussion, see Sachdev (2011).

7.5 *Appendix:* **Matsubara frequency sums**

In order to perform the basic Matsubara frequency sums for bosons and fermions, consider

$$g_\pm(z) = \frac{1}{e^{\mp\beta\hbar z} - 1}\frac{e^{-\tau z}}{z - \omega}, \quad g'_\pm(z) = \frac{1}{e^{\mp\beta\hbar z} + 1}\frac{e^{-\tau z}}{z - \omega}. \tag{7.212}$$

The meromorphic functions $g_\pm(z)$ and $g'_\pm(z)$ have poles at $z = i\omega_n$, with respectively the bosonic and fermionic Matsubara frequencies $\omega_n = 2n\pi/\beta\hbar$ and $\omega_n = (2n+1)\pi/\beta\hbar$, with integer $n = 0, \pm1, \pm2, \dots$. Let us choose as integration contour a circle in the complex z plane with radius R. Recalling that $|\tau| < \beta\hbar$, it is then easy to establish that the contributions to the integral from this contour vanish as $R \to \infty$, provided we select $g_+^{()}(z)$ for $\tau > 0$ and $g_-^{()}(z)$ for $\tau < 0$. The residue theorem consequently gives $\sum_n \text{Res}\left[g_\pm^{()}(z), i\omega_n\right] = -\text{Res}\left[g_\pm^{()}(z), \omega\right]$, and thus yields

$$\frac{1}{\beta\hbar}\sum_n \frac{e^{-i\omega_n\tau}}{-i\omega_n + \omega} = \frac{e^{-\omega\tau}}{1 \mp e^{-\beta\hbar\omega}}\Theta(\tau) \pm \frac{e^{-\omega\tau}}{e^{\beta\hbar\omega} \mp 1}\Theta(-\tau) \tag{7.213}$$

for the bosonic and fermionic cases. By direct summation, one then obtains

$$\frac{1}{\beta\hbar}\sum_n \frac{e^{-i\omega_n\tau}}{\omega_n^2 + \omega^2} = \frac{1}{2\omega}\frac{e^{\beta\hbar\omega}e^{-\omega|\tau|} \pm e^{\omega|\tau|}}{e^{\beta\hbar\omega} \mp 1}, \tag{7.214}$$

whence at $\tau = 0$

$$\frac{1}{\beta\hbar}\sum_n \frac{1}{\omega_n^2 + \omega^2} = \frac{1}{2\omega}\frac{e^{\beta\hbar\omega} \pm 1}{e^{\beta\hbar\omega} \mp 1}. \tag{7.215}$$

Further sum formulas follow through taking derivatives with respect to ω, for example

$$\frac{1}{\beta\hbar}\sum_n \frac{\omega_n^2 - \omega^2}{(\omega_n^2 + \omega^2)^2} = -\frac{\beta\hbar}{4\sinh^2(\beta\hbar\omega/2)} \tag{7.216}$$

for bosons (see Problem 7.8), and similarly for fermions, with the hyperbolic sine replaced by the hyperbolic cosine function.

Problems

7.1 *Particle number fluctuations in coherent and Fock states*

    (a) Determine the mean-square particle number fluctuation $(\Delta N)^2$ for non-interacting bosons or fermions in terms of the Bose–Einstein/Fermi–Dirac distributions, and compare with a classical ideal gas.

(b) Find the probability $P(n_\alpha)$ of finding the single-particle states $\{\alpha\}$ occupied with n_α particles in a bosonic coherent state $|\phi\rangle$, and compute the mean particle number $\langle N\rangle_\phi$ as well as $(\Delta N)^2$ in this coherent state.

7.2 *Overlap and closure relation for bosonic coherent states*

(a) Confirm the inner product/overlap (7.14) between coherent states.

(b) Employ polar coordinates in the complex plane $\phi_\alpha = \rho_\alpha e^{i\theta_\alpha}$ to verify the (over-)completeness relation (7.17) via direct integration.

7.3 *Propagator for an ideal Bose gas*

Use the discretized action and path integral representation (7.24) to compute the single-particle Green function (propagator) $G_{0\alpha}(\tau)$ for an ideal Bose gas, confirming Eq. (7.40).

7.4 *Closure relation and operator trace for fermionic coherent states*

(a) By evaluating its matrix elements with arbitrary N-particle fermion states $|\alpha_1 \ldots \alpha_N\rangle$, demonstrate that the left-hand side of Eq. (7.64) indeed represents the identity operator.

(b) Thus confirm the fermionic operator trace formula (7.65).

7.5 *Gaussian ensemble with Grassmann variables*

For a Gaussian 'action' of Grassmann variables, $S_0[\xi^*, \xi] = \sum_\alpha a_\alpha \xi_\alpha^* \xi_\alpha$, establish that in accord with Wick's theorem for fermions,

(a) $\langle \xi_\alpha \xi_\beta^* \rangle_0 = a_\alpha^{-1} \delta_{\alpha\beta}$, and

(b) $\langle \xi_\alpha \xi_\beta \xi_\gamma^* \xi_\delta^* \rangle_0 = \langle \xi_\alpha \xi_\delta^* \rangle_0 \langle \xi_\beta \xi_\gamma^* \rangle_0 - \langle \xi_\alpha \xi_\gamma^* \rangle_0 \langle \xi_\beta \xi_\delta^* \rangle_0$, where

$$\langle \xi_{\alpha_1} \cdots \xi_{\alpha_N}^* \rangle_0 = \frac{\int \prod_\alpha d\xi_\alpha^* \, d\xi_\alpha \, \xi_{\alpha_1} \cdots \xi_{\alpha_N}^* \, e^{-S_0[\xi^*,\xi]}}{\int \prod_\alpha d\xi_\alpha^* \, d\xi_\alpha \, e^{-S_0[\xi^*,\xi]}}.$$

7.6 *Coherent-state path integral action in the superfluid phase*

Confirm that with the parametrization (7.121), the action (7.99) becomes $S[\pi, \Theta] = -n_0^2 V_0 \beta \hbar V/2 + S_0[\pi, \Theta] + S_{int}[\pi, \Theta]$, with the harmonic and non-linear fluctuation contributions given respectively by Eqs. (7.123) and (7.124).

7.7 *One-loop self-energy for the quantum non-linear sigma model*

Compute the one-loop fluctuation corrections for the two-point vertex function, confirming the results given in Eq. (7.188).

7.8 *Matsubara frequency sums*

Confirm the Matsubara frequency sum formulas (7.213)–(7.216) listed in the Appendix (Section 7.5).

References

Amit, D. J., 1984, *Field Theory, the Renormalization Group, and Critical Phenomena*, Singapore: World Scientific, chapter II 6.

Ceperley, D. M., 1995, Path integrals in the theory of condensed helium, *Rev. Mod. Phys.* **67**, 279–355.

Chaikin, P. M. and T. C. Lubensky, 1995, *Principles of Condensed Matter Physics*, Cambridge: Cambridge University Press, chapter 9.

Chakravarty, S., B. I. Halperin, and D. R. Nelson, 1989, Two-dimensional quantum Heisenberg antiferromagnet at low temperatures, *Phys. Rev. B* **39**, 2344–2371.

Coldea, R., D. A. Tennant, E. M. Wheeler, *et al.*, 2010, Quantum criticality in an Ising chain: experimental evidence for emergent E_8 symmetry, *Science* **327**, 177–180.

Continentino, M. A., 1994, Quantum scaling in many-body systems, *Phys. Rep.* **239**, 179–213.

Fetter, A. L. and J. D. Walecka, 1971, *Quantum Theory of Many-particle Systems*, New York: McGraw-Hill, chapter 14.

Fisher, D. S. and M. P. A. Fisher, 1988, Onset of superfluidity in random media, *Phys. Rev. Lett.* **61**, 1847–1850.

Fisher, D. S., 1989, Universality, low-temperature properties, and finite-size scaling in quantum antiferromagnets, *Phys. Rev. B* **39**, 11 783–11 792.

Fisher, M. P. A., P. B. Weichman, G. Grinstein, and D. S. Fisher, 1989, Boson localization and the superfluid–insulator transition, *Phys. Rev. B* **40**, 546–570.

Hertz, J. A., 1976, Quantum critical phenomena, *Phys. Rev. B* **14**, 1165–1184.

Kamenev, A., 2011, *Field Theory of Non-equilibrium Systems*, Cambridge: Cambridge University Press, chapters 1, 5, 7–9.

Meier, H. and M. Wallin, 2012, Quantum critical dynamics simulation of dirty boson systems, *Phys. Rev. Lett.* **108**, 055701-1–4.

Negele, J. W. and J. Orland, 1988, *Quantum Many-particle Systems*, Redwood City: Addison–Wesley, chapter 1.

Nelson, D. R., 2002, *Defects and Geometry in Condensed Matter Physics*, Cambridge: Cambridge University Press, chapter 2.

Onsager, L., 1944, Crystal statistics I. a two-dimensional model with an order-disorder transition, *Phys. Rev.* **65**, 117–149.

Popov, V. N., 1987, *Functional Integrals and Collective Excitations*, Cambridge: Cambridge University Press.

Sachdev, S., 2011, *Quantum Phase Transitions*, Cambridge: Cambridge University Press, 2nd edn.

Shankar, R., 1994, Renormalization-group approach to interacting fermions, *Rev. Mod. Phys.* **66**, 129–192.

Schwabl, F., 2008, *Advanced Quantum Mechanics*, Berlin: Springer, 4th edn., chapter 1.

Sondhi, L., S. M. Girvin, J. P. Carini, and D. Shahar, 1997, Continuous quantum phase transitions, *Rev. Mod. Phys.* **69**, 315–333.

Täuber, U. C. and D. R. Nelson, 1997, Superfluid bosons and flux liquids: disorder, thermal fluctuations, and finite-size effects, *Phys. Rep.* **289**, 157–233.

Van Vliet, C. M., 2010, *Equilibrium and Non-equilibrium Statistical Mechanics*, New Jersey: World Scientific, 2nd edn., chapter 9.

Vojta, M., 2003, Quantum phase transitions, *Rep. Prog. Phys.* **66**, 2069–2110.

Vojta, T., 2008, Computing quantum phase transitions, *Reviews in Computational Chemistry* **26**, 167–221.

Young, A. P., 1994, Bosons in a random potential, in: *Fundamental Problems in Statistical Mechanics*, Vol. VIII, eds. H. van Beijeren and M. H. Ernst, 27–47, Amsterdam: Elsevier Science.

Further reading

Belitz, D., T. R. Kirkpatrick, and T. Vojta, 2005, How generic scale invariance influences classical and quantum phase transitions, *Rev. Mod. Phys.* **77**, 579–632.

Fradkin, E., 2013, *Field Theories of Condensed Matter Physics*, Cambridge: Cambridge University Press, 2nd edn.

Huse, D.A. and L.R. Radzihovsky, 1994, Statistical mechanics of vortices in type-II superconductors, in: *Fundamental Problems in Statistical Mechanics*, Vol. VIII, eds. H. van Beijeren and M. H. Ernst, 49–82, Amsterdam: Elsevier Science.

Kirkpatrick, T. R., D. Belitz, and J. V. Sengers, 2002, Long-time tails, weak localization, and classical and quantum critical behavior, *J. Stat. Phys.* **109**, 373–405.

Lawrie, I. D., 1993, Critical phenomena in field theories at finite temperature, *J. Phys. A: Math. Gen.* **26**, 6825–6846.

Mercaldo, M. T., L. De Cesare, I. Rabuffo, and A. Caramico D'Auria, 2007, Unified static renormalization-group treatment of finite-temperature crossovers close to a quantum critical point, *Phys. Rev. B* **75**, 014105-1–14.

Tsvelik, A. M., 1995, *Quantum Field Theory in Condensed Matter Physics*, Cambridge: Cambridge University Press.

Tyč, S., B. I. Halperin, and S. Chakravarty, 1989, Dynamic properties of a two-dimensional Heisenberg antiferromagnet at low temperatures, *Phys. Rev. Lett.* **62**, 835–838.

Vojta, T., 2000, Quantum phase transitions in electronic systems, *Ann. Phys. (Leipzig)* **9**, 403–440.

Part II

Scale invariance in non-equilibrium systems

Introduction

In the second part of this book, we consider the dynamics of systems far away from thermal equilibrium. This departure from equilibrium may be caused by an *external driving* force, as is the case for *driven diffusive systems* or *growing interfaces* considered in Chapter 11; there also exist fundamentally *open* athermal systems which never reach equilibrium, as is true for some of the *reaction-diffusion systems* considered in Chapter 9. In both instances, the constraints imposed by detailed balance and the ensuing fluctuation-dissipation theorem on the form of phenomenological equations describing the temporal evolution of such systems are absent. Even at the fundamental level of quantum mechanics, the dynamical description of such *open dissipative systems* is still quite poorly understood. Generally, it may thus seem a hopeless task to derive coarse-grained equations of motion for only a few mesoscopic degrees of freedom from such an unsatisfactory foundation.

Fortunately, this conclusion is too pessimistic, at least if we are interested in systems whose *non-equilibrium steady state* is tuned close to a critical point, or displays generic *scale invariance*. For, in these situations, we may appeal to the concept of *universality* to allow us to constrain through basic symmetry and conservation arguments the terms which must be retained in an *effective* dynamical description. Similarly, we may hope that the universal properties of drastically simplified models which happen to be exactly solvable may extend to more realistic and technologically relevant systems.

Indeed, statistical mechanics far from equilibrium has become a vigorous research area over the past two decades, specifically owing to dramatic progress on exact solutions of many interesting problems in one dimension, and to the enormous growth in computing power that now allows extensive numerical simulation studies. It is certainly not the intent of this text to provide a comprehensive survey of the fast expanding field of non-equilibrium physics, with its obvious and promising ramifications on biological systems, ranging from biochemistry and microbiology to the macroscopic realm of population dynamics, ecology, and even sociology.

Rather, the focus here will be on those problems in critical dynamics which may be tackled using straightforward modifications of the basic toolkit developed in Part I of this exposition, with the goal of understanding the origin of scaling and universality by means of the field-theoretic renormalization group machinery. Indeed, several non-trivial paradigmatic model systems fall into this category.

In Chapter 8, we begin the discussion of universal features in critical dynamics away from thermal equilibrium. Here, we consider the relaxation kinetics starting from non-equilibrated random initial conditions, and briefly discuss phase ordering, coarsening, aging, and persistence phenomena. We also study the effects of violating the detailed-balance conditions on the familiar dynamical universality classes, and derive important general non-equilibrium work and fluctuation theorems for the relaxational models.

Diffusion-limited 'chemical' reactions are the subject of Chapter 9. It explains how the stochastic master equation for interacting particle systems can be represented through a non-Hermitean time evolution operator in Fock space with an intriguing Hamiltonian structure. Subsequently, this stochastic particle dynamics is mapped onto a coherent-state path integral by means of the tools developed in Chapter 7. We utilize this field theory framework to elucidate the effect of depletion zones on the universal algebraic decay towards the empty state for simple annihilation reactions. Generalizations to multi-species systems emphasize the crucial role of conservation laws connected with particle segregation into inert domains separated by sharp reaction fronts. We also address strongly fluctuating spatio-temporal structures in stochastic Lotka–Volterra models for predator–prey competition.

The following Chapter 10 is devoted to prominent universality classes for genuine non-equilibrium continuous phase transitions from active stationary to inactive, absorbing states. It includes detailed RG treatments of directed and dynamic isotropic percolation, several variants thereof, and of branching and annihilating random walks and certain higher-order reaction processes.

In the final Chapter 11, we turn to non-equilibrium steady states in externally driven systems that display generic scale invariance. We start with driven diffusive systems which encompass simple exclusion processes. Next we add nearest-neighbor interactions in order to address the non-equilibrium phase transition in driven Ising lattice gases. We then analyze the scaling properties of the important Kardar–Parisi–Zhang model for driven interfaces or growing surfaces away from thermal equilibrium, also explaining and utilizing its connection with the noisy Burgers equation and the statistical mechanics of directed lines in disordered environments.

8

Non-equilibrium critical dynamics

We now begin our exposition of universal dynamic scaling behavior that emerges under non-equilibrium conditions. We are mainly concerned with the relaxational models A and B here, but also touch on other dynamic universality classes. First, we consider non-equilibrium critical relaxation from a disordered initial state, and compute the associated universal scaling exponents within the dynamic renormalization group framework. Related phenomena in the early-time regime, prior to reaching temporally translationally invariant asymptotics, are 'aging' and a non-equilibrium fluctuation-dissipation ratio, both of which have enjoyed prominence in the literature on 'glassy' kinetics. In this context, we derive the critical initial slip exponent for the order parameter, and briefly discuss interesting persistence properties. Second, we explore the long-time scaling laws in phase ordering kinetics and coarsening for the relaxational models, following a fast temperature 'quench' into the ordered phase, in systems with either non-conserved or conserved order parameters. Aside from a few explicit computations in the spherical model limit (wherein the number of order parameter components $n \to \infty$), we largely employ phenomenological considerations and scaling theory, stressing the importance of topological defects for energy dissipation during coarsening. Next we address the question how violations of Einstein's relation that links relaxation coefficients and noise correlations might affect the asymptotic dynamic critical scaling behavior. Whereas in systems with non-conserved order parameter typically the equilibrium scaling laws are recovered in the vicinity of the critical point, in contrast genuinely novel universality classes may emerge in the case of a conserved parameter, provided it is driven out of equilibrium in a spatially anisotropic fashion. Quite similar anisotropic scaling properties emerge in driven diffusive systems (see Chapter 11). Finally, we invoke the dynamic response functional to derive general fluctuation and work theorems, and discuss important implications.

8.1 Non-equilibrium critical relaxation and 'aging'

In our preceding considerations of critical dynamics, we have tacitly taken the long-time limit, and asserted that the *initial* conditions did not matter. In the early stages of the temporal evolution, however, the kinetics is not yet stationary, and the dynamic correlation functions are not translationally invariant in time. At a continuous phase transition, this non-stationary regime extends to all times, and it turns out that there exists in general one additional critical exponent that describes how the system 'remembers' its initial preparation.[1]

8.1.1 Relaxational models with random initial conditions

Including the effect of random initial conditions on critical dynamics within the field theory framework, and the renormalization group computation of the critical *initial-slip exponent* proceeds in a manner rather analogous to the analysis of 'ordinary' equilibrium statical surface critical phenomena.[2] The 'surface' here consists of the d-dimensional space at the initial time $t = 0$, where we assume the order parameter fields to take the values $S^\alpha(x, 0) = a^\alpha(x)$. Allowing for a statistical variation for these initial conditions, we may describe this situation by means of the Gaussian weight

$$e^{-\mathcal{H}_i[S]} = \exp\left(-\frac{\Delta}{2}\int d^d x \sum_\alpha [S^\alpha(x, 0) - a^\alpha(x)]^2\right), \qquad (8.1)$$

which implies for the fluctuations of the initial configurations

$$\overline{[S^\alpha(x, 0) - a^\alpha(x)][S^\beta(x', 0) - a^\beta(x')]} = \frac{1}{\Delta}\delta(x - x')\delta^{\alpha\beta}, \qquad (8.2)$$

where the overline indicates an average taken over the Gaussian ensemble (8.1) of initial conditions.

To be specific, let us consider the relaxational models A and B. Upon including the initial weight (8.1), the Gaussian action becomes

$$\mathcal{A}_0[\tilde{S}, S] = \int d^d x \sum_\alpha \left(\int_0^\infty dt \left[\tilde{S}^\alpha(x, t)\left(\frac{\partial}{\partial t} + D(i\nabla)^a(r - \nabla^2)\right) S^\alpha(x, t)\right.\right.$$

$$\left.\left. - \tilde{S}^\alpha(x, t) D(i\nabla)^a \tilde{S}^\alpha(x, t)\right] + \frac{\Delta}{2}[S^\alpha(x, 0) - a^\alpha(x)]^2\right), \qquad (8.3)$$

---

[1] This section largely follows the original work by Janssen, Schaub, and Schmittmann (1989); as well as the overviews by Janssen (1992); Calabrese and Gambassi (2005); Henkel and Pleimling (2010).

[2] See, e.g., Diehl (1986).

to be supplemented with the non-linear vertex (4.15) with the same temporal integration boundaries $0 \leq t \leq \infty$. The scaling dimension of the initial configuration distribution width is $[\Delta] = \mu^2$. It is therefore a *relevant* parameter, and under the renormalization group flow one expects $\Delta \to \infty$, i.e., asymptotically the system should be aptly described by a sharp distribution on the initial time sheet corresponding to *Dirichlet boundary conditions*.

We shall see that associated with the relevant coupling Δ, i.e., the initial fluctuations $S^\alpha(x, 0)$, is a critical exponent θ that specifically governs the universal short-time behavior. More precisely, it describes the temporal window between any microscopic time scales t_0 and the asymptotic long-time regime that is characterized by the dynamic exponent z. Because the initial conditions of course break time translation invariance, the two-point correlation functions will explicitly depend on *two* time variables. Let us take $t \gg t' \gg t_0$, with both t and t' large compared to the microscopic time scale t_0; in this *'aging scaling'* limit one finds universal behavior

$$\chi(q; t, t'/t \to 0) = D q^a \, G(q; t, t'/t \to 0)$$

$$= D \, |q|^{z-2+\eta} \left(\frac{t}{t'} \right)^\theta \, \hat{\chi}_0(q\xi, |q|^z Dt) \, \Theta(t) \qquad (8.4)$$

for the dynamic susceptibility; t' is often referred to as *waiting time*. Although the fluctuation-dissipation theorem (2.34) does not hold in this non-equilibrium regime, it turns out that the short-time scaling for the correlation function is nevertheless governed by the same exponents:

$$C(q; t, t'/t \to 0) = |q|^{-2+\eta} \left(\frac{t}{t'} \right)^{\theta-1} \, \hat{C}_0(q\xi, |q|^z Dt) \, . \qquad (8.5)$$

For the temporal evolution of the order parameter, one obtains

$$\langle S^n(t) \rangle = S_0 \, t^{\theta'} \, \hat{S}\left(S_0 \, t^{\theta'+\beta/zv} \right) , \qquad (8.6)$$

with $\hat{S}(y) \sim y^{-1}$ as its argument $y \to \infty$, and where respectively for models A and B with non-conserved and conserved order parameter,

$$a = 0 : \quad \theta' = \theta - 1 + \frac{2 - \eta}{z} \, , \quad a = 2 : \quad \theta' = \theta = 0 \, . \qquad (8.7)$$

For a non-conserved order parameter, $\langle S^n(t) \rangle \sim t^{\theta'}$ initially grows algebraically, characterized by the *initial-slip* exponent θ', before eventually crossing over to the long-time decay $\langle S^n(t) \rangle \sim t^{-\beta/zv}$. Note that here and in the following the brackets $\langle \ldots \rangle$ entail averages over *both* the initial configurations as well as thermal noise histories.

8.1.2 *Gaussian models and fluctuation-dissipation ratio*

In order to determine θ via field-theoretic methods, we first need the Gaussian propagators. They are conveniently determined via solving the classical field equations for the generating functional $\mathcal{Z}_0[\tilde{\jmath}, j]$ associated with the action (8.3) and subject to the boundary conditions $\tilde{S}^\alpha(x, 0) = \Delta\,[S^\alpha(x, 0) - a^\alpha(x)]$ and $\tilde{S}^\alpha(x, t \to \infty) = 0$ in Fourier space (see Problem 8.1),

$$\tilde{S}^\alpha(q, t) = \int_0^\infty e^{Dq^a(r+q^2)(t-t')}\,\Theta(t'-t)\,j^\alpha(q, t')\,dt'\,, \tag{8.8}$$

$$S^\alpha(q, t) = \int_0^\infty e^{-Dq^a(r+q^2)(t-t')}\,\Theta(t-t')\big[\tilde{\jmath}^\alpha(q, t') + 2Dq^a\,\tilde{S}^\alpha(q, t')$$

$$+ \big(a^\alpha(q) + \Delta^{-1}\tilde{S}^\alpha(q, 0)\big)\delta(t')\big]\,dt\,. \tag{8.9}$$

Thus one finds that the harmonic response propagator is not affected by the initial conditions,

$$G_0(q; t, t') = G_0(q; t-t') = e^{-Dq^a(r+q^2)(t-t')}\,\Theta(t-t')\,, \tag{8.10}$$

and remains translationally invariant in time. However, the initial preparation is distinctly reflected in the form of the correlation propagator

$$C_0(q; t, t') = C_0(q; t', t) = C_D(q; t, t') + \frac{1}{\Delta}\,G_0(q, t)\,G_0(q, t')\,. \tag{8.11}$$

Asymptotically, as $\Delta \to \infty$, this reduces to the *Dirichlet correlator*

$$C_D(q; t, t') = \frac{1}{r+q^2}\big(e^{-Dq^a(r+q^2)|t-t'|} - e^{-Dq^a(r+q^2)(t+t')}\big)\,, \tag{8.12}$$

which corresponds to sharp initial conditions with vanishing fluctuations, $C_D(q; t > 0, 0) = 0$. We may thus view the effects of finite Δ as corrections to the leading scaling behavior. In the short-time scaling limit, we have

$$C_D(q; t \gg t' \to 0) = \frac{1}{r+q^2}\,e^{-Dq^a(r+q^2)t}\,2\sinh\big[Dq^a(r+q^2)t'\big]$$

$$\to e^{-Dq^a(r+q^2)t}\,2Dq^a t \left(\frac{t}{t'}\right)^{-1}\,. \tag{8.13}$$

Since $\eta = 0$ and $z = 2 + a$ in the harmonic approximation, comparison with Eqs. (8.4) and (8.5) yields $\theta = 0$, and $\hat{\chi}_0(y, \tilde{t}) = \exp[-\tilde{t}\,(1 + y^{-2})]$, $\hat{C}_0(y, \tilde{t}) = 2\tilde{t}\,\exp[-\tilde{t}\,(1 + y^{-2})]$. A non-trivial critical initial-slip exponent can clearly be obtained only by taking anharmonic fluctuations into account.

In this context, one often studies the *fluctuation-dissipation ratio*

$$X(q;t > t', t') = k_B T \frac{\chi(q;t > t', t')}{dC(q;t, t')/dt'}, \tag{8.14}$$

which according to the fluctuation-dissipation theorem (2.36) acquires the value 1 in thermal equilibrium. For the Gaussian relaxational models with random initial conditions, Eqs. (8.10)–(8.12) yield (with $k_B T = 1$)

$$X_0(q;t > t', t')^{-1} = 1 + e^{-2Dq^a(r+q^2)t'}[1 - (r + q^2)/\Delta], \tag{8.15}$$

independent of the larger 'measurement' time t. For $q \neq 0$ indeed $X_0(q;t > t', t') \to 1$ as $t' \to \infty$. More precisely, the system reaches equilibrium for $t' \gg t_c/2$, with the characteristic relaxation time scale $t_c = [Dq^a(r + q^2)]^{-1}$. However, at $T = T_c^0$ ($r = 0$) the relaxation time diverges, and the $q = 0$ Fourier component *never* reaches equilibrium, reflected in the ratio $X_0(0;t > t', t') = 1/2$ for *all* t'. In model B, as a consequence of the conservation law the zero-wavevector order parameter mode cannot relax even away from the critical point, whence the fluctuation-dissipation ratio at $q = 0$ acquires the non-trivial but *non-universal* value $X_0(0;t > t', t')^{-1} = 2 - r/\Delta$.

It is a straightforward task (Problem 8.2) to compute the Gaussian fluctuation-dissipation ratio in real space. For model A, the result in the asymptotic limit $\Delta \to \infty$ is

$$X_0(x;t > t', t')^{-1} = \frac{dC_D(x;t, t')/dt'}{\chi(x;t > t', t')}$$

$$= 1 + \left(\frac{t - t'}{t + t'}\right)^{d/2} \exp\left[-2Dt'\left(r - \frac{x^2}{4D^2(t^2 - t'^2)}\right)\right]. \tag{8.16}$$

For $t, t' \to \infty$, but with the ratio $s = t'/t$ held fixed, we obtain again $X_0(x;t > t', t') \to 1$ for $r > 0$ for any value of s. At the critical point $r = 0$, however, one observes a slow crossover from the equilibrium value $X_0(x;t, t) = 1$ (i.e., $s = 1$) to $X_0(x;t > 0, 0) = 1/2$ ($s = 0$). For model B at criticality and $x = 0$, one finds similarly

$$T = T_c : \quad X_0(x = 0;t > t', t')^{-1} = 1 + \left(\frac{t - t'}{t + t'}\right)^{(d+2)/4}, \tag{8.17}$$

which displays qualitatively the same crossover behavior, but is characterized by a different temporal exponent $(d + 2)/4$ that replaces $d/2$ for a non-conserved order parameter. In either case, the terms $\propto 1/\Delta$ constitute corrections to the leading power laws, see Problem 8.2.

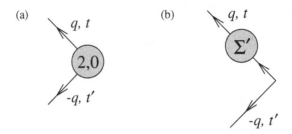

Fig. 8.1 Diagrammatic representation for the correlation function $C(q;t,t')$ with fluctuation contributions from: (a) the noise vertex $\Gamma^{(2,0)}$, and (b) the one-particle reducible propagator self-energy Σ'.

8.1.3 Perturbation theory and renormalization

In the graphical representation of the perturbation expansion, we must now treat the response and correlation propagators as independent, and therefore resort to the diagrammatics outlined in Section 4.3.4. Since specifying the initial conditions breaks temporal translation invariance, we have to draw the Feynman diagrams and perform all computations in the time rather than the frequency domain, as explained in Section 4.4.4. In addition we cannot employ the vertex functions, but need to directly renormalize the one-particle reducible correlation functions. The 'bulk' UV singularities for $t > 0$ are of course taken care of by the renormalization constants determined in Section 5.2. However, we may encounter new infrared divergences in the fluctuation corrections to the correlation propagator $C(q;t,t')$ that originate from the second term in the Dirichlet correlator (8.12) on the initial time 'surface' $t + t' = 0$. In the limit $\Delta \to \infty$, we encounter Dirichlet boundary conditions implying $\langle S^\alpha(x,t)S^\beta(x',0)\rangle = 0$. Thus, novel singularities can only be contained in $\partial\langle S^\alpha(x,t)S^\beta(x',t')\rangle/\partial t'|_{t'=0}$.

Let us now look at the diagrammatic representation of the perturbation expansion in the time domain, where the end points of the two full propagator lines carry the labels t and t', respectively; see Fig. 8.1. We observe that at time $t' \to 0$, causality eliminates any contribution from the noise vertex renormalization $\Gamma^{(2,0)}$, shown in Fig. 8.1(a), for these would have to occur *prior* to the initial time, whereas there is no such constraint for the diagrams of type Fig. 8.1(b). The initial correlation function derivative $\partial\langle S^\alpha(x,t)S^\beta(x',t')\rangle/\partial t'|_{t'=0}$ is thus renormalized by the very diagrams that contribute to the response propagator, namely with the one-particle *reducible* 'bubbles' Σ' of Fig. 8.1(b), cf. Figs. 4.3 and 4.11. Consequently no distinct renormalization for the initial auxiliary field $\tilde{S}^\alpha(x,0)$ is required. In fact, it follows from these considerations that *within* correlation functions one may identify

$$\left.\frac{\partial S^{\alpha}(x,t)}{\partial t}\right|_{t=0} \simeq 2D(i\nabla)^{a}\,\tilde{S}^{\alpha}(x,0)\,. \tag{8.18}$$

From the structure of the Feynman diagrams, we infer that the response propagator can be written quite generally as a convolution of the *equilibrium* response propagator, computed with only the first contribution in the correlator (8.12), and the one-particle reducible self-energy Σ' of Fig. 8.1(b),

$$\langle S^{\alpha}(-q,t)\,\tilde{S}^{\beta}(q,0)\rangle = \int_{0}^{t}\sum_{\gamma}\langle S^{\alpha}(-q,t)\,\tilde{S}^{\gamma}(q,t')\rangle_{\rm eq}\,\Sigma'^{\gamma\beta}(q,t')\,dt'\,. \tag{8.19}$$

The corresponding one-loop graph is just a Hartree loop, as in Fig. 4.9(b), but evaluated with the Dirichlet correlator:

$$\langle S^{\alpha}(-q,t)\,\tilde{S}^{\beta}(q,0)\rangle = \left[G_{0}(q,t) - \frac{n+2}{6}\,u\,Dq^{a}\int_{0}^{t}dt'\,G_{0}(q,t-t')\right.$$

$$\left. \times\int_{k}C_{\rm D}(k;t',t')\,G_{0}(q,t') + O(u^{2})\right]\delta^{\alpha\beta}\,. \tag{8.20}$$

Explicitly, this yields

$$\Sigma'(q,t) = \delta(t) - \frac{n+2}{6}\,u\,Dq^{a}\,e^{-Dq^{a}(r+q^{2})t}\,\Theta(t)$$

$$\times\int_{k}\frac{1}{r+k^{2}}\left[1-e^{-2Dk^{a}(r+k^{2})t}\right] + O(u^{2})\,, \tag{8.21}$$

and for its Fourier transform

$$\Sigma'(q,\omega) = 1 - \frac{n+2}{3}\,u\,\frac{Dq^{a}}{i\omega+Dq^{a}(r+q^{2})}$$

$$\times\int_{k}\frac{Dk^{a}}{i\omega+Dq^{a}(r+q^{2})+2Dk^{a}(r+k^{2})} + O(u^{2})\,. \tag{8.22}$$

For model B with conserved order parameter dynamics, the wavevector dependence of the four-point vertex implies that all loop contributions to the response propagator and hence to the correlation function vanish as $q \to 0$, just as in Eq. (4.84) for the 'bulk'. For model A we obtain

$$\Sigma'(0,\omega) = 1 - \frac{n+2}{6}\,\frac{u}{i\omega/D+r}\int_{k}\frac{1}{i\omega/2D+3r/2+k^{2}} + O(u^{2})\,. \tag{8.23}$$

We next define a distinct renormalization for the initial response field $\widetilde{S}_0^\alpha(x) = \widetilde{S}^\alpha(x, 0)$ with the convention

$$\widetilde{S}_{0R}^\alpha = (Z_0 \, Z_{\widetilde{S}})^{1/2} \, \widetilde{S}_0^\alpha , \qquad (8.24)$$

which according to the definition (8.19) absorbs the initial-time ultraviolet divergences in the self-energy Σ' in Z_0,

$$\Sigma'_R(q, \omega) = Z_0^{1/2} \, \Sigma'(q, \omega) . \qquad (8.25)$$

Evaluating Eq. (8.23) at the convenient normalization point $i\omega/2D = \mu^2, \tau = 0 = q$ gives

$$Z_0 = 1 - \frac{n+2}{6} \frac{u A_d \, \mu^{-\epsilon}}{\epsilon} + O(u^2) = 1 - \frac{n+2}{6} \frac{u_R}{\epsilon} + O(u_R^2) \qquad (8.26)$$

for model A, while no new renormalization constant is required for the initial time behavior of model B, i.e., $Z_0 = 1$ to all orders in the perturbation expansion.

Cumulants of N fields S^α, $\widetilde{N}$ response fields $\widetilde{S}^\alpha$, and $\widetilde{N}_0$ initial response fields $\widetilde{S}_0^\alpha$ become multiplicatively renormalized with the Z factors

$$G_R^{(N,\widetilde{N},\widetilde{N}_0)} = Z_S^{N/2} Z_{\widetilde{S}}^{(\widetilde{N}+\widetilde{N}_0)/2} Z_0^{\widetilde{N}_0/2} \, G^{(N,\widetilde{N},\widetilde{N}_0)} . \qquad (8.27)$$

Therefore, the renormalization group equation with explicit consideration of the initial fields can be obtained from the corresponding one for the 'bulk', see Eq. (5.41), by adding the term $\frac{1}{2}\widetilde{N}_0(\gamma_{\widetilde{S}} + \gamma_0)$ to $\frac{1}{2}N\gamma_S + \frac{1}{2}\widetilde{N}\gamma_{\widetilde{S}}$. Here the additional Wilson flow function

$$\gamma_0 = \mu \left. \frac{\partial}{\partial \mu} \right|_0 \ln Z_0 = \frac{n+2}{6} u_R + O(u_R^2) \qquad (8.28)$$

has been introduced. Consequently, in the vicinity of an infrared-stable fixed point, we obtain the leading scaling behavior

$$G_R^{(N,\widetilde{N},\widetilde{N}_0)}(\mu, D_R, \tau_R, u_R, \{x_i\}, \{t_i\})$$

$$= (\mu \ell)^{-\bar{d}_{N,\widetilde{N},\widetilde{N}_0}} \widetilde{G}_R^{(N,\widetilde{N},\widetilde{N}_0)}\left(\tau_R \ell^{\gamma_\tau^*}, u^*, \{x_i \mu \ell\}, \{D_R t_i (\mu \ell)^{2+a+\gamma_D^*}\}\right), \qquad (8.29)$$

compare with Eq. (5.60), and where, using Eq. (5.7),

$$\bar{d}_{N,\widetilde{N},\widetilde{N}_0} = -\frac{N}{2}(d-2-\gamma_S^*) - \frac{\widetilde{N}+\widetilde{N}_0}{2}(d+2-\gamma_{\widetilde{S}}^*) + \frac{\widetilde{N}_0}{2}\gamma_0^* . \qquad (8.30)$$

Setting $\widetilde{N} = 0$, $N = 1 = \widetilde{N}_0$, and applying the matching condition $\mu \ell = 1/\xi$, one obtains for the dynamic susceptibility at large times $t \gg t' = 0$:

$$\chi_R(x; t, 0) = \xi^{-\lambda_R} \, \widetilde{\chi}_R(x/\xi, t/\xi^z) = t^{-\lambda_R/z} \, \hat{\chi}_R(x/\xi, x\, t^{-1/z}) . \qquad (8.31)$$

Thus, $\chi_R(0; t, 0) \sim t^{-\lambda_R/z}$ with the *autoresponse exponent*

$$\lambda_R = d + a - z\theta', \qquad \theta' = \frac{\gamma_{\tilde{S}}^* + \gamma_{\tilde{\tilde{S}}}^* + \gamma_0^*}{2z}. \qquad (8.32)$$

Using the relations (5.46) and (5.68) for the critical exponents, we obtain for model A with non-conserved order parameter

$$a = 0 : \quad \lambda_R = d - z\theta' = d - 2 + \eta + z - \frac{\gamma_0^*}{2}. \qquad (8.33)$$

Inserting the Heisenberg fixed point (5.70), we find to first order in the dimensional ϵ expansion

$$\theta' = \frac{(n+2)}{4(n+8)}\epsilon + O(\epsilon^2), \quad \lambda_R = 4 - \frac{3(n+6)}{2(n+8)}\epsilon + O(\epsilon^2). \qquad (8.34)$$

For model B with conserved order parameter, the identities $\gamma_0 = 0$ and $\gamma_S = -\gamma_{\tilde{S}}$ lead to the *exact* results

$$a = 2 : \quad \theta' = 0, \quad \lambda_R = d + 2. \qquad (8.35)$$

8.1.4 Critical initial-slip and persistence exponents

We must expect the above scaling functions $\tilde{G}_R^{(N,\tilde{N},\tilde{N}_0)}$ to become singular as any temporal argument approaches the initial time surface. Therefore, in order to be able to infer the universal scaling properties in the initial-slip regime where $t \gg t' > 0$, we require additional knowledge about the behavior of the fields at early times $t' \to 0$. To make progress, we invoke a formal *short-time* (more generally, *operator product*) *expansion* that essentially assumes regularity in the ultraviolet regime,

$$\tilde{S}_0^\alpha(x, t') = \tilde{\sigma}(t')\tilde{S}_0^\alpha(x) + \cdots, \qquad (8.36)$$

$$S_0^\alpha(x, t') = 2Dt'\,\tilde{\sigma}(t')\tilde{S}_0^\alpha(x) + \cdots. \qquad (8.37)$$

In (8.37) we have exploited the (asymptotic) Dirichlet boundary condition $S_0^\alpha(x) = 0$ and the relation (8.18). Upon comparing Eq. (8.29) for $N = 0 = \tilde{N}_0$, $\tilde{N} = 1$ and $N = 0 = \tilde{N}$, $\tilde{N}_0 = 1$, we obtain via matching $\mu\ell = (Dt')^{-1/z}$ in the scaling limit for $t' \to 0$

$$\tilde{\sigma}(t') = (Dt')^{-\theta}\,\hat{\sigma}(t/\xi^z), \qquad \theta = \frac{\gamma_0^*}{2z}. \qquad (8.38)$$

In conjunction with the short-time expansions (8.36), (8.37), we thereby arrive at the initial-slip scaling laws (8.4) and (8.5) for the dynamic response and correlation functions. Here, the critical initial-slip exponent θ reads explicitly for models A and B:

$$a = 0 : \quad \theta = \frac{(n+2)}{4(n+8)}\epsilon + O(\epsilon^2), \quad a = 2 : \quad \theta = 0 . \tag{8.39}$$

Next we address the order parameter initial scaling behavior. To this end, we introduce a generating functional $\ln Z[\tilde{j}_0, \tilde{j}, j]$ with external sources $\tilde{j}_0^\alpha$ that couple to the initial fields $\tilde{S}_0^\alpha$, see Eq. (4.58). Expanding with respect to a homogeneous source $\tilde{j}_0 = S_0$, we may write

$$\langle S^n(x,t)\rangle = \frac{\delta \ln \mathcal{Z}[S_0, \tilde{j}, j]}{\delta j(x,t)}\bigg|_{\tilde{j}=j=0}$$

$$= \sum_{\tilde{N}_0=1}^{\infty} \frac{S_0^{\tilde{N}_0}}{\tilde{N}_0!} \int \prod_{i=1}^{\tilde{N}_0} d^d x_i \, G^{(1,0,\tilde{N}_0)}(\tau, \{x_i\}, t) . \tag{8.40}$$

After renormalization, and upon exploiting the asymptotic scaling (8.29) that follows from the solution of the RG equation near a stable fixed point, one directly recovers Eq. (8.6), recalling that $2\beta/\nu = d - 2 + \eta$, and with the exponent θ' defined in (8.32) (Problem 8.3).

Thus, for model A with non-conserved order parameter, the critical relaxation from random initial conditions obeys the power laws listed in Section 8.1.1, with novel universal initial-slip exponents that have to be determined independently (via γ_0^*) from the 'bulk' critical exponents. This is true also for model C wherein the order parameter is coupled to a conserved scalar field, albeit with different values for γ_0^*. In the diffusive relaxational critical dynamics of a conserved order parameter (model B), however, both the dynamic exponent z and the initial scaling laws can be expressed entirely through static critical exponents; in fact, $\gamma_0^* = 0$ *exactly* in this case.[3] For critical dynamics with reversible mode-coupling terms, the situation changes yet again. For model J capturing the critical dynamics of isotropic ferromagnets, one finds, for example,[4]

$$\theta = \frac{z - 4 + \eta}{z} = -\frac{6 - d - \eta}{d + 2 - \eta} . \tag{8.41}$$

In systems where a *non-conserved* order parameter is dynamically coupled to other conserved modes, the initial-slip exponent θ is actually *not* a universal number, but depends on the width of the initial distribution. Yet the conserved field response function still displays universal scaling features.

Further interesting and non-trivial features emerge when one considers *persistence* properties in critical dynamics.[5] For example, the probability $P(t, t')$ that the

---

[3] Oerding and Janssen (1993a).   [4] Oerding and Janssen (1993b).
[5] Majumdar, Bray, Cornell, and Sire (1996); Oerding, Cornell, and Bray (1997).

global magnetization does not change sign in the time interval $[t', t]$ follows a power law $P(t, t') \sim (t'/t)^\mu$ in the limit $t'/t \to 0$. Under the assumption of Markovian kinetics, one may derive the approximate scaling relation $\mu\, z = \lambda_C - d + 1 - \eta/2$, where λ_C denotes an *autocorrelation exponent* defined in analogy with (8.31): $\langle S^\alpha(x, t)\, S^\beta(x, 0)\rangle \sim t^{-\lambda_C/z}$. In equilibrium $\lambda_C = \lambda_R$, whence $\mu = 1/2 + O(\epsilon)$. It has been demonstrated that the persistence exponent μ is also accessible within the dynamic renormalization group formalism, although its actual computation poses a quite formidable task. For the critical persistence exponent of model A, one finds that non-Markovian corrections appear only to second order in $\epsilon = 4 - d$, and moreover disappear in the spherical limit $n \to \infty$,

$$\mu = \frac{\lambda_C - d + 1 - \eta/2}{z}\left[1 + \frac{3(n+2)}{4(n+8)^2}\,\alpha\,\epsilon^2 + O(\epsilon^3)\right], \quad \alpha \approx 0.2716\,.$$

(8.42)

In contrast, for model C deviations from the Markovian scaling relation already emerge to first order in ϵ.

8.2 Coarsening and phase ordering

Let us next consider a system that is instantaneously 'quenched' from the high-temperature, disordered phase into its ordered phase. Starting from its spatially random initial state, the order parameter cannot immediately assume its equilibrium value. Rather, in the broken-symmetry phase regions with different values of the order parameter are competing. Since the longest relaxation time scale diverges with system size as a consequence of the broken ergodicity in the ordered phase, in the thermodynamic limit thermal equilibrium is *not* attained within any finite time. The coarsening of ordered domains proceeds through universal power laws and is governed by intriguing scaling phenomena.[6]

8.2.1 Scaling hypothesis for phase ordering

In analogy with time-dependent critical phenomena, we invoke a *dynamic scaling hypothesis* whereupon the growing ordered domains are characterized by a typical length scale $L(t)$ that grows algebraically with time,

$$L(t) \sim t^{1/z}\,.$$

(8.43)

[6] For overviews, see, e.g., Bray (1993, 1994); Bray and Rutenberg (1994); Biroli (2005); Henkel and Pleimling (2010).

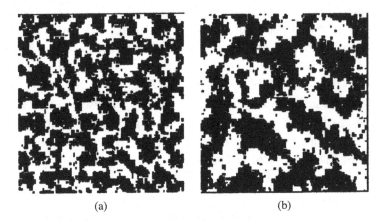

(a)   (b)

Fig. 8.2 Two-dimensional snapshots of a Monte Carlo simulation for the three-dimensional Ising model with Glauber kinetics following a quench from the high-temperature into the ordered phase, after (a) 10^3 and (b) 10^5 Monte Carlo steps. [Figure reproduced with permission from: G. Biroli, *J. Stat. Mech.*, P05014 (2005); DOI: 10.1088/1742-5468/2005/05/P05014; copyright (2005) by Institute of Physics Publ.]

Systems at different times are then presumed to be statistically similar after appropriate rescaling of lengths with the single scale $L(t)$; this is indeed borne out by experiments and numerical simulations, see Fig. 8.2. Regions with identical order parameter values thus grow through *coarsening*. Clearly, the dynamics for a non-conserved order parameter must be manifestly different from the conserved case. In the latter situation, the system can only coarsen through *phase separation*; for fluids, this process is referred to as *spinodal decomposition*, whereas it is known as *Ostwald ripening* in binary alloy metallurgy. In addition, the energetics of the domain walls or topological defects that separate the locally ordered regions must play a crucial role.

In either case, one postulates the two-point correlation function to assume the general scaling form

$$\langle S^n(x, t)\, S^n(x', t') \rangle = f\left(\frac{|x - x'|}{L(t)}, \frac{L(t')}{L(t)}\right),\qquad (8.44)$$

with $f(0, 1) = $ const. Consequently, the (connected) *autocorrelation* function ($x = x'$) must be a function of the ratio $L(t')/L(t)$ only; in fact, in the aging scaling limit $t \gg t'$, i.e., $L(t) \gg L(t')$, it becomes a pure power law

$$C(0; t, t')_{\mathrm{c}} \sim [L(t')/L(t)]^\lambda \sim (t'/t)^{\lambda/z}.\qquad (8.45)$$

The *equal-time* correlations ($t = t'$) are governed by the single length scale $L(t)$. Correspondingly their Fourier transform, the dynamic order parameter *structure factor*, should obey

$$C(q, t)_c = L(t)^d\, g(q\, L(t)) \,. \tag{8.46}$$

While it is suggestive that a zero-temperature fixed point should govern the universal coarsening kinetics, a consistent RG treatment of this situation has remained elusive to date. There is also numerical evidence that the scaling functions in coarsening depend on the symmetries and other features of the underlying lattice, and are in this sense less universal than those at critical points. In the following, we shall derive the above scaling laws for the relaxational models A and B in the spherical limit, wherein the number of order parameter components $n \to \infty$. We then provide a set of illuminating scaling arguments to obtain the growth laws for both non-conserved and conserved order parameters with arbitrary n.

8.2.2 Aging and coarsening in the spherical models A/B

Our goal is to analyze the partition function $Z[h = 0]$ for the n-component Landau–Ginzburg–Wilson Hamiltonian (4.12) in the spherical model limit $n \to \infty$. To this end, we 'linearize' the quartic non-linearity by means of a Gaussian Hubbard–Stratonovich transformation, through introducing an auxiliary field $\Psi(x)$,

$$\int d(i\Psi)\, e^{-(\Psi\, \vec{S}^2 - 3\Psi^2/u)/2} = \sqrt{\frac{2\pi u}{3}}\, e^{-u(\vec{S}^2)^2/4!} \,.$$

Thus $Z[h = 0] \propto \int \mathcal{D}[\vec{S}] \int \mathcal{D}[i\Psi]\, e^{-\widetilde{\mathcal{H}}[S,\Psi]}$, with the augmented Hamiltonian

$$\widetilde{\mathcal{H}}[S, \Psi] = -\frac{1}{2}\int d^d x \left([r + \Psi(x)]\, \vec{S}(x)^2 + [\nabla \vec{S}(x)]^2 - \frac{3}{u}\Psi(x)^2 \right) , \tag{8.47}$$

which is Gaussian in the order parameter fluctuations $S^\alpha(x)$, and diagonal in its components $\alpha = 1, \ldots, n$. Upon integrating out the original order parameter fields, compare with Eq. (8.116) in the Appendix (Section 8.5), one arrives at

$$Z[h = 0] \propto \int \mathcal{D}[i\Psi]\exp\left[\frac{3}{2u}\int d^d x\, \Psi(x)^2 - \frac{n}{2}\, \mathrm{Tr}\, \ln \frac{G_\Psi(x, x')^{-1}}{2\pi} \right], \tag{8.48}$$

with the inverse Green function

$$G_\Psi(x, x')^{-1} = [r + \Psi(x) - \nabla^2]\delta(x - x') \,, \tag{8.49}$$

which becomes in Fourier space

$$G_\Psi(q, q')^{-1} = (r + q^2)\,(2\pi)^d\, \delta(q + q') + \Psi(q + q') \,. \tag{8.50}$$

Recall that the RG fixed point (5.70) for the non-linear coupling $u_H^* \propto \epsilon/(n+2)$; for large n, it is therefore useful to set $u = u'/n$, where u' becomes independent of the number of components n as $n \to \infty$. Consequently $Z[h = 0] \propto \int \mathcal{D}[i\Psi] e^{-n\,\Phi[\Psi]}$, with

$$\Phi[\Psi] = -\frac{3}{2u'} \int d^d x\, \Psi(x)^2 + \frac{1}{2} \mathrm{Tr}\ln \frac{G_\Psi(x, x')^{-1}}{2\pi} . \tag{8.51}$$

In the limit $n \to \infty$, the steepest-descent approximation to the partition function becomes exact, i.e., we merely require the solution of the stationarity condition $\delta\Phi[\Psi]/\delta\Psi(x) = 0$. We expect the minimizer to be a homogeneous solution, $\Psi(x) = \Psi$, whence

$$\mathrm{Tr}\ln \frac{G_\Psi(x, x')^{-1}}{2\pi} = \int d^d x \int \frac{d^d q}{(2\pi)^d} \ln \frac{r + \Psi + q^2}{2\pi} , \tag{8.52}$$

and the stationarity condition yields a self-consistent equation for Ψ:

$$\Psi = \frac{u'}{6} \int \frac{d^d q}{(2\pi)^d} \frac{1}{r + \Psi + q^2} . \tag{8.53}$$

The integral here is just the bare correlation function $C_0(x = 0)$, with a shifted temperature parameter $r \to r + \Psi$. By means of the self-consistent equation (8.52), one may readily infer the static critical exponents of the spherical model: $\eta = 0$ and $\gamma = 2\nu = 2/(d-2)$ for $d < d_c = 4$, while of course $\gamma = 2\nu = 1$ for $d > 4$, see Problem 8.4(a). These results naturally coincide with the values found in Sections 5.3.3 and 7.4.3 when the limit $n \to \infty$ is taken, see Eqs. (5.71), (5.72), and (7.203), (7.205), respectively.

In essence, the spherical limit for the Hamiltonian (4.12) is obtained through the decoupling of the quartic non-linearity $u\,(\vec{S}^2)^2 \to 6\,\Psi\,\vec{S}^2$, accompanied by the shift $\tau \to \tau + \Psi$. This prescription obviously carries over to the purely relaxational dynamics of models A and B, yielding the mean-field dynamic exponents $z = 2 + a$. We can also see this directly by considering the perturbational expansion. Since $u = u'/n$, Feynman diagrams require an independent closed loop $\propto n$ for each appearing non-linear coupling in order to survive the limit $n \to \infty$. Consequently, among the response propagator self-energy graphs (to two-loop order depicted in Fig. 4.13), only the products of Hartree loops will contribute (see also Section 5.4.2). For example, the third diagram in Fig. 4.13 contains two loops to second order in u, but these share a propagator line and are hence not independent, leading to an overall factor $\propto 1/n$. These facts are confirmed by the explicit analytic results (4.82).

With these observations in mind, we may now write down the dynamical response and correlation functions in the spherical model limit as the appropriate generalizations of Eqs. (3.29) and (3.31) to a time-dependent 'mass' term $\tau(t) = \tau + \Psi(t)$. The response propagator reads

$$G_\infty(q;t,t') = \Theta(t-t') \exp\left(-Dq^a \int_{t'}^{t} [\tau(\tilde{t}) + q^2]\, d\tilde{t}\right), \qquad (8.54)$$

while the two-time correlation function becomes a temporal convolution

$$C_\infty(q;t,t') = 2Dq^a \int_0^\infty G_\infty(q;t,\tilde{t})\, G_\infty(q;t',\tilde{t})\, d\tilde{t}. \qquad (8.55)$$

Both these expressions are to be supplemented with the self-consistency condition

$$\tau(t) = \tau + \frac{u'}{6} \int \frac{d^d q}{(2\pi)^d} C_\infty(q;t,t)$$

$$= \tau_\infty + \frac{u'}{6} \int \frac{d^d q}{(2\pi)^d} [C_\infty(q;t,t) - C_\infty(q)], \qquad (8.56)$$

where $\tau_\infty = \tau(t \to \infty)$ and $C_\infty(q) = \lim_{t \to \infty} C_\infty(q;t,t) = (\tau_\infty + q^2)^{-1}$ is the static correlation function of the spherical model. Equations (8.54) and (8.55) can also be formally established using the techniques of Problem 8.1.

Next we consider the critical scaling of the spherical relaxational models in the aging regime. Thus we set $\tau_\infty = 0$ and seek for solutions of the equal-time correlation function in the scaling form $C_\infty(q;t=t') = q^{-2} \hat{C}_\infty (q^{2+a} Dt)$, using $\eta = 0$ and $z = 2 + a$. Since $[\Psi(t)] = [\tau] = \mu^2$, dimensional analysis tells us that $\tau(t) = c_a (2Dt)^{-2/(2+a)}$ with $c_a = $ const. For model A with non-conserved order parameter, inserting $\tau(t) = c_0(2Dt)^{-1}$ into (8.54) yields

$$G_\infty(q;t,t') = \Theta(t-t')(t/t')^{-c_0/2}\, e^{-Dq^2(t-t')}, \qquad (8.57)$$

and comparison with Eq. (8.4) establishes that $c_0 = -2\theta$. After investing a bit of algebra, see Problem 8.4(c), the condition (8.56) fixes this constant, resulting in

$$\theta = \frac{4-d}{4}, \qquad (8.58)$$

which indeed agrees with the one-loop result (8.39) in the limit $n \to \infty$. For model B, we have in contrast $\tau(t) = c_2(2Dt)^{-1/2}$, and obtain

$$G_\infty(q;t,t') = \Theta(t-t')\, e^{-c_2\sqrt{2D}q^2(\sqrt{t}-\sqrt{t'})-Dq^4(t-t')}, \qquad (8.59)$$

which is not a power law in t/t', confirming that $\theta = 0$ for relaxational dynamics with conserved order parameter.

A very different physical situation is encountered when we rapidly quench the system from a random initial state (effectively at $T = \infty$) into the ordered phase ($T < T_c$). The system's temporal evolution should then be governed by fluctuations in the initial state rather than thermal noise. Upon omitting the stochastic forcing and with the spherical model factorization, the ensuing *deterministic* equations of motion for the order parameter components become

$$\frac{\partial S^\alpha(x, t)}{\partial t} = -D(i\nabla)^a [\tau(t) - \nabla^2] S^\alpha(x, t),$$

(8.60)

and are formally solved in Fourier space by

$$S^\alpha(q, t) = S^\alpha(q, 0) \exp\left[-Dq^a \left(q^2 t + \int_0^t \tau(t') dt'\right)\right].$$

(8.61)

We then exploit the initial correlations given by Eq. (8.2) with $a^\alpha(x) = 0$, i.e., $\langle S^\alpha(q, 0) S^\beta(q', 0)\rangle = \Delta^{-1}(2\pi)^d \delta(q + q') \delta^{\alpha\beta}$, to evaluate

$$\langle S^\alpha(x, t)^2\rangle_\infty = \frac{1}{\Delta} \int \frac{d^d q}{(2\pi)^d} \exp\left[-2Dq^a \left(q^2 t + \int_0^t \tau(t') dt'\right)\right].$$

(8.62)

But the self-consistency condition (8.56) demands $\tau(t) = \tau + u'\langle S^\alpha(x, t)^2\rangle/6$, and thus, since $\lim_{t\to\infty}\langle S^\alpha(x, t)^2\rangle = 6|\tau|/u'$ for $T < T_c$, that $\tau_\infty = 0$.

We may use (8.62) along with this limiting condition on the left-hand side to determine the asymptotic scaling law for the order parameter correlations as $t \to \infty$. Setting first $a = 0$, the wavevector integral gives $(8\pi Dt)^{-d/2}$, and we obtain for $t > t_0 \gg 0$ (with appropriately chosen t_0) that $2D \int_{t_0}^t \tau(t') dt' \approx -\frac{d}{2} \ln(t/t_0)$, i.e., $\tau(t) \approx -d/4Dt$. From Eq. (8.61) consequently $S^\alpha(q, t) \approx S^\alpha(q, 0) (6|\tau|\Delta/u')^{1/2} (8\pi Dt)^{d/4} e^{-Dq^2 t}$, whence Fourier backtransform finally yields for $t, t' \to \infty$

$$\langle S^\alpha(x, t) S^\beta(x', t)\rangle_\infty \approx \frac{6|\tau|}{u'} \left(\frac{4t t'}{(t + t')^2}\right)^{d/4} e^{-(x-x')^2/4D(t+t')} \delta^{\alpha\beta}.$$

(8.63)

The order parameter correlation function for purely relaxational coarsening with non-conserved order parameter thus acquires the scaling form (8.44) with $L(t) \sim \sqrt{Dt}$, whence we infer the spherical model A scaling exponents

$$a = 0: \quad z = 2, \quad \lambda = \frac{d}{2}.$$

(8.64)

In the case of model B with conserved order parameter, similar considerations merely lead to subleading corrections in the exponential of (8.61) rather than genuine power laws, akin to those in Eq. (8.59) for the critical response function. Therefore, for quenches to $T < T_c$ the characteristic coarsening length scales as $L(t) \sim (Dt)^{1/4}$, and we arrive at

$$a = 2: \quad z = 4, \quad \lambda = 0$$

(8.65)

in the spherical model limit $n \to \infty$.

8.2.3 Phase ordering kinetics, coarsening growth laws

For a finite number of order parameter components, we resort to two versions of dynamical scaling theory. First, we consider model B with diffusive relaxation, i.e., the stochastic dynamics governed by Eq. (3.16) or (4.13) with $a = 2$, and specifically exploit the fact that the diffusion coefficient D does not renormalize independently as a consequence of the conservation law. We may then simply invoke scale transformations, without actually performing any explicit momentum shell elimination or other renormalization steps, in the spirit of Section 1.4. Aside from rescaling space coordinates according to $x \to x' = x/b$ (with scale parameter $b > 1$), according to the dynamic scaling hypothesis we require $t \to t' = t/b^z$, and $D \to D' = D$. In addition we implement the scale transformations $S^\alpha(x) \to S'^\alpha(x') = b^\zeta S^\alpha(x)$ for the order parameter, and $\mathcal{H}[S^\alpha] \to b^{-\Theta} \mathcal{H}'[S'^\alpha]$ for the Hamiltonian. After rescaling, the left-hand side of the equation of motion (4.13) acquires an overall scale factor $b^{-\zeta-z}$, whereas the right-hand side picks up the term $b^{-2+\zeta+\Theta-d}$ (keeping in mind that the functional derivative involves a spatial delta function). Demanding that both sides of the rescaled equations should transform identically, we immediately obtain the scaling relation

$$z = d + 2 - 2\zeta - \Theta . \tag{8.66}$$

Upon properly rescaling of the noise terms and invoking the noise correlator (3.17), one may confirm the temperature rescaling $T \to T' = T/b^\Theta$, which is also evident on dimensional grounds.

At the critical point $T = T_c$, the Hamiltonian should be scale-invariant, i.e., $\Theta = 0$; as we established earlier from the power law decay of the order parameter correlation function that $2\zeta = d - 2 + \eta$, we recover the exact result for the model B dynamic critical exponent $z = 4 - \eta$. In contrast, in the ordered phase $C(x) \to$ const. as $|x| \to \infty$ rather than obeying a power law decay, and consequently $\zeta = 0$, $z = d + 2 - \Theta$. Moreover, near the zero-temperature fixed point that supposedly controls the coarsening dynamics, $\Theta > 0$ such that indeed $T \to 0$ under successive RG transformations. In fact, in this regime the energy scaling should be determined by the scale dependence of the domain walls or other generic topological defects in the system. We should therefore identify the energy scaling exponent Θ with the stiffness exponent defined in Eq. (1.80), $\Theta = d - 1$ in the case of Ising symmetry ($n = 1$), whereas $\Theta = d - 2$ for vector order parameters with $n \geq 2$. Hence the scaling relation (8.66) yields the *model B dynamic coarsening exponents*

$$a = 2, \, n = 1 : \quad z = 3, \quad n \geq 2 : \quad z = 4, \tag{8.67}$$

which includes the spherical limit (8.65).

(a)   (b)   (c)

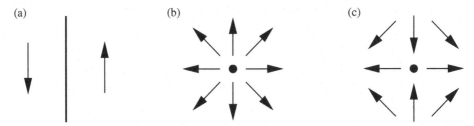

Fig. 8.3 Topological defects in the two-dimensional $O(n)$ model: (a) domain wall
for $n = 1$; (b) vortex for $n = 2$; and (c) anti-vortex for $n = 2$. The arrows indicate
the orientation of the order parameter field.

Our second scaling approach involves a more detailed analysis of the energy
dissipation induced by the topological defects that dominate the low-temperature
kinetics.[7] Let us consider a general $O(n)$-symmetric Hamiltonian for an n-
component vector order parameter in d spatial dimensions,

$$\mathcal{H}[\vec{S}] = \int d^d x \left[\frac{1}{2} [\nabla \vec{S}(x)]^2 + V(\vec{S}(x)) \right] , \qquad (8.68)$$

where the potential $V(\vec{S}) = v(|\vec{S}|)$ depends only on the magnitude of the order
parameter, but not on its orientation. *Topological defects* are given by the solution
of the associated Landau–Ginzburg equation $\nabla^2 \vec{S}(x) = dV/d\vec{S}(x)$ with appropriate
boundary conditions. In the scalar case ($n = 1$), we demand that the order parameter
asymptotically assume its opposite homogeneous low-temperature values $\phi_{\pm}$ in
different half-spaces, which are thus separated by a *domain wall*, i.e., a $(d - 1)$-
dimensional surface where $S(x) = 0$, see Fig. 8.3(a). (An explicit calculation of
the Ising domain wall profile is the subject of Problem 1.4.) For an n-component
order parameter, the solution of the Landau–Ginzburg equation yields a surface of
codimension $d - n$. Topological defects may therefore exist provided $n \leq d$; for
$n < d$, they are spatially extended. The different possibilities in two dimensions
are depicted in Fig. 8.3, where the arrows indicate the local orientation of the
order parameter field. Aside from the linear domain wall for $n = 1$, for $n = 2$ one
encounters point-like *vortex* and *anti-vortex* solutions, with the order parameter
orientation angle respectively changing by $+2\pi$ and -2π upon traversing a closed
loop encircling the topological defect. In three dimensions, the domain walls of the
scalar model become planar defects. For $n = 2$, one encounters vortex and anti-
vortex *line* defects with cylindrical symmetry: Figs. 8.3(b) and (c) then represent
just the cuts perpendicular to the cylinder axis. Finally, for $n = 3$ there exist
spherically symmetric point defects, often named *monopoles* or *hedgehogs*; these

---

[7] This discussion follows Bray and Rutenberg (1994).

are just the three-dimensional extension of the two-dimensional vortex shown in Fig. 8.3(b); the anti-monopole simply has all the arrows reversed.

The presence of topological defects induces specific power laws in the spatial order parameter correlation function at distances short compared to the coarsening scale $L(t)$, but large compared to the defect core size that is determined by the correlation length ξ, i.e., for $\xi \ll |x| \ll L$, and correspondingly for wavevectors in the range $L^{-1} \ll |q| \ll \xi^{-1}$. Indeed, according to (8.44), the equal-time correlations should be only a function of the ratio $|x - x'|/L(t)$. In a volume $L(t)^d$, topological defects with codimension $d - n$ will induce (anti)-correlations proportional to their volume fraction $1/L(t)^n$. Thus one expects $C(x - x', t)_c \sim [|x - x'|/L(t)]^n$ as the leading singular contribution,[8] which in turn yields in Fourier space

$$C(q, t)_c \sim L(t)^{-n} |q|^{-(d+n)} . \tag{8.69}$$

The power law tails for the order parameter structure factor (8.46) in the ordered phase of the $O(n)$ model are known as (generalized) *Porod's law*. For spatially extended topological defects ($n < d$), coarsening proceeds via the shrinkage of regions with large curvature (small domain bubbles or vortex loops), and concomitantly the reduction of defect surface areas. For point defects ($n = d$), instead the long-time dynamics is governed by the mutual annihilation of defect–anti-defect pairs. In either situation, Porod's law (8.69) in conjunction with the dynamic scaling hypothesis permits the determination of the growth laws for the coarsening scale $L(t)$.

To this end, we evaluate the energy dissipation as the system relaxes toward its equilibrium state, which should be dominated by the kinetics of the topological defects (when they exist), and hence governed by structures at the characteristic scale $L(t)$. By means of the deterministic equation of motion $\partial S^\alpha(q, t)/\partial t = -Dq^a \, \delta \mathcal{H}[S^\alpha]/\delta S^\alpha(-q, t)$ for the order parameter components of the relaxational models in Fourier space, we obtain

$$\frac{d \langle \mathcal{H}[S^\alpha](t) \rangle}{dt} = \int \frac{d^d q}{(2\pi)^d} \sum_{\alpha=1}^{n} \left\langle \frac{\delta \mathcal{H}[S^\alpha]}{\delta S^\alpha(q, t)} \frac{\partial S^\alpha(q, t)}{\partial t} \right\rangle$$

$$= -\int \frac{d^d q}{(2\pi)^d} \sum_{\alpha=1}^{n} \frac{1}{Dq^a} \left\langle \frac{\partial S^\alpha(q, t)}{\partial t} \frac{\partial S^\alpha(-q, t)}{\partial t} \right\rangle$$

$$= -V \frac{\partial^2}{\partial t \, \partial t'} \int \frac{d^d q}{(2\pi)^d} \sum_{\alpha=1}^{n} \frac{C^{\alpha\alpha}(q; t, t')}{Dq^a} \bigg|_{t=t'} . \tag{8.70}$$

---

[8] For even n, a more detailed quantitative analysis gives a logarithmic correction $C(x, t)_c \sim [|x|/L(t)]^n \ln[|x|/L(t)]$ that however does not modify the result (8.69).

Next we determine the scaling behavior of the energy dissipation in terms of $L(t)$ (following similar procedures as in Section 5.1). For the $O(n)$-symmetric models (8.68), we have

$$\langle \mathcal{H}[\vec{S}](t) \rangle \sim \left\langle \int d^d x \, [\nabla \vec{S}(x)]^2 \right\rangle = V \int \frac{d^d q}{(2\pi)^d} \sum_{\alpha=1}^{n} q^2 \, C^{\alpha\alpha}(q;t,t) \,, \qquad (8.71)$$

and, provided the integral converges, the scaling assumption (8.46) yields immediately $\langle \mathcal{H}(t) \rangle \sim V/L(t)^2$. However, if the wavevector integral in Eq. (8.71) diverges, we must impose an ultraviolet cutoff, naturally provided by the defect core size $\approx \xi$. In that case, the integral is dominated by the large momentum tails in Porod's law (8.69), and we obtain

$$\langle \mathcal{H}(t) \rangle \sim \frac{V}{L(t)^n} \int_0^{1/\xi} q^{1-n} \, dq \sim \begin{cases} V \, \xi^{n-2}/L(t)^n, & n < 2 \,, \\ V \, L(t)^{-2} \, \ln[L(t)/\xi], & n = 2 \,, \end{cases} \qquad (8.72)$$

where $1/L(t)$ has been employed as an appropriate infrared cutoff in the marginal case $n = 2$. We also infer that the integral is finite for $n > 2$. The right-hand side of Eq. (8.70) is analyzed in a similar manner. Exploiting the scaling hypothesis gives the scaling $\sim -(V/D) \, L(t)^{a-2} \, [dL(t)/dt]^2$, if the wavevector integral is convergent, whereas otherwise (i.e., for $n + a \leq 2$)

$$\frac{d \, \langle \mathcal{H}(t) \rangle}{dt} \sim -\frac{V}{D \, L(t)^n} \left(\frac{dL(t)}{dt} \right)^2 \int_0^{1/\xi} q^{1-a-n} \, dq$$

$$\sim \begin{cases} V \, \xi^{n+a-2} \, [dL(t)/dt]^2/DL(t)^n, & n + a < 2 \,, \\ (V/D) \, L(t)^{a-2} \, [dL(t)/dt]^2 \, \ln[L(t)/\xi], & n + a = 2 \,. \end{cases} \qquad (8.73)$$

Upon self-consistently equating the scaling in (8.73) with the explicit time derivative of Eq. (8.72), we are led to a differential equation for $L(t)$, whose integration yields the desired growth law. Consider first the case $n < 2$ and $n + a > 2$, i.e., $-nV \, \xi^{n-2}/L(t)^{n+1} dL(t)/dt \sim -(V/D) \, L(t)^{a-2} \, [dL(t)/dt]^2$ or $dL(t)/dt \sim L(t)^{1-n-a}$. This is readily integrated to $t \sim L^{n+a}$ or $L(t) \sim t^{1/(n+a)}$, which recovers the conserved dynamics scalar domain wall growth law $L(t) \sim t^{1/3}$ for $a = 2$, $n = 1$. For $n > 2$, we merely need to formally replace n with 2 on the left-hand side, and thus arrive at $L(t) \sim t^{1/(2+a)}$, covering both the conserved and non-conserved dynamics for vector order parameters with more than two components. For the scalar model A ($a = 0$, $n = 1$), one finds $dL(t)/dt \sim 1/L(t)$, which again gives $L(t) \sim t^{1/2}$. This leaves the two marginal situations: In the case of the two-component model A, to leading order the logarithms cancel, whence

$$a = 0: \quad z = 2 \tag{8.74}$$

for all n. Yet since the microscopic cutoffs for the integrals on either side of Eq. (8.71) will differ, one would expect a logarithmic correction to scaling for $n = 2$ of the form $L(t) \sim t^{1/2} [1 + O(1/\ln Dt)]$. Finally, for $n = 2$ and $a > 0$, $dL(t)/dt \sim 1/[L(t)\ln L(t)]$, whose asymptotic solution

$$a > 0, \, n = 2: \quad L(t) \sim (t \ln Dt)^{1/(2+a)} \tag{8.75}$$

refines our previous scaling result (8.67) for a two-component conserved order parameter.

This concludes our scaling derivation of coarsening growth laws for the $O(n)$ model. Notice that the dynamic scaling exponents z are independent of dimension; but of course the above analysis always assumed the existence of an ordered phase, and hence pertains to $d > d_{lc}$. The preceding analysis can in fact be extended to a computation of the exact amplitudes of the leading growth power laws. However, a consistent analytic calculation of the full scaling functions in coarsening and phase separation poses a formidable problem. An intriguing recent development generalizes the global scaling hypothesis to postulating *local scale invariance*.[9] This local symmetry constraint then essentially fixes the corresponding scaling functions in the aging regime, and yields results that in many situations compare favorably with numerical simulation data.

8.3 Effects of violating the detailed balance conditions

In this section, we proceed to explore the effects of violating the detailed balance conditions on the critical dynamics (and statics) in the vicinity of a second-order phase transition.[10] We begin with *isotropic* deviations from the Einstein relation in the purely relaxational models A and B, and then comment on the effects of similar perturbations away from thermal equilibrium for model J with reversible mode couplings, as well as models C, D, E, G (or rather, their n-component generalization, namely the SSS model), and H. Next we consider *spatially anisotropic* detailed balance violations in systems with *conserved* order parameter, first in the case of purely diffusive kinetics which asymptotically leads to the anisotropic critical scaling behavior of the *two-temperature model B* (or *randomly driven diffusive dynamics*), and then in systems that contain additional slow modes.

---

[9] See Henkel and Pleimling (2010).
[10] See Schmittmann and Zia (1991); Schmittmann (1993); and the concise overview by Täuber, Akkineni, and Santos (2002).

8.3.1 Isotropic detailed balance perturbations

Fur purely relaxational kinetics, the condition that the canonical probability distribution (3.11) should be attained in the long-time limit is encoded in the Einstein relation for the relaxation constant D and the noise correlator (3.19). Yet even if we modify the noise strength $D \to \widetilde{D}$, it is immediately obvious that we can reinterpret this as a mere rescaling of the temperature $k_B T \to k_B T' = \widetilde{D}/D$. But the absolute value of T_c, a non-universal quantity, does not enter the description of the *universal* critical properties at all. We therefore expect that any such isotropic detailed balance violation does *not* change the critical exponents and asymptotic scaling functions.

This assertion can be formally verified by means of the response functional, which now reads (with $k_B T$ absorbed into the noise strength $\widetilde{D}$)

$$
\mathcal{A}[\widetilde{S}, S] = \int d^d x \int dt \sum_{\alpha} \left(\widetilde{S}^{\alpha} \left[\frac{\partial}{\partial t} + D(i\nabla)^a (r - \nabla^2) \right] S^{\alpha} \right.
$$

$$
\left. - \widetilde{D}\, \widetilde{S}^{\alpha} (i\nabla)^a \widetilde{S}^{\alpha} + D \frac{u}{6} \sum_{\beta} \widetilde{S}^{\alpha} (i\nabla)^a S^{\alpha} S^{\beta} S^{\beta} \right) . \tag{8.76}
$$

Upon rescaling the fluctuating fields $\widetilde{S}^{\alpha} \to \widetilde{S}'^{\alpha} = (\widetilde{D}/D)^{1/2} \widetilde{S}^{\alpha}$, $S^{\alpha} \to S'^{\alpha} = (D/\widetilde{D})^{1/2} S^{\alpha}$, the noise strength and relaxation constant become identical, and in terms of $\widetilde{S}'^{\alpha}$ and S'^{α} the dynamic response functional assumes its equilibrium form, albeit with a modified non-linear coupling

$$
u \to \tilde{u} = \frac{\widetilde{D}}{D} u . \tag{8.77}
$$

The universal static and dynamic critical properties are, however, governed by the Heisenberg fixed point u_H^* and the RG flow in its vicinity. The non-universal modification (8.77) thus has no bearing on the critical behavior of either model A or B.

For the critical dynamics of isotropic ferromagnets as captured in model J, see Section 6.2, one may proceed in the very same manner. Rescaling the order parameter and response fields as above recovers essentially the equilibrium action, just with the modified $u \to \tilde{u}$ of (8.77) as well as an altered mode-coupling strength $g \to \tilde{g} = (\widetilde{D}/D)^{1/2} g$ in the vertex (6.55). These changes do of course not affect the scaling properties governed by the infrared-stable fixed point with universal values u_H^* and f_J^*.

Such a simple reinterpretation in terms of an effective temperature T' cannot be accomplished for those dynamic universality classes that entail a coupling to additional slow modes. In models C, D (see Section 6.1), E, G (or more generally the n-component SSS model, Section 6.3) and H (Section 6.4), one has the freedom to also alter the noise correlation for the conserved non-critical fields, $\lambda \to \tilde{\lambda}$. This is reflected in the *non-equilibrium parameter*

$$\Theta = \tilde{\lambda} D / \lambda \tilde{D} , \tag{8.78}$$

i.e., the *ratio* of the effective temperatures of the heat baths for the conserved quantities and the order parameter. Upon appropriate rescaling of the fluctuating fields, the relaxation rates and noise strength can be readily set equal; however, the *dynamic* couplings in the Langevin equations for the order parameter and the non-critical modes then assume different values, typically by factors of Θ. Whereas the Einstein relations can thus be formally satisfied for either variable, the second equilibrium condition (3.58), that the probability current be divergence-free, is not in general fulfilled, except at the equilibrium fixed point $\Theta_{\rm eq}^* = 1$.

A detailed renormalization group analysis reveals that (at least to one-loop order) there exist at most two additional non-equilibrium fixed points, namely $\Theta_0^* = 0$ and $\Theta_\infty^* = \infty$. Holding the order parameter temperature fixed at $T_{\rm c}$, these respectively correspond to vanishing and infinite effective temperatures of the heat bath for the conserved non-critical modes, and hence unidirectional energy flow either from or into the order parameter subsystem. It is then intuitively obvious that when $\Theta = 0$, the order parameter fluctuations affect the dynamics of the conserved quantities, but there is no feedback from the latter into the Langevin equation for the critical modes; and conversely for $\Theta = \infty$. Consequently, one would expect the order parameter dynamics at Θ_0^* to be described by the model A/B universality class, with anomalous power laws governing the relaxation of the conserved modes. In contrast, at Θ_∞^* (provided this fixed point exists) genuinely novel non-equilibrium critical properties may ensue in the order parameter sector, whereas the non-critical fields just relax diffusively.

Precisely these different scenarios emerge for the non-equilibrium SSS model with n-component non-conserved order parameter.[11] Yet the equilibrium fixed point $\Theta_{\rm eq}^* = 1$ turns out to be stable. For the scalar model C ($n = 1$), to one-loop order there exists no non-equilibrium fixed point at all; for $n \geq 2$, the coupling to the energy density vanishes asymptotically anyway.[12] In model D, there is at best an effect of the order parameter on the conserved fields (namely for $n < 4$), but the above non-equilibrium perturbations can be removed entirely in the asymptotic limit where $w \to w_{\rm D}^* = 0$ through integrating out the energy fluctuations ρ and $\tilde{\rho}$. In the non-equilibrium model H with conserved order parameter, the only possible additional fixed point is at $\Theta_0^* = 0$, but it is unstable.[13]

In summary, one finds that the large fluctuations in the vicinity of the phase transition effectively *restore* detailed balance. The ensuing static and dynamic

---

[11] Täuber and Rácz (1997).   [12] Akkineni and Täuber (2004.)   [13] Santos and Täuber (2002).

critical scaling laws are consequently those of the corresponding equilibrium system. This remains true even if we permit spatially anisotropic noise for the conserved *non-critical* fluctuations, $-\tilde{\lambda}\, \nabla^2 \to -\tilde{\lambda}_\parallel\, \nabla_\parallel^2 - \tilde{\lambda}_\perp\, \nabla_\perp^2$, which introduces the temperature ratio of the longitudinal and transverse spatial sectors as an additional degree of freedom (holding the order parameter temperature fixed, say, near T_c). Yet again at the stable fixed points, both in model C and the SSS model, the flow leads back to the equilibrium ratio 1. Models with *non-conserved* order parameter thus turn out to be remarkably robust against non-equilibrium perturbations in the form of detailed balance violations.

8.3.2 Conserved systems with anisotropic random drive

For models with *conserved* order parameter, we may in addition investigate the possibility of spatially anisotropic violation of detailed balance in the noise correlations for the critical fields.[14] In general, this implies that we need to both split the isotropic relaxation coefficient according to $-D\, \nabla^2 \to -D_\parallel\, \nabla_\parallel^2 - D_\perp\, \nabla_\perp^2$ and impose anisotropic stochastic noise correlations

$$\langle \zeta(x, t)\zeta(x', t')\rangle = -2\left(\tilde{D}_\parallel \nabla_\parallel^2 + \tilde{D}_\perp \nabla_\perp^2\right) \delta(x - x')\delta(t - t'), \qquad (8.79)$$

where $\tilde{D}_\parallel/D_\parallel \neq \tilde{D}_\perp/D_\perp$. This may be interpreted as coupling the longitudinal and transverse spatial sectors, with respective dimensions $d_\parallel$ and $d_\perp = d - d_\parallel$, to heat baths with distinct temperatures $T_\parallel$ and $T_\perp$. If we take $T_\perp < T_\parallel$, the transverse sector will soften first, while the longitudinal order parameter sector remains non-critical. In the vicinity of $r_\perp \propto T_\perp - T_c = 0$, one may then disregard non-linear longitudinal order parameter fluctuations. This becomes manifest if we apply the following *anisotropic* scaling and power counting appropriate for this situation,

$$[q_\perp] = \mu, \quad [q_\parallel] = \mu^2, \quad [\omega] = \mu^4, \quad [\tilde{D}_\perp] = [D_\perp] = \mu^0, \qquad (8.80)$$

which implies the negative scaling dimensions

$$[\tilde{D}_\parallel] = \mu^{-2} = [D_\parallel]. \qquad (8.81)$$

Hence the longitudinal noise strength and relaxation constant are irrelevant in the renormalization group sense, and can be set to zero asymptotically.

Let us study the ensuing critical behavior for the purely relaxational model B. Upon rescaling the fields as above, i.e., $\tilde{S}^\alpha \to (\tilde{D}_\perp/D_\perp)^{1/2}\tilde{S}^\alpha$, $S^\alpha \to (D_\perp/\tilde{D}_\perp)^{1/2}S^\alpha$, and setting

$$r = r_\perp, \quad D = D_\perp, \quad c = \frac{D_\parallel}{D_\perp}r_\parallel, \quad \tilde{u} = \frac{\tilde{D}_\perp}{D_\perp}u \qquad (8.82)$$

[14] Täuber, Santos, and Rácz (1999).

(recall that $r_\| \propto T_\| - T_c > 0$), we arrive at the Langevin equation

$$\frac{\partial S^\alpha(x, t)}{\partial t} = D\left[c\nabla_\|^2 + \nabla_\perp^2(r - \nabla_\perp^2)\right] S^\alpha(x, t)$$

$$+ D\frac{\tilde{u}}{6}\nabla_\perp^2 S^\alpha(x, t) \sum_\beta [S^\beta(x, t)]^2 + \zeta^\alpha(x, t), \qquad (8.83)$$

with noise correlator

$$\langle \zeta^\alpha(x, t)\zeta^\beta(x', t')\rangle = -2D\nabla_\perp^2 \delta(x - x')\delta(t - t')\delta^{\alpha\beta}. \qquad (8.84)$$

The effective dynamics (8.83) with noise (8.84) is referred to as model B with *anisotropic random drive*, or as the *two-temperature* model B.

Remarkably, this non-equilibrium version of purely diffusive critical dynamics is actually equivalent to an *equilibrium* model, however with spatially long-range correlations. The stochastic equation of motion (8.83) can be inferred from the purely relaxational kinetics

$$\frac{\partial S^\alpha(x, t)}{\partial t} = D\nabla_\perp^2 \frac{\delta\mathcal{H}_{\text{eff}}[S]}{\delta S^\alpha(x, t)} + \zeta^\alpha(x, t), \qquad (8.85)$$

derived from an effective Hamiltonian

$$\mathcal{H}_{\text{eff}}[S] = \int \frac{d^d q}{(2\pi)^d} \sum_\alpha \frac{c q_\|^2 + q_\perp^2(r + q_\perp^2)}{2q_\perp^2} S^\alpha(q) S^\alpha(-q)$$

$$+ \frac{\tilde{u}}{4!}\int d^d x \sum_{\alpha,\beta} [S^\alpha(x)]^2 [S^\beta(x)]^2. \qquad (8.86)$$

This Hamiltonian describes, e.g., a class of elastic structural phase transitions, where phonon modes soften only in certain wavevector sectors; or, for $d_\perp = 1$ and $n = 1$, uniaxial dipolar Ising magnets. It is characterized by the anisotropic Gaussian correlation function

$$C_0(q_\|, q_\perp) = \frac{q_\perp^2}{c q_\|^2 + q_\perp^2(r + q_\perp^2)}. \qquad (8.87)$$

We shall later encounter the same structure for the driven Ising lattice gas near criticality (Section 11.2). Longitudinal fluctuations are suppressed, whence we expect the upper critical dimension to be *reduced* as compared to the isotropic case. Indeed, power counting for fields and the non-linear coupling yields

$$[S^\alpha(x, t)] = \mu^{-1+d_\|+d_\perp/2}, \quad [\tilde{u}] = \mu^{4-2d_\|-d_\perp}, \qquad (8.88)$$

and hence $d_c = 4 - d_\|$.

As a consequence of the field rescaling, the Einstein relation between the transverse diffusion coefficient and noise strength is fulfilled. The anisotropic model B described by Eqs. (8.85), (8.86), and (8.84) satisfies the equilibrium fluctuation-dissipation theorem (2.34). We merely have to allow for different scaling in the longitudinal and transverse wavevector sectors. Thus, the general scaling form for the dynamic response function becomes

$$\chi(\tau, q_\|, q_\perp, \omega) = |q_\perp|^{-2+\eta} \, \hat{\chi}\left(\frac{\tau}{|q_\perp|^{1/\nu}}, \frac{\sqrt{c}\,q_\|}{|q_\perp|^{1+\Delta}}, \frac{\omega}{D\,|q_\perp|^z}\right), \tag{8.89}$$

where we have introduced a new scaling exponent Δ originating in the intrinsic system anisotropy. In the mean-field approximation or for $d > d_c$, we see from (8.87) that $\Delta = 1$, corresponding to the naive scaling (8.80). Alternatively, we could have defined a new set of distinct longitudinal scaling exponents $\nu_\| = \nu(1 + \Delta)$ and $z_\| = z/(1 + \Delta)$.

As in the isotropic model B, the fluctuation-dissipation relation and the momentum dependence of the non-linear vertex stemming from the conservation law imply the identity $\gamma_D = \gamma_S$, and consequently

$$z = 4 - \eta \,. \tag{8.90}$$

Moreover, for $q_\perp = 0$ to all orders in the perturbation expansion

$$\Gamma^{(1,1)}(q_\|, q_\perp = 0, \omega) = i\omega + D\,c\,q_\|^2 \,, \tag{8.91}$$

since the vertex $\propto \tilde{u}$ only supports transverse external momenta. This absence of any renormalization of fluctuations in the non-critical longitudinal sector implies $Z_c = Z_{D_\perp}^{-1}$ or, with $[c] = \mu^0$,

$$\gamma_c = -\gamma_{D_\perp} = -\gamma_S \,. \tag{8.92}$$

As we further readily identify $1 + \Delta = 2 - \gamma_c^*/2$ in the vicinity of a stable RG fixed point, we are led to the *exact* scaling relations

$$\Delta = 1 - \frac{\gamma_c^*}{2} = 1 - \frac{\eta}{2} = \frac{z}{2} - 1 \tag{8.93}$$

for the two-temperature model B, and therefore

$$\nu_\| = \nu(1 + \Delta) = \frac{\nu}{2}(4 - \eta) \,, \quad z_\| = \frac{z}{1 + \Delta} = 2 \,. \tag{8.94}$$

This leaves us again with only the two independent static critical exponents η and ν. To one-loop order, the UV poles in the integrals for the singular vertex functions $\Gamma^{(1,1)}$ and $\Gamma^{(1,3)}$ are identical (if evaluated in minimal subtraction), as we have seen earlier for the isotropic model. Thus, the critical exponents are simply determined by the associated combinatorial factors, which are unaltered by the anisotropies. Formally, to this order therefore the exponents look precisely as in equilibrium,

$$\eta = 0 + O(\epsilon'^2), \quad \frac{1}{\nu} = 2 - \frac{n+2}{n+8}\epsilon' + O(\epsilon'^2), \tag{8.95}$$

albeit with a modified $\epsilon' = d_c - d = 4 - d - d_\parallel$. (The straightforward explicit calculation is left to Problem 8.5.) To higher loop orders, the anisotropic propagator and noise correlation render the contributions from the nested integrals different from the isotropic model.

This analysis is readily extended to a two-temperature version of model D, generated from the standard equilibrium model D by allowing for anisotropic noise correlations (8.79), and in the same manner as for the conserved field.[15] Following the above procedure, only the transverse energy fluctuations are relevant near criticality, which leads to

$$\frac{\partial \rho(x, t)}{\partial t} = \lambda \nabla_\perp^2 \frac{\delta \mathcal{H}_{\text{eff}}[S, \rho]}{\delta \rho(x, t)} + \eta(x, t) \tag{8.96}$$

in addition to Eq. (8.85), with (3.48) added to the long-range Hamiltonian (8.86). From here, one proceeds just as in Section 6.1 for the equilibrium model D. To one-loop order, the specific heat exponent reads as in Eq. (6.40), but with the dimensional expansion parameter ϵ'. For $\alpha > 0$ ($n < 4$ to one-loop order), the two-temperature model B dynamic critical exponent $z_S = 4 - \eta$ for the order parameter is consequently supplemented with $z_\rho = 2 + \alpha/\nu$, with the new static exponents pertaining to the elastic Hamiltonian (8.86), whereas $z_\rho = 2$ for $\alpha \leq 0$ ($n \geq 4$).

While the relaxational models A and C with non-conserved order parameter are not affected by detailed balance violation, we see that spatially anisotropic noise drives the related models B and D with conserved diffusive dynamics to a different universality class characterized by *long-range* correlations already in the disordered phase, and concomitantly reduced upper critical dimension. The effects of spatial anisotropies in the conserved order parameter noise correlations appear to be even more drastic in models J and H. Here, the presence of reversible mode-coupling terms precludes the ensuing dynamics to correspond to an effective equilibrium system, since the divergence-free condition (3.58) for the probability current cannot be satisfied. Within a one-loop calculation, one finds no stable RG fixed point in the two-temperature models J and H.[16] Instead, the RG flows 'run away' to some perturbatively inaccessible strong-coupling regime, which might

---

[15] Akkineni and Täuber (2004).   [16] Täuber, Santos, and Rácz (1999); Santos and Täuber (2002).

indicate that simple scaling laws do not adequately capture the long-wavelength and large-time limits in these systems anymore.

8.4 Non-equilibrium work and fluctuation theorems

To conclude this chapter, we employ the response functional formalism to derive several important and general non-equilibrium fluctuation and work theorems specifically for the relaxational models A and B. These include the famous Jarzynski work theorem and Crooks' symmetry relation, as well as non-equilibrium generalizations of the fluctuation-dissipation theorem.[17]

8.4.1 Jarzynski's work theorem and Crooks' relation

We set out to investigate the following scenario. For $t \leq t_0 = 0$, a system described by the Hamiltonian $\mathcal{H}_0$ is prepared in its equilibrium stationary state at temperature T, governed by the canonical probability distribution (1.51). At $t = 0$, a time-dependent external field $h^\alpha(x, t)$ is switched on that drives the system away from equilibrium and induces non-trivial non-equilibrium stochastic dynamics. After a finite time $t = t_f$, the external field is held fixed, whence the system would relax to a new thermal equilibrium state with generally different Hamiltonian $\mathcal{H}_1$, provided one allows a sufficiently long time interval to elapse subsequently. In addition, we shall consider the effect of the *time-reversed protocol* for the external driving field $\hat{h}^\alpha(x, t) = h^\alpha(x, t_f - t)$.

To be concrete, we consider the $O(n)$-symmetric Landau–Ginzburg–Wilson Hamiltonian (4.12); for simplicity, we invoke purely relaxational kinetics for the non-conserved ($a = 0$) or conserved ($a = 2$) order parameter field $S^\alpha(x, t)$. For $t \leq 0$, the external field $h_0^\alpha(x)$ is maintained constant in time (but not necessarily uniform), and the associated Hamiltonian $\mathcal{H}_0[S]$ is stationary. In the finite time interval $[0, t_f]$, however, it acquires an *explicit* time dependence through the externally prescribed protocol $h^\alpha(x, t)$,

$$\frac{\partial \mathcal{H}[S](t)}{\partial t} = -\int d^d x \sum_{\alpha=1}^n \frac{\partial h^\alpha(x, t)}{\partial t} S^\alpha(x) . \qquad (8.97)$$

For $t \geq t_f$, the external field components are frozen at the values $h_1^\alpha(x) = h^\alpha(x, t_f)$, resulting again in a time-independent Hamiltonian $\mathcal{H}_1[S]$.

Following Section 4.1.2, the statistical average of an observable $A[S]$ with respect to the equilibrium canonical ensemble for $t < 0$ and the stochastic evolution

---

[17] This section closely follows the original work by Mallick, Moshe, and Orland (2011).

for $0 \leq t \leq t_f$ can be expressed as a functional integral

$$\langle A[S] \rangle = \int \mathcal{D}[S_0] \frac{1}{Z_0(T)} e^{-\mathcal{H}_0[S_0]/k_B T} \int \mathcal{D}[S_1]$$

$$\times \int \mathcal{D}[i\tilde{S}] \int_{S(x,0)=S_0}^{S(x,t_f)=S_1} \mathcal{D}[S] A[S] e^{-A[\tilde{S},S]} , \qquad (8.98)$$

with the Janssen–De Dominicis functional

$$A[\tilde{S}, S] = \int d^d x \int_0^{t_f} dt \sum_\alpha \tilde{S}^\alpha(x, t) \left[\frac{\partial S^\alpha(x, t)}{\partial t} + D(i\nabla)^a \frac{\delta \mathcal{H}[S](t)}{\delta S^\alpha(x, t)} \right.$$

$$\left. - k_B T D(i\nabla)^a \tilde{S}^\alpha(x, t) \right] , \qquad (8.99)$$

where the temperature factor in the noise correlator has been reinstated.

We now apply a variable transformation to new Martin–Siggia–Rose auxiliary fields

$$\bar{S}^\alpha(x, t) = -\tilde{S}^\alpha(x, t) + \frac{1}{k_B T} \frac{\delta \mathcal{H}[S]}{\delta S^\alpha(x, t)} , \qquad (8.100)$$

which leaves the functional integration measure invariant, but (for $a = 2$, after twice integrating by parts) transforms the response functional to

$$A[\bar{S}, S] = \int d^d x \int_0^{t_f} dt \sum_\alpha \left(\bar{S}^\alpha(x, t) \left[\frac{\partial S^\alpha(x, t)}{\partial(-t)} + D(i\nabla)^a \frac{\delta \mathcal{H}[S](t)}{\delta S^\alpha(x, t)} \right. \right.$$

$$\left. \left. - k_B T D(i\nabla)^a \bar{S}^\alpha(x, t) \right] + \frac{1}{k_B T} \frac{\delta \mathcal{H}[S]}{\delta S^\alpha(x, t)} \frac{\partial S^\alpha(x, t)}{\partial t} \right) . \qquad (8.101)$$

The last term here may be rewritten as follows:

$$\int d^d x \sum_\alpha \frac{\delta \mathcal{H}[S]}{\delta S^\alpha(x, t)} \frac{\partial S^\alpha(x, t)}{\partial t} = \int_0^{t_f} \left(\frac{d\mathcal{H}[S](t)}{dt} - \frac{\partial \mathcal{H}[S](t)}{\partial t} \right) dt$$

$$= \mathcal{H}_1[S_1] - \mathcal{H}_0[S_0] - W_J[S] , \qquad (8.102)$$

with *Jarzynski's non-equilibrium work* defined as

$$W_J[S] = \int_0^{t_f} \frac{\partial \mathcal{H}[S](t)}{\partial t} dt ; \qquad (8.103)$$

through (8.97) it contains all the information about the external field protocol that drives the system through a series of non-equilibrium states.

Upon time inversion, the action (8.101) reverts to its original form (8.99), aside from the contributions (8.102). One may therefore directly relate the expectation value (8.98) to its counterpart for the observable $\hat{A}[S^\alpha(x, t)] = A[S^\alpha(x, t - t_f)]$

pertaining to the time-reversed field protocol $\hat{h}^\alpha(x,t)$:

$$\langle A[S]\rangle = \frac{Z_1(T)}{Z_0(T)} \int \mathcal{D}[S_0] \int \mathcal{D}[S_1] \frac{1}{Z_1(T)} e^{-\mathcal{H}_1[S_1]/k_B T}$$

$$\times \int \mathcal{D}[i\bar{S}] \int_{S(x,t_f)=S_1}^{S(x,0)=S_0} \mathcal{D}[S]\, \hat{A}[S]\, e^{\tilde{W}_J[S]-\mathcal{A}[\bar{S},S]}.$$

With $\Delta F = F_1 - F_0 = k_B T \ln[Z_0(T)/Z_1(T)]$ and noting that Jarzynski's work is *odd* under time reversal, this yields the basic relation

$$\langle A[S]\rangle = e^{-\Delta F/k_B T} \left\langle \hat{A}[S]\, e^{-W_J[S]/k_B T} \right\rangle_R, \tag{8.104}$$

or equivalently, setting $A'[S] = A[S]\, e^{-W_J[S]/k_B T}$,

$$\left\langle A[S]\, e^{-W_J[S]/k_B T} \right\rangle = e^{-\Delta F/k_B T} \langle \hat{A}[S]\rangle_R. \tag{8.105}$$

Setting $A[S] = 1$, one arrives at *Jarzynski's work theorem*

$$\left\langle e^{-W_J[S]/k_B T} \right\rangle = e^{-\Delta F/k_B T}, \tag{8.106}$$

which provides a remarkable connection between the free energy difference of the equilibrium configurations for the initial and final Hamiltonians (although the latter in fact need never be reached) with the statistical average of the exponential of the non-equilibrium work (8.103). Since $e^{-x} \geq 1 - x$ (concavity of the exponential function), the theorem (8.106) implies the inequality

$$\Delta F = -k_B T \ln \left\langle e^{-W_J[S]/k_B T} \right\rangle \geq \langle W_J[S]\rangle \tag{8.107}$$

that constrains the extractable non-equilibrium work by the equilibrium free energy difference, a standard corollary of the second law of thermodynamics.

Next let $A[S] = e^{-\lambda W_J[S]}$ with arbitrary real λ; Eq. (8.104) then gives

$$\left\langle e^{-\lambda W_J[S]} \right\rangle = e^{-\Delta F/k_B T} \left\langle e^{\lambda W_J[S]}\, e^{-W_J[S]/k_B T} \right\rangle_R. \tag{8.108}$$

This can be reformulated as a symmetry property for the *characteristic function* associated with the probability distribution $\mathcal{P}(W_J)$ for the non-equilibrium work:

$$\chi(\lambda) = \int \mathcal{P}(W_J)\, e^{-\lambda W_J} dW_J = \left\langle e^{-\lambda W_J[S]} \right\rangle = e^{-\beta\Delta F} \chi(\beta - \lambda)_R, \tag{8.109}$$

with $\beta = 1/k_B T$. Laplace backtransform of Eq. (8.109) immediately yields *Crooks' relation* for the work probability distributions of the forward and reverse processes

$$\frac{\mathcal{P}(W_J)}{\mathcal{P}_R(-W_J)} = e^{(W_J-\Delta F)/k_B T}. \tag{8.110}$$

Jarzynski's and Crooks' theorems are quite general and can be derived in a variety of different settings, ranging from classical Hamiltonian to Markovian stochastic dynamics. For example, allowing for reversible mode-coupling terms in the Langevin equation merely adds a cumbersome contribution to the non-equilibrium work (8.103), see Problem 8.6, but does not otherwise alter the fundamental results (8.104)–(8.110).

8.4.2 Non-equilibrium FDT generalizations

Through functional derivatives of the fundamental relation (8.105) with respect to the external field $h^\alpha(x, t)$, one may derive a series of relations between dynamic response and correlation functions that apply to the above non-equilibrium scenario induced by the explicitly time-dependent protocol. They all represent generalizations of the equilibrium fluctuation-dissipation theorem, which naturally is recovered by formally setting $\Delta F = 0 = W_J[S]$.

From (8.103) and (8.97) one obtains $\delta W_J[S]/\delta h^\beta(x', t') = \partial S^\beta(x', t')/\partial t'$ after integrating by parts, whence the functional derivative of Eq. (8.105) leads to

$$\frac{\delta}{\delta h^\beta(x', t')}\langle A[S]\,e^{-W_J[S]/k_B T}\rangle - e^{-\Delta F/k_B T}\,\frac{\delta\langle\hat{A}[S]\rangle_R}{\delta\hat{h}^\beta(x', t')}$$

$$= \frac{1}{k_B T}\left\langle A[S]\,\frac{\partial S^\beta(x', t')}{\partial t'}\,e^{-W_J[S]/k_B T}\right\rangle. \tag{8.111}$$

On the left-hand side, the functional derivative only affects the observable $A[S]$, i.e., refers to fixed external protocol and hence $W_J[S]$. While the physical implications of (8.111) and similar general identities are perhaps not immediately transparent, they clearly contain the equilibrium FDT as a special case: just replacing the observable with $S^\alpha(x, t)$ and letting $\Delta F = 0 = W_J[S]$ yields the fluctuation-dissipation theorem (4.25) in the form

$$\frac{\delta\langle S^\alpha(x, t)\rangle}{\delta h^\beta(x', t')} - \frac{\delta\langle S^\alpha(x, t_f - t)\rangle}{\delta h^\beta(x', t_f - t')} = \frac{1}{k_B T}\frac{\partial}{\partial t'}\langle S^\alpha(x, t)S^\beta(x', t')\rangle. \tag{8.112}$$

8.5 Appendix: general Gaussian fluctuations

Let $S_s^\alpha(x)$ denote the solution of the zero-field Landau–Ginzburg equation, i.e., the classical field equation obtained through the saddle point approximation, say for the Hamiltonian (4.12),

$$0 = \frac{\delta\mathcal{H}_{h=0}[S]}{\delta S^\alpha(x)}\bigg|_{S=S_s} = \left[r - \nabla^2 + \frac{u}{6}\sum_{\beta=1}^{n} S_s^\beta(x)^2\right]S_s^\alpha(x). \tag{8.113}$$

We wish to compute the partition function in the Gaussian approximation, and therefore expand the Hamiltonian to quadratic order in the fluctuations $\delta S^\alpha(x) = S^\alpha(x) - S_s^\alpha(x)$:

$$\mathcal{H}[S] = \mathcal{H}[S_s] - \int d^d x \sum_\alpha h^\alpha(x)\, \delta S^\alpha(x)$$

$$+ \frac{1}{2} \int d^d x \int d^d x' \sum_{\alpha,\beta} \delta S^\alpha(x)\, G_0^{\alpha\beta}(x,x')^{-1}\, \delta S^\beta(x') + \cdots . \qquad (8.114)$$

Here we have introduced

$$G_0^{\alpha\beta}(x,x')^{-1} = \frac{\delta^2 \mathcal{H}_{h=0}[S]}{\delta S^\alpha(x)\, \delta S^\beta(x')}\bigg|_{S=S_s}$$

$$= \left(\left[r - \nabla^2 + \frac{u}{6} \sum_{\beta=1}^n S_s^\beta(x)^2 \right] \delta^{\alpha\beta} + \frac{u}{3} S_s^\alpha(x)\, S_s^\beta(x) \right) \delta(x - x') ,$$

$$(8.115)$$

which is the inverse Green function for the differential operator in brackets.

Next we recall the Gaussian integral formula for general bilinear coupling between the vector integration variables S^α,

$$\int \prod_{\alpha=1}^n dS^\alpha\, e^{-\frac{1}{2} \sum_{\alpha,\beta} S^\alpha\, C^{\alpha\beta}\, S^\beta + \sum_\alpha j^\alpha\, S^\alpha} = \sqrt{\frac{(2\pi)^n}{\det C}}\, e^{\frac{1}{2} \sum_{\alpha,\beta} j^\alpha\, C^{-1\,\alpha\beta}\, j^\beta} .$$

This assumes that the matrix $C^{\alpha\beta}$ can be diagonalized; after transforming to its eigenvectors and completing the square, one has n independent Gaussian integrals, and the product of the eigenvalues gives the determinant $\det C$. Upon rewriting $\ln \sqrt{(2\pi)^n / \det C} = -\frac{1}{2} \operatorname{Tr} \ln(C/2\pi)$, we then formally obtain for the free energy in our general Gaussian approximation

$$F_0[h] = \mathcal{H}[S_s] + \frac{k_B T}{2} \operatorname{Tr} \ln \frac{G_0^{\alpha\beta}(x,x')^{-1}}{2\pi k_B T}$$

$$- \frac{1}{2} \int d^d x \int d^d x' \sum_{\alpha,\beta} h^\alpha(x)\, G_0^{\alpha\beta}(x,x')\, h^\beta(x') . \qquad (8.116)$$

The trace operation here should be understood as follows:

$$\operatorname{Tr} C^{\alpha\beta}(x,x') = \int d^d x \sum_\alpha C^{\alpha\alpha}(x,x) = V \int \frac{d^d q}{(2\pi)^d} \sum_\alpha C^{\alpha\alpha}(q) ,$$

where the final expression is valid for translationally invariant systems.

For example, in the Landau–Ginzburg equation we have $S_s^\alpha = 0$ for $r > 0$, and $G_0^{\alpha\beta}(q, q') = (r + q^2)^{-1} \delta^{\alpha\beta} (2\pi)^d \delta(q + q')$, whence

$$F_0[h] = -V k_B T \frac{n}{2} \int \frac{d^d q}{(2\pi)^d} \ln \frac{2\pi k_B T}{r + q^2} - \frac{1}{2} \int \frac{d^d q}{(2\pi)^d} \sum_{\alpha=1}^{n} \frac{|h^\alpha(q)|^2}{r + q^2}, \quad (8.117)$$

compare with Eq. (1.61). In the low-temperature phase ($r < 0$), we have $S_s^\alpha = \sqrt{6|r|/u}\, \delta^{\alpha n}$, and consequently $G_0^{nn}(q, q') = (2|r| + q^2)^{-1} (2\pi)^d \delta(q + q')$ and $G_0^{\alpha\beta}(q, q') = q^{-2} \delta^{\alpha\beta} (2\pi)^d \delta(q + q')$ for the massless transverse fluctuations with $\alpha = 1, \ldots, n - 1$. This yields, with the Landau free energy contribution (1.47), see Problem 1.5,

$$F_0[h] = -V \frac{3r^2}{2u} - V \frac{k_B T}{2} \int \frac{d^d q}{(2\pi)^d} \left[\ln \frac{2\pi k_B T}{2|r| + q^2} + (n - 1) \ln \frac{2\pi k_B T}{q^2} \right]$$

$$- \frac{1}{2} \int \frac{d^d q}{(2\pi)^d} \left[\frac{|h^n(q)|^2}{2|r| + q^2} + \sum_{\alpha=1}^{n-1} \frac{|h^n(q)|^2}{q^2} \right]. \quad (8.118)$$

Problems

8.1 *Gaussian propagators with specified initial conditions*
Write down and solve the classical field equations $\delta \mathcal{Z}_0 / \delta \tilde{S}^\alpha = 0 = \delta \mathcal{Z}_0 / \delta S^\alpha$ for the harmonic generating functional $\mathcal{Z}_0[\tilde{j}, j]$ corresponding to the action (8.3). Thus, obtain the Gaussian propagators (8.10), (8.11) for the relaxational models with initial conditions given by (8.1).

8.2 *Gaussian fluctuation-dissipation ratios in real space*
(a) Using the Dirichlet correlator (8.12), confirm the result (8.16) for the model A Gaussian fluctuation-dissipation ratio in real space.
(b) Compute $X_0(x = 0; t > t', t')$ for the Gaussian models A/B at T_c with finite Δ, and thus show that the terms $\propto 1/\Delta$ represent corrections to the leading scaling behavior given in Eqs. (8.16) and (8.17).

8.3 *Order parameter initial-slip scaling*
Starting from the expansion (8.40), derive the initial-slip scaling law (8.6) for the order parameter.

8.4 *Spherical model critical exponents*
(a) Use Eq. (8.49) for the propagator and the self-consistency condition (8.53) to compute the static critical exponents for the spherical model. (The fluctuation-induced critical temperature shift must be determined first; the result also shows that the lower critical dimension is $d_{lc} = 2$.)
(b) Determine the logarithmic corrections for the susceptibility at the upper critical dimension $d_c = 4$.

(c) For the spherical relaxational models A and B, confirm that the dynamic critical exponent is $z = 2 + a$.

(d) Establish that $\theta = 1 - d/4$ for the spherical model A.

8.5 *One-loop integrals and exponents for the two-temperature model B*

For model B with conserved order parameter and anisotropic noise, as defined by Eqs. (8.83) and (8.84), explicitly evaluate the UV-divergent integrals for $\Gamma^{(1,1)}$ and $\Gamma^{(1,3)}$, and thereby confirm the result (8.95).

8.6 *Modified non-equilibrium work induced by reversible mode couplings*

Consider general Langevin equations (3.60) that contain both relaxational and reversible mode-coupling terms (3.59), see Section 3.3.2.

(a) Establish the relation (8.104) for this more general situation, where Jarzynski's work (8.103) picks up the additional contribution

$$\Delta W_J[\psi] = \int d^d x \int_0^{t_f} dt \sum_\alpha \Big[(L^\alpha)^{-1} F_{\text{rev}}^\alpha [\psi](x, t) \Big] \frac{\partial \psi^\alpha(x, t)}{\partial t} \, .$$

Here, $L^\alpha = D^\alpha (i\nabla)^{a_\alpha}$ denote the Onsager coefficients, and $a_\alpha = 0, 2$ for non-conserved and conserved fields, respectively.

(b) Derive generalizations of Eq. (8.111) and the equilibrium FDT (8.112) for non-equilibrium dynamics involving mode-coupling terms.

References

Akkineni, V. K. and U. C. Täuber, 2004, Non-equilibrium critical dynamics of the relaxational models C and D, *Phys. Rev. E* **69**, 036113-1–25.

Biroli, G., 2005, A crash course on ageing, *J. Stat. Mech.*, P05014-1–19.

Bray, A. J., 1993, Domain growth and coarsening, in: *Phase Transitions and Relaxation in Systems with Competing Energy Scales*, eds. T. Riste and D. Sherrington, NATO ASI Series C **415**, Dordrecht: Kluwer Academic, 405–436.

Bray, A. J., 1994, Theory of phase ordering kinetics, *Adv. Phys.* **43**, 357–459.

Bray, A. J. and A. D. Rutenberg, 1994, Growth laws for phase ordering, *Phys. Rev. E* **49**, R27–R30.

Calabrese, P. and A. Gambassi, 2005, Ageing properties of critical systems, *J. Phys. A: Math. Gen.* **38**, R133–R193.

Diehl, H. W., 1986, Field-theoretic approach to critical behaviour at surfaces, in: *Phase Transitions and Critical Phenomena*, Vol. 10, eds. C. Domb and J. L. Lebowitz, London: Academic Press, 75–267.

Henkel, M. and M. Pleimling, 2010, *Non-equilibrium Phase Transitions, Vol. 2: Ageing and Dynamical Scaling Far from Equilibrium*, Dordrecht: Springer.

Janssen, H. K., 1992, On the renormalized field theory of nonlinear critical relaxation, in: *From Phase Transitions to Chaos*, eds. G. Györgyi, I. Kondor, L. Sasvári, and T. Tél, Singapore: World Scientific, 68–91.

Janssen, H. K., B. Schaub, and B. Schmittmann, 1989, New universal short-time scaling behaviour of critical relaxation processes, *Z. Phys. B Cond. Matt.* **73**, 539–549.

Majumdar, S. N., A. J. Bray, S. J. Cornell, and C. Sire, 1996, Global persistence exponent for nonequilibrium critical dynamics, *Phys. Rev. Lett.* **77**, 3704–3707.

Mallick, K., M. Moshe, and H. Orland, 2011, A field-theoretic approach to non-equilibrium work identities, *J. Phys. A: Math. Theor.* **44**, 095002-1–19.
Oerding, K. and H. K. Janssen, 1993a, Non-equilibrium critical relaxation with coupling to a conserved quantity, *J. Phys. A: Math. Gen.* **26**, 3369–3381.
Oerding, K. and H. K. Janssen, 1993b, Non-equilibrium critical relaxation with reversible mode coupling, *J. Phys. A: Math. Gen.* **26**, 5295–5303.
Oerding, K., S. J. Cornell, and A. J. Bray, 1997, Non-Markovian persistence and nonequilibrium critical dynamics, *Phys. Rev. E* **56**, R25–R28.
Santos, J. E. and U. C. Täuber, 2002, Non-equilibrium behavior at a liquid-gas critical point, *Eur. Phys. J. B* **28**, 423–440.
Schmittmann, B., 1993, Fixed point Hamiltonian for a randomly driven diffusive system, *Europhys. Lett.* **24**, 109–114.
Schmittmann, B. and R. K. P. Zia, 1991, Critical properties of a randomly driven diffusive system, *Phys. Rev. Lett.* **66**, 357–360.
Täuber, U. C., V. K. Akkineni, and J. E. Santos, 2002, Effects of violating detailed balance on critical dynamics, *Phys. Rev. Lett.* **88**, 045702-1–4.
Täuber, U. C. and Z. Rácz, 1997, Critical behavior of $O(n)$-symmetric systems with reversible mode-coupling terms: stability against detailed-balance violation, *Phys. Rev. E* **55**, 4120–4136.
Täuber, U. C., J. E. Santos, and Z. Rácz, 1999, Non-equilibrium critical behavior of $O(n)$-symmetric systems: effect of reversible mode-coupling terms and dynamical anisotropy, *Eur. Phys. J. B* **7**, 309–330; err. *Eur. Phys. J.* **9**, 567–568.

Further reading

Aron, C., G. Biroli, and L. F. Cugliandolo, 2010, Symmetries of generating functionals of Langevin processes with colored multiplicative noise, *J. Stat. Mech.*, P11018-1–40.
Baumann, F. and M. Henkel, 2007, Kinetics of phase separation in the critical spherical model and local scale invariance, *J. Stat. Mech.*, P07012-1–27.
Bonart, J., L. F. Cugliandolo, and A. Gambassi, 2011, Critical Langevin dynamics of the $O(N)$ Ginzburg–Landau model with correlated noise, *J. Stat. Mech.*, P01014-1–41.
Bray, A. J., 1989, Exact renormalization-group results for domain-growth scaling in spinodal decomposition, *Phys. Rev. Lett.* **62**, 2841–2844.
Bray, A. J., 1990, Renormalization-group approach to domain-growth scaling, *Phys. Rev. B* **41**, 6724–6732.
Calabrese, P. and A. Gambassi, 2002, Two-loop critical fluctuation-dissipation ratio for the relaxational dynamics of the $O(N)$ Landau–Ginzburg Hamiltonian, *Phys. Rev. E* **66**, 066101-1–12.
Calabrese, P. and A. Gambassi, 2007, Slow dynamics in critical ferromagnetic vector models relaxing from a magnetized initial state, *J. Stat. Mech.*, P07001-1–40.
Calabrese, P., A. Gambassi, and F. Krzakala, 2006, Critical ageing of Ising ferromagnets relaxing from an ordered state, *J. Stat. Mech.*, P06016-1–35.
Caracciolo, S., A. Gambassi, M. Gubinelli, and A. Pelissetto, 2005, Critical behavior of the two-dimensional randomly driven lattice gas, *Phys. Rev. E* **72**, 056111-1–4.
Cardy, J. L., 1992, Random initial conditions and nonlinear relaxation, *J. Phys. A: Math. Gen.* **25**, 2765–2790.
Coniglio, A. and M. Zanetti, 1990, Renormalization group for growth kinetics in the large-N limit, *Phys. Rev. B* **42**, 6873–6876.
Das, S. K., S. Roy, S. Majumder, and S. Ahmad, 2012, Finite-size effects in dynamics: critical vs. coarsening phenomena, *EPL* **97**, 66006-1–6.

Fedorenko, A. A. and S. Trimper, 2006, Critical aging of a ferromagnetic system from a completely ordered state, *Europhys. Lett.* **74**, 89–95.

Godréche, C. and J. M. Luck, 2013, Asymmetric Langevin dynamics for the ferromagnetic spherical model, *J. Stat. Mech.*, P05006-1–43.

Henkel, M., M. Pleimling, and R. Sanctuary (eds.), 2007, *Ageing and the Glass Transition*, Lecture Notes in Physics **716**, Berlin: Springer.

Kissner, J. G. and A. J. Bray, 1992, Dynamic correlations in phase ordering: the $1/n$-expansion reconsidered, *J. Phys. A: Math. Gen.* **26**, 1571–1588.

Mazenko, G. F. and M. Zanetti, 1984, Growth of order in a system with continuous symmetry, *Phys. Rev. Lett.* **53**, 2106–2109.

Mazenko, G. F. and M. Zanetti, 1985, *Instability, spinodal decomposition, and nucleation in a system with continuous symmetry*, *Phys. Rev. B* **32**, 4565–4575.

Nam, K., B. Kim, and S. J. Lee, 2011, Coarsening kinetics of a two-dimensional $O(2)$ Ginzburg–Landau model: the effect of reversible mode coupling, *J. Stat. Mech.*, P03013-1–15.

Newman, T. J. and A. J. Bray, 1990a, New exponent for dynamic correlations in domain growth, *J. Phys. A: Math. Gen.* **23**, L279–L284.

Newman, T. J. and A. J. Bray, 1990b, Dynamic correlations in domain growth: a $1/n$ expansion, *J. Phys. A: Math. Gen.* **23**, 4491–4507.

Newman, T. J., A. J. Bray, and M. A. Moore, 1990, Growth of order in vector spin systems and self-organized criticality, *Phys. Rev. B* **42**, 4514–4523.

Patashinski, A., 2001, Nonequilibrium critical phenomena, *Physica A* **292**, 452–464.

Paul, R., A. Gambassi, and G. Schehr, 2007, Dynamic crossover in the global persistence at criticality, *Europhys. Lett.* **78**, 10007-1–5.

Prudnikov, P. V., V. V. Prudnikov, and I. A. Kalashnikov, 2008, Renormalization-group description of nonequilibrium critical short-time relaxation processes: a three-loop approximation, *JETP* **106**, 1095–1101.

Sciolla, B. and G. Biroli, 2013, Quantum quenches, dynamical transitions, and off-equilibrium quantum criticality, *Phys. Rev. B* **88**, 20110(R)-1–5.

Sieberer, L. M., S. D. Huber, E. Altman, and S. Diehl, 2013, Dynamical critical phenomena in driven-dissipative systems, *Phys. Rev. Lett.* **110**, 195301-1–5.

Zanetti, M. and G. F. Mazenko, 1987, Anomalous time correlations in quenched systems with continuous symmetry, *Phys. Rev. B* **35**, 5043–5047.

9

Reaction-diffusion systems

This chapter addresses the stochastic dynamics of interacting particle systems, specifically reaction-diffusion models that, for example, capture chemical reactions in a gel such that convective transport is inhibited. Generic reaction-diffusion models are in fact utilized to describe a multitude of phenomena in various disciplines, ranging from population dynamics in ecology, competition of bacterial colonies in microbiology, dynamics of magnetic monopoles in the early Universe in cosmology, equity trading on the stock market in economy, opinion exchange in sociology, etc. More concrete physical applications systems encompass excitons kinetics in organic semconductors, domain wall interactions in magnets, and interface dynamics in growth models. Yet most of our current knowledge in this area stems from extensive computer simulations, and actual experimental realizations allowing accurate quantitative analysis are still deplorably rare. We begin with a brief review of mean-field and scaling arguments including Smoluchowski's self-consistent treatment of diffusion-limited binary annihilation. The main goal of this chapter is to demonstrate how one may systematically proceed from a microscopic master equation for interacting particles, which perhaps represents the most straightforward description of a system far from equilibrium, to a non-Hermitean bosonic 'quantum' many-body Hamiltonian, and thence to a continuum field theory representation that permits subsequent perturbative expansions and renormalization group treatment. The ensuing rich physics is illustrated with simple examples that include the annihilation reactions $k A \rightarrow l A$ $(l < k)$ and $A + B \rightarrow \emptyset$, their generalization to multiple particle species, as well as reversible recombination $A + A \rightleftharpoons B$. We also explore the crucial role of conservation laws and the emergence and scaling properties of reaction fronts. Finally, we address the spontaneous formation of complex spatio-temporal structures and erratic population oscillations in stochastic Lotka–Volterra models for predator–prey coexistence.

9.1 Rate equations and scaling theory

We turn our attention to *stochastic interacting particle systems*, whose microscopic dynamics is ultimately defined through a (classical) master equation. Neglecting spatio-temporal correlations and fluctuations, one arrives at the associated *mean-field rate equations* that incorporate the chemical law of mass action as a consequence of the approximate factorization of correlation functions into density products. This chapter also demonstrates how Smoluchowski's insightful physical intuition led to a refined self-consistent mean-field approximation that adequately captures dynamically generated particle anticorrelations for single-species pair annihilation processes. We shall later see how a controlled, systematic renormalization group treatment elegantly confirms the underlying assumptions of Smoluchowski's and related scaling approaches.

9.1.1 Chemical reaction kinetics

We set out to investigate systems of interacting 'particles' $A, B, \ldots$ that propagate through hopping to nearest neighbors on a d-dimensional lattice, or, in the continuum limit, via diffusion. Either spontaneously or upon encounter, these particles may undergo species transformations, annihilate, or produce offspring with prescribed reaction rates. Such basic stochastic processes find broad application in various academic fields; here, we will interchangeably use the language of chemical reactions and population dynamics.[1] Already two fundamentally different regimes must be distinguished owing to the competition of diffusive spreading and local reactivity. At sufficiently large particle densities, the characteristic time scales of the kinetics will be governed by the reaction rates, and the system is termed *reaction-limited*. In contrast, at low densities, any reactions that require at least two particles to be in proximity become rare and *diffusion-limited*; the basic time scale is in this case set by the hopping rate or diffusion coefficient.

For an initial approach to the dynamics of such 'chemical' reactions, let us assume homogeneous mixing of each species. We then attempt to capture the kinetics in terms of *rate equations* for each particle concentration or mean density. Note that such a description neglects any spatial fluctuations and correlations in the system, constituting in essence a *mean-field* approximation. For concreteness, consider the generic *reversible* on-site reaction scheme $k\,A + l\,B \rightleftharpoons m\,C$ with rates λ and σ assigned to the forward 'fusion' and backward 'fission' processes, respectively. The system is fully characterized by the local occupation numbers $n_\alpha = 0, 1, \ldots$ for species $\alpha = A, B, C$ (we omit the lattice site index for now).

---

[1] For general overviews, see Kuzovkov and Kotomin (1988); Ovchinnikov, Timashev, and Belyy (1989); Krapivsky, Redner, and Ben-Naim (2010).

We may then construct the *master equation* for the configurational probability $P(n_A, n_B, n_C; t)$ to have n_α particles of species α present at time t, subject to the above stochastic chemical reaction kinetics (see Section 2.3.2):

$$\frac{\partial P(n_A, n_B, n_C; t)}{\partial t} = \lambda \frac{(n_A + k)!}{n_A!} \frac{(n_B + l)!}{n_B!} P(n_A + k, n_B + l, n_C - m; t)$$

$$+ \sigma \frac{(n_C + m)!}{n_C!} P(n_A - k, n_B - l, n_C + m; t)$$

$$- \left[\lambda \frac{n_A!}{(n_A - k)!} \frac{n_B!}{(n_B - l)!} + \sigma \frac{n_C!}{(n_C - m)!} \right] P(n_A, n_B, n_C; t),$$

$$(9.1)$$

with the understanding that $P(n_A, n_B, n_C; t) = 0$ if any $n_\alpha < 0$, and where the factorials account for the number of different possibilities of picking the *uncorrelated* reactants (in a given order) from the available particle pools.[2]

For the temporal evolution of the average particle numbers $\langle n_\alpha(t) \rangle = \sum_{n_A, n_B, n_C = 0}^{\infty} n_\alpha P(n_A, n_B, n_C; t)$, the master equation (9.1) yields

$$R(t) = -\frac{1}{k} \frac{d\langle n_A(t) \rangle}{dt} = -\frac{1}{l} \frac{d\langle n_B(t) \rangle}{dt} = -\frac{1}{m} \frac{d\langle n_C(t) \rangle}{dt}$$

$$= \lambda \langle [n_A(n_A - 1) \cdots (n_A - k + 1) n_B(n_B - 1) \cdots (n_B - l + 1)](t) \rangle$$

$$- \sigma \langle [n_C(n_C - 1) \cdots (n_C - m + 1)](t) \rangle, \qquad (9.2)$$

which implies the obvious exact conservation laws (for integer $k, l, m > 0$)

$$l \langle n_A(t) \rangle - k \langle n_B(t) \rangle = \text{const.}, \qquad (9.3)$$

$$m \langle n_A(t) \rangle + k \langle n_C(t) \rangle = \text{const.}, \quad m \langle n_B(t) \rangle + l \langle n_C(t) \rangle = \text{const.} \qquad (9.4)$$

Equation (9.2) states that the changes in the particle concentrations are determined by non-linear correlators. As is common in this type of problem, solving an infinite hierarchy of coupled differential equations for N-point correlation functions would be required to fully describe the system's dynamics. A massive simplification, in the spirit of mean-field theory, is fashioned by brute-force factorization of these higher-order correlations into simple density products. For large particle numbers $\langle n_\alpha(t) \rangle \gg k, l, m$, the reaction terms on the right-hand side of Eq. (9.2) thus become replaced by the expression $R(t) \approx \lambda \langle n_A(t) \rangle^k \langle n_B(t) \rangle^l - \sigma \langle n_C(t) \rangle^m$, resulting in a

---

[2] If instead of this convention arbitrary reactant orderings are allowed, the factorial quotients in Eq. (9.1) become replaced with binomial factors, which in turn amounts to a mere rescaling of the reaction rates, $\lambda \to \lambda/k!l!$ and $\sigma \to \sigma/m!$. In our convention, the rates λ, σ indicate the actual number of reactions occurring per unit time, see Eq. (9.2).

closed set of coupled *rate equations* for the mean densities, henceforth denoted as $a(t) = \langle n_A(t) \rangle$, etc. After a sufficient amount of time $t \gg 1/\lambda$, $1/\sigma$ has elapsed, the system equilibrates to a stationary state $R(t \to \infty) \to 0$ wherein the stochiometric density product adjusts itself to a ratio

$$\frac{c_\infty^m}{a_\infty^k \, b_\infty^l} = \frac{\lambda}{\sigma} \tag{9.5}$$

determined by the reaction rates. This is precisely the statement of the classical *law of mass action*; in thermal equilibrium, the logarithm of (9.5) may be related to the net reaction enthalpy. Increasing the forward rate λ naturally shifts the chemical equilibrium in favor of the C concentration, whereas conversely a larger fission rate σ enhances the density product of the A and B species. If either reaction rate is set to zero, the reaction becomes *irreversible*, irretrievably approaching a state with disappearing C species for $\lambda = 0$, or conversely $a(t)\, b(t) \to 0$ if $\sigma = 0$.

9.1.2 Mean-field rate equations

As a first specific example, let us study the single-species *annihilation* kinetics in the irreversible kth-order reaction $k\, A \to l\, A$, with $k > l$ and fixed reaction rate λ. The corresponding rate equation for the average particle number $a(t) = \langle n_A(t) \rangle$ reads

$$\frac{da(t)}{dt} = -(k - l)\lambda \, a(t)^k . \tag{9.6}$$

This ordinary differential equation is readily solved, with the result

$$k = 1: \ a(t) = a_0 \, e^{-\lambda t} , \tag{9.7}$$

$$k \geq 2: \ a(t) = \left[a_0^{1-k} + (k - l)(k - 1)\lambda \, t \right]^{-1/(k-1)} , \tag{9.8}$$

where $a_0 = a(0)$. Simple 'radioactive' decay ($k = 1$) of course gives an exponential time dependence, indicative of statistically independent events. For pair ($k = 2$) and higher-order ($k \geq 3$) processes, algebraic long-time behavior ensues, $a(t) \to [(k - 1)(k - 1)\lambda \, t]^{-1/(k-1)}$, with an amplitude that ultimately becomes independent of the initial density a_0 (but still depends on the stochiometric numbers k and l). The absence of a characteristic time scale hints at cooperative effects, and the crucial question arises whether and under which circumstances correlations might qualitatively affect the asymptotic long-time power laws, and cause deviations from the rate equation predictions. As in equilibrium critical phenomena, the influence of spatial fluctuations is expected to diminish with increasing dimensionality, and we shall see that the notion of an upper critical dimension d_c separating qualitatively mean-field behavior at $d > d_c$ from distinctly correlation-dominated scaling laws in low dimensions $d \leq d_c$ carries over as well (Section 9.3).

Particle annihilation processes invariably generate zones of depleted density as long as diffusive spreading remains ineffective at maintaining a homogeneous well-mixed state. In low dimensions, the ensuing particle *anti*correlations dominate the system's dynamics. For *two-species* pair annihilation $A + B \to \emptyset$ (without mixing), particle *segregation* into A- or B-rich domains further complicates and enriches the kinetics (Section 9.4). Regions dominated by either species become largely inert, and the annihilation processes are confined to rather sharp *reaction fronts*. These distinctly spatial effects are of course absent in the simplest mean-field rate equation description, which assumes homogeneous particle mixing and ignores correlations.

For two-species pair annihilation, one obtains two coupled rate equations,

$$\frac{da(t)}{dt} = \frac{db(t)}{dt} = -\lambda \, a(t) \, b(t) \,. \tag{9.9}$$

Crucially, however, the reaction $A + B \to \emptyset$ preserves the difference of particle numbers (even locally), i.e., there exists a *conservation law* for $c(t) = a(t) - b(t) = c_0$, see Eq. (9.3) for $k = 1 = l$. In the special situation of exactly *equal* initial densities $a_0 = b_0$, i.e., $c_0 = 0$, both rate equations (9.9) become identical, and are solved by

$$a(t) = \frac{a_0}{1 + a_0 \lambda t} = b(t) \,, \tag{9.10}$$

which is just the single-species pair annihilation mean-field power law (9.8). For *unequal* initial densities $c_0 = a_0 - b_0 > 0$, say, the majority species eventually reaches a saturation value $a(t \to \infty) \to a_\infty = c_0 > 0$, while the minority population disappears: $b(t \to \infty) \to 0$. From Eqs. (9.9), one deduces an exponential asymptotic approach towards $a_\infty = c_0$ and $b_\infty = 0$,

$$a(t) - c_0 \sim e^{-c_0 \lambda t} \sim b(t) \,. \tag{9.11}$$

The effective decay rate here is $c_0 \lambda$; hence $c_0 = 0$ is clearly discernible as a special critical point for this stochastic process.

Competition between particle decay and production processes, e.g., in the reactions $A \to \emptyset$ (with rate κ), $A \rightleftharpoons A + A$ (with forward and backward rates σ and λ, respectively), leads to even richer scenarios, as can already be inferred from the associated rate equation

$$\frac{da(t)}{dt} = (\sigma - \kappa) \, a(t) - \lambda \, a(t)^2 \,. \tag{9.12}$$

For $\sigma < \kappa$, obviously $a(t \to \infty) \sim e^{-(\kappa - \sigma)t} \to 0$. The system eventually enters an *inactive* state, which even in the fully stochastic model remains *absorbing*: since once there is no particle left, none of the allowed processes can drive the system out of the empty state again. On the other hand, for $\sigma > \kappa$, we encounter an *active* state with $a(t \to \infty) \to a_\infty = (\sigma - \kappa)/\lambda$, approached again exponentially with rate $\sim \sigma - \kappa$. We have thus identified a *continuous non-equilibrium phase transition* at $\sigma_c = \kappa$.

In fact, as in thermal equilibrium, this critical point is governed by characteristic power laws, with the reaction rates serving as appropriate control parameters. For example, the asymptotic particle density behaves as

$$a_\infty \sim (\sigma - \sigma_c)^\beta , \quad \sigma > \sigma_c , \tag{9.13}$$

with the mean-field exponent $\beta = 1$. Recall that in any *finite* system with an absorbing state, the latter is inevitably reached at sufficiently long times, and hence represents the unique stationary configuration. As in equilibrium critical systems, the existence of a genuine phase transition first requires taking the thermodynamic limit, and Eq. (9.13) in addition relies on the long-time limit $t \to \infty$ *before* the external control parameters are adjusted. At criticality, the exponential approaches toward stationarity in both the absorbing and active phases are replaced by the critical density decay

$$a_c(t) \sim t^{-\alpha} , \quad \sigma = \sigma_c ; \tag{9.14}$$

the rate equation (9.8) predicts $\alpha = 1$ ($k = 2$).

Augmenting Eq. (9.12) with a diffusive spreading term, we arrive at the following *reaction-diffusion model* for the *local* particle density $a(x, t)$:

$$\frac{\partial a(x, t)}{\partial t} = (\sigma - \kappa + D\nabla^2)\, a(x, t) - \lambda\, a(x, t)^2 . \tag{9.15}$$

As this coarse-grained description still discards non-trivial correlations, it remains mean-field in character; however, it qualitatively captures important spatial features, such as the existence of soliton-like invasion fronts from an active into an empty region (Problem 9.1). In addition, we may extract from Eq. (9.15) the typical mean-field values for the correlation length and dynamic critical exponents, $\nu = 1/2$ and $z = 2$. Naturally then, the following questions arise (to be at least partially answered in Chapter 10). What are the *critical exponent values* once statistical fluctuations, specifically the internal reaction noise, are properly included in the analysis? Is it possible, as for dynamical critical phenomena in equilibrium, to identify and characterize distinct *universality classes*? Which *mesoscopic or global features* determine this classification scheme and set the associated *critical dimensions*?

The above competing set of reactions may also be viewed as a (rather crude) model for the *population dynamics* of a single species subject to spontaneous birth and death processes, and binary competition.[3] Indeed, Eq. (9.15) is known

---

[3] For introductions, see, e.g., Haken (1983), Hofbauer and Sigmund (1998), and Murray (2002).

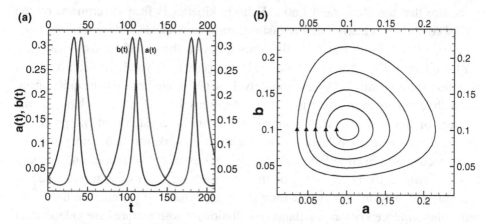

Fig. 9.1 Numerical solutions of the coupled Lotka–Volterra mean-field rate equations (9.16). (a) Non-linear predator $a(t)$ (grey) and prey $b(t)$ (black) population density oscillations; (b) periodic orbits in the a–b phase plane. For small amplitudes the oscillations become circular orbits, with harmonic oscillation frequency $\omega_0 = \sqrt{\kappa\,\sigma}$. [Figure reproduced with permission from: M. J. Washenberger, M. Mobilia, and U. C. Täuber, *J. Phys. Condens. Matter* **19**, 065139 (2007); DOI: 10.1088/0953-8984/19/6/065139; copyright (2007) by Institute of Physics Publ.]

as the *Fisher–Kolmogorov equation* in this context. In the same language, we may formulate a stochastic version of the classical *Lotka–Volterra predator–prey competition model*: In the absence of prey, the 'predators' A are left to die out, $A \rightarrow \varnothing$ with rate κ; if left alone, the prey in contrast spontaneously reproduce, $B \rightarrow B + B$ with rate σ, an unimpeded branching proliferation that inevitably causes a Malthusian population explosion. The predators are kept alive and the prey population under control through a mutual binary *predation* interaction, in the simplest scenario represented by the reaction $A + B \rightarrow A + A$. With rate λ, a prey is devoured by a predator; incited by the favorable food supply, the predator simultaneously produces a single offspring. The coupled kinetic rate equations for this system read

$$\frac{da(t)}{dt} = \lambda\,a(t)\,b(t) - \kappa\,a(t)\,, \qquad \frac{db(t)}{dt} = \sigma\,b(t) - \lambda\,a(t)\,b(t)\,. \qquad (9.16)$$

Taking their ratio and separating variables, one obtains $(-\lambda + \sigma/a)\,da = (\lambda - \kappa/b)\,db$, which is readily integrated to demonstrate that the quantity

$$K(t) = \lambda[a(t) + b(t)] - \sigma \ln a(t) - \kappa \ln b(t) \qquad (9.17)$$

constitutes a constant of motion for the coupled system of ordinary differential equations (9.16), i.e., $K(t) = K_0$. As a consequence, the system is governed by regular *population oscillations*, as depicted in Fig. 9.1.

Notice that the mean-field Lotka–Volterra kinetics is fully determined by the *initial* conditions, clearly not a realistic feature ecologically. Moreover, the existence of neutral non-linear periodic orbits relies on the specific conservation law for the quantity $K(t)$ that has no physical underpinning beyond the mean-field rate equation approximation. Correspondingly, Eqs. (9.16) are known to be quite unstable with respect to even minor model modifications. Indeed, if one includes spatial degrees of freedom and takes account of the full stochasticity of the processes involved, the system's behavior turns out to be remarkably rich, and genuinely more robust (see Section 9.5.2). In the species coexistence phase, one encounters for sufficiently large values of the predation rate an incessant sequence of *'pursuit and evasion'* waves that form complex dynamical patterns. In finite systems, these induce *erratic* population oscillations whose features are independent of the specific initial configuration; in the thermodynamic limit, the randomized phases for the localized oscillators average out the overall oscillation amplitude. In addition, if the prey 'carrying capacity' is restricted, there appears an *extinction threshold* for the predator population that separates the absorbing state of a system filled with prey from the active coexistence regime through a continuous phase transition.[4]

Just as mean-field theory in thermal equilibrium is a useful tool to arrive at a qualitative understanding of phase diagrams, rate equations for chemical kinetics or population dynamics usually constitute the essential starting point for further analysis. Indeed, the coupled non-linear rate equations for multi-component systems typically describe quite complex dynamical behavior including chaotic regimes, spontaneous pattern formation, as, e.g., in the Belousov–Zhabotinski reaction, and morphological Turing instabilities. All these examples call for a systematic approach to including stochastic fluctuations in the mathematical description of interacting reaction-diffusion systems that would be conducive to the application of the powerful field theory machinery, including the dynamic renormalization group, specifically in scale-invariant regimes. In the following, we shall describe such a general method which allows a representation of the classical master equation for interacting or reacting particle systems in terms of non-Hermitean bosonic many-particle pseudo-Hamiltonians, turning in the continuum limit into coherent-state path integral actions by means of the techniques detailed in Chapter 7. Yet before we proceed with this program, it is useful to demonstrate how incorporating the essential physics into a refined self-consistent mean-field treatment manages to aptly capture the implications of depletion zones on the kinetics of pair annihilation processes.

---

[4] For a recent overview, see Mobilia, Georgiev, and Täuber (2007). This volume's cover illustrates the spatio-temporal patterns in a two-dimensional stochastic lattice Lotka–Volterra model.

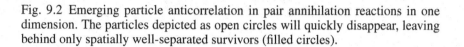

Fig. 9.2 Emerging particle anticorrelation in pair annihilation reactions in one dimension. The particles depicted as open circles will quickly disappear, leaving behind only spatially well-separated survivors (filled circles).

9.1.3 Smoluchowski's self-consistent theory

Let us return to the single-species binary reaction $A + A \rightarrow \emptyset$, and explore where the mean-field approximation of neglecting correlations is bound to fail. Adjacent particle pairs become annihilated quickly, inducing local particle *anti*correlations, as illustrated in Fig. 9.2. These may develop into extended density *depletion zones*, if at long times the surviving particles remain well separated from each other. Since we are concerned with diffusive propagation, we may now invoke the fundamental *recurrence property of random walks*. In low dimensions $d \leq 2$, two random walkers will eventually meet again with certainty, while they will escape from each other with a finite probability for $d > 2$. In high dimensions $d > d_c = 2$ therefore, diffusive spreading should be sufficient to maintain a spatially well-mixed configuration on mesoscopic scales, and the rate equation prediction (9.8) with $k = 2$ for the overall density decay $a(t) \sim (\lambda t)^{-1}$ essentially correct, aside perhaps from a slight renormalization of the effective annihilation rate.

For $d \leq d_c = 2$, however, particle anticorrelations should persist, and render the rate equation approximation inadequate. Indeed, as any particle pair will assuredly meet again, the long-time kinetics is bound to become completely *independent* of the reactivity λ. The effective rate is instead limited by the time it takes to traverse a depletion zone emptied by previous annihilations. Thus the characteristic diffusion length $x_D \sim \sqrt{Dt}$ sets the relevant length scale, and our basic scaling argument predicts that asymptotically

$$n(t) \sim x_D^{-d} \sim (Dt)^{-d/2} \, , \tag{9.18}$$

slower than the mean-field decay in dimensions $d \leq 2$. Note that the diffusion-limited power law (9.18) coincides with the rate equation prediction at the (upper) critical dimension $d_c = 2$, but with a different amplitude.

Smoluchowski's self-consistent approach refines the above simple scaling theory, adding quantitative details. Consider the diffusive flux for particles with asymptotic density n_∞ to reach a reference reactant located at the origin. In a continuum representation of the problem, one needs to assign a finite sphere with radius b to the reactive center, within which the annihilation reaction occurs with probability one, see Fig. 9.3. The idea is to self-consistently determine the stationary density profile $n(r)$ with the boundary conditions $n(b) = 0$ and $n(r) \rightarrow n_\infty$ as $r \rightarrow \infty$. With these

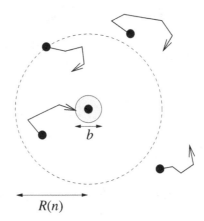

Fig. 9.3 Schematic picture illustrating Smoluchowski's approach to single-species pair annihilation reactions $A + A \to \emptyset$ (or A). The reaction radius is labeled b, and the average interparticle spacing, $R(n) \sim n^{-1/d}$, indicated as a dashed circle around the reference particle or reactive center.

assumptions, the diffusion equation reduces to the radial Laplace equation

$$\nabla^2 n(r) = \frac{\partial^2 n(r)}{\partial r^2} + \frac{d-1}{r}\frac{\partial n(r)}{\partial r} = 0 \, , \tag{9.19}$$

whose solution in $d > 2$ dimensions is

$$n(r) = n_\infty [1 - (b/r)^{d-2}] \, . \tag{9.20}$$

With the relative diffusivity $2D$, one then obtains the total flux across the reaction surface $2DK_d b^{d-1}(\partial n/\partial r)|_{r=b} = 2DK_d(d-2)b^{d-2}n_\infty$. Dividing by n_∞, this gives an effective reaction rate λ_{eff} with proper scaling dimension (length)^d/time, which can be inserted self-consistently into the rate equation

$$\frac{dn(t)}{dt} = -\lambda_{\mathrm{eff}}\, n(t)^2 \, , \quad \lambda_{\mathrm{eff}} = 2(d-2)K_d\, b^{d-2}D \, . \tag{9.21}$$

While Eq. (9.21) accounts for the region close to a given reference particle becoming depleted by earlier annihilation events, it still yields the mean-field decay law $n(t) \sim (\lambda_{\mathrm{eff}}t)^{-1}$, but with an effective reactivity $\lambda_{\mathrm{eff}} \propto D$ that vanishes as $d \downarrow 2$.

Indeed, the solution (9.20) is no longer valid in low dimensions. For $d < 2$, one must impose a modified boundary condition on Eq. (9.19), namely that $n(r) \approx n_\infty$ when r approaches the typical spacing $R(n) = (nK_d/d)^{-1/d}$, beyond which the particles should be distributed homogeneously. This gives

$$n(r) = n_\infty \frac{(r/b)^{2-d} - 1}{[R(n)/b]^{2-d} - 1} \, , \tag{9.22}$$

which for $R(n) \gg b$ leads to a crucially altered rate equation

$$\frac{dn(t)}{dt} = -\lambda_{\text{eff}}(n)\, n(t)^2\,, \quad \lambda_{\text{eff}}(n) = 2DK_d(2-d)\, R(n)^{d-2}\,. \tag{9.23}$$

The dependence on the reaction radius, which incorporates the original reaction rate and served as ultraviolet cutoff in the continuum theory for $d > 2$, has now been replaced by the infrared regulator $R(n)$. It imposes a density dependence on the effective reactivity $\lambda_{\text{eff}}(n) \sim D\, n^{(2-d)/d}$, and (9.23) is consequently solved by the decelerated density decay law (9.18). At $d_c = 2$, Eqs. (9.20) and (9.22) must be replaced by a logarithmic density profile, which results in

$$n(t) \sim (Dt)^{-1} \ln(Dt/b^2)\,, \tag{9.24}$$

see Problem 9.2(a), reminiscent of the logarithmic corrections at the upper critical dimension for continuous phase transitions.

As we shall see below, the functional forms for the pair annihilation density decay obtained with the Smoluchowski approach are actually confirmed by a field-theoretic renormalization group analysis. Yet the numerical values for the associated amplitudes in fact disagree with universal RG results and with exact solutions available in one dimension, which hints at the limitations of this method. Neither can Smoluchowski's approach, while straightforward to apply, be improved upon in a systematic manner. A controlled inclusion of fluctuation corrections is afforded by casting the problem as a field theory and applying renormalization group methods.

9.2 Field theory for stochastic interacting particle systems

This section details the mapping from the classical master equation for interacting particle systems to a non-Hermitean 'quantum' many-body Hamiltonian dynamics.[5] General features of the formalism are summarized, before finally the coherent-state path integral representation is utilized to arrive at a continuum field theory description that faithfully incorporates the intrinsic reaction noise and discreteness of the original stochastic processes.

9.2.1 Master equation and Fock space representation

A systematic inclusion of fluctuations and correlation effects for reaction-diffusion models and similar stochastic interacting particle systems faces two fundamental

---

[5] This formalism was originally devised by Doi (1976a, b); Grassberger and Scheunert (1980); and Peliti (1985). For recent reviews, see Cardy (1997); Mattis and Glasser (1998); Täuber, Howard, and Vollmayr-Lee (2005); and Cardy (2008). An illustrative alternative derivation is presented in Andreanov, Biroli, Bouchaud, and Lefèvre (2006).

difficulties. First, as is true generally in situations far away from thermal equilibrium, the choice of appropriate mesoscopic variables is not obvious. Second, one cannot easily apply the Langevin methodology of supplementing generalized mean-field rate equations for coarse-grained particle densities with random forcings. The functional form of the noise correlations is not constrained by any kind of fluctuation-dissipation theorem or Einstein relation; and as one is often concerned with the presence of absorbing states wherein fluctuations cease entirely, mere additive noise clearly does not suffice, but some kind of multiplicative random forces coupled to the density fields must appear.

One therefore must resort to the basic underlying stochastic description of such processes in terms of a master equation that provides a balance of gain and loss terms for the temporal evolution of the system's configurational probability (see Section 2.2). Interacting and reacting particle systems on a d-dimensional lattice with sites i are fully characterized by the set of integer occupation numbers $n_i = 0, 1, 2, \ldots$ for each particle species, and the configurational probability as well as any observable are functions of this string of integers $\{n_i\}$. Moreover, any stochastic process involving particle motion or chemical reactions is encapsulated in changing these occupation number variables by *integer* amounts. The fundamental linearity and the complete characterization of all possible configurations by integer numbers and their changes in time then call for an adaptation of the 'second-quantized' *Fock space representation* familiar from the quantum-mechanical description of indistinguishable particles, as outlined in Section 7.1.

In order to track the integer occupation number changes by means of a suitable mathematical counting device, we introduce for each participating particle species $A, B, \ldots$ a *bosonic ladder operator algebra*

$$[a_i, a_j] = 0 = \left[a_i^\dagger, a_j^\dagger\right], \quad \left[a_i, a_j^\dagger\right] = \delta_{ij}, \tag{9.25}$$

and similarly for b_i, $b_i^\dagger$, etc. The operators pertaining to different particle species of course commute, $[a_i, b_j] = 0 = [a_i, b_j^\dagger]$. From these commutation relations one establishes in the standard manner that $a_i, b_i, \ldots$ and $a_i^\dagger, b_i^\dagger, \ldots$ respectively constitute annihilation and creation operators, whence particle number eigenstates $|n_i\rangle$ can be constructed that satisfy

$$a_i |n_i\rangle = n_i |n_i - 1\rangle, \quad a_i^\dagger |n_i\rangle = |n_i + 1\rangle, \quad a_i^\dagger a_i |n_i\rangle = n_i |n_i\rangle \tag{9.26}$$

with integer eigenvalues $n_i = 0, 1, \ldots$, as desired. Notice that for convenience a different normalization convention has been applied here than is customary in many-particle quantum mechanics.

An instantaneous configuration with n_i particles of species A, m_i particles of species B, etc. residing on lattice sites i is then constructed from the empty vacuum state $|0\rangle$, defined through $0 = a_i |0\rangle = b_i |0\rangle \ldots$, as the *Fock product state*

$$|\{n_i\}, \{m_i\}, \ldots\rangle = \prod_i \left(a_i^\dagger\right)^{n_i} \left(b_i^\dagger\right)^{m_i} \cdots |0\rangle . \qquad (9.27)$$

To make contact with the time-dependent configuration probability and generate temporal evolution, one next introduces the formal *state vector*

$$|\Phi(t)\rangle = \sum_{\{n_i\}} P(\{n_i\}, \{m_i\}, \ldots ; t) \, |\{n_i\}, \{m_i\}, \ldots\rangle \qquad (9.28)$$

as a superposition of all possible occupation number configurations, weighted with their corresponding probabilities $P(\{n_i\}, \{m_i\}, \ldots ; t)$. The linear time evolution governed by the original (classical) master equation is thus transported to an *'imaginary-time' Schrödinger equation*

$$\frac{\partial |\Phi(t)\rangle}{\partial t} = -H \, |\Phi(t)\rangle , \qquad (9.29)$$

with a *stochastic pseudo-Hamiltonian H*, or rather the Liouville operator, see Eq. (2.75). For on-site reaction processes, it must be a sum of *local* and *normal-ordered* contributions, $H = \sum_i H_i(a_i^\dagger, b_i^\dagger, \ldots, a_i, b_i, \ldots)$; since particles are destroyed and/or produced on each site i, captured through the presence of corresponding annihilation and creation operators, H will generally *not* be Hermitean. Recall from Section 2.2.2 that non-Hermitean Liouville operators indicate genuine non-equilibrium kinetics.

An explicit example illustrates how this straightforward translation from master equation to stochastic pseudo-Hamiltonian works. Consider the on-site single-species reaction $k A \to l A$ (rate λ). For each lattice point i, the corresponding master equation becomes

$$\frac{\partial P(n_i; t)}{\partial t} = \lambda \left[\frac{(n_i + k - l)!}{(n_i - l)!} P(n_i + k - l; t) - \frac{n_i!}{(n_i - k)!} P(n_i; t) \right], \qquad (9.30)$$

cf. Eq. (9.1). The associated site state vector contains a sum over all values of n_i; shifting the summation index $n_i \to n_i + k - l$ in the gain term gives

$$\frac{\partial |\Phi_i(t)\rangle}{\partial t} = \lambda \sum_{n_i = \max(0, k-l)}^{\infty} P(n_i; t) \, n_i (n_i - 1) \cdots (n_i - k + 1) \, |n_i - k + l\rangle$$

$$- \lambda \sum_{n_i = k}^{\infty} P(n_i; t) \, n_i (n_i - 1) \cdots (n_i - k + 1) \, |n_i\rangle .$$

Noting that $a_i^k |n_i\rangle = n_i(n_i - 1)\cdots(n_i - k + 1)|n_i - k\rangle$, this can be written in the form of Eq. (9.29), with the *on-site annihilation Hamiltonian*

$$H_i = \lambda\big[(a_i^\dagger)^k - (a_i^\dagger)^l\big]a_i^k . \tag{9.31}$$

For the case of *binary annihilation* $A + A \to \emptyset$ (rate λ) and *coagulation* (fusion process) $A + A \to A$ (rate λ'), this becomes

$$H_i = \lambda\left(a_i^{\dagger 2} - 1\right)a_i^2 + \lambda'\left(a_i^\dagger - 1\right)a_i^\dagger a_i^2 . \tag{9.32}$$

Each reaction process is thus represented through two contributions in the pseudo-Hamiltonian, which may be interpreted physically as follows.

- The first, positive contribution originates from the loss term in the master equation, and encodes the *'order'* k of each reaction, i.e., the number operator $a_i^\dagger a_i$ raised to the kth power, but in normal-ordered form $(a_i^\dagger)^k a_i^k$.
- The second, negative contribution stems from the balance gain, and directly reflects the reaction process, namely how many particles are removed and subsequently (re-)created, in a ladder operator language.

The combination of spontaneous decay $A \to \emptyset$ (rate κ), branching $A \to A + A$ (rate σ), and pair coagulation (rate λ) thus immediately translates to

$$H_i = (a_i^\dagger - 1)(\kappa - \sigma\, a_i^\dagger + \lambda a_i^\dagger a_i)a_i . \tag{9.33}$$

These procedures are readily generalized for reactions with multiple particle species. For example, the master equation (9.1) for the reversible on-site processes $k A + l B \rightleftharpoons m C$ are encoded in the local pseudo-Hamiltonian

$$H_i = \big[(a_i^\dagger)^k (b_i^\dagger)^l - (c_i^\dagger)^m\big](\lambda a_i^k b_i^l - \sigma c_i^m) , \tag{9.34}$$

see Problem 9.3(a). Specializing to $k = 1 = l$, $m = 0$, this yields for irreversible two-species pair annihilation $A + B \to \emptyset$ with rate λ,

$$H_i = \lambda\left(a_i^\dagger b_i^\dagger - 1\right) a_i b_i , \tag{9.35}$$

while the Lotka–Volterra predator–prey interactions are captured by

$$H_i = \kappa(a_i^\dagger - 1)a_i + \sigma b_i^\dagger(1 - b_i^\dagger)b_i + \lambda a_i^\dagger(b_i^\dagger - a_i^\dagger)a_i b_i . \tag{9.36}$$

In a similar manner, non-local processes are easily accounted for. One may just regard interactions involving distinct lattice sites i as pertaining to different species A_i. Unbiased hopping between neighboring lattice sites $\langle ij\rangle$ is thus represented in this formalism through the pseudo-Hamiltonian

$$H_{\text{dif}} = D \sum_{\langle ij \rangle} \left(a_i^\dagger - a_j^\dagger \right)(a_i - a_j) , \qquad (9.37)$$

see Problem 9.3(b).

This bosonic annihilation and creation operator representation tacitly assumed that there exist no restrictions on the particle occupation numbers n_i on each site. If, instead, there is a maximum number of particles $n_i \leq 2S + 1$ permitted per lattice site, referred to as a finite *carrying capacity* in the ecology literature, one may instead employ a representation in terms of spin S operators. For example, particle exclusion systems with $n_i = 0$ or 1 can thus be mapped onto non-Hermitean spin $S = \frac{1}{2}$ 'quantum' systems. Specifically in one dimension, such representations in terms of integrable spin chains have proved a fruitful tool to procure various remarkable *exact* results.[6] An alternative approach that is more amenable to a field-theoretic treatment employs the bosonic theory, but encodes the site occupation restrictions through appropriate exponentials in the number operators.[7]

9.2.2 General features and Hamiltonian structure

For notational simplicity, consider for now a single particle species A. As initial configuration for the master equation describing stochastic particle reactions, one may for example assume a random, uncorrelated Poisson distribution $P(\{n_i\}; 0) = \prod_i P_0(n_i)$, where $P_0(n_i) = e^{-\bar{n}_0} \bar{n}_0^{n_i} / n_i!$ with mean density $\bar{n}_0$. With Eqs. (9.27) and (9.28) this yields the initial state

$$|\Phi(0)\rangle = e^{\bar{n}_0 \sum_i \left(a_i^\dagger - 1 \right)} |0\rangle , \qquad (9.38)$$

and akin to quantum mechanics, Eq. (9.29) is then formally solved by

$$|\Phi(t)\rangle = e^{-Ht} |\Phi(0)\rangle , \qquad (9.39)$$

cf. Eq. (2.77). Various mathematical techniques familiar from quantum theory can thus be adapted to stochastic many-particle systems (see Section 2.2.2); e.g., the late-time dynamics is obviously dominated by those *right* eigenstates of H whose eigenvalues have the lowest real parts. Yet there are crucial differences. The linear superposition (9.28) entails real-valued configurational probabilities, *not* complex probability amplitudes, whence no interference phenomena arise. Also, the standard quantum-mechanical expression $\langle A(t) \rangle_\psi = \langle \psi(t)|A|\psi(t)\rangle$ for the *expectation value* of an observable A in state $|\psi(t)\rangle$ clearly cannot apply here, since it would be a bilinear form in $P(\{n_i\}; t)$.

---

[6] For overviews, see, e.g., Alcaraz, Droz, Henkel, and Rittenberg (1994); Henkel, Orlandini, and Santos (1997); Schütz (2001); Stinchcombe (2001).
[7] van Wijland (2001).

Instead, for an observable $A(\{n_i\})$, necessarily expressible as a function of the lattice occupation numbers n_i, we wish to compute

$$\langle A(t) \rangle = \sum_{\{n_i\}} A(\{n_i\}) \, P(\{n_i\}; t) . \tag{9.40}$$

To this end, we introduce the *projection state* $\langle \mathcal{P} | = \langle 0 | \prod_i e^{a_i}$, which satisfies $\langle \mathcal{P} | 0 \rangle = 1$ and $\langle \mathcal{P} | a_i^\dagger = \langle \mathcal{P} |$, since $[e^{a_i}, a_j^\dagger] = e^{a_i} \delta_{ij}$. For the desired *statistical averages* of observables, we then obtain the prescription

$$\langle A(t) \rangle = \langle \mathcal{P} | A(\{a_i^\dagger a_i\}) | \Phi(t) \rangle = \langle \mathcal{P} | A(\{a_i^\dagger a_i\}) \, e^{-H(\{a_i^\dagger\}, \{a_i\}) t} \, | \Phi(0) \rangle , \tag{9.41}$$

as is readily verified by commuting the factors e^{a_i} from the projection state to the right, and recalling that $\langle 0 | a_i^\dagger = 0$. When acting on the number eigenstates $|\{n_i\}\rangle$, the operator-valued function $A(\{a_i^\dagger a_i\})$ is replaced by the function $A(\{n_i\})$ of their integer eigenvalues n_i.

Let us next explore the consequences of *probability conservation*, setting $A = 1$: $1 = \langle \mathcal{P} | \Phi(t) \rangle = \langle \mathcal{P} | e^{-Ht} | \Phi(0) \rangle$ for any value of t. With proper normalization of the initial state $\langle \mathcal{P} | \Phi(0) \rangle = 1$, this requires $\langle \mathcal{P} | H(\{a_i^\dagger\}, \{a_i\}) = 0$. Upon commuting $e^{\sum_i a_i}$ with the pseudo-Hamiltonian, effectively all creation operators become shifted: $a_i^\dagger \to 1 + a_i^\dagger$; probability conservation is consequently guaranteed if $H(\{a_i^\dagger \to 1\}; \{a_i\}) = 0$. All explicit examples in the preceding Section 9.2.1 of course fulfill this condition. This very same commutation procedure may be utilized to rewrite the observable averages (9.41) as follows. First, one needs to normal-order the ladder operators occurring in $A(\{a_i^\dagger a_i\})$, resulting in the rearrangement $\widetilde{A}(\{a_i^\dagger\}, \{a_i\})$; subsequently $\widetilde{A}(\{a_i^\dagger \to 1\}, \{a_i\})$ results from the action on the vacuum state on the left. For example, for the particle density on site i simply $a_i^\dagger a_i \to a_i$, whereas for the two-point operator $a_i^\dagger a_i a_j^\dagger a_j \to a_i \delta_{ij} + a_i a_j$. Next the shift $a_i^\dagger \to 1 + a_i^\dagger$ is to be performed in the creation operators of the pseudo-Hamiltonian. Finally, the initial state vector becomes modified to

$$|\widetilde{\Phi}(0)\rangle = e^{\sum_i a_i} |\Phi(0)\rangle = e^{\bar{n}_0 \sum_i a_i^\dagger} |0\rangle , \tag{9.42}$$

where the last expression holds for an uncorrelated Poissonian initial distribution (9.38). In a slightly symbolic notation, we arrive at

$$\langle A(t) \rangle = \langle 0 | \widetilde{A}(\{1\}, \{a_i\}) \, e^{-H(\{1 + a_i^\dagger\}, \{a_i\}) t} \, |\widetilde{\Phi}(0)\rangle . \tag{9.43}$$

A direct connection with the generating function formalism introduced in Section 2.3.2 sheds additional light on the underlying Hamiltonian structure.[8] Consider

---

[8] Elgart and Kamenev (2004, 2006).

on-site reactions of a single particle species, e.g., again $k\,A \to l\,A$, and construct the generating function (2.107): $g(x, t) = \sum_n x^n\,P_n(t)$ with the auxiliary variable x, recalling that $\langle n(t)\rangle = \partial g(x, t)/\partial x|_{x=1}$, etc. By basically the same steps that led from the master equation (9.30) to Eq. (9.31), one derives the partial differential equation for the temporal evolution of g,

$$\frac{\partial g(x, t)}{\partial t} = -H(x, p)\,g(x, t)\,, \tag{9.44}$$

where $p = \partial/\partial x$ represents the 'momentum' operator conjugate to x, and

$$H(x, p) = \lambda\,(x^k - x^l)\,p^k\,. \tag{9.45}$$

With the correspondences $x \leftrightarrow a^\dagger$ and $p \leftrightarrow a$, this is precisely the pseudo-Hamiltonian (9.31); indeed, $[p, x] = 1$. Note that Eq. (2.112) in Section 2.3.2 is just Eq. (9.44) for the first-order processes $A \to \emptyset$ and $A \to A + A$ with $H = (x - 1)(\kappa - \sigma\,x)\,p$, see Eq. (9.33).

Probability conservation is encoded in the boundary condition $g(1, t) = 1$, which is automatically satisfied if $H(1, p) = 0$. The 'classical trajectories' associated with the stochastic reaction dynamics are the solutions of Hamilton's equations of motion familiar from classical mechanics,

$$\frac{dx(t)}{dt} = \frac{\partial H(x, p)}{\partial p}\,, \qquad \frac{dp(t)}{dt} = -\frac{\partial H(x, p)}{\partial x}\,; \tag{9.46}$$

we have already encountered an explicit example in Eq. (2.115). Due to probability conservation, the first of these differential equations is always solved by $x = 1$; identifying $\langle n(t)\rangle = p(t)$, the second Hamilton's equation then simply becomes the corresponding rate equation. For an *absorbing state*, $dp(t)/dt = 0$ if $p = 0$, whence $H(x, 0) = 0$. For example, for the kth order annihilation process encoded in (9.45), one has $dx(t)/dt = k\lambda(x^k - x^l)p^{k-1}$ and $dp(t)/dt = -\lambda(kx^{k-1} - lx^{l-1})p^k$; inserting $x = 1$ recovers Eq. (9.6).

Generally, the 'phase space' trajectories $x(t)$ and $p(t)$ are confined to constant 'energy' hypersurfaces $H = H(x(0), p(0))$, as the pseudo-Hamiltonian carries no explicit time dependence. Trajectories can intersect only on the special $H = 0$ surfaces (note that these include both probability-conserving and absorbing-state trajectories), where they turn into separatrices dividing phase space into disjoint sectors. This Hamiltonian formulation has found various fruitful applications, specifically as a starting point for 'semi-classical' treatments in the spirit of the Wentzel–Kramers–Brillouin method, e.g., for the computation of fluctuation effects on extinction times as a stochastic system approaches an absorbing state.

9.2.3 *Coherent-state path integral representation*

We may now follow the direct route established for identical bosons in Section 7.1 and proceed from the pseudo-Hamiltonian for stochastic interacting particle systems towards a *path integral representation* based on *coherent states*.[9] Since the master and many-body Schrödinger equations are both *linear*, we may carry out strictly analogous computations. In order to evaluate *expectation values* subject to the 'imaginary-time' evolution (9.29), we discretize $t_l = l\,\tau$ ($l = 0, 1, \ldots, M$), identifying $t_M = t_f$. Inserting the closure relation (7.17) at all $M + 1$ time steps, and using the overlap integral (7.15) for the ensuing matrix elements, one obtains

$$\langle A(t) \rangle = \lim_{\tau \to 0} \int \prod_i \prod_{l=0}^{M} \frac{d\phi_i^*(t_l)\,d\phi_i(t_l)}{2\pi i}\, e^{\sum_i \phi_i(t_f)}$$

$$\times e^{-\tau \sum_{l=1}^{M} \sum_i \phi_i^*(t_l) \frac{\phi_i(t_l)-\phi_i(t_{l-1})}{\tau} + H(\{\phi_i(t_l)\},\{\phi_i(t_{l-1})\})}$$

$$\times e^{\sum_i (-|\phi_i(0)|^2 + \bar{n}_0 [\phi_i^*(0)-1])} . \tag{9.47}$$

Taking the temporal continuum limit, these expectation values (9.40) are represented by a functional integral

$$\langle A(t) \rangle \propto \int \prod_i \mathcal{D}[\phi_i^*, \phi_i]\, \widetilde{A}(\{1\}, \{\phi_i\})\, e^{-\mathcal{A}[\phi_i^*, \phi_i]} , \tag{9.48}$$

with the *Doi–Peliti action* for *local* (on-site) reactions (of one set of particles)

$$\mathcal{A}[\phi_i^*, \phi_i] = \int_0^{t_f} \left[\sum_i \phi_i^*(t)\, \frac{\partial \phi_i(t)}{\partial t} + H\big(\phi_i^*(t), \phi_i(t)\big) \right] dt$$

$$- \sum_i \phi_i(t_f) + \sum_i \left(|\phi_i(0)|^2 - \bar{n}_0 [\phi_i^*(0) - 1] \right) . \tag{9.49}$$

The site indices here assume the role of the single-particle state labels in (7.20) and (7.21). The temporal boundary terms at $t = t_f$ and $t = 0$ are accounted for in the second line. The final-time contribution stems from the projection state in Eq. (9.41). The last term in the action (9.49) encodes the initial Poisson distribution (9.38). Indeed, performing the functional integral over the variables $\phi_i^*(0)$ just yields the constraints $\phi_i(0) = \bar{n}_0$ for each lattice site. In the pseudo-Hamiltonian, the creation and annihilation operators $a_i^\dagger$ and a_i are once again simply replaced by the complex numbers ϕ_i^* and ϕ_i, respectively. Note that in Eq. (9.48) the projection state exponentials have been commuted through the operator-valued function $A(\{a_i^\dagger\, a_i\})$ to finally leave behind the reduced function $\widetilde{A}(\{1\}, \{\phi_i\})$, compare with

---

[9] The fundamental relations (7.11), (7.15), and (7.17) are not affected by the modified Fock state normalization in Eqs. (9.26) and (9.27).

(9.43). It should be emphasized that the path integral representation (9.48) with the Doi–Peliti action (9.49) faithfully mirrors the master equation for the *microscopic* stochastic process under consideration, without invoking any approximation or resorting to additional assumptions. It consequently properly encodes the discreteness in site occupation numbers and their changes due to hopping, particle generation, annihilation, transformation, etc. and fully incorporates the associated *internal* reaction noise. For example, the bulk contributions of the action governing the on-site reactions $k A \to l A$ (rate λ) combined with unbiased nearest-neighbor hopping (rate D) are

$$A[\phi_i^*, \phi_i] = \int_0^{t_f} dt \left(\sum_i \phi_i^*(t) \frac{\partial \phi_i(t)}{\partial t} + \lambda \sum_i [\phi_i^*(t)^k - \phi_i^*(t)^l] \phi_i(t)^k \right.$$

$$\left. + D \sum_{\langle ij \rangle} [\phi_i^*(t) - \phi_j^*(t)][\phi_i(t) - \phi_j(t)] \right). \qquad (9.50)$$

The generalization to an arbitrary number of particle species is straightforward, with complex coherent-state eigenvalues $\phi_{\alpha i}$ carrying an additional internal index connected with species α on site i, and the temporal derivative and boundary terms in (9.49) replicated for each α.

In order to arrive at a genuine field theory as the natural basis for further coarse-graining and renormalization group analysis, we take the *spatial continuum limit*, i.e., $\sum_i \to a_0^{-d} \int d^d x$ (where a_0 is the original lattice constant), and let $\phi_i^*(t) \to \hat{\psi}(x, t)$ remain dimensionless, whence $\phi_i(t) \to a_0^d \psi(x, t)$. The continuum versions of Eqs. (9.48) and (9.49) become

$$\langle A(t) \rangle \propto \int \mathcal{D}[\hat{\psi}, \psi] \, \tilde{A}(\{1\}, \{\psi\}) \, e^{-A[\hat{\psi}, \psi]}, \qquad (9.51)$$

$$A[\hat{\psi}, \psi] = \int d^d x \left(\int_0^{t_f} \left[\hat{\psi}(x, t) \frac{\partial \psi(x, t)}{\partial t} + \mathcal{H}(\hat{\psi}(x, t), \psi(x, t)) \right] dt \right.$$

$$\left. - \psi(x, t_f) + \hat{\psi}(x, 0) \psi(x, 0) - \bar{n}_0 [\hat{\psi}(x, 0) - 1] \right), \qquad (9.52)$$

where $\mathcal{H}$ indicates the 'pseudo-Hamiltonian density' associated with the stochastic processes under investigation: e.g., for $k A \to l A$:

$$\mathcal{H}(\hat{\psi}(x, t), \psi(x, t)) = \tilde{\lambda}[\hat{\psi}(x, t)^k - \hat{\psi}(x, t)^l] \psi(x, t)^k, \qquad (9.53)$$

with $\tilde{\lambda} = a_0^{(k-1)d} \lambda$. To lowest order in a gradient expansion, the nearest-neighbor hopping terms in the second line of Eq. (9.50) transform into

$$\mathcal{H}_{\text{dif}}(\hat{\psi}(x, t), \psi(x, t)) = \tilde{D} \, \nabla \hat{\psi}(x, t) \, \nabla \psi(x, t) \to -\tilde{D} \, \hat{\psi}(x, t) \, \nabla^2 \psi(x, t) \qquad (9.54)$$

after integration by parts, with $\tilde{D} = a_0^2 D$. Together with the time derivative term in (9.52), this yields the standard diffusion propagator.

Let us next have a look at the *classical field equations* for the action (9.52). The first one, $\delta A[\hat{\psi}, \psi]/\delta\psi(x, t) = 0$, is always solved by $\hat{\psi}(x, t) = 1$, reflecting probability conservation, see Section 9.2.2. Inserting this uniform solution into $\delta A[\hat{\psi}, \psi]/\delta\hat{\psi} = 0$ then leads to the associated rate equation, or, if the diffusive propagation is retained, the corresponding *reaction-diffusion equation* for the local particle density $\psi(x, t)$. For example, the classical field equations associated with (9.53) and (9.54) in this manner reduce to

$$\frac{\partial\psi(x, t)}{\partial t} = \tilde{D}\,\nabla^2\,\psi(x, t) - (k - l)\,\tilde{\lambda}\,\psi(x, t)^k ,\qquad (9.55)$$

see Eq. (9.6). The field theory action (9.52), derived directly from the master equation that we take to define the stochastic particle processes in question, now obviously provides a means of systematically improving upon mean-field approximations and properly including fluctuations and correlation effects.

In certain situations it can be useful and instructive to perform a shift in the field $\hat{\psi}$ about the probability-conserving mean-field solution, $\hat{\psi}(x, t) = 1 + \tilde{\psi}(x, t)$, whereupon the boundary contributions in the Doi–Peliti action assume a simpler form,

$$A[\tilde{\psi}, \psi] = \int d^d x \left(\int_0^{t_f} \left[\tilde{\psi}(x, t) \frac{\partial\psi(x, t)}{\partial t} + \mathcal{H}'(\tilde{\psi}(x, t), \psi(x, t)) \right] dt \right.$$

$$\left. + \tilde{\psi}(x, 0)\,[\psi(x, 0) - \bar{n}_0] \right),\qquad (9.56)$$

where $\mathcal{H}'(\tilde{\psi}, \psi) = \mathcal{H}(1 + \tilde{\psi}, \psi)$; note that integrating out $\tilde{\psi}(x, 0)$ just returns the initial condition $\psi(x, 0) = \bar{n}_0$. Consider the binary annihilation and coagulation or fusion pseudo-Hamiltonian (9.32), for which the shifted Hamiltonian density reads (with $\tilde{\lambda} = a_0^d \lambda$ and $\tilde{\lambda}' = a_0^d \lambda'$)

$$\mathcal{H}'(\tilde{\psi}, \psi) = (2\tilde{\lambda} + \tilde{\lambda}')\,\tilde{\psi}\,\psi^2 + (\tilde{\lambda} + \tilde{\lambda}')\,\tilde{\psi}^2\,\psi^2 .\qquad (9.57)$$

The annihilation ($\propto \tilde{\lambda}$) and coagulation ($\propto \tilde{\lambda}'$) processes thus yield essentially *identical* terms, except for one factor 2. Aside from non-universal amplitudes, these binary reactions must therefore be characterized by the same asymptotic scaling behavior in the long-time limit.

Just as in the Janssen–De Dominicis functional for Langevin equations (Chapter 4) and the coherent-state path integral in many-particle quantum dynamics (Chapter 7), the Doi–Peliti field theory action encodes the stochastic master equation kinetics through *two independent* fields. In fact, provided one is concerned with mere first- or second-order reactions, immediate connection can be made with a response functional, since the fields $\hat{\psi}$ and $\tilde{\psi}$ respectively enter at most quadrat-ically in the pseudo-Hamiltonians of (9.52) and (9.56), whence one may interpret them as reflecting Gaussian white noise. The Langevin dynamics associated with the action $\mathcal{A}[\tilde{\psi}, \psi]$ turns out to represent the reaction-diffusion equations with additional stochastic forcing. As in the path integral (4.8), the integration con-tour for the field $\tilde{\psi}$ then has to be taken along the imaginary axis to guarantee convergence.

The single-species reactions $k A \rightarrow l A$ with $k, l = 0, 1, 2$ illuminate this intriguing connection. Applying the shift $\hat{\psi} = 1 + \tilde{\psi}$ to the pseudo-Hamiltonian (9.53) results in

$$\mathcal{H}'(\tilde{\psi}, \psi) = \tilde{\lambda} \left[(k - l) \tilde{\psi} + \frac{k(k-1) - l(l-1)}{2} \tilde{\psi}^2 \right] \psi^k . \tag{9.58}$$

In conjunction with the diffusion term (9.54), the bulk action (9.56) may be viewed as the response functional associated with the Langevin equation

$$\frac{\partial \psi(x, t)}{\partial t} = \tilde{D} \nabla^2 \psi(x, t) - (k - l) \tilde{\lambda} \psi(x, t)^k + \zeta(x, t) , \tag{9.59}$$

i.e., in $\mathcal{H}'(\tilde{\psi}, \psi)$ we identify the contribution linear in $\tilde{\psi}$ with the systematic force term $-\tilde{\psi} F[\psi]$, and the parts quadratic in this auxiliary field $-\tilde{\psi}^2 L[\psi]$ as originating from the functional $L[\psi]$ that encapsulates the second moment for the Gaussian white noise ζ, with $\langle \zeta(x, t) \rangle = 0$; in our specific example:

$$L[\psi] = \tilde{\lambda} \frac{l(l-1) - k(k-1)}{2} \psi^k . \tag{9.60}$$

If $k = l$ we have of course just diffusive propagation and no reactions take place; notice that Brownian random walk noise (see Section 2.4) is *not* captured by this formalism, since $L = 0$. The intrinsic reaction noise correlator also van-ishes for the *linear* processes $A \rightarrow \emptyset$ $(k = 1, l = 0)$ and $\emptyset \rightarrow A$ $(k = 0, l = 1)$; the reaction-diffusion equation (9.55) hence describes the large-scale long-time kinet-ics accurately. Spontaneous pair production $\emptyset \rightarrow A + A$ $(k = 0, l = 2)$ is governed by genuine *additive noise*,

$$\langle \zeta(x, t) \zeta(x', t') \rangle = 2 \tilde{\lambda} \delta(x - x') \delta(t - t') . \tag{9.61}$$

In contrast, the *branching process* $A \rightarrow A + A$ $(k = 1, l = 2)$ has an empty absorb-ing state, which is reflected in $L[\psi] = \tilde{\lambda} \psi \rightarrow 0$ as $\psi \rightarrow 0$. Somewhat sloppily, this *multiplicative noise* is often written as

$$\langle \zeta(x, t) \zeta(x', t') \rangle = 2 \tilde{\lambda} \psi(x, t) \delta(x - x') \delta(t - t') , \tag{9.62}$$

'equating' a noise history average with an instantaneous local stochastic vari-able. Replacing the additive stochastic force in the Langevin equation (9.59) with

$\zeta(x, t) = \sqrt{\psi(x, t)}\, \eta(x, t)$ recovers the correlator (9.61) for the *'square-root' noise* $\eta(x, t)$. This is, however, problematic as well, since the field $\psi(x, t)$ is in general a complex function, whence the square-root operation induces a branch cut;[10] the original definition through the noise functional inside the path integral is in fact the least ambiguous.

This statement applies even more for the case $k = 2$, i.e., either pair annihilation ($l = 0$) or coagulation ($l = 1$), see (9.57), for which $L[\psi] = -\tilde{\lambda}\,\psi^2$ is strictly negative if ψ is real. In loose notation again, we have

$$\langle \zeta(x, t)\,\zeta(x', t') \rangle = -2\tilde{\lambda}\,\psi(x, t)^2\,\delta(x - x')\,\delta(t - t')\,, \qquad (9.63)$$

indicating *'imaginary' multiplicative noise*, since reinstating (9.61) would necessitate $\zeta(x, t) = i\,\psi(x, t)\,\eta(x, t)$. This Langevin description for pair processes is thus not truly well-defined; physically, the negative sign in (9.63) reflects the emerging particle *anti*correlations, see Section 9.1.3. It is important to recall that the field $\psi(x, t)$ is *not* to be identified with the fluctuating local particle density $n(x, t)$: whereas $\langle \psi(x, t) \rangle = n(t) = \langle n(x, t) \rangle$, one has $\langle n(x, t)n(x', 0) \rangle = n(t)\delta(x - x') + \langle \psi(x, t)\psi(x', 0) \rangle$ and similarly for higher moments, cf. Section 9.2.2. The negative noise correlator clearly renders the second term negative as well, but of course $(\Delta n)^2 > 0$. In contrast, the local branching process $A \to A + A$ generates clusters of particles, naturally governed by positive correlations.

The noise correlator for pair annihilation and coagulation can actually be rendered positive through a non-linear *Cole–Hopf transformation*

$$\hat{\psi}(x, t) = e^{\tilde{\rho}(x,t)}\,, \quad \psi(x, t) = e^{-\tilde{\rho}(x,t)}\,\rho(x, t)\,, \qquad (9.64)$$

applied to the unshifted action (9.52).[11] The Jacobian for this variable transformation is 1, and the local particle density $\hat{\psi}\,\psi = \rho$. Furthermore, $\hat{\psi}\,\frac{\partial \psi}{\partial t} = \frac{\partial}{\partial t}[(1 - \tilde{\rho})\rho] + \tilde{\rho}\,\frac{\partial \rho}{\partial t}$, and, omitting boundary contributions, $-\tilde{D}\,\hat{\psi}\,\nabla^2\psi = -\tilde{D}\,(\nabla^2\hat{\psi})\,\psi = -\tilde{D}\,[\nabla^2\tilde{\rho} + (\nabla\tilde{\rho})^2]\,\rho$ for the diffusion term. The pair annihilation reaction pseudo-Hamiltonian becomes $\tilde{\lambda}\,(\hat{\psi}^2 - 1)\,\psi^2 = \tilde{\lambda}\,(1 - e^{-2\tilde{\rho}})\,\rho^2 = 2\tilde{\lambda}\,\tilde{\rho}\,\rho^2 - 2\tilde{\lambda}\,\tilde{\rho}^2\,\rho^2 \ldots$, if we expand the exponential, noticing that $\tilde{\rho} = 0$ for $\hat{\psi} = 1$. Similarly, for coagulation/fusion $\tilde{\lambda}'\,(\hat{\psi} - 1)\,\hat{\psi}\,\psi^2 = \tilde{\lambda}'\,(1 - e^{-\tilde{\rho}})\,\rho^2 = \tilde{\lambda}'\,\tilde{\rho}\,\rho^2 - \frac{\tilde{\lambda}'}{2}\,\tilde{\rho}^2\,\rho^2 \ldots$. Hence in either case the quadratic term in the response field $\tilde{\rho}$ has assumed the opposite sign to before, and therefore corresponds to 'real' rather than imaginary noise (for real ρ). However, the truncation of this expansion at

---

[10] An efficient numerical algorithm for Langevin equations with 'square-root' multiplicative noise is described in Dornic, Chaté, and Muñoz (2005).

[11] Janssen (2001); we are loosely treating the stochastic fields $\hat{\psi}$, ψ, $\tilde{\rho}$, and ρ as continuous differentiable functions. A more precise approach would be to take recourse to the original discretized version of the functional integral.

second order is not generally justifiable, and a consistent description of the annihilation kinetics in terms of the fields ρ and $\tilde{\rho}$ comes at the price of having to
incorporate non-linear 'diffusion noise' $-\tilde{D} (\nabla \tilde{\rho})^2 \rho$. For third- and higher-order
reactions, a direct mesoscopic representation in terms of Langevin equations that in
a straightforward manner extend the kinetic rate equations is at least not obviously
possible.

9.3 Diffusion-limited annihilation: depletion zones

The Doi–Peliti field theory representation of the stochastic master equation for
interacting particle systems affords us with a proper framework to employ the perturbative renormalization group tools developed in Chapters 4 and 5, transferred
here to a genuine *non-equilibrium* context. We next explore in some detail the
approach to the absorbing state for single-species annihilation reactions, where
dynamically emerging depletion zones crucially affect the long-time density decay
in low dimensions $d \leq d_c$. The renormalization group treatment for pair annihilation or coagulation/fusion affirms and explains the validity of Smoluchowski's
self-consistent theory (see Section 9.1.3).

9.3.1 Single-species annihilation: RG approach

We begin by analyzing diffusion-limited single-species annihilation reactions
$k A \to l A$, with $k > l$, $k \geq 2$.[12] Note that, microscopically, the final absorbing
state is reached after a finite time, and may consist of $l, \dots, k - 1$ particles, with
the outcome possibly depending on their initial number N_0. For example, for
$A + A \to \emptyset$ one ends up with a single particle if N_0 is odd, and an empty system
if N_0 is even, whereas for $A + A \to A$ there always remains one residual particle.
The field theory, however, addresss the continuous thermodynamic limit $N_0 \to \infty$,
wherein these different absorbing states are all indistinctly described by a vanishing
particle density $a(t) \to 0$, reached only asymptotically as $t \to \infty$. With (9.53) and
(9.54), the corresponding unshifted Doi–Peliti action (9.52) reads

$$\mathcal{A}[\hat{\psi}, \psi] = \int d^d x \left(-\psi(x, t_f) + \int_0^{t_f} dt \left[\hat{\psi}(x, t) \left(\frac{\partial}{\partial t} - D \nabla^2 \right) \psi(x, t) \right. \right.$$

$$\left. + \lambda [\hat{\psi}(x, t)^k - \hat{\psi}(x, t)^l] \psi(x, t)^k \right]$$

$$\left. + \hat{\psi}(x, 0) \psi(x, 0) - \bar{n}_0 [\hat{\psi}(x, 0) - 1] \right). \tag{9.65}$$

---

[12] Peliti (1986); Lee (1994); Sections 9.3.1 and 9.3.2 rather closely follow the presentation in Täuber, Howard,
and Vollmayr-Lee (2005).

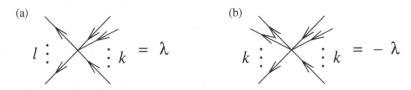

Fig. 9.4 Annihilation vertices for the kth order annihilation reaction $kA \to lA$, $k > l$: (a) annihilation 'sink' $\Gamma^{(l,k)}$; (b) scattering vertex $\Gamma^{(k,k)}$.

(We shall henceforth omit the '~' that was used in Section 9.2 to carefully distinguish coarse-grained continuum from microscopic reaction rates.) The harmonic part of the action (9.65) naturally contains the diffusion propagator, or Gaussian model A propagator (4.33), (3.29) at criticality $r = 0$,

$$G_0(q, \omega) = (-i\omega + Dq^2)^{-1} , \quad G_0(q, t) = e^{-Dq^2 t} \Theta(t) . \qquad (9.66)$$

Introducing a momentum scale $[q] = \mu$, or $[x] = \mu^{-1}$, and the associated diffusive temporal scaling $[t] = \mu^{-2}$, $[\omega] = \mu^2$, i.e., $[D] = \mu^0$, cf. Eq. (5.3) for $a = 0$, one obtains for the *scaling dimensions* of the fields (recall that we elected $\hat{\psi}$ to be dimensionless) and reaction rate:

$$[\hat{\psi}(x, t)] = \mu^0 , \quad [\psi(x, t)] = \mu^d , \quad [\lambda] = \mu^{2-(k-1)d} . \qquad (9.67)$$

The non-linear coupling λ is thus *relevant* in the RG sense in low dimensions $d < d_c(k)$, and becomes *marginal* at the *upper critical dimension*

$$d_c(k) = \frac{2}{k - 1} . \qquad (9.68)$$

For $d > d_c(k)$, one expects the mean-field power laws to be valid, e.g., $a(t) \sim (\lambda t)^{-1/(k-1)}$ for the density decay, see Eq. (9.8). For $k = 2$, indeed $d_c(2) = 2$ as anticipated earlier from the recurrence properties of random walks, and non-classical power laws induced by particle anticorrelations ensue only in one dimension. At $d_c(2) = 2$ for pair processes and $d_c(3) = 1$ for triplet reactions, the rate equation predictions become altered by logarithmic corrections. Since $d_c(k > 3) < 1$, mean-field theory should be essentially correct in any physical dimension for quartic or higher-order annihilation reactions. The action (9.65) contains two 'bulk' vertices, namely the 'annihilation sink' $\propto \lambda$ with $k \geq 2$ incoming and $l < k$ outgoing lines, and the 'scattering' vertex $\propto -\lambda$ with both k incoming and outgoing lines, shown in Fig. 9.4. No combination of these vertices can possibly generate any Feynman diagram contributing to propagator renormalization, which would require a single incoming line. Hence $Z_{\hat{\psi}} = Z_\psi = Z_D = 1$ to *all* orders in the perturbation expansion, and the annihilation field theory remains strictly massless and in that sense *critical* with bare diffusion propagator (9.66). Consequently,

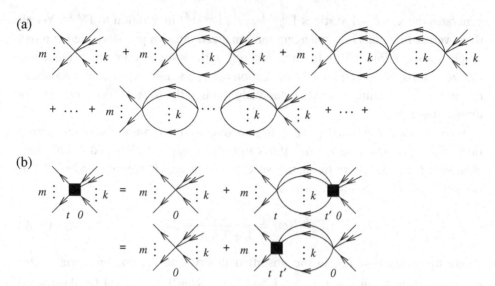

Fig. 9.5 Vertex renormalization for diffusion-limited annihilation processes $k A \to l A$. (a) Series of concatenated $(k-1)$-loop diagrams for $\Gamma^{(m,k)}$, with $m = l, k$ in the unshifted, $m = l+1, \ldots, k$ in the shifted field theory; (b) equivalent self-consistent Bethe–Salpeter equation for the full vertex $\Gamma^{(m,k)}$, indicated by the filled square; compare with the diagrammatic representation of Dyson's equation in Fig. 4.12.

one deduces the *exact* scaling exponents $\eta = 0$ and $z = 2$. Indeed, Smoluchowski's approach relies precisely on purely diffusive particle spreading, unaltered by the reaction kinetics even at long time scales.

This leaves us with the renormalization of the reaction rate λ, which constitutes the non-linear coupling in the action (9.65). In the unshifted field theory, we may consider either the renormalization of the annihilation vertex $\Gamma^{(l,k)}$ or the scattering vertex $\Gamma^{(k,k)}$. Probability conservation naturally demands that both vertex functions renormalize in the same manner, since they just represent the gain and loss contributions in the original master equation. Indeed, an identical set of loop diagrams describes the fluctuation corrections for both renormalized vertex functions, as depicted in Fig. 9.5(a). These strings of diagrams offer a straightforward physical interpretation: the bare annihilation vertex $\propto \lambda$ is proportional to the rate that k particles annihilate to l survivors when they meet, ignoring their earlier history. This must be corrected by subtracting off the probability that these particles have met at least once in the past and reacted; but that is overcounting, since again, the particles may have just scattered off each other before, so the third term needs to be added on, and so forth. One may of course equivalently work with the shifted action (9.56) with $\hat{\psi} = 1 + \tilde{\psi}$; the binomial expansion of the reaction pseudo-Hamiltonian (9.53) then eliminates the annihilation sink $\Gamma^{(l,k)}$, but instead generates the $k - l - 1$ vertices $\Gamma^{(l+1,k)}, \ldots, \Gamma^{(k-1,k)}$ in addition to $\Gamma^{(k,k)}$. Yet all these vertex functions become again renormalized through precisely the Feynman graphs shown in Fig. 9.5(a). In direct analogy with Dyson's equation for the renormalized propagator, see Fig. 4.12 in Section 4.4.1, one may equivalently represent this infinite diagrammatic series recursively through the *Bethe–Salpeter equations* drawn in Fig. 9.5(b).

We may set the external wavevectors to zero, and determine the renormalized rate at finite frequency $i\omega = D\mu^2$, thus carefully avoiding the infrared singularities as $\omega \to 0$ (or $t \to \infty$). In frequency space, the concatenated graphs in Fig. 9.5(a) obviously sum to a geometric series,

$$\Gamma_R^{(m,k)}(\omega) = \frac{c_{km}\lambda}{1 + \lambda I_{kd}(\omega)} \, , \tag{9.69}$$

where the index $m = l, k$ in the unshifted theory, and the combinatorial factor $c_{km} = k!\, m!$, while $m = l + 1, \ldots, k$ and $c_{km} = k!\, m! \left[\binom{k}{m} - \binom{l}{m} \right]$ for the shifted variant; $I_{kd}(\omega)$ denotes the $(k-1)$-loop integral of the lowest-order fluctuation correction. In the time domain, the second Bethe–Salpeter graph in Fig. 9.5(b) translates to the self-consistent equation

$$\Gamma_R^{(m,k)}(t) = c_{km}\lambda\, \delta(t) - \lambda \int_0^t \Gamma_R^{(m,k)}(t - t')\, I_{kd}(t')\, dt' \, . \tag{9.70}$$

Fourier or rather Laplace transform, as $\Gamma_R^{(m,k)}(t) \propto \Theta(t)$, together with the convolution theorem then also yields (9.69). The loop integral is most conveniently evaluated in the time domain, by means of appropriate linear transformations of the momentum variables (Problem 9.4):

$$I_{kd}(t) = k! \int \prod_{i=1}^{k} \frac{d^d p_i}{(2\pi)^d}\, e^{-D(p_1^2 + \cdots + p_k^2)t}\, (2\pi)^d \delta(p_1 + \cdots + p_k)\, \Theta(t)$$

$$= \frac{k!}{k^{d/2}(4\pi Dt)^{(k-1)d/2}}\, \Theta(t) \, . \tag{9.71}$$

Back in frequency space, the resulting gamma function displays the expected singularity at the upper critical dimension $d_c(k) = 2/(k-1)$:

$$I_{kd}(\omega) = \frac{C_{kd}/D}{d_c(k) - d} \left(\frac{i\omega}{D} \right)^{-1+d/d_c(k)} , \qquad C_{kd} = \frac{k!\,\Gamma(2 - d/d_c(k))\, d_c(k)}{k^{d/2}\,(4\pi)^{d/d_c(k)}} \, . \tag{9.72}$$

If we define the renormalized effective coupling conveniently according to

$$g_R = Z_g\, \frac{\lambda}{D}\, C_{kd}\, \mu^{-2[1-d/d_c(k)]} \, , \tag{9.73}$$

we obtain at the normalization point $i\omega/D = \mu^2$

$$Z_g^{-1} = 1 + \frac{\lambda}{D} \frac{C_{kd}\, \mu^{-2[1-d/d_c(k)]}}{d_c(k) - d} , \qquad (9.74)$$

to *all* orders in the perturbation expansion. Consequently, the associated RG beta function becomes

$$\beta_g = \mu \left.\frac{\partial}{\partial \mu}\right|_0 g_R = \frac{2g_R}{d_c(k)} \left[d - d_c(k) + g_R\right] , \qquad (9.75)$$

displaying two fixed points: the Gaussian fixed point $g_0^* = 0$ is infrared-stable for $d > d_c(k)$, whereas in low dimensions $d < d_c(k)$, the RG flow approaches the non-trivial *annihilation fixed point*

$$g^* = d_c(k) - d , \qquad (9.76)$$

which governs *universal* asymptotic behavior, independent of specific reaction rate values. Note that Eqs. (9.73) and (9.74) can be inverted exactly to relate the bare ratio of reactivity and diffusion coefficient λ/D with the dimensionless renormalized coupling g_R:

$$\frac{\lambda}{D} = \frac{[d_c(k) - d]\, g_R}{d_c(k) - d - g_R} C_{kd}^{-1} \mu^{2[1-d/d_c(k)]} . \qquad (9.77)$$

Physical rates therefore correspond to $g_R < g^*$; at the annihilation fixed point itself, $\lambda/D \to \infty$, indicating that the reaction processes indeed become entirely diffusion-limited.

With the quadratic beta function (9.75), one may readily solve the RG flow equation for the running coupling

$$\ell\frac{d\tilde{g}(\ell)}{d\ell} = \beta_g(\ell) , \quad \tilde{g}(1) = g_R , \qquad (9.78)$$

to obtain an algebraic function for $d \neq d_c(k)$,

$$\tilde{g}(\ell) = \frac{[d_c(k) - d]\, g_R}{g_R + [d_c(k) - d - g_R]\, \ell^{2[1-d/d_c(k)]}} , \qquad (9.79)$$

which is replaced by a logarithmic decay at the critical dimension $d = d_c(k)$:

$$\tilde{g}(\ell) = \frac{g_R}{1 - [2g_R/d_c(k)]\ln \ell} . \qquad (9.80)$$

Since no other quantities acquire any renormalizations, one may anticipate that inserting the scale-dependent effective reactivity $\tilde{\lambda}(\ell)$ into the rate equation should yield the correct density decay. For $d > d_c(k)$, Eq. (9.80) gives $\tilde{g}(\ell) \sim g_R\, \ell^{2[d-d_c(k)]/d_c(k)} \to 0$, whence according to (9.77) the effective reaction rate $\tilde{\lambda}(\ell)/D \sim \tilde{g}(\ell) C_{kd}^{-1}(\mu\ell)^{2(d_c-d)/d_c} \to$ const., as implicitly assumed in the mean-field approach, and the power laws (9.8) are recovered. For $d < d_c(k)$, we match the flow parameter with the diffusive time scale according to $\mu\ell = (Dt)^{-1/2}$; as $\tilde{g} \to g^*$, we obtain the asymptotic temporal decrease $\tilde{\lambda}(t) \sim D(Dt)^{-1+d/d_c(k)}$. The rate equation $\partial a(t)/\partial t = -\tilde{\lambda}(t)a(t)^k$ is then solved by $a(t) \sim (Dt)^{-d/2}$ as anticipated in (9.18) by simple scaling arguments; this may be turned into a density-dependent reaction rate $\tilde{\lambda}(a) \sim a^{2[d_c(k)-d]/d_c(k)d}$, in accord with Smoluchowski's theory for pair processes with $d_c(2) = 2$, see Eq. (9.23). At $d_c(k)$, $\tilde{\lambda}(t) \sim D/\ln(Dt)$ according to (9.80), which yields the logarithmic corrections $a(t) \sim [(Dt)^{-1}\ln(Dt)]^{1/(k-1)}$ and $\tilde{\lambda}(a) \sim D/\ln(1/a)$, compare with Problem 9.2(a). This heuristic approach is confirmed by a careful renormalization group analysis of the particle density.

9.3.2 RG analysis of the particle density decay

Following the procedures of Section 5.3.1, we construct the renormalization group equation for the particle density, retaining its dependence on the initial value a_0. To this end, we note that the particle density has scaling dimension $[a] = \mu^d$, whence $a_R(\mu, D_R, a_0, g_R) = \mu^d \, \hat{a}_R(D_R, a_0\,\mu^{-d}, g_R)$; recalling that neither the fields nor the diffusion constant D pick up any renormalization for the pure annihilation model, $\gamma_D = 0$ and $\gamma_{a_0} = -d$, the RG equation for the density becomes

$$\left[d - d\,a_0 \frac{\partial}{\partial a_0} + \beta_g \frac{\partial}{\partial g_R} \right] \hat{a}_R\left(a_0\,\mu^{-d}, g_R\right) = 0\,, \qquad (9.81)$$

compare with Eq. (5.41). With the characteristics set equal to $\ell = (D\mu^2 t)^{-1/2}$, its solution in the vicinity of the IR-stable RG fixed point $g^* > 0$ reads

$$a_R(a_0, t) \propto (D\mu^2 t)^{-d/2}\, \hat{a}_R\left(a_0\,(D\mu^2 t)^{d/2}, g^*\right)\,, \qquad (9.82)$$

and thus yields the expected diffusion-limited decay law $a(t) \sim (Dt)^{-d/2}$ for $d < d_c(k)$. However, the first argument in (9.82) flows to infinity under the RG. One therefore has to establish that the scaling function $\hat{a}(y) = \hat{a}_R(y, g^*)$ remains finite as its argument $y \to \infty$.

For convenience, we use the shifted Doi–Peliti action (9.56) to first set up a perturbation expansion with respect to the initial density a_0, and then sum it to *all* orders in a_0. The contributing Feynman diagrams are depicted in Fig. 9.6, for clarity just for pair annihilation or coagulation/fusion, up to third order in both the reaction rate λ and the initial density a_0. First consider the tree diagrams, shown up to second order in the annihilation vertex λ in Fig. 9.6. As demonstrated in Fig. 9.7, the entire set of tree-level graphs is iteratively produced through a Dyson-like structure that contains the bare vertex $\Gamma_0^{(1,k)}(0, 0) = -(k - l)\,\lambda$, see (9.58). The

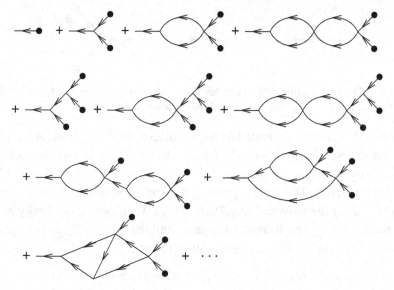

Fig. 9.6 Feynman diagrams contributing to the density for pair annihilation, up to third order in both the vertex coupling λ (two-loop) and the source strength n_0.

corresponding self-consistent analytical expression for the tree-level density reads

$$a_{\text{tr}}(t) = a_0 - (k - l)\lambda \int_0^t a_{\text{tr}}(t')^k \, dt' , \tag{9.83}$$

which is precisely the integral version of the mean-field rate equation (9.6), whose solution is manifestly finite in the limit $a_0 \to \infty$ (for $k > 1$). Through the RG equation, the solution (9.8) becomes transported into the diffusion-limited regime that is dominated by emerging depletion zones,

$$a_{\text{tr}}(t) = (\mu\ell)^d \frac{\tilde{a}_0(\ell)\mu^{-d}}{\left[1 + [\tilde{a}_0(\ell)\mu^{-d}]^{2/d_c(k)}(k - l)(k - 1)\, C_{kd}^{-1}\tilde{g}(\ell)(\mu\ell)^2 Dt\right]^{d_c(k)/2}} ,$$

where $\tilde{a}_0(\ell) = a_0\ell^{-d}$. At the annihilation fixed point, this basically amounts to the replacement $\lambda t \to C_{kd}^{-1}g^*(Dt)^{d/d_c(k)}$ in the rate equation solution,

$$a_{\text{tr}}(t) \sim \frac{a_0}{\left[1 + a_0^{2/d_c(k)}(k - l)(k - 1)\, C_{kd}^{-1}g^*(Dt)^{d/d_c(k)}\right]^{d_c(k)/2}}$$

$$\to A_{kl}(d)\,(Dt)^{-d/2} \tag{9.84}$$

as $Dt = x_D^2 \gg a_0^{-2/d}$, with a *universal* (independent of a_0 and λ) amplitude

$$A_{kl}(d) = \left[(k - l)(k - 1)\, C_{kd}^{-1}[d_c(k) - d]\right]^{-1/(k-1)} . \tag{9.85}$$

The tree approximation for the particle density is therefore equivalent to renormalized simple mean-field theory, i.e., with bare parameters replaced by their renormalized running counterparts. Inspecting the loop corrections in Fig. 9.6, we

Fig. 9.7 Dyson equation that generates the entire series of tree diagrams for the particle density a_{tr} in diffusion-limited annihilation reactions.

observe that these are all generated through replacing the bare propagators with full renormalized response functions. At fixed order in a_0, for $d \leq d_c(k)$ each vertex asymptotically contributes a term $\propto g^*$, leaving the overall temporal power law in Eq. (9.84) unaltered. The loop expansion thus transforms into a power series in $\epsilon = d_c(k) - d$ for the universal *amplitude* $A_{kl}(\epsilon)$. With (9.85), we finally arrive at the asymptotic long-time behavior for pair annihilation ($l = 0$) and coagulation/ fusion ($l = 1$) below the critical dimension $d_c(2) = 2$,

$$a(t) \rightarrow A_{2l}(\epsilon)(D\,t)^{-d/2} , \qquad A_{2l}(\epsilon) = [(2-l)\,2\pi\,\epsilon]^{-1} + O(1) . \qquad (9.86)$$

At the critical dimension $d_c(k)$, the slow decay (9.80) yields instead the still diffusion-limited logarithmic correction

$$a(t) \rightarrow \left[B_{kl}(Dt)^{-1} \ln(\mu^2 Dt) \right]^{1/(k-1)} , \qquad B_{kl} = \frac{k!}{4\pi(k-l)(k-1)k^{1/(k-1)}} ; \quad (9.87)$$

i.e., $a(t) \rightarrow [2\pi(2-l)Dt]^{-1} \ln(\mu^2 Dt)$ for pair processes ($l = 0, 1$) at $d_c(2) = 2$, and $a(t) \rightarrow ([4\pi(3-l)Dt/\sqrt{3}]^{-1} \ln(\mu^2 Dt))^{1/2}$ for triplet reactions ($l = 0, 1, 2$) at $d_c(3) = 1$, see Problem 9.2(b).

Instead of particle propagation via nearest-neighbor hopping, one may also study faster long-range *Lévy flights* with superdiffusive mean-square displacement $\langle x(t)^2 \rangle \sim (D_L t)^{1/\sigma}$, where $0 < \sigma < 2$.[13] The critical dimension for pair annihi- lation or fusion processes is lowered to $d_c(\sigma) = \sigma$, and for $d < \sigma$ one finds the asymptotic density decay $a(t) \sim (D_L t)^{-d/\sigma}$ (Problem 9.5).

Recent experiments have indeed confirmed the anomalous power law decay $n(t) \sim t^{-1/2}$ in one dimension for exciton recombination kinetics $A + A \rightarrow A$ in molecular wires, semiconducting $N(CH_3)_4 MnCl_3$ (TMMC) polymer chains, and carbon nanotubes.[14] Particularly convincing are the data obtained by Allam *et al.* for the exciton statistics in single-walled carbon nanotubes. For high excitation levels, Fig. 9.8(a) shows non-universal mean-field decay $\sim t^{-1}$ and subsequent crossover to the universal behavior $\sim t^{-1/2}$. When the absorption is fully saturated, the plot in Fig. 9.8(b) of the exciton concentration against $t + t_0 + t_1$ allows the determination of characteristic times t_0 and t_1, respectively indicating the onset

---

[13] Hinrichsen and Howard (1999); Vernon (2003).
[14] Kopelman (1988); Kroon, Fleurent, and Sprik (1993); Russo *et al.* (2006); Allam *et al.* (2013).

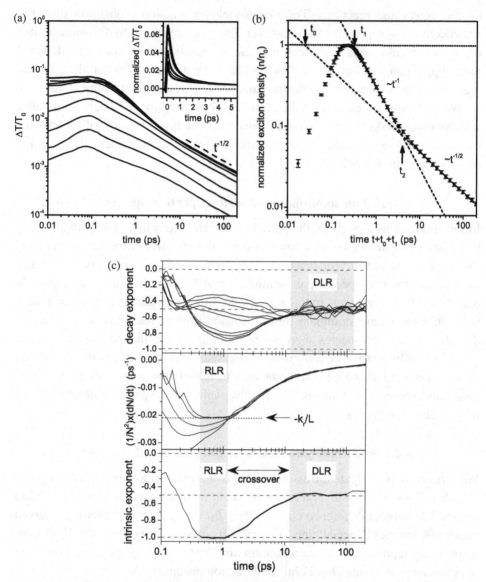

Fig. 9.8 Experimental data for exciton recombination in single-walled carbon nanotubes showing the crossover from reaction- to diffusion-limited scaling: (a) decay of the exciton density, proportional to the differential transmission $\Delta T/T_0$ at various laser pulse energies (bottom to top: 0.19, 0.48, 1.1, 4.0, 12, 40, 60, 80, and 104 nJ; inset: same data normalized to the amplitude at long time); (b) normalized exciton concentration n/n_0 plotted against the scaling time $t + t_0 + t_1$; and (c) upper panel: evolution of the effective decay exponent (for the same pulse energies as in (a) from top to bottom); middle panel: $n(t)^{-2}\mathrm{d}n(t)/\mathrm{d}t$ for the highest excitation levels (decreasing from top to bottom); lower panel: intrinsic exponent given by the measured exponent multiplied by $(t + t_0 + t_1)/t$, with the heavy line indicating the experimentally determined crossover function [Figures adapted with permission from: J. Allam, M. T. Sajjad, R. Sutton, *et al.*, *Phys. Rev. Lett.* **111**, 197401 (2013).]

of encounters and reactions. Their ratio yields a reaction probability of ≈ 0.2 per encounter; t_2 represents the crossover time from reaction- to diffusion-limited decay. An unambiguous identification of the asymptotic decay exponent is depicted in the upper panel of Fig. 9.8(c). The middle panel demonstrates that the effective reaction rate is constant only in the intermediate reaction-limited regime (RLR), but evolves with time (or density) in the diffusion-limited regime (DLR), in accordance with Smoluchowski theory and the RG analysis. The lower panel finally shows the experimentally determined crossover function from the mean-field decay to the universal asymptotic behavior.

9.4 Pair annihilation of distinct particle species

In low dimensions $d \leq d_c(k)$, the particle anticorrelations induced by annihilation processes generate depletion zones that cause the effective renormalized reaction rate to decay with time (or density). This mechanism also crucially decelerates the kinetics for two-species pair annihilation $A + B \to \emptyset$, if the initial particle densities differ. For *equal* initial species densities, correlations of a fundamentally different nature dominate these diffusion-limited reactions in dimensions $d \leq d_s = 4$, namely species *segregation* into coarsening inert domains.[15] These are delimited by sharp *reaction fronts*, whose scaling properties can be determined *exactly*. These basic spatial correlation effects, namely the emergence of depletion zones and segregated domains, are finally explored in general multiple-species pair annihilation processes.

9.4.1 Renormalized kinetics for unequal initial densities

We proceed to investigate diffusion-limited pair annihilation among distinct particles $A + B \to \emptyset$. Crucially, no reactions occur among either A or B species, and the particle number difference $c(t) = a(t) - b(t) = c_0$ remains strictly conserved under this kinetics. Specializing to equal diffusivities for both A and B species, respectively represented in the continuum limit by the fields ψ and φ, and omitting the boundary terms, the Doi–Peliti 'bulk' action becomes

$$\mathcal{A}[\hat{\psi}, \hat{\varphi}, \psi, \varphi] = \int d^d x \int dt \left[\hat{\psi}(x, t)\left(\frac{\partial}{\partial t} - D\nabla^2 \right)\psi(x, t) \right.$$

$$+ \hat{\varphi}(x, t)\left(\frac{\partial}{\partial t} - D\nabla^2 \right)\varphi(x, t)$$

$$\left. + \lambda[\hat{\psi}(x, t)\,\hat{\varphi}(x, t) - 1]\psi(x, t)\,\varphi(x, t) \right], \qquad (9.88)$$

---

[15] Toussaint and Wilczek (1983); Lee and Cardy (1994, 1995).

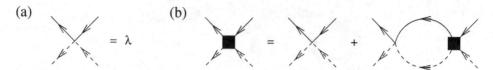

Fig. 9.9 Diffusion-limited pair annihilation $A + B \to \emptyset$: (a) scattering vertex (species A/B: full/dashed lines); and (b) vertex renormalization Bethe–Salpeter equation which takes the same form as for single-species pair annihilation, see Fig. 9.5.

cf. Eq. (9.35). When written in terms of the fields $\tilde{\psi}(x, t) = \hat{\psi}(x, t) - 1$ and $\tilde{\varphi}(x, t) = \hat{\varphi}(x, t) - 1$, the resulting shifted action can be interpreted as a Janssen–De Dominicis functional that encodes the Langevin equations

$$\frac{\partial \psi(x, t)}{\partial t} = D\nabla^2 \psi(x, t) - \lambda \psi(x, t) \varphi(x, t) + \zeta(x, t), \tag{9.89}$$

$$\frac{\partial \varphi(x, t)}{\partial t} = D\nabla^2 \varphi(x, t) - \lambda \psi(x, t) \varphi(x, t) + \eta(x, t), \tag{9.90}$$

with $\langle \zeta(x, t) \rangle = 0 = \langle \eta(x, t) \rangle$, and the noise correlations

$$\langle \zeta(x, t) \zeta(x', t') \rangle = 0 = \langle \eta(x, t) \eta(x', t') \rangle, \tag{9.91}$$

$$\langle \zeta(x, t) \eta(x', t') \rangle = -\lambda \, \psi(x, t) \varphi(x, t) \delta(x - x')\delta(t - t'). \tag{9.92}$$

The stochastic partial differential equations (9.89) and (9.90) constitute the mean-field reaction-diffusion equations supplemented with multiplicative noise that captures the emerging anticorrelations between different particles.

As in the single-species case, the action (9.88) allows no propagator renormalization, and the string of Feynman diagrams for the renormalized annihilation vertex, or equivalently, the corresponding Bethe–Salpeter equation, are of precisely the form as for $A + A \to \emptyset$, see Fig. 9.9. Consequently we deduce that $A-B$ anticorrelations and depletion zones should qualitatively influence the long-time statistics in dimensions $d \leq d_c = 2$, the associated upper critical dimension. Let us first consider the more generic situation of initially *unequal* particle densities, $c_0 > 0$. Following the procedure in the preceding section, we study the renormalized mean-field theory, with the RG flow for the annihilation rate properly accounted for. For $d > 2$, this recovers the basic mean-field result (9.11). For $d < 2$, we merely need to replace λt in the exponent with $(Dt)^{d/2}$, leading to the *stretched exponential decay*

$$\ln[a(t) - a_\infty] \sim \ln b(t) \sim -(Dt)^{d/2}; \tag{9.93}$$

at $d_c = 2$, one similarly obtains the logarithmically slowed-down kinetics

$$\ln[a(t) - a_\infty] \sim \ln b(t) \sim -Dt / \ln(Dt). \tag{9.94}$$

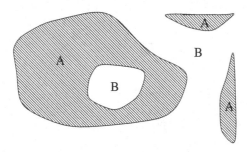

Fig. 9.10 Typical late-time state for two-species pair annihilation with homogeneous random initial conditions. Almost pure regions of typical size $x_D \sim (Dt)^{1/2}$ are separated by narrow reaction zones.

9.4.2 Equal initial densities: segregation, reaction fronts

For *equal* initial densities $a_0 = b_0$, the mean-field decay rate $c_0 \lambda$ vanishes, indicating a special scenario, where fluctuations in the *initial* particle distributions become crucial. A simple scaling argument captures the essential physics. Starting from a random (Poissonian) particle distribution, nearby A–B pairs will quickly annihilate; following this fast initial regime, density fluctuations on distances shorter than $x_D \sim (Dt)^{1/2}$ will be smoothed out after time t, leaving behind essentially homogeneous regions with the respective local majority population, see Fig. 9.10. The two particle species hence *segregate* asymptotically, and reactions can only occur at the sharp boundaries separating the A- and B-rich regions. Since the diffusion length x_D sets the characteristic scale, one expects the local density excess to decay like $c(t) \sim x_D(t)^{-d/2} \sim (Dt)^{-d/4}$, which is *slower* than the rate equation prediction $\sim (\lambda t)^{-1}$ in dimensions $d < d_s = 4$, and even supersedes the slowing down due to rate renormalization $\sim (Dt)^{-d/2}$ for $d \leq d_c = 2$.

At least for $d > 2$, this scenario can be captured quantitatively by means of the classical field or reaction-diffusion requations (9.89), (9.90) with vanishing noise $\zeta = 0 = \eta$. The local density difference $c(x, t) = \psi(x, t) - \varphi(x, t)$ then becomes a purely *diffusive mode*, governed by the propagator (9.66). One may consequently employ the spatial Green function in d dimensions,

$$G_0(x, t) = \frac{1}{(4\pi Dt)^{d/2}} \, e^{-x^2/4Dt} \, \Theta(t) \,, \tag{9.95}$$

see Eq. (2.104), to solve the general initial value problem,

$$c(x, t) = \int d^d x' \, G_0(x - x', t) \, c(x', 0) \,. \tag{9.96}$$

We assume a Poisson distribution for the initial density fluctuations; indicating averages over initial conditions by an overbar, therefore $\overline{\psi(x, 0) \, \psi(x', 0)} = a_0^2 + a_0 \, \delta(x - x') = \overline{\varphi(x, 0) \, \varphi(x', 0)}$ and $\overline{\psi(x, 0) \, \varphi(x', 0)} = a_0^2$, which implies

$\overline{c(x,0)\,c(x',0)} = 2a_0\,\delta(x-x')$. Averaging the squared density excess (9.96) at time $t > 0$ over the initial distribution yields

$$\overline{c(x,t)^2} = 2a_0 \int d^d x'\, G_0(x-x',t)^2 = \frac{2a_0}{(8\pi Dt)^{d/2}}\,; \qquad (9.97)$$

since the distribution for $c(x,t)$ will be a Gaussian, we obtain for the root-mean-square *local density excess* originating in a random initial fluctuation:

$$\overline{|c(x,t)|} = \sqrt{\frac{2}{\pi}\,\overline{c(x,t)^2}} = \sqrt{\frac{a_0}{\pi}}\,\frac{2}{(8\pi Dt)^{d/4}}\,. \qquad (9.98)$$

Since this decay is slower than the mean-field prediction $a(t) = b(t) \sim (\lambda t)^{-1}$ in dimensions $d < d_s = 4$, the particle distribution of neither species can be uniform. Instead, locally either $a(x,t) \gg b(x,t)$ or vice versa, whence (9.98) in fact describes the density scaling of the majority species in their respective domains. Summing over these segregated regions then yields that the average A and B densities follow the algebraic decay (9.98), $a(t) \sim b(t) \sim (Dt)^{-d/4}$. The three-dimensional non-classical asymptotic power law $\sim t^{-3/4}$ has indeed been observed in a calcium/fluorophore system.[16] Note that there are no logarithmic corrections in four dimensions.

For very specially chosen initial states, the situation can be different. For example, consider hard-core particles (or equivalently, reactivity $\lambda \to \infty$) on a one-dimensional chain that are regularly arranged in an alternating manner $\ldots ABABABABAB \ldots$. The reactions $A + B \to \emptyset$ actually preserve this arrangement, whence the distinction between A and B particles becomes meaningless, and their densities indeed decay according to the $t^{-1/2}$ power law from the single-species pair annihilation reaction.

Let us return to the generic case, with spatially segregated regions in dimensions $d < 4$ that are separated by sharp reactive boundaries, where pair annihilation processes happen, Fig. 9.10, and study the scaling properties of these *reaction fronts*, sketched in Fig. 9.11. We begin with a simple *steady-state* approximation for $d > d_c = 2$ dimensions, wherein a fixed particle *current* J is set up along (the arbitrarily chosen) x_1 direction:

$$\left(\frac{\partial a(x)}{\partial x_1},\frac{\partial b(x)}{\partial x_1}\right) \to \begin{cases} (-J,0)\,, & x_1 \to -\infty\,, \\ (0,J)\,, & x_1 \to \infty\,, \end{cases}$$

i.e., $c(x) = a(x) - b(x) = -Jx_1$ from the stationary diffusion equation for $c(x)$. This allows us to eliminate one of the densities from the coupled reaction-diffusion

---

[16] Monson and Kopelman (2004).

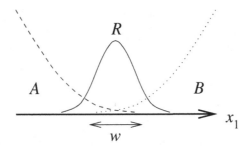

Fig. 9.11 Sketch of a reaction front cross section. The density of A (dashed)/B (dotted) particles decreases almost linearly on the left/right, and vanishes rapidly on the right/left. The solid curve indicates the reaction zone density $R(x)$ of width w.

equations, for example

$$D \frac{\partial^2 a(x)}{\partial x_1^2} - \lambda J x_1 \, a(x) - \lambda a(x)^2 = 0 . \tag{9.99}$$

It is now easily seen that the ordinary differential equation (9.99) can be rendered dimensionless by introducing rescaled variables $y = (\lambda J/D)^{1/3} x_1$ and $\rho = (DJ^2/\lambda)^{1/3} a$; in other words, the solution of Eq. (9.99) must assume the *scaling form* ($x_\perp$ denotes the coordinates perpendicular to the front),

$$a(x) = (DJ^2/\lambda)^{1/3} \, \hat{a}\big((\lambda J/D)^{1/3} x_1, x_\perp\big) , \tag{9.100}$$

whence we obtain for the *steady-state reaction front density*

$$R(x) = \lambda \, a(x) \, b(x) = J^{\beta'} \hat{R}\big(J^{\alpha'} x_1, x_\perp\big) \tag{9.101}$$

with the mean-field exponents $\alpha' = 1/3$ and $\beta' = 4/3$. The argument in the scaling function may be interpreted as $x_1/w(J)$ with the *reaction front width* $w(J) \sim J^{-\alpha'}$.

Remarkably, this straightforward analysis can be extended to dimensions $d \leq 2$, and one can even obtain *exact* scaling laws. This is possible due to the absence of field renormalization, and the fact that since $\tilde{g}(\ell) \to g^*$, asymptotically the dependence on the reaction rate λ must drop out. With the unaltered scaling dimensions $[R(x)] = \mu^{d+2}$ and $[J] = \mu^{d+1}$, thus necessarily

$$\alpha' = \frac{1}{d+1} , \quad \beta' = \frac{d+2}{d+1} . \tag{9.102}$$

For the associated temporal scaling, one may invoke time scale separation, and argue that the local relaxation of fronts to their stationary state proceeds fast compared with actual front motion. This allows the replacement of the stationary current J with its quasi-stationary time-dependent counterpart $J(t) \sim a(t)/x_D(t)$. Upon assembling the characteristic density decay and diffusion scaling, one thus arrives at

$$J(t) \sim (Dt)^{-\lambda/\alpha'}, \quad w(t) \sim (Dt)^{\lambda}, \quad \lambda = \frac{d+2}{4(d+1)}. \tag{9.103}$$

Relative to the growing size $\sim (Dt)^{1/2}$ of the segregated areas, the reaction zones sharpen with time $\sim (Dt)^{-d/4(d+1)}$. In two dimensions, one finds logarithmic corrections to the mean-field power laws, see Problem 9.6.

Finally, we briefly address the generalization to *q-species pair annihilation* processes, $A_i + A_j \to \emptyset$, $1 \le i < j \le q$, with *equal* initial densities $a_i(0)$ as well as uniform diffusion and reaction rates.[17] For $q > 2$, there exists *no* conservation law in the stochastic system, and one may argue, based on the study of fluctuations in the associated Fokker–Planck equation, that particle segregation happens only for $d < d_s(q) = 4/(q-1)$. In any physical dimension $d \ge 2$, one should therefore obtain the same behavior as for the single-species processes $A + A \to \emptyset$; in fact, this is obvious in the limit $q \to \infty$, since in this case the probability for particles of the same species to ever meet tends to zero, whence the species labeling becomes irrelevant. In one dimension, with its special topology, segregation does occur, facilitated by the fact that asymptotically no particles can pass through each other, as the bare reaction rate $\lambda \to \infty$ at the annihilation fixed point. For generic initial conditions one may derive the decay law

$$a_i(t) \sim t^{-\alpha(q)} + C\,t^{-1/2}, \quad \alpha(q) = \frac{q-1}{2q}, \tag{9.104}$$

which correctly reproduces $\alpha(2) = 1/4$ and $\alpha(\infty) = 1/2$. Note the sub-leading contribution $\sim t^{-1/2}$ which originates from the rate renormalization caused by the appearance of depletion zones. As shown in the Monte Carlo simulation data in Fig. 9.12, these and other corrections to the leading power laws generally cause a rather slow convergence of the effective density decay exponent $\alpha_{\text{eff}}(q, t)$ to its asymptotic value given by (9.104). The reaction zone width grows as

$$w(t) \sim t^{\lambda(q)}, \quad \lambda(q) = \frac{2q-1}{4q}; \tag{9.105}$$

for $q = 2$ this gives $\lambda(2) = 3/8$, which coincides with (9.103) for $d = 1$.

Once again, one may construct special situations, e.g., the alignment $\dots ABCDABCDABCD\dots$ for $q = 4$, for which single-species scaling ensues. There are also intriguing *cyclic* variants, for example if for four species only

---

[17] Deloubrière, Hilhorst, and Täuber (2002); Hilhorst, Deloubrière, Washenberger, and Täuber (2004); Hilhorst, Washenberger, and Täuber (2004).

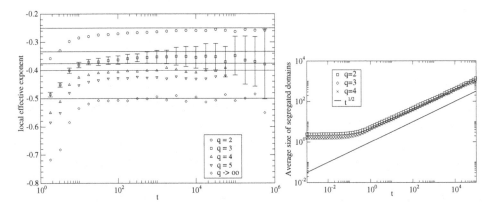

Fig. 9.12 Monte Carlo results for q-species pair annihilation reactions, obtained with random initial conditions on one-dimensional lattices with 10^5 sites, averaged over 50 simulation runs. (a) Local effective density decay exponent $-\alpha_{\text{eff}}(q, t)$ for $q = 2, 3, 4, 5$ different species, and for $q \to \infty$ (i.e., $A + A \to \emptyset$); (statistical) error bars are indicated for $q = 3$. Note the rather slow approach to the asymptotic predictions (9.104), shown as straight lines, except for the single-species annihilation reaction. (b) Growth $\sim t^{1/2}$ of the average domain size of the single-species segregated regions for $q = 2, 3$, and 4 in one dimension. [Figures reproduced with permission from: H. J. Hilhorst, O. Deloubrière, M. J. Washenberger, and U. C. Täuber, *J. Phys. A: Math. Gen.* **37**, 7063 (2004); DOI: 10.1088/0305-4470/37/28/001; copyright (2004) by Inst. of Physics Publ.]

the reactions $A + B \to \emptyset$, $B + C \to \emptyset$, $C + D \to \emptyset$, and $D + A \to \emptyset$ are permitted. We may then obviously identify $A = C$ and $B = D$, reducing the problem to two-species pair annihilation. Generally, within essentially mean-field theory one finds for cyclic multi-species annihilation processes $a_i(t) \sim t^{-\alpha(q,d)}$ for $d < d_s(q)$ where $\alpha(q, d) = d/d_s(q)$ and $d_s = 4$ for even $q = 2, 4, 6, \ldots$, while $d_s(q) = 4\cos(\pi/q) > 2$ for odd $q = 3, 5, 7, \ldots$. Curiously, for five species this yields the borderline dimension $d_s(5) = 1 + \sqrt{5}$ for segregation to occur, and non-trivial decay exponents $\alpha(5, 2) = \frac{1}{2}(\sqrt{5} - 1)$ in $d = 2$ and $\alpha(5, 3) = \frac{3}{4}(\sqrt{5} - 1)$ in $d = 3$ that involve the golden ratio.

9.5 Fluctuation effects in two-species binary reactions

Applying the basic concepts and analytical methods developed in the previous sections allows us to study the intriguing non-trivial effects of fluctuations and reaction-induced correlations in two other binary processes that involve two particle species. First, we consider the emergence of *generic scale invariance* in *reversible recombination* reactions, caused by the presence of a conserved quantity. Second, we employ our field theory tools to explain the presence of strong fluctuations associated with persistent activity fronts that in turn induce local stochastic population oscillations in the predator–prey coexistence phase of the Lotka–Volterra model.

9.5.1 Reversible recombination reactions

Another instructive example of a simple stochastic reaction-diffusion process with non-trivial power law scaling, for which a conservation law plays a crucial role, is the *reversible recombination* reaction $A + A \rightleftharpoons B$ with forward rate λ and backward rate σ.[18] We assume again for simplicity that all particles have the same diffusivities D, and that the initial conditions are random but homogeneous with mean densities a_0 and b_0, respectively. The reaction processes conserve the quantity $c(t) = a(t) + 2b(t) = c_0$. Upon setting $k = 2$, $l = 0$, and $m = 1$ in (9.34), taking the continuum limit, and shifting $\hat{\psi}(x,t) = 1 + \tilde{\psi}(x,t)$, $\hat{\varphi}(x,t) = 1 + \tilde{\varphi}(x,t)$, one arrives at the pseudo-Hamiltonian density

$$\mathcal{H}'[\tilde{\psi}, \tilde{\varphi}, \psi, \varphi] = [2\tilde{\psi}(x,t) - \tilde{\varphi}(x,t) + \tilde{\psi}(x,t)^2][\lambda\psi(x,t)^2 - \sigma\varphi(x,t)] \,,$$

(9.106)

where the fields ψ and φ represent the A and B species. With the same diffusion terms as in (9.88) added, one may interpret the resulting Doi–Peliti action as a Janssen–De Dominicis response functional, which yields the equivalent Langevin description

$$\frac{\partial\psi(x,t)}{\partial t} = D\nabla^2\psi(x,t) - 2\lambda\psi(x,t)^2 + 2\sigma\varphi(x,t) + \zeta(x,t) \,, \qquad (9.107)$$

$$\frac{\partial\varphi(x,t)}{\partial t} = D\nabla^2\varphi(x,t) + \lambda\psi(x,t)^2 - \sigma\varphi(x,t) \,, \qquad (9.108)$$

with $\langle \zeta(x,t) \rangle = 0$, and the noise correlation

$$\langle \zeta(x,t)\zeta(x',t') \rangle = 2[\sigma\varphi(x,t) - \lambda\psi(x,t)^2]\delta(x-x')\delta(t-t') \,. \quad (9.109)$$

Discarding the noise and the diffusion terms yields the associated mean-field rate equations, which describe the system's evolution into a stable non-trivial stationary state with non-zero densities a_∞ and b_∞ that are related through the condition $\lambda a_\infty^2 = \sigma b_\infty$. Since in addition $a_\infty + 2b_\infty = c_0$, the asymptotic densities are completely determined as functions of the reaction rates λ, σ, and the initial density c_0,

$$a_\infty = \sqrt{\frac{\sigma^2}{16\lambda^2} + \frac{c_0\sigma}{2\lambda}} - \frac{\sigma}{4\lambda} \,, \qquad b_\infty = \frac{c_0 - a_\infty}{2} \,. \qquad (9.110)$$

[18] Rey and Cardy (1999).

Linearizing about these steady-state densities, one finds an exponentially decaying eigenmode with relaxation rate

$$\gamma = \sigma + 4\lambda a_\infty = \sqrt{\sigma(\sigma + 8c_0\lambda)}\,, \tag{9.111}$$

whereas the other eigenvalue is exactly zero. The appearance of this zero mode reflects the presence of the conserved density c, which induces slow relaxational dynamics, causing fluctuations in the initial conditions to die away algebraically rather than exponentially fast. Reversible recombination reactions constitute a prototypical example for a non-equilibrium system displaying the phenomenon of *generic scale invariance*.

Returning to the Langevin description (9.107)–(9.109), we note that the equation of motion for the field φ representing the B species is deterministic, since the dissociation process $B \to A + A$ is of first order. The reaction noise correlator (9.109) is positive in the dissociation-dominated regime, but turns negative when recombination prevails; it vanishes in the stationary state (a_∞, b_∞). (Recall that diffusion noise is not properly incorporated in this formalism.) Consider the field $c(x, t) = \psi(x, t) + 2\varphi(x, t) - c_0$, which satisfies the noisy diffusion equation

$$\frac{\partial c(x, t)}{\partial t} = D\nabla^2 c(x, t) + \zeta(x, t)\,. \tag{9.112}$$

Its solution follows with the aid of the diffusion Green function (9.95):

$$c(x, t) = \int d^d x' \int dt'\, G_0(x - x', t - t')\zeta(x', t')$$

$$+ \int d^d x'\, G_0(x - x', t)\, c(x', 0)\,. \tag{9.113}$$

The second term here turns out to be unimportant unless the initial state is prepared with manifest spatial correlations. The first term implies, by the central-limit theorem, that the density fluctuations $c(x, t)$ should be asymptotically governed by a Gaussian distribution with zero mean, and that in particular with the noise correlator (9.109)

$$\langle c(x, t)^2 \rangle \approx 2 \int d^d x' \int dt'\, G_0(x - x', t - t')^2 [\sigma\varphi(x', t') - \lambda\psi(x', t')^2]$$

$$= 2 \int d^d x' \int dt'\, G_0(x - x', t - t')^2 \left(D\nabla'^2 - \frac{\partial}{\partial t'} \right)\varphi(x', t')\,, \tag{9.114}$$

by means of the deterministic equation (9.108). This crucial substitution allows us to average over the initial conditions, inserting (9.95) and performing the spatial

Gaussian integral, to conclude that

$$\overline{\langle c(x,t)^2\rangle} \approx -2\int_0^t \frac{1}{[8\pi D(t-t')]^{d/2}} \frac{\partial \overline{\varphi(t')}}{\partial t'}\,dt'\,.$$

For large times t, we may replace $(t-t')^{d/2}$ in the denominator by $t^{d/2}$; the integral over t' then yields the remarkable result that the amplitude of the late-time fluctuations in the conserved field are set entirely by the difference in the initial and asymptotic values of the B species density:

$$\overline{\langle c(x,t)^2\rangle} \approx \frac{2(b_0 - b_\infty)}{(8\pi Dt)^{d/2}}\,. \tag{9.115}$$

Now we can once again use Eq. (9.108), writing it in terms of the fields $c(x,t)$ and $\varphi(x,t)$:

$$\left(\frac{\partial}{\partial t} - D\nabla^2\right)\varphi(x,t) = \lambda[c(x,t) + c_0 - 2\varphi(x,t)]^2 - \sigma\varphi(x,t)\,. \tag{9.116}$$

Given $c(x,t)$ and the initial conditions, see (9.113), this in principle fully determines $\varphi(x,t)$ and thus $\psi(x,t) = c(x,t) + c_0 - 2\varphi(x,t)$. We know that if we ignore the conserved quantity c, $b(t)$ will approach its asymptotic value b_∞ exponentially fast; yet $c(x,t)$ typically only decays algebraically with time. Therefore, the leading terms on the right-hand side of Eq. (9.108) must cancel each other. This may be mathematically justified by regarding the left-hand side as a perturbation and showing that it produces sub-leading terms in the final solution. Physically, this procedure is equivalent to asserting that the balance requirement $\lambda\psi(x,t)^2 \approx \sigma\varphi(x,t)$ is even *locally* satisfied, before the fluctuations of the conserved mode allow the system to relax to full equilibrium. Since $c(x,t)$ is typically small, we can find a solution by setting the right-hand side to zero, expanding $\varphi(x,t) \approx b_\infty + k_1 c(x,t) + k_2 c(x,t)^2 + \cdots$. After straightforward algebra, the coefficients are found in terms of the input parameters λ, σ, and c_0: $k_1 = 2\lambda a_\infty/\gamma$, $k_2 = \lambda(1 - 2k_1)^2/\gamma = \sigma^2\lambda/\gamma^3$, etc. Taking averages over the initial conditions, noting that $\overline{\langle c(x,t)\rangle} = 0$, and using (9.115), one eventually arrives at

$$\overline{\langle a(t)\rangle} - a_\infty \approx -2(\overline{\langle b(t)\rangle} - b_\infty) \approx \frac{4\sqrt{\sigma}\lambda\,(b_\infty - b_0)}{(\sigma + 8c_0\lambda)^{3/2}(8\pi Dt)^{d/2}}\,; \tag{9.117}$$

i.e., monotonic power law decay $\sim (Dt)^{-d/2}$ in *any* dimension, with a *non-universal* amplitude whose sign is determined by the system's initiation through the difference $b_\infty - b_0$.

The *reversible two-species recombination* reactions $A + B \rightleftharpoons C$ are characterized by *two* conserved densities $a(t) - b(t)$ and $a(t) + b(t) + 2c(t)$. For uncorrelated initial conditions, the approach to the stationary state is asymptotically described by the very same algebraic decay as in (9.117), but with a different amplitude; this is true even for unequal diffusivities, including even the extreme situation of immobile C particles. In contrast to the upper critical dimension $d_c(k)$ for the importance of depletion zones in annihilation reactions, and the boundary d_s demarkating the regime of particle segregation in multiple-species pair annihilation, the generically scale-invariant algebraic decay in reversible recombination reactions is apparently *not* associated with a finite borderline dimension.

9.5.2 Lotka–Volterra predator–prey competition

As mentioned at the end of Section 9.1.1, implementations of the Lotka–Volterra model for competing predator–prey populations on one-, two-, or three-dimensional lattices (with or without site exclusion) are characterized by anomalously strong fluctuations, associated with prominent spatio-temporal structures (featured in this book's cover illustration). Specifically, deep in the species coexistence phase, Monte Carlo simulations observe radially expanding fronts of prey closely followed by predators that periodically sweep the system. Any small surviving prey clusters then serve as nucleation centers for new population waves that subsequently interact and for large population densities eventually merge.[19] From equal-time density correlation functions, one infers a spatial width of ~ 10–20 lattice sites of the spreading activity regions. Compared with predictions from the mean-field reaction-diffusion equations, the front speed is found to be enhanced by a factor up to ~ 2–3. Averaging over the weakly coupled and periodically emerging structures yields long-lived but damped population oscillations. As the system size increases, one observes the relative oscillation amplitudes to decrease. In the thermodynamic limit, the quasi-periodic population fluctuations eventually terminate; locally, however, density oscillations persist. Prominent peaks in the Fourier-transformed concentration signals allow the measurement of the characteristic oscillation frequencies, which turn out to be reduced by factors ~ 2–6 in the stochastic spatially extended system relative to the mean-field rate equation prediction. One suspects this considerable downward renormalization to be likely caused by reaction-induced spatio-temporal correlations; indeed, a one-loop analysis of the associated Doi–Peliti field theory in the predator–prey coexistence phase uncovers the mechanism behind those anomalously strong fluctuation corrections.[20]

The pseudo-Hamiltonian (9.36) captures the on-site Lotka–Volterra reactions $A \rightarrow \emptyset$ (predator death, rate κ), $B \rightarrow B + B$ (prey reproduction, rate σ), and $A + B \rightarrow A + A$ (non-linear predation interaction, rate λ). Adding particle diffusion (with equal diffusivities D for both species), the bulk part of the shifted continuum

---

[19] The text summarizes Monte Carlo simulation results from Mobilia, Georgiev, and Täuber (2007); Washenberger, Mobilia, and Täuber (2007); Dobramysl and Täuber (2008, 2013).

[20] Täuber (2012).

Doi–Peliti action (9.56) becomes

$$A[\tilde{\psi}, \tilde{\varphi}; \psi, \varphi] = \int d^d x \int dt \left[\tilde{\psi}(x, t) \left(\frac{\partial}{\partial t} - D\nabla^2 + \kappa \right) \psi(x, t) \right.$$

$$+ \tilde{\varphi}(x, t) \left(\frac{\partial}{\partial t} - D\nabla^2 - \sigma \right) \varphi(x, t) - \sigma \tilde{\varphi}(x, t)^2 \varphi(x, t)$$

$$\left. - \lambda [1 + \tilde{\psi}(x, t)][\tilde{\psi}(x, t) - \tilde{\varphi}(x, t)]\psi(x, t)\varphi(x, t) \right], \quad (9.118)$$

where the fields ψ and φ represent the predator and prey populations, respectively. Through interpretation as Janssen–De Dominicis response functional, this action encodes the coupled stochastic Langevin equations

$$\frac{\partial \psi(x, t)}{\partial t} = (D\nabla^2 - \kappa)\psi(x, t) + \lambda \psi(x, t)\varphi(x, t) + \zeta(x, t), \quad (9.119)$$

$$\frac{\partial \varphi(x, t)}{\partial t} = (D\nabla^2 + \sigma)\varphi(x, t) - \lambda \psi(x, t)\varphi(x, t) + \eta(x, t), \quad (9.120)$$

with Gaussian white noise $\langle \zeta(x, t) \rangle = 0 = \langle \eta(x, t) \rangle$, and (cross-)correlations

$$\langle \zeta(x, t) \zeta(x', t') \rangle = 2\lambda \, \psi(x, t) \, \varphi(x, t) \, \delta(x - x')\delta(t - t'), \quad (9.121)$$

$$\langle \zeta(x, t) \eta(x', t') \rangle = -\lambda \, \psi(x, t) \, \varphi(x, t) \, \delta(x - x')\delta(t - t'), \quad (9.122)$$

$$\langle \eta(x, t) \eta(x', t') \rangle = 2\sigma \, \varphi(x, t) \, \delta(x - x')\delta(t - t'), \quad (9.123)$$

describing multiplicative noise terms that vanish with the particle densities, as appropriate for the absorbing state at $a = 0 = b$.

In order to evaluate fluctuation corrections in the predator–prey coexistence phase, one needs to introduce new fluctuating fields $\bar{\psi}(x, t) = \psi(x, t) - \langle \psi \rangle$ and $\bar{\varphi}(x, t) = \varphi(x, t) - \langle \varphi \rangle$ with zero mean:

$$\psi(x, t) = \frac{\sigma}{\lambda}(1 + C_A) + \bar{\psi}(x, t), \quad \varphi(x, t) = \frac{\kappa}{\lambda}(1 + C_B) + \bar{\varphi}(x, t). \quad (9.124)$$

Here, the mean-field values $a_\infty = \sigma/\lambda$ and $b_\infty = \kappa/\lambda$ are already accounted for, such that the *counterterms* C_A and C_B, which are naturally determined by the conditions $\langle \bar{\psi} \rangle = 0 = \langle \bar{\varphi} \rangle$, contain only fluctuation corrections; compare with Section 5.4.1. It is convenient to separate the resulting action $\mathcal{A} = \mathcal{A}_0 + \mathcal{A}_{ct} + \mathcal{A}_{int}$ into three contributions: the bilinear part

$$\mathcal{A}_0[\tilde{\psi}, \tilde{\varphi}; \bar{\psi}, \bar{\varphi}] = \int d^d x \int dt \left[\tilde{\psi} \left(\frac{\partial}{\partial t} - D\nabla^2 - \kappa \, C_B \right) \bar{\psi} \right.$$

$$+ \kappa(1 + C_B)\tilde{\varphi}\,\tilde{\psi} - \sigma(1 + C_A)\tilde{\psi}\,\tilde{\varphi}$$

$$\left. + \tilde{\varphi} \left(\frac{\partial}{\partial t} - D\nabla^2 + \sigma \, C_A \right) \bar{\varphi} \right], \quad (9.125)$$

the source terms, both linear and quadratic in the response fields,

$$A_{sr}[\tilde{\psi}, \tilde{\varphi}] = -\frac{\kappa\sigma}{\lambda} \int d^d x \int dt \left[(1 + C_A)C_B \, \tilde{\psi} - C_A(1 + C_B)\tilde{\varphi} \right.$$

$$\left. + (1 + C_A)(1 + C_B)\, \tilde{\psi}\,(\tilde{\psi} - \tilde{\varphi}) + (1 + C_B)\tilde{\varphi}^2 \right], \quad (9.126)$$

and finally the non-linear vertices

$$A_{int}[\tilde{\psi}, \tilde{\varphi}; \bar{\psi}, \bar{\varphi}] = - \int d^d x \int dt \left[\kappa(1 + C_B)\, \tilde{\psi}\,(\tilde{\psi} - \tilde{\varphi})\, \bar{\psi} \right.$$

$$\left. + \sigma(1 + C_A)\, \tilde{\psi}\,(\tilde{\psi} - \tilde{\varphi})\bar{\varphi} + \sigma\,\tilde{\varphi}^2\bar{\varphi} + \lambda(1 + \tilde{\psi})(\tilde{\psi} - \tilde{\varphi})\,\bar{\psi}\,\bar{\varphi} \right].$$

$$(9.127)$$

The next step is to diagonalize the harmonic action (9.125) by means of the linear field transformations

$$\tilde{\psi}(x, t) = \frac{\tilde{\phi}_+(x, t) - \tilde{\phi}_-(x, t)}{i\sqrt{2\sigma}}, \quad \tilde{\varphi}(x, t) = \frac{\tilde{\phi}_+(x, t) + \tilde{\phi}_-(x, t)}{\sqrt{2\kappa}}, \quad (9.128)$$

$$\bar{\psi}(x, t) = \frac{\phi_+(x, t) + \phi_-(x, t)}{\sqrt{2\kappa}}, \quad \bar{\varphi}(x, t) = \frac{\phi_+(x, t) - \phi_-(x, t)}{i\sqrt{2\sigma}}, \quad (9.129)$$

which gives

$$A_0[\tilde{\phi}_+, \tilde{\phi}_-; \phi_+, \phi_-]$$

$$= \frac{1}{i\omega_0} \int d^d x \int dt \left[\tilde{\phi}_+ \left(\frac{\partial}{\partial t} - D\nabla^2 + i\omega_0 \right)\phi_+ - \tilde{\phi}_- \left(\frac{\partial}{\partial t} - D\nabla^2 - i\omega_0 \right)\phi_- \right.$$

$$+ \tilde{\phi}_+ \left(\frac{\sigma + i\omega_0}{2} C_A - \frac{\kappa - i\omega_0}{2} C_B \right)\phi_+ - \tilde{\phi}_+ \left(\frac{\sigma + i\omega_0}{2} C_A + \frac{\kappa - i\omega_0}{2} C_B \right)\phi_-$$

$$\left. + \tilde{\phi}_- \left(\frac{\sigma - i\omega_0}{2} C_A + \frac{\kappa + i\omega_0}{2} C_B \right)\phi_+ + \tilde{\phi}_- \left(\frac{\sigma - i\omega_0}{2} C_A - \frac{\kappa + i\omega_0}{2} C_B \right)\phi_- \right],$$

$$(9.130)$$

where $\omega_0 = \sqrt{\kappa\sigma}$ denotes the population oscillation frequency in the Gaussian approximation; the last four terms contain only counterterms $\propto C_A, C_B$. Thus we obtain the Gaussian propagators in the diagonalized action,

$$\langle \tilde{\phi}_\pm(q, \omega)\phi_\pm(q', \omega') \rangle_0 = \frac{\pm i\omega_0}{-i\omega + Dq^2 \pm i\omega_0 + \gamma} (2\pi)^{d+1}\delta(q + q')\delta(\omega + \omega').$$

$$(9.131)$$

Akin to spin waves in ferro- or antiferromagnets, the poles of the propagators (9.131) describe (anti-)clockwise propagating waves with frequency ω_0 and relaxation rate (damping) γ, with additional diffusive relaxation $\propto Dq^2$. Note that $\gamma = 0$

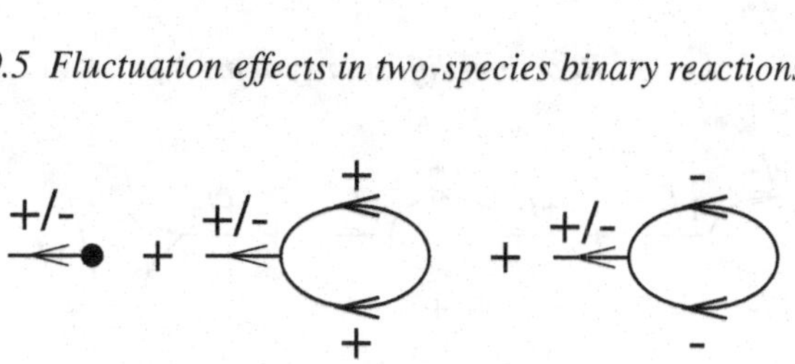

Fig. 9.13 Feynman diagrams for the expectation values $\langle \varphi_\pm \rangle$ up to one-loop order, where the '•' symbol in the first diagram represents the counterterms.

in the bare harmonic action (9.130); in Eq. (9.131) we anticipate that fluctuations will dynamically generate a 'mass' term, i.e., a non-zero imaginary part of the self-energy $\pm \omega_0 \, \mathrm{Im} \, \Sigma_\pm (q = 0, \omega = 0)$.

For our subsequent one-loop calculation, we merely need the lowest-order counterterms,

$$\mathcal{A}_{ct}[\tilde{\phi}_+, \tilde{\phi}_-] = \sqrt{\frac{\sigma}{2}} \frac{1}{i\lambda} \int d^d x \int dt \left[(i\omega_0 C_A - \kappa C_B) \tilde{\phi}_+ \right.$$

$$\left. + (i\omega_0 C_A + \kappa C_B) \tilde{\phi}_- + O(\lambda^2) \right], \quad (9.132)$$

and may omit the counterterms $\propto \lambda$ in the Langevin noise contributions

$$\mathcal{A}_{sr}[\tilde{\phi}_+, \tilde{\phi}_-] = -\frac{1}{2\lambda} \int d^d x \int dt \left[(i\omega_0 - \kappa + \sigma) \tilde{\phi}_+^2 + 2(\kappa + \sigma) \tilde{\phi}_+ \tilde{\phi}_- \right.$$

$$\left. - (i\omega_0 + \kappa - \sigma) \tilde{\phi}_-^2 + O(\lambda) \right], \quad (9.133)$$

as well as in the three-point vertices with two response fields,

$$\mathcal{A}_{int}[\tilde{\phi}_+, \tilde{\phi}_-; \phi_+, \phi_-] = \frac{1}{2\sqrt{2\kappa} \, \omega_0^2} \int d^d x \int dt \left[[(i\omega_0 - \kappa)^2 + i\omega_0 \sigma] \tilde{\phi}_+^2 \phi_+ \right.$$

$$+ [\kappa(\kappa + \sigma) - i\omega_0 \sigma] \tilde{\phi}_+^2 \phi_- + 2[\kappa(i\omega_0 - \kappa) + i\omega_0 \sigma] \tilde{\phi}_+ \tilde{\phi}_- \phi_+$$

$$- 2[\kappa(i\omega_0 + \kappa) + i\omega_0 \sigma] \tilde{\phi}_+ \tilde{\phi}_- \phi_- + [\kappa(\kappa + \sigma) + i\omega_0 \sigma] \tilde{\phi}_-^2 \phi_+$$

$$+ [(i\omega_0 + \kappa)^2 - i\omega_0 \sigma] \tilde{\phi}_-^2 \phi_- + O(\lambda)$$

$$- \lambda(i\omega_0 - \kappa) \tilde{\phi}_+ (\phi_+^2 - \phi_-^2)$$

$$\left. - \lambda(i\omega_0 + \kappa) \tilde{\phi}_- (\phi_+^2 - \phi_-^2) \right]; \quad (9.134)$$

the four-point vertices do not enter one-loop Feynman graphs. The action remains invariant under label exchange $+ \leftrightarrow -$, aside from complex conjugation, reflecting the symmetry between the two propagating eigenmodes.

We can now determine the counterterms C_A and C_B to first order in λ from the conditions $\langle \phi_\pm \rangle = 0$. The contributing Feynman graphs up to one-loop order are shown in Fig. 9.13, cf. Fig. 5.2 in Section 5.4.1; they yield analytically (see

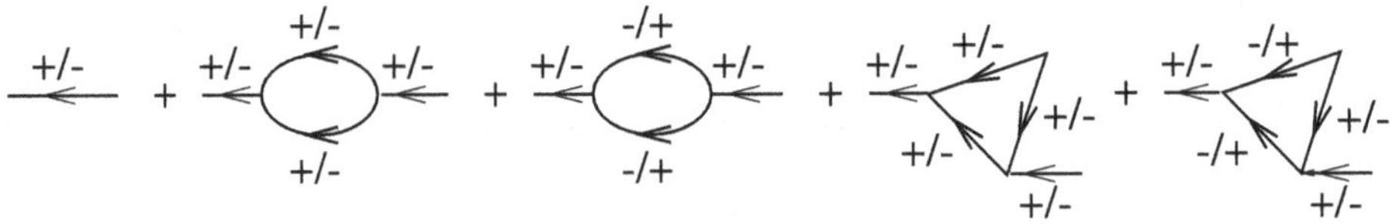

Fig. 9.14 One-loop Feynman graphs for the two-point vertex functions $\Gamma^{(\pm;\pm)}(q,\omega)$ or propagator self-energies $\Sigma_\pm(q,\omega)$.

Problem 9.7)

$$0 = \sqrt{\frac{\sigma}{2}}\frac{i}{\lambda}(i\omega_0 C_A \mp \kappa C_B)$$

$$+ \frac{i\omega_0 \mp \kappa}{4\sqrt{2\kappa}}\int\frac{d^d k}{(2\pi)^d}\left(\frac{\kappa - \sigma - i\omega_0}{i\omega_0 + \gamma + Dk^2} - \frac{\kappa - \sigma + i\omega_0}{-i\omega_0 + \gamma + Dk^2}\right).$$

These two coupled conditions are readily solved, with the result

$$C_A = C_B = \frac{\lambda}{2}\int\frac{d^d k}{(2\pi)^d}\frac{\kappa - \sigma + \gamma + Dk^2}{\omega_0^2 + (\gamma + Dk^2)^2} + O(\lambda^2),\qquad (9.135)$$

a necessary ingredient for the evaluation of the fluctuation corrections to the propagator (9.131), see (9.130). The one-loop Feynman graphs for the two-point vertex functions $\Gamma^{(\pm;\pm)}(q,\omega)$ are depicted in Fig. 9.14. Performing the internal frequency integrals, and collecting alike terms, one eventually arrives at the explicit expressions

$$\Gamma^{(\pm;\pm)}(q,\omega) = \frac{1}{\pm i\omega_0}\left[i\omega \pm i\omega_0 + \gamma + Dq^2 + \left(\frac{\sigma - \kappa}{2} \pm i\omega_0\right)C_A\right]$$

$$+ \frac{\lambda(\pm i\omega_0 - \kappa)}{8\omega_0^2\kappa}[\kappa(\sigma - \kappa \pm 2i\omega_0) \mp i\omega_0\sigma]\int_k\frac{1}{\frac{i\omega}{2} \pm i\omega_0 + \gamma + D(\frac{q^2}{4} + k^2)}$$

$$+ \frac{\lambda(\pm i\omega_0 - \kappa)}{8\omega_0^2\kappa}[\kappa(\sigma + \kappa) \pm i\omega_0\sigma]\int_k\frac{1}{\frac{i\omega}{2} \mp i\omega_0 + \gamma + D(\frac{q^2}{4} + k^2)}$$

$$\pm \frac{\lambda}{8i\omega_0}(\sigma - \kappa \pm 2i\omega_0)(\sigma - \kappa \pm i\omega_0)\int_k\frac{1}{\pm i\omega_0 + \gamma + D(\frac{q}{2} + k)^2}$$

$$\times \frac{1}{\pm i\omega_0 + \gamma_0 + D(\frac{q}{2} - k)^2}\frac{\pm i\omega_0 + \gamma + D(\frac{q^2}{4} + k^2)}{\frac{i\omega}{2} \pm i\omega_0 + \gamma + D(\frac{q^2}{4} + k^2)}$$

$$\mp \frac{\lambda(\sigma + \kappa)^2}{8i\omega_0}\int_k\frac{1}{\gamma + D(\frac{q}{2} + k)^2}\frac{1}{\gamma + D(\frac{q}{2} - k)^2}$$

$$\times \frac{\gamma + D(\frac{q^2}{4} + k^2)}{\frac{i\omega}{2} \mp i\omega_0 + \gamma + D(\frac{q^2}{4} + k^2)},\qquad (9.136)$$

where the last two terms have been symmetrized with respect to the external wavevector q (Problem 9.7).

Upon writing the renormalized two-point vertex functions in the form

$$\Gamma_R^{(\pm;\pm)}(q, \omega) = 1 \pm \frac{\gamma_R}{i\omega_R} \pm \frac{\omega}{\omega_R} \pm \frac{D_R q^2}{i\omega_R} , \qquad (9.137)$$

and comparing with the long-wavelength, low-frequency expansion

$$\Gamma^{(\pm;\pm)}(q, \omega) = 1 + \text{Re}\,\Gamma_1^{(\pm;\pm)}(0, 0) \pm \frac{\gamma}{i\omega_0} + i\,\text{Im}\,\Gamma_1^{(\pm;\pm)}(0, 0) \pm \frac{Dq^2}{i\omega_0}$$

$$+ q^2 \left.\frac{\partial \Gamma_1^{(\pm;\pm)}(q, 0)}{\partial q^2}\right|_{q=0} \pm \frac{\omega}{\omega_0} + i\omega \left.\frac{\partial \Gamma_1^{(\pm;\pm)}(0, \omega)}{\partial i\omega}\right|_{\omega=0} + \cdots ,$$

where $\Gamma_1^{(\pm;\pm)}(q, \omega)$ indicate the one-loop fluctuation corrections, one identifies the renormalized frequency ω_R, relaxation rate or damping γ_R, and diffusivity D_R via

$$\omega_R = \frac{\omega_0 \left[1 + \text{Re}\,\Gamma_1^{(\pm;\pm)}(0, 0)\right]}{1 \mp \omega_0\,\text{Im}\left[\partial\Gamma_1^{(\pm;\pm)}(0, \omega)/\partial i\omega\right]_{\omega=0}} , \qquad (9.138)$$

$$\gamma_R = \frac{\gamma \mp \omega_0\,\text{Im}\,\Gamma_1^{(\pm;\pm)}(0, 0)}{1 \mp \omega_0\,\text{Im}\left[\partial\Gamma_1^{(\pm;\pm)}(0, \omega)/\partial i\omega\right]_{\omega=0}} , \qquad (9.139)$$

$$D_R = \frac{D \mp \omega_0\,\text{Im}\left[\partial\,\Gamma_1^{(\pm;\pm)}(q, 0)/\partial q^2\right]_{q=0}}{1 \mp \omega_0\,\text{Im}\left[\partial\Gamma_1^{(\pm;\pm)}(0, \omega)/\partial i\omega\right]_{\omega=0}} . \qquad (9.140)$$

The common denominators originate from field renormalization. Note that a negative 'damping' $\gamma_R < 0$ signals an instability towards a spatially inhomogeneous configuration at wavenumber $q_c = \sqrt{|\gamma_R|/D_R}$ or characteristic wavelength $\lambda_c = 2\pi\sqrt{D/|\gamma_R|} + O(\lambda^2)$.

It is a straightforward yet laborious task to evaluate the renormalized parameters (9.138)–(9.140) from the one-loop vertex function (9.136). The required integrals are all UV-convergent in dimensions $d < 4$, and can be carried out with the aid of the fundamental integral (5.10), decomposition into partial fractions, and taking parametric derivatives with respect to γ and/or ω_0. One finally obtains the following explicit expressions:

$$\omega_R = \omega_0 + \lambda \frac{\Gamma(1-d/2)}{2^{d+4}\pi^{d/2}} \left(\frac{\omega_0}{D}\right)^{d/2} \left[\left(\frac{\sigma}{\kappa} + \frac{\kappa}{\sigma} + 2\right)\mathrm{Re}\left(\frac{\gamma}{\omega_0} + i\right)^{-1+d/2}\right.$$

$$+ 4\sqrt{\frac{\sigma}{\kappa}}\,\mathrm{Im}\left(\frac{\gamma}{\omega_0} + i\right)^{-1+d/2} - \left(\frac{\sigma}{\kappa} + \frac{\kappa}{\sigma} + 2\right)\left(\frac{\gamma}{\omega_0}\right)^{-1+d/2}\Big]$$

$$+ \lambda \frac{\Gamma(2-d/2)}{2^{d+4}\pi^{d/2}} \left(\frac{\omega_0}{D}\right)^{d/2} \left[4\left(\sqrt{\frac{\sigma}{\kappa}} - \sqrt{\frac{\kappa}{\sigma}}\right)\mathrm{Re}\left(\frac{\gamma}{\omega_0} + i\right)^{-2+d/2}\right.$$

$$+ \left(\frac{\sigma}{\kappa} + \frac{\kappa}{\sigma} - 4\right)\mathrm{Im}\left(\frac{\gamma}{\omega_0} + i\right)^{-2+d/2}\Big]$$

$$+ \lambda \frac{\Gamma(3-d/2)}{2^{d+5}\pi^{d/2}} \left(\frac{\omega_0}{D}\right)^{d/2} \left[\left(\frac{\sigma}{\kappa} + \frac{\kappa}{\sigma} - 4\right)\mathrm{Re}\left(\frac{\gamma}{\omega_0} + i\right)^{-3+d/2}\right.$$

$$- 3\left(\sqrt{\frac{\sigma}{\kappa}} - \sqrt{\frac{\kappa}{\sigma}}\right)\mathrm{Im}\left(\frac{\gamma}{\omega_0} + i\right)^{-3+d/2}\Big] + O(\lambda^2), \tag{9.141}$$

$$\gamma_R = \gamma + \lambda \frac{\Gamma(1-d/2)}{2^{d+3}\pi^{d/2}} \left(\frac{\omega_0}{D}\right)^{d/2} \left(\frac{\sigma}{\kappa} + \frac{\kappa}{\sigma}\right)\mathrm{Im}\left(\frac{\gamma}{\omega_0} + i\right)^{-1+d/2}$$

$$+ \lambda \frac{\Gamma(2-d/2)}{2^{d+3}\pi^{d/2}} \left(\frac{\omega_0}{D}\right)^{d/2} \left[\left(\frac{\sigma}{\kappa} + \frac{\kappa}{\sigma} - 4\right)\mathrm{Re}\left(\frac{\gamma}{\omega_0} + i\right)^{-2+d/2}\right.$$

$$- 3\left(\sqrt{\frac{\sigma}{\kappa}} - \sqrt{\frac{\kappa}{\sigma}}\right)\mathrm{Im}\left(\frac{\gamma}{\omega_0} + i\right)^{-2+d/2}\Big] + O(\lambda^2), \tag{9.142}$$

$$D_R = D + \lambda \frac{\Gamma(1-d/2)}{d \cdot 2^{d+3}\pi^{d/2}} \left(\frac{\omega_0}{D}\right)^{-1+d/2} \left(\frac{\sigma}{\kappa} + \frac{\kappa}{\sigma} + 2\right)\mathrm{Im}\left(\frac{\gamma}{\omega_0} + i\right)^{d/2}$$

$$- \lambda \frac{\Gamma(1-d/2)}{2^{d+4}\pi^{d/2}} \left(\frac{\omega_0}{D}\right)^{-1+d/2} \left(\frac{\sigma}{\kappa} + \frac{\kappa}{\sigma} + 2\right)\mathrm{Re}\left(\frac{\gamma}{\omega_0} + i\right)^{-1+d/2}$$

$$- \lambda \frac{\Gamma(2-d/2)}{2^{d+5}\pi^{d/2}} \left(\frac{\omega_0}{D}\right)^{-1+d/2} \left[2\left(\sqrt{\frac{\sigma}{\kappa}} - \sqrt{\frac{\kappa}{\sigma}}\right)\mathrm{Re}\left(\frac{\gamma}{\omega_0} + i\right)^{-2+d/2}\right.$$

$$+ \left(\frac{\sigma}{\kappa} + \frac{\kappa}{\sigma} - 4\right)\mathrm{Im}\left(\frac{\gamma}{\omega_0} + i\right)^{-2+d/2}\Big]$$

$$+ \lambda \frac{\Gamma(3-d/2)}{3 \cdot 2^{d+5}\pi^{d/2}} \left(\frac{\omega_0}{D}\right)^{-1+d/2} \left[\left(\frac{\sigma}{\kappa} + \frac{\kappa}{\sigma} - 4\right)\mathrm{Re}\left(\frac{\gamma}{\omega_0} + i\right)^{-3+d/2}\right.$$

$$- 3\left(\sqrt{\frac{\sigma}{\kappa}} - \sqrt{\frac{\kappa}{\sigma}}\right)\mathrm{Im}\left(\frac{\gamma}{\omega_0} + i\right)^{-3+d/2}\Big] + O(\lambda^2). \tag{9.143}$$

The effective expansion parameter in this perturbation series turns out to be $(\lambda/\omega_0)(\omega_0/D)^{d/2}$; naturally, when diffusion is fast compared to the oscillation period, the system becomes well-mixed and correlations irrelevant. In dimensions $d < 2$, when we let $\gamma \to 0$, the leading fluctuation correction to the population oscillation frequency diverges $\sim (\omega_0/\gamma)^{1-d/2}$; it is negative, and symmetric under formal reaction rate exchange $\kappa \leftrightarrow \sigma$, cf. the last term in the second line in (9.141). This infrared singularity originates from destructive interference of clockwise and anti-clockwise propagating internal modes in the fluctuation loop of the last Feynman graph in Fig. 9.14. As a consequence, the imaginary 'mass' terms $\pm i\omega_0$ cancel each other in the final contribution to (9.136). We may interpret γ in the above equations as a small, self-consistently determined damping, e.g., in the spirit of mode-coupling theory. The infrared singularity then disappears, and one finds, in remarkable qualitative agreement with Monte Carlo observations, that spatio-temporal fluctuations and correlations induced by the stochastic reaction processes cause a strong downward renormalization of the oscillation frequency, with very similar functional dependence on the rates κ and σ. Note that $d_c = 2$ can be viewed as (upper) critical dimension for the appearance of IR-singular fluctuation contributions, which resemble the dynamic coexistence anomalies induced by Goldstone modes in systems with spontaneously broken continuous order parameter symmetry (Section 5.4).

For $d \leq 4$, the fluctuation-induced contribution to γ_R is always negative, indicating an instability towards spatially inhomogeneous structures. The diffusion coefficient D_R is enhanced through the one-loop fluctuation corrections, inducing faster spreading of population fronts than predicted by the mean-field reaction-diffusion equations. Infrared singularities in the limit $\gamma \to 0$ appear only in the renormalized frequency. Specifically in two dimensions, where Monte Carlo simulations are typically carried out, the renormalized oscillation frequency, relaxation rate, and diffusivity become

$$
\omega_R = \omega_0 - \frac{\lambda}{32\pi}\frac{\omega_0}{D}\ln\frac{\omega_0}{\gamma}\cdot\left[1+\frac{1}{2}\left(\frac{\sigma}{\kappa}+\frac{\kappa}{\sigma}\right)\right]
$$

$$
+ \frac{3\lambda}{32\pi}\frac{\omega_0}{D}\left[1-\frac{\pi}{3}\sqrt{\frac{\sigma}{\kappa}}-\frac{1}{4}\left(\frac{\sigma}{\kappa}+\frac{\kappa}{\sigma}\right)\right]+O(\lambda^2)\,, \tag{9.144}
$$

$$
\gamma_R = \gamma + \frac{\lambda}{64}\frac{\omega_0}{D}\left[\frac{6}{\pi}\left(\sqrt{\frac{\sigma}{\kappa}}-\sqrt{\frac{\kappa}{\sigma}}\right)-\left(\frac{\sigma}{\kappa}+\frac{\kappa}{\sigma}\right)\right]+O(\lambda^2)\,, \tag{9.145}
$$

$$
D_R = D + \frac{\lambda}{96\pi}\left[1+2\left(\frac{\sigma}{\kappa}+\frac{\kappa}{\sigma}\right)\right]+O(\lambda^2)\,. \tag{9.146}
$$

Note the strong negative logarithmic correction to the frequency (9.144).

Problems

9.1 *Fisher–Kolmogorov invasion wave*

Confirm that the rescaled one-dimensional Fisher–Kolmogorov partial differential equation

$$\frac{\partial u(x, t)}{\partial t} = \frac{\partial^2 u(x, t)}{\partial x^2} + u(x, t)\,[1 - u(x, t)]$$

permits a traveling wave solution of the form $u(x, t) = v(x \mp ct)$ for the specific front speed $c = 5/\sqrt{6}$, which reads explicitly

$$v(z) = (1 + C\,e^{\pm z/\sqrt{6}})^{-2}\,,$$

with arbitrary amplitude $C > 0$, interpolating between the empty, absorbing state $u = 0$ as $z \to \pm\infty$ and the active phase $u = 1$ as $z \to \mp\infty$. Such invasion fronts actually exist for all wave velocities $c > 2$, with their shape uniquely determined by c.

9.2 *Pair annihilation in two dimensions*

(a) Apply Smoluchowski's self-consistent approach to the binary annihilation reaction $A + A \to \emptyset$ at $d_c = 2$ to derive the effective reaction rate $\lambda_{\mathrm{eff}}(n) = 4\pi D/\ln[R(n)/b]$. Through iterative solution of the resulting rate equation, obtain the asymptotic density decay (9.24).

(b) Use the renormalization group approach for the kth order annihilation processes $k A \to l A$ ($l < k$) to evaluate the logarithmic corrections to the mean-field density decay at the critical dimension $d_c(k)$, confirming Eq. (9.87). For pair annihilation, compare with the result in (a).

9.3 *Pseudo-Hamiltonians for reversible reactions and hopping transport*

(a) Starting from the master equation (9.1) for the reversible processes $k A + l B \rightleftharpoons m C$, derive the associated pseudo-Hamiltonian (9.34).

(b) Confirm Eq. (9.37) for unbiased particle hopping.

(c) Construct the Doi–Peliti action (9.49) for the reactions in (a) including hopping terms as in (b). Via the classical field equations in the continuum limit, derive the associated reaction-diffusion equations.

(d) Find the equivalent Langevin description for $k + l, m = 0, 1, 2$.

9.4 *Annihilation loop integral*

Evaluate the one-loop fluctuation integral for binary and triplet annihilations in the frequency and time domain to confirm (9.72) for $k = 2, 3$.

9.5 *Lévy flights in reactive systems: pair annihilation/coagulation*

In *Lévy flights*, nearest-neighbor hopping is replaced by typically long-range jumps that follow a distance distribution $P_L(x) \propto |x|^{-d-\sigma}$ in d dimensions, with Lévy exponent $0 < \sigma < 2$ (see Section 10.2.2). This yields a *superdiffusive* mean-square displacement $\langle x(t)^2 \rangle \sim (D_L t)^{1/\sigma}$; the associated propagator is $G_0(q, \omega) = (-i\omega + D_L|q|^\sigma)^{-1}$.

    (a) Establish that for Lévy flight dispersion, the critical dimension for binary annihilation or coagulation/fusion $A + A \to \emptyset$, A is $d_c(\sigma) = \sigma$.

    (b) Compute the annihilation fixed point, and determine the asymptotic particle density decay $a(t) \sim (D_L t)^{-d/\sigma}$.

9.6 *Reaction fronts for $A + B \to \emptyset$ in dimensions $d \le 2$*

    (a) Derive the reaction front scaling exponents (9.102) for two-species annihilation by inserting the renormalized reactivity $\lambda(a) \sim Da^{-1+2/d}$ into the mean-field equation (9.99) and the front density (9.101).

    (b) Use the scaling form (9.100) and the RG flow (9.80) to predict the logarithmic corrections to the front width and height at $d_c = 2$.

9.7 *One-loop fluctuation corrections for the Lotka–Volterra propagator*

Confirm Eqs. (9.135) and (9.136) for the counterterm $C_{A/B}$ and the propagator self-energy $\Sigma_\pm(q, \omega) = -\Gamma^{(\pm;\pm)}(q, \omega)$ in the predator–prey coexistence phase of the Lotka–Volterra model to one-loop order.

References

Alcaraz, F. C., M. Droz, M. Henkel, and V. Rittenberg, 1994, Reaction-diffusion processes, critical dynamics, and quantum chains, *Ann. Phys. (NY)* **230**, 250–302.

Allam, J., M. T. Sajjad, R. Sutton, *et al.*, 2013, Measurement of a reaction–diffusion crossover in exciton–exciton recombination inside carbon nanotubes using femtosecond optical absorption, *Phys. Rev. Lett.* **111**, 197401-1–5.

Andreanov, A., G. Biroli, J.-P. Bouchaud, and A. Lefèvre, 2006, Field theories and exact stochastic equations for interacting particle systems, *Phys. Rev. E* **74**, 030101-1–4.

Cardy, J., 1997, Renormalisation group approach to reaction-diffusion problems, in: *Proceedings of Mathematical Beauty of Physics*, ed. J.-B. Zuber, Adv. Ser. in Math. Phys. **24**, 113–128.

Cardy, J., 2008, Reaction-diffusion processes, in: *Non-equilibrium Statistical Mechanics and Turbulence*, London Math. Soc. Lecture Note Ser. **355**, Cambridge: Cambridge University Press, 108–161.

Deloubrière, O., H. J. Hilhorst, and U. C. Täuber, 2002, Multispecies pair annihilation reactions, *Phys. Rev. Lett.* **89**, 250601-1-4.

Dobramysl, U. and U. C. Täuber, 2008, Spatial variability enhances species fitness in stochastic predator-prey interactions, *Phys. Rev. Lett.* **101**, 258102-1–4.

Dobramysl, U. and U. C. Täuber, 2013, Environmental versus demographic variability in two-species predator-prey models, *Phys. Rev. Lett.* **110**, 048105-1–5.

Doi, M., 1976a, Second quantization representation for classical many-particle systems, *J. Phys. A: Math. Gen.* **9**, 1465–1477.

Doi, M., 1976b, Stochastic theory of diffusion-controlled reactions, *J. Phys. A: Math. Gen.* **9**, 1479–1495.

Dornic, I., H. Chaté, and M. A. Muñoz, 2005, Integration of Langevin equations with multiplicative noise and the viability of field theories for absorbing phase transitions, *Phys. Rev. Lett.* **94**, 100601-1–4.

Elgart, V. and A. Kamenev, 2004, Rare event statistics in reaction-diffusion systems, *Phys. Rev. E* **70**, 041106-1–12.

Elgart, V. and A. Kamenev, 2006, Classification of phase transitions in reaction-diffusion models, *Phys. Rev. E* **74**, 041101-1–16.

Grassberger, P. and M. Scheunert, 1980, Fock–space methods for identical classical objects, *Fortschr. Phys.* **28**, 547–578.

Haken, H., 1983, *Synergetics – an Introduction*, Berlin: Springer, chapters 9, 10.

Henkel, M., E. Orlandini, and J. Santos, 1997, Reaction-diffusion processes from equivalent integrable quantum chains, *Ann. Phys. (NY)* **259**, 163–231.

Hilhorst, H. J., O. Deloubrière, M. J. Washenberger, and U. C. Täuber, 2004, Segregation in diffusion-limited multispecies pair annihilation, *J. Phys. A: Math. Gen.* **37**, 7063–7093.

Hilhorst, H. J., M. J. Washenberger, and U. C. Täuber, 2004, Symmetry and species segregation in diffusion-limited pair annihilation, *J. Stat. Mech.* P10002-1–19.

Hinrichsen, H. and M. Howard, 1999, A model for anomalous directed percolation, *Eur. Phys. J. B* **7**, 635–644.

Hofbauer, J. and K. Sigmund, 1998, *Evolutionary Games and Population Dynamics*, Cambridge: Cambridge University Press.

Janssen, H. K., 2001, Directed percolation with colors and flavors, *J. Stat. Phys.* **103**, 801–839.

Kopelman, R., 1988, Fractal reaction kinetics, *Science* **241**, 1620–1626.

Krapivsky, P. K., S. Redner, and E. Ben-Naim, 2010, *A Kinetic View of Statistical Physics*, Cambridge: Cambridge University Press, chapters 12, 13.

Kroon, R., H. Fleurent, and R. Sprik, 1993, Diffusion-limited exciton fusion reaction in one-dimensional tetramethylammonium manganese trichloride (TMMC), *Phys. Rev. E* **47**, 2462–2472.

Kuzovkov, V. and E. Kotomin, 1988, Kinetics of bimolecular reactions in condensed media: critical phenomena and microscopic self-organisation, *Rep. Prog. Phys.* **51**, 1479–1523.

Lee, B. P., 1994, Renormalization group calculation for the reaction $kA \to \emptyset$, *J. Phys. A: Math. Gen.* **27**, 2633–2652.

Lee, B. P. and J. Cardy, 1994, Scaling of reaction zones in the $A + B \to \emptyset$ diffusion-limited reaction, *Phys. Rev. E* **50**, R3287–R3290.

Lee, B. P. and J. Cardy, 1995, Renormalization group study of the $A + B \to \emptyset$ diffusion-limited reaction, *J. Stat. Phys.* **80**, 971–1007.

Mattis, D. C. and M. L. Glasser, 1998, The uses of quantum field theory in diffusion-limited reactions, *Rev. Mod. Phys.* **70**, 979–1002.

Mobilia, M., I. T. Georgiev, and U. C. Täuber, 2007, Phase transitions and spatio-temporal fluctuations in stochastic lattice Lotka–Volterra models, *J. Stat. Phys.* **128**, 447–483. [Various Monte Carlo simulation movies are accessible at http://www.phys.vt.edu/~tauber/PredatorPrey/movies/].

Monson, E. and R. Kopelman, 2004, Nonclassical kinetics of an elementary $A + B \to C$ reaction-diffusion system showing effects of a speckled initial reactant distribution and eventual self-segregation: experiments, *Phys. Rev. E* **69**, 021103-1–12.

Murray, J. D., 2002, *Mathematical Biology*, Vols. I and II, New York: Springer, 3rd edn.

Ovchinnikov, A. A., S. F. Timashev, and A. A. Belyy, 1989, *Kinetics of Diffusion-controlled Chemical Processes*, New York: Nova Science.

Peliti, L., 1985, Path integral approach to birth-death processes on a lattice, *J. Phys. (Paris)* **46**, 1469–1482.

Peliti, L., 1986, Renormalisation of fluctuation effects in the $A + A \rightarrow A$ reaction, *J. Phys. A: Math. Gen.* **19**, L365–367.

Rey, P.-A. and J. Cardy, 1999, Asymptotic form of the approach to equilibrium in reversible recombination reactions, *J. Phys. A: Math. Gen.* **32**, 1585–1603.

Russo, R. M., E. J. Mele, C. L. Kane, I. V. Rubtsov, M. J. Therien, and D. E. Luzzi, 2006, One-dimensional diffusion-limited relaxation of photoexcitations in suspensions of single-walled carbon nanotubes, *Phys. Rev. B* **74**, 041405(R)-1–4.

Schütz, G. M., 2001, Exactly solvable models for many-body systems far from equilibrium, in: *Phase Transitions and Critical Phenomena*, Vol. 19, eds. C. Domb and J. L. Lebowitz, London: Academic Press.

Stinchcombe, R., 2001, Stochastic nonequilibrium systems, *Adv. Phys.* **50**, 431–496.

Täuber, U. C., 2012, Population oscillations in spatial stochastic Lotka–Volterra models: a field-theoretic perturbational analysis, *J. Phys. A: Math. Theor.* **45**, 405002-1–34.

Täuber, U. C., M. J. Howard, and B. P. Vollmayr-Lee, 2005, Applications of field-theoretic renormalization group methods to reaction-diffusion problems, *J. Phys. A: Math. Gen.* **38**, R79–R131.

Toussaint, D. and F. Wilczek, 1983, Particle-antiparticle annihilation in diffusive motion, *J. Chem. Phys.* **78**, 2642–2647.

van Wijland, F., 2001, Field theory for reaction-diffusion processes with hard-core particles, *Phys. Rev. E* **63**, 022101-1–4.

Vernon, D., 2003, Long range hops and the pair annihilation reaction $A + A \rightarrow \emptyset$: renormalization group and simulation, *Phys. Rev. E* **68**, 041103-1–4.

Washenberger, M. J., M. Mobilia, and U. C. Täuber, 2007, Influence of local carrying capacity restrictions on stochastic predator-prey models, *J. Phys. Cond. Matt.* **19**, 065139-1–14.

Further reading

Barkema, G. T., M. J. Howard, and J. L. Cardy, 1996, Reaction-diffusion front for $A + B \rightarrow \emptyset$ in one dimension, *Phys. Rev. E* **53**, R2017–R2020.

Cardy, J., 1995, Proportion of unaffected sites in a reaction-diffusion process, *J. Phys. A: Math. Gen.* **28**, L19–L24.

Cardy, J. and M. Katori, 2003, Families of vicious walkers, *J. Phys. A: Math. Gen.* **36** 609–630.

Chen, L. and M. W. Deem, 2002, Reaction, Lévy flights, and quenched disorder, *Phys. Rev. E* **65** 011109-1–6.

Deem, M. W. and J.-M. Park, 1998a, Effect of static disorder and reactant segregation on the $A + B \rightarrow \emptyset$ reaction, *Phys. Rev. E* **57**, 2681–2685.

Deem, M. W. and J.-M. Park, 1998b, Reactive turbulent flow in low-dimensional, disordered media, *Phys. Rev. E* **58**, 3223–3228.

Dobrinevski, A. and E. Frey, 2012, Extinction in neutrally stable stochastic Lotka–Volterra models, *Phys. Rev. E* **85**, 051903-1–12.

Droz, M. and L. Sasvári, 1993, Renormalization-group approach to simple reaction-diffusion phenomena, *Phys. Rev. E* **48**, R2343–R2346.

Emmerich, T., A. Bunde, and S. Havlin, 2012, Diffusion, annihilation, and chemical reactions in complex networks with spatial constraints, *Phys. Rev. E* **86**, 046103-1–5.

Hnatič, M. and J. Honkonen, 2000, Velocity-fluctuation-induced anomalous kinetics of the $A + A \rightarrow \emptyset$ reaction, *Phys. Rev. E* **61**, 3904–3911.

Hnatič, M., J. Honkonen, and T. Lučivjanský, 2013, Two-loop calculation of anomalous kinetics of the reaction $A + A \to \emptyset$ in randomly stirred fluid, *Eur. Phys. J. B* **86**, 214-1–16.

Honkonen, J., 1991, Renormalization group analysis of superdiffusion in random velocity fields, *J. Phys. A: Math. Gen.* **24**, L1235–L1242.

Honkonen, J., 2012, Functional methods in stochastic systems, *Lecture Notes in Computer Science* **7125**, 66–78.

Howard, M., 1996, Fluctuation kinetics in a multispecies reaction-diffusion system, *J. Phys. A: Math. Gen.* **29**, 3437–3460.

Howard, M. J. and G. T. Barkema, 1996, Shear flows and segregation in the reaction $A + B \to \emptyset$, *Phys. Rev. E* **53**, 5949–5956.

Howard, M. and J. Cardy, 1995, Fluctuation effects and multiscaling of the reaction-diffusion front for $A + B \to \emptyset$, *J. Phys. A: Math. Gen.* **28**, 3599–3622.

Howard, M. and C. Godrèche, 1998, Persistence in the Voter model: continuum reaction-diffusion approach, *J. Phys. A: Math. Gen.* **31**, L209–L216.

Howard, M. J. and U. C. Täuber, 1997, 'Real' vs 'imaginary' noise in diffusion-limited reactions, *J. Phys. A: Math. Gen.* **30**, 7721–7731.

Itakura, K., J. Ohkubo, and S.-i. Sasa, 2010, Two Langevin equations in the Doi–Peliti formalism, *J. Phys. A: Math. Theor.* **43**, 125001-1–14.

Konkoli, Z., 2004, Application of Bogolyubov's theory of weakly nonideal Bose gases to the $A + A$, $A + B$, $B + B$ reaction-diffusion system, *Phys. Rev. E* **69**, 011106-1–16.

Konkoli, Z., H. Johannesson, and B. P. Lee, 1999, Fluctuation effects in steric reaction-diffusion systems, *Phys. Rev. E* **59**, R3787–R3790.

Konkoli, Z. and H. Johannesson, 2000, Two-species reaction-diffusion system with equal diffusion constants: anomalous density decay at large times, *Phys. Rev. E* **62**, 3276–3280.

Krishnamurthy, S., R. Rajesh, and O. Zaboronski, 2003, Persistence properties of a system of coagulating and annihilating random walkers, *Phys. Rev. E* **68**, 046103-1–12.

Murthy, K. P. N. and G. M. Schütz, 1998, Aging in two- and three-particle annihilation processes, *Phys. Rev. E* **57**, 1388–1394.

Oerding, K., 1996, The $A + B \to \emptyset$ annihilation reaction in a quenched random velocity field, *J. Phys. A: Math. Gen.* **29**, 7051–7065.

Ohkubo, J., 2012, One-parameter extension of the Doi–Peliti formalism and its relation with orthogonal polynomials, *Phys. Rev. E* **86**, 042102-1–4.

Park, J.-M. and M. W. Deem, 1998, Disorder-induced anomalous kinetics in the $A + A \to \emptyset$ reaction, *Phys. Rev. E* **57**, 3618–3621.

Peruani, F. and C. F. Lee, 2013, Fluctuations and the role of collision duration in reaction-diffusion systems, *EPL* **102**, 58001-1–6.

Rajesh, R. and O. Zaboronski, 2004, Survival probability of a diffusing test particle in a system of coagulating and annihilating random walkers, *Phys. Rev. E* **70**, 036111-1–9.

Rey, P.-A. and M. Droz, 1997, A renormalization group study of a class of reaction-diffusion models, *J. Phys. A: Math. Gen.* **30**, 1101–1114.

Richardson, M. J. E. and J. Cardy, 1999, The reaction process $A + A \to \emptyset$ in Sinai disorder, *J. Phys. A: Math. Gen.* **32** 4035–4046.

Santos, R. V. dos, and R. Dickman, 2013, Survival of the scarcer in space, e-print arXiv:1304.5956.

Sasamoto, T., S. Mori, and M. Wadati, 1997, Universal properties of the $mA + nB \to \emptyset$ diffusion-limited reaction, *Physica A* **247**, 357–378.

Schütz, G. M., 1995, Reaction-diffusion processes of hard-core particles, *J. Stat. Phys.* **79**, 243–264.

Schütz, G. M., 1997, Diffusion-limited annihilation in inhomogeneous environments, *Z. Phys. B Cond. Matt.* **104**, 583–590.

Winkler, A. and E. Frey, 2012, Validity of the law of mass action in three-dimensional coagulation processes, *Phys. Rev. Lett.* **108**, 108301-1–5.

Winkler, A. and E. Frey, 2013, Long-range and many-body effects in coagulation processes, *Phys. Rev. E* **87**, 022136-1–13.

Zaboronski, O., 2001, Stochastic aggregation of diffusive particles revisited, *Phys. Lett. A* **281**, 119–125.

10

Active to absorbing state transitions

Continuous phase transitions from active to inactive, absorbing states represent prime examples of genuine non-equilibrium processes whose properties are strongly influenced by fluctuations. They arise in a broad variety of macroscopic phenomena, ranging from extinction thresholds in population dynamics and epidemic spreading models to certain diffusion-limited chemical reactions, and even turbulent kinetics in magnetic fluids. Intriguingly, the generic universality class for such active to absorbing phase transitions is intimately related to the scaling properties of critical directed percolation clusters. After elucidating this remarkable connection of stochastic kinetics with an originally geometric problem through mappings of both a specific microscopic interacting particle model and a more general mesoscopic Langevin description onto the corresponding Reggeon field theory action, we exploit the mathematical and conceptual techniques developed in previous chapters to compute the associated critical exponents to lowest non-trivial order in a dimensional ϵ expansion about the upper critical dimension $d_c = 4$. We then set out to explore generalizations to systems with multiple particle species, and to investigate the dynamic percolation model variant that generates isotropic critical percolation clusters in the quasi-static limit. Particle spreading via long-range Lévy flights rather than nearest-neighbor hopping and coupling to an additional conserved field that may cause a fluctuation-induced first-order transition are also discussed. Motivated by the domain wall kinetics in non-equilibrium Ising systems, we address more general stochastic reaction systems of branching and annihilating random walks, and study the ensuing non-equilibrium phase diagrams and continuous transitions, including the parity-conserving universality class. This chapter concludes with remarks on absorbing state transitions that involve higher-order reactions, whose critical properties are as yet incompletely understood, and which therefore still await satisfactory classification.

10.1 The directed percolation universality class

In this section, we establish the important statement that active to absorbing state transitions are *generically* governed by the universality class of directed percolation (DP).[1] We begin by employing the Doi–Peliti formalism to map the stochastic field theory for two simple population dynamics models near their respective extinction thresholds onto the effective Reggeon field theory action. We then show that, more generally, universal features of simple epidemic processes are captured through equivalent mesoscopic Langevin equations, before we proceed to construct the perturbation expansion for this problem. Finally, we employ the dynamic renormalization group tools introduced in previous chapters to derive the critical dynamic scaling laws and compute scaling exponents to first order in a dimensional expansion about the upper critical dimension $d_c = 4$.

10.1.1 Extinction threshold and directed percolation

Let us return to the competing reactions $A \to \emptyset$ with spontaneous decay rate κ, $A \to A + A$ with branching rate σ, and density-limiting pair coagulation or fusion $A + A \to A$ with rate λ, introduced in Section 9.1.2. Adding diffusive spreading, these reactions also describe the dynamics of a single migrating population species that incorporates an extinction threshold (when $\sigma = \kappa$ in the rate equation approximation); the empty state is absorbing, since all stochastic reactions require the presence of at least one particle or agent A. The associated reaction-diffusion equation (9.15) is known as *Fisher–Kolmogorov equation* in the ecology literature. As we saw earlier in Section 9.1, it directly implies the mean-field critical exponents $\alpha = 1 = \beta$, $\nu = 1/2$, and $z = 2$. In order to properly account for spatio-temporal fluctuations, one may resort to the stochastic pseudo-Hamiltonian (9.33) for this set of reactions, which faithfully encodes the internal reaction noise.

Before proceeding with taking the continuum limit and performing a detailed field-theoretic analysis for this system, let us draw a typical realization, say on a tilted square ('diamond') lattice of the stochastic processes associated with these competing reactions, starting from a single particle 'seed', as depicted in Fig. 10.1. As usual, time flows from right to left here, and particle moves are indicated by arrows. The combination of nearest-neighbor hopping, decay, branching, and fusion processes generates a cluster of directed links between the lattice sites. It terminates at finite length if species A goes extinct after time $t = L_\parallel$. The inactive to active phase transition as $L_\parallel \to \infty$ and in the infinite transverse system size limit $L_\perp \to \infty$ corresponds to a *percolation threshold*: in the active state, there

---

[1] This chapter partially follows the expositions in Janssen and Täuber (2005); Täuber, Howard, and Vollmayr-Lee (2005); for excellent overviews over absorbing phase transitions in general, see Hinrichsen (2001); Ódor (2004); Henkel, Hinrichsen, and Lübeck (2008).

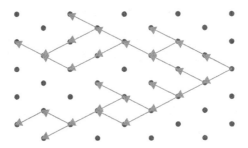

Fig. 10.1 Elementary birth, spreading, and coagulation/fusion processes initiated by a single active seed generate a directed percolation cluster; bonds connecting the lattice sites may only be formed by advancing along the forward 'time' direction from right to left. [Figure reproduced with permission from: H K. Janssen and U. C. Täuber, *Ann. Phys. (NY)* **315**, 147 (2005); DOI: 10.1016/j.aop.2004.09.011; copyright (2004) by Elsevier Inc.]

exists at least one spanning cluster that traverses the entire system. In contrast with standard isotropic bond percolation, in this *directed percolation* (DP) problem links between adjacent lattice sites can only be formed in the forward direction in one special 'time' coordinate.[2] This induces *anisotropic scaling* at the percolation (or extinction) threshold; e.g., for the temporal 'correlation length' as a function of the distance τ from the critical point,

$$\xi_\|(\tau) \sim \xi(\tau)^z \sim |\tau|^{-\nu_\|} , \qquad \nu_\| = z\,\nu . \tag{10.1}$$

The dynamic critical exponent z thus constitutes an *anisotropy exponent* in the geometric interpenetration in $d+1$ dimensions, indicating the ratio of the correlation length exponents $\nu_\|$ along the 'temporal' direction and $\nu_\perp = \nu$ in the d transverse coordinates.

Just as in Section 7.3 for quantum critical phenomena, anisotropic scaling induces the modification $d \to d+z$ in hyperscaling relations, e.g.,

$$a_\infty(\tau) = a(t \to \infty) \sim |\tau|^\beta , \qquad \beta = \frac{\nu}{2}(d+z-2+\eta) \tag{10.2}$$

for the order parameter exponent characterizing the growth of the asymptotic particle density on the active side of the absorbing state transition. At finite time but still in the critical regime, we may combine (10.1) and (10.2) to the scaling ansatz for the particle density

$$a(\tau, t) = a_0^{-d} |\tau|^\beta \, \hat{a}\big(Dt/a_0^2 |\tau|^{-\nu_\|}\big) \tag{10.3}$$

with the microscopic diffusivity D and lattice constant a_0, cf. Eq. (3.5). Eliminating τ at the extinction threshold yields the critical density decay

$$a(0, t) \sim t^{-\alpha} , \qquad \alpha = \beta/\nu_\| = \beta/z\,\nu . \tag{10.4}$$

---

[2] For overviews, see Kinzel (1983); Hinrichsen (2001).

(a)

(b)

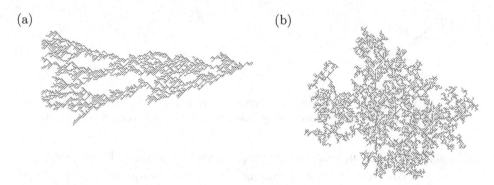

Fig. 10.2 (a) Directed (time t flows from right to left) and (b) isotropic percolation clusters, generated from Monte Carlo simulations on a two-dimensional square lattice. [Figures reproduced with permission from: E. Frey, U. C. Täuber, and F. Schwabl, *Phys. Rev. E* **49**, 5058 (1994); DOI: 10.1103/PhysRevE.49.5058; copyright (1994) by The American Physical Society.]

A simulated critical directed percolation cluster at the percolation threshold, originating from a single active seed, is shown in Fig. 10.2(a). Its strongly anisotropic features become apparent in comparison with the critical isotropic percolation cluster depicted in Fig. 10.2(b).

By the methods developed in Section 9.2, one may construct the coherent-state path integral in the continuum limit for the competing stochastic processes $A \to \emptyset$ and $A \rightleftharpoons A + A$. The pseudo-Hamiltonian that captures the associated master equation is given in Eq. (9.33). Adding diffusive spreading, we thus obtain for the 'bulk' part of the Doi–Peliti action (9.52):[3]

$$A[\hat{\psi}, \psi] = \int d^d x \int dt \left[\hat{\psi}(x, t) \left(\frac{\partial}{\partial t} - D\nabla^2 \right) \psi(x, t) \right.$$

$$\left. + [\hat{\psi}(x, t) - 1][\kappa - \sigma \, \hat{\psi}(x, t) + \lambda \, \hat{\psi}(x, t) \, \psi(x, t)]\psi(x, t) \right]. \quad (10.5)$$

Upon shifting the field $\hat{\psi}$ about its stationary value 1 and rescaling according to $\hat{\psi}(x, t) = 1 + \sqrt{\lambda/\sigma} \, \tilde{S}(x, t)$ and $\psi(x, t) = \sqrt{\sigma/\lambda} \, S(x, t)$, the action becomes

$$A[\tilde{S}, S] = \int d^d x \int dt \left[\tilde{S}(x, t) \left[\frac{\partial}{\partial t} + D \left(r - \nabla^2 \right) \right] S(x, t) \right.$$

$$\left. - u \, \tilde{S}(x, t)[\tilde{S}(x, t) - S(x, t)]S(x, t) + \lambda \, \tilde{S}(x, t)^2 \, S(x, t)^2 \right], \quad (10.6)$$

where $r = (\kappa - \sigma)/D$. Notice that the field theory (10.6) remains form-invariant under field rescaling $\tilde{S}(x, t) \to K^{-1} \tilde{S}(x, t)$, $S(x, t) \to K \, S(x, t)$, with an amplitude K that carries an arbitrary scaling dimension. It is convenient to choose this

---

[3] We henceforth omit the distinction between reaction rates in discrete and continuum systems.

Fig. 10.3 Directed percolation (DP) three-point vertices: (a) multiplicative noise source; (b) non-linear vertex stemming from pair coagulation/fusion $A + A \to A$.

redundant parameter such that both three-point vertices acquire identical strengths $u = \sqrt{\sigma}\,\lambda$, whose square indeed represents the effective coupling in the perturbation expansion, see Fig. 10.4.

In terms of the momentum scale $[q] = \mu$, $[x] = \mu^{-1}$, $t = \mu^{-2}$, one determines from (10.6) the *scaling dimensions*

$$[\tilde{S}(x, t)] = \mu^{d/2} = [S(x, t)],\tag{10.7}$$

$$[D] = \mu^0,\quad [r] = \mu^2,\quad [u] = \mu^{2-d/2},\tag{10.8}$$

whence we infer the upper critical dimension $d_c = 4$ for the non-linear coupling u. The four-point vertex $\propto \lambda$, with scaling dimension $[\lambda] = \mu^{2-d}$, is therefore *irrelevant* in the renormalization group sense, and may be dropped for the computation of universal, asymptotic scaling properties. The *effective action* describing the critical behavior of the *directed percolation universality class* then becomes[4]

$$\mathcal{A}_{\mathrm{eff}}[\tilde{S}, S] = \int d^d x \int dt\, \left[\tilde{S}(x, t)\left[\frac{\partial}{\partial t} + D\left(r - \nabla^2\right)\right] S(x, t) \right.$$

$$\left. -u\, \tilde{S}(x, t)[\tilde{S}(x, t) - S(x, t)]S(x, t)\right],\tag{10.9}$$

and is known as *Reggeon field theory.*[5] It consists of the model A (response) propagator $G_0(q, \omega)^{-1} = -i\omega + D(r + q^2)$ and the two three-point vertices displayed in Fig. 10.3. Both actions (10.6) and (10.9) are invariant under so-called *rapidity reversal* that entails time inversion in addition to exchange of the fields S and $\tilde{S}$, see Problem 10.3(a),

$$S(x, t) \leftrightarrow -\tilde{S}(x, -t).\tag{10.10}$$

Finally, we may interpret (10.9) as a Janssen–De Dominicis response functional, and extract a formally equivalent non-linear Langevin equation describing

---

[4] Obukhov (1980); Cardy and Sugar (1980); Janssen (1981).
[5] This name originates from the analysis of universal scaling in high-energy scattering amplitudes in particle physics; see Moshe (1978).

a stochastic process with multiplicative noise ζ, with $\langle \zeta \rangle = 0$:

$$\frac{\partial S(x,t)}{\partial t} = -D\left(r - \nabla^2\right) S(x,t) - u\, S(x,t)^2 + \zeta(x,t)\,, \qquad (10.11)$$

$$\langle \zeta(x,t)\,\zeta(x',t') \rangle = 2u\, S(x,t)\, \delta(x - x')\delta(t - t')\,. \qquad (10.12)$$

Naturally, Eq. (10.11) represents a noisy Fisher–Kolmogorov equation, albeit for a *complex* field $S(x,t)$. The noise correlator (10.12) ensures that the fluctuations indeed cease in the absorbing state where $\langle S \rangle = 0$.

10.1.2 Simple epidemic processes: universal features

Indeed, this non-linear Langevin equation, or equivalently, the Reggeon field theory action (10.9) for directed percolation, encodes a remarkably general class of stochastic processes that display phase transitions from an active to an inactive, absorbing state with identical universal scaling properties. In order to address such generic situations, we need to pursue a complementary approach to the direct mapping of a specific particle reaction master equation onto a coherent-state path integral, namely a phenomenological construction of a *mesoscopic* continuum theory for a local density subject to Markovian stochastic dynamics. We formulate the problem as a *simple epidemic process* (SEP), i.e., a spreading infectious disease with recovery.[6]

(i) A *'susceptible' medium* becomes locally *'infected'*, depending on the local density $n(x,t)$ of neighboring *'sick'* individuals. The infected regions may recover after a brief time interval.

(ii) The state $n = 0$, which describes the *extinction* of the epidemic, is *absorbing*.

(iii) The disease spreads *diffusively* via the short-range infection (i) of adjacent susceptible regions.

(iv) Microscopic, fast degrees of freedom are incorporated via local *noise* or stochastic forces that respect the absorbing state condition (ii). Stochastic noise by itself cannot regenerate the disease.

These four defining ingredients are captured by the following coarse-grained Langevin equation for the local density $n(x,t)$ of infected agents:

$$\frac{\partial n(x,t)}{\partial t} = D\nabla^2 n(x,t) - R[n(x,t)]\, n(x,t) + \eta(x,t)\,, \qquad (10.13)$$

[6] Murray (2002).

with diffusion constant D, an appropriate reaction functional $R[n]$, and multiplicative stochastic noise $\langle \eta \rangle = 0$ with correlator

$$\langle \eta(x,t)\,\eta(x',t')\rangle = 2L[n(x,t)]\,\delta(x-x')\delta(t-t'), \quad L[n] = n\,N[n]. \quad (10.14)$$

Notice that the absorbing-state constraint (ii) is explicitly implemented through factoring out a density field from both reaction and noise functionals. Near the extinction threshold, we may expand $R[n] = Dr + u'n + \cdots$ and $N[n] = v + \cdots$, omitting higher-order terms as well as powers of gradients of the density field, since simple power counting yields that all these additional contributions are in fact irrelevant in the RG sense (Problem 10.1). Straightforward rescaling then recovers the non-linear Langevin equation (10.11) with noise correlation (10.12).

The above considerations thus establish the Janssen–Grassberger *DP conjecture*.[7] The asymptotic critical properties for continuous non-equilibrium phase transitions from active to inactive, absorbing states described by a scalar order parameter field governed by Markovian stochastic dynamics that is decoupled from any other slow variables and not subject to the influence of quenched randomness, are generically captured by the universality class of directed percolation.

10.1.3 Perturbation expansion and renormalization

We now proceed with the explicit perturbational analysis of the singular vertex functions of Reggeon field theory (10.9) to one-loop order. To this end, we may closely follow the perturbative loop expansion in Chapter 4 for model A with the identical propagator $G_0(q,\omega)$, yet with the two three-point vertices $\propto u$ rather than the two-point noise source and four-point anharmonicity. By virtue of the associated Feynman rules, the single one-loop graph for the two-point vertex function $\Gamma^{(1,1)}(q,\omega)$ or propagator self-energy in Fig. 10.4(a) yields the analytic expression

$$\Gamma^{(1,1)}(q,\omega) = i\omega + D\left(r+q^2\right) + 2u^2 \int_k \frac{1}{i\omega + 2D\left(r + \frac{q^2}{4} + k^2\right)} + O(u^4).$$

$$(10.15)$$

First, we impose the standard *criticality condition* $\Gamma^{(1,1)}(0,0) = 0$ at $r = r_c$ to determine the fluctuation-induced shift of the percolation threshold (*additive* renormalization) as the self-consistent solution of the implicit equation

$$r_c = -\frac{u^2}{D^2}\int_k \frac{1}{r_c + k^2} + O(u^4). \quad (10.16)$$

[7] Janssen (1981); Grassberger (1982).

(a)   (b)   (c)

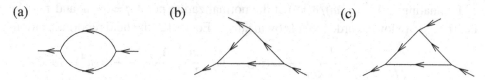

Fig. 10.4 One-loop Feynman diagrams for directed percolation: (a) two-point vertex function (self-energy) $\Gamma^{(1,1)}(q, \omega)$; (b) three-point noise vertex $\Gamma^{(2,1)}(q, \omega)$; and (c) three-point non-linear vertex $\Gamma^{(1,2)}(q, \omega)$.

With the integral (5.16) this implies an essential singularity as $d \uparrow 4$,

$$|r_c| = \left[\frac{2A_d}{(d-2)(4-d)}\frac{u^2}{D^2}\right]^{2/(4-d)}, \tag{10.17}$$

precisely as in (5.17) for the relaxational models or the static Φ^4 theory.

Defining the distance from the true phase transition via $\tau = r - r_c$, and inserting (10.16) into Eq. (10.15), one obtains to order u^2:

$$\Gamma^{(1,1)}(q, \omega) = i\omega + D(\tau + q^2) - \frac{u^2}{D}\int_k \frac{1}{k^2}\frac{i\omega + 2D(\tau + \frac{q^2}{4})}{i\omega + 2D(\tau + \frac{q^2}{4} + k^2)} + O(u^4). \tag{10.18}$$

Ultraviolet divergences here appear in dimensions $d \geq d_c = 4$ in the derivatives with respect to ω, q^2, and τ, which will be absorbed into multiplicatively renormalized fields and parameters D_R and τ_R. The only remaining UV-singular vertex functions are the three-point functions $\Gamma^{(1,2)}$ and $\Gamma^{(2,1)}$. From the associated one-loop graphs, depicted in Fig. 10.4(b) and (c), one infers at vanishing external wavevectors and frequencies $\underline{k}_i = (q_i, \omega_i) = (0, 0)$,

$$\Gamma^{(1,2)}(\{\underline{0}\}) = -\Gamma^{(2,1)}(\{\underline{0}\}) = -2u\left[1 - \frac{2u^2}{D^2}\int_k \frac{1}{(\tau + k^2)^2} + O(u^4)\right]. \tag{10.19}$$

Rapidity reversal symmetry (10.10) implies the vertex function identity

$$\Gamma^{(\tilde{N},N)}(\{q_i\}, \{\omega_i\}) = (-1)^{\tilde{N}+N}\Gamma^{(N,\tilde{N})}(\{q_i\}, \{-\omega_i\}), \tag{10.20}$$

which is of course borne out by the explicit results (10.15) and (10.19). Consequently only a single Z factor is required to renormalize both response and ordinary fields, $Z_{\tilde{s}} = Z_S$, whence Eq. (5.22) reduces to

$$\Gamma^{(\tilde{N},N)}_R = Z_S^{-(\tilde{N}+N)/2}\Gamma^{(\tilde{N},N)}. \tag{10.21}$$

For the renormalized parameters, we follow our standard conventions (5.23),

$$D_R = Z_D D, \quad \tau_R = Z_\tau \tau \mu^{-2}, \quad u_R = Z_u u A_d^{1/2}\mu^{(d-4)/2}. \tag{10.22}$$

Evaluating $\partial\Gamma^{(1,1)}(q,\omega)/\partial(i\omega)$ at the normalization point $q=0=\omega$ and $\tau_R=1$, i.e., $\tau=\mu^2$ to lowest order, yields by means of Eq. (5.12) the field renormalization

$$Z_S = 1 - \frac{u^2}{2D^2}\int_k \frac{1}{(\mu^2+k^2)^2} + O(u^4) = 1 - \frac{u^2}{2D^2}\frac{A_d\,\mu^{-\epsilon}}{\epsilon} + O(u^4)\,, \quad (10.23)$$

with $\epsilon=4-d$. Next from $\partial\Gamma^{(1,1)}(q,0)/\partial q^2$ and $\partial\Gamma^{(1,1)}(0,0)/\partial\tau$ one obtains

$$Z_S Z_D = 1 - \frac{u^2}{4D^2}\frac{A_d\,\mu^{-\epsilon}}{\epsilon} + O(u^4)\,, \quad Z_S Z_D Z_\tau = 1 - \frac{u^2}{D^2}\frac{A_d\,\mu^{-\epsilon}}{\epsilon} + O(u^4)\,,$$

whence one infers

$$Z_D = 1 + \frac{u^2}{4D^2}\frac{A_d\,\mu^{-\epsilon}}{\epsilon} + O(u^4)\,, \quad Z_\tau = 1 - \frac{3u^2}{4D^2}\frac{A_d\,\mu^{-\epsilon}}{\epsilon} + O(u^4)\,. \quad (10.24)$$

Finally, Eq. (10.19) immediately gives $Z_S^{3/2}Z_u$ at the normalization point, and consequently with (10.23)

$$Z_u = 1 - \frac{5u^2}{4D^2}\frac{A_d\,\mu^{-\epsilon}}{\epsilon} + O(u^4)\,. \quad (10.25)$$

10.1.4 DP scaling laws and critical exponents

From these renormalization constants (10.23)–(10.25) we deduce Wilson's RG flow functions

$$\gamma_S = \mu\left.\frac{\partial}{\partial\mu}\right|_0 \ln Z_S = \frac{v_R}{2} + O\left(v_R^2\right)\,, \quad (10.26)$$

$$\gamma_D = \mu\left.\frac{\partial}{\partial\mu}\right|_0 \ln\frac{D_R}{D} = -\frac{v_R}{4} + O\left(v_R^2\right)\,, \quad (10.27)$$

$$\gamma_\tau = \mu\left.\frac{\partial}{\partial\mu}\right|_0 \ln\frac{\tau_R}{\tau} = -2 + \frac{3v_R}{4} + O\left(v_R^2\right)\,, \quad (10.28)$$

$$\gamma_u = \mu\left.\frac{\partial}{\partial\mu}\right|_0 \ln\frac{u_R}{u} = \frac{d-4}{2} + \frac{5v_R}{4} + O\left(v_R^2\right)\,, \quad (10.29)$$

where we have introduced the renormalized dimensionless effective coupling

$$v_R = \frac{Z_u^2}{Z_D^2}\frac{u^2}{D^2}A_d\,\mu^{d-4}\,. \quad (10.30)$$

Its renormalization group beta function reads to this order

$$\beta_v = \mu\left.\frac{\partial}{\partial\mu}\right|_0 v_R = 2v_R(\gamma_u-\gamma_D) = v_R\left[-\epsilon + 3v_R + O\left(v_R^2\right)\right]\,. \quad (10.31)$$

Its graphs in the various dimensional regimes look just like Fig. 5.1: for $d > d_c = 4$ ($\epsilon < 0$), the Gaussian fixed point $v_0^* = 0$ is stable, and one recovers the mean-field critical exponents. Yet for $\epsilon = 4 - d > 0$, one finds the non-trivial infrared-stable RG fixed point

$$v_{DP}^* = \frac{\epsilon}{3} + O(\epsilon^2) \tag{10.32}$$

that controls the critical fluctuations at the directed percolation threshold. Setting up and solving the renormalization group equation for the vertex functions proceeds exactly in Section 5.3.1, and results in

$$\left[\mu \frac{\partial}{\partial \mu} + \frac{\tilde{N} + N}{2} \gamma_S + \gamma_D \, D_R \frac{\partial}{\partial D_R} + \gamma_\tau \, \tau_R \frac{\partial}{\partial \tau_R} + \beta_v \frac{\partial}{\partial v_R} \right]$$

$$\times \Gamma_R^{(\tilde{N}, N)}(\mu, D_R, \tau_R, v_R) = 0 \,. \tag{10.33}$$

After solving the RG equation with the method of characteristics, near the fixed point v_{DP}^* the two-point function assumes the form

$$G_R^{(1,1)}(\mu, D_R, \tau_R, v_R, q, \omega)^{-1} = \Gamma_R^{(1,1)}(\mu, D_R, \tau_R, v_R, q, -\omega)$$

$$\propto D_R \, \mu^2 \ell^{2 + \gamma_S^* + \gamma_D^*} \, \tilde{\Gamma}_R^{(1,1)} \left(\tau_R \, \ell^{\gamma_\tau^*}, v_{DP}^*, \frac{q}{\mu \ell}, \frac{\omega}{D_R \, \mu^2 \ell^{2 + \gamma_D^*}} \right), \tag{10.34}$$

compare with Eqs. (5.65) and (5.66) for the relaxational models. Upon matching $\ell = |q|/\mu$, this yields the scaling law

$$G_R^{(1,1)}(\tau_R, q, \omega)^{-1} = D_R \, q^{2-\eta} \, \hat{\Gamma} \left(\tau_R \, |q|^{-1/\nu}, \omega/D_R \, |q|^z \right), \tag{10.35}$$

where we identify the three independent *critical exponents* for directed percolation to first order in ϵ:[8]

$$\eta = -\gamma_S^* - \gamma_D^* = -\frac{\epsilon}{12} + O(\epsilon^2), \tag{10.36}$$

$$\frac{1}{\nu} = -\gamma_\tau^* = 2 - \frac{\epsilon}{4} + O(\epsilon^2), \tag{10.37}$$

$$z = 2 + \gamma_D^* = 2 - \frac{\epsilon}{12} + O(\epsilon^2). \tag{10.38}$$

Alternatively, setting $\ell = |\tau_R|^\nu$, one finds $G_R^{(1,1)}(\tau_R, 0, 0) \sim |\tau_R|^{-\gamma}$, with

$$\gamma = \nu(2 - \eta) = 1 + \frac{\epsilon}{6} + O(\epsilon^2). \tag{10.39}$$

---

[8] The two-loop results for the DP critical exponents to order ϵ^2 in the perturbative dimensional expansion are, e.g., derived and listed in Janssen and Täuber (2005).

Table 10.1 *Comparison of the DP critical exponents from series expansions and Monte Carlo simulations (numbers in brackets indicate the uncertainties in the last digits) with the results from the ϵ expansion.*

exponent	$d = 1$	$d = 2$	$d = 3$	$d = 4 - \epsilon$
ν	1.096854(4)	0.7333(75)	0.584(5)	$1/2 + \epsilon/16 + O(\epsilon^2)$
z	1.580745(10)	1.7660(16)	1.901(5)	$2 - \epsilon/12 + O(\epsilon^2)$
γ	2.277730(5)	1.5948(148)	1.237(23)	$1 + \epsilon/6 + O(\epsilon^2)$
β	0.276486(8)	0.5834(30)	0.813(9)	$1 - \epsilon/6 + O(\epsilon^2)$
α	0.159464(6)	0.4505(10)	0.732(4)	$1 - \epsilon/4 + O(\epsilon^2)$
θ	0.313686(8)	0.2295(10)	0.114(4)	$\epsilon/12 + O(\epsilon^2)$

[Numerical values adapted from: M. Henkel, H. Hinrichsen, and S. Lübeck, *Non-equilibrium Phase Transitions, Vol. 1: Absorbing Phase Transitions*, Springer (Dordrecht, 2008), p. 159, table 4.3.]

In the same manner, the solution of the RG equation for the *order parameter* in the vicinity of v_{DP}^* reads, recalling (10.7),

$$\langle S_R(\tau_R, t) \rangle = \mu^{d/2} \ell^{(d - \gamma_S^*)/2} \hat{S}\left(\tau_R \ell^{\gamma_\tau^*}, v_R^*, D_R \mu^2 \ell^{2 + \gamma_D^*} t\right). \tag{10.40}$$

Inserting the matching condition $\ell = |\tau_R|^\nu$ again confirms the scaling relations (10.2) and (10.4), and yields the explicit exponent values

$$\beta = 1 - \frac{\epsilon}{6} + O(\epsilon^2), \quad \alpha = 1 - \frac{\epsilon}{4} + O(\epsilon^2). \tag{10.41}$$

Another observable that is easily accessible to computer simulations is the number of active sites at time t generated by a seed at time $t = 0$ at the origin, which is just

$$N(\tau, t) = \int d^d x \, G^{(1,1)}(x, t) = G^{(1,1)}(q = 0, t). \tag{10.42}$$

The temporal Fourier transform of its renormalized counterpart follows immediately from (10.35), and is readily seen to take the scaling form

$$N_R(\tau_R, t) = t^\theta \, \hat{N}(\tau_R \, t^{1/z\nu}),$$

$$\theta = \frac{2 - \eta}{z} - 1 = \frac{d}{z} - 2\alpha = \frac{\epsilon}{12} + O(\epsilon^2). \tag{10.43}$$

At the percolation threshold, the number of active sites grows algebraically, with an exponent θ that plays the role of θ' for the order parameter initial slip in Eq. (8.6), compare with (8.7) for model A. Indeed, note that no new singularities

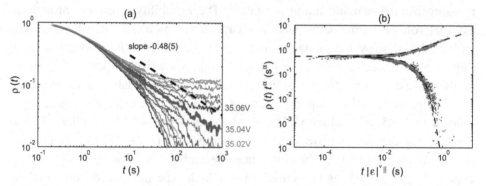

Fig. 10.5 Evidence for DP critical behavior in two-dimensional turbulent liquid crystals: (a) order parameter decay in the vicinity of the critical voltage $V_c \approx$ 35.04V; and (b) scaling collapse with DP critical exponents $\alpha \approx 0.48$ and $\nu_\parallel \approx$ 1.29; the dashed lines depict the corresponding scaling functions in the active (upper curve) and absorbing (lower curve) states as measured in Monte Carlo simulations for the contact process. [Figures reproduced with permission from: K. A. Takeuchi, M. Kuroda, H. Chaté, and M. Sano, *Phys. Rev. E* **80**, 051116 (2009); DOI: 10.1103/PhysRevE.80.051116; copyright (2009) by The American Physical Society.]

can be generated on the initial-time surface owing to the absence of correlation propagators in the DP field theory, which again leads to the scaling relation (10.43).

The scaling exponents for critical directed percolation are known for a plethora of physical quantities (but unfortunately a variety of different conventions defining and naming them are used in the literature). Table 10.1 compares the exponent values obtained in the dimensional expansion to order ϵ with high-accuracy results from Monte Carlo computer simulations in dimensions $d = 2$ and $d = 3$, and precision series expansions in one dimension. It is evident that the trends in the dimension dependence encoded already in the one-loop approximation are impressively reflected in the actual measured data.

Regrettably, though, only a few real experiments have to date confirmed the DP conjecture and actually measured the scaling exponents for this prominent non-equilibrium universality class. This may be at least partially due to the instability of the DP fixed point and associated scaling laws with respect to quenched disorder. A notable exception is the identification of several DP critical exponents in studies of spatio-temporal intermittency of ferrofluidic spikes.[9] Convincing evidence for directed-percolation scaling was found at the phase transition between two different turbulent states of electrohydrodynamic convection in thin, effectively

---

[9] Rupp, Richter, and Rehberg (2003).

two-dimensional nematic liquid crystals.[10] The instability is driven through an external voltage V; the active state is characterized by a large density of disclinations that destroy long-range nematic order, and detectable through reduced light transmittance. Takeuchi *et al.* extract an impressive twelve independent exponents and five scaling functions from their experimental data. As an example, Fig. 10.5(a) shows the temporal decay of the order parameter near the critical voltage $V_c \approx 35.04$ V, which yields the exponent $\alpha \approx 0.48(5)$. Excellent scaling collapse for different square voltage deviations $\varepsilon = (V^2 - V_c^2)/V_c^2$ is demonstrated in Fig. 10.5(b) for both the inactive and active phases, if the scaling exponent $\nu_\parallel = z\nu \approx 1.29(11)$ is used. In fact, both experimental scaling functions agree remarkably well with those obtained in numerical simulations of the two-dimensional contact process, which represents a particularly simple microscopic realization of the DP universality class.

10.2 DP variants and other percolation processes

We now turn our attention to several illuminating variants of directed percolation processes. First, we study extensions to systems with multiple particle species that also allow for special multi-critical points. The second modification adds long-range Lévy flight propagation to diffusive spreading. Next, immune 'debris' and concomitant memory is introduced into a more general epidemic process; in its quasi-static limit, the associated absorbing phase transition is governed by the critical exponents of isotropic percolation. Finally, we briefly address intriguing scenarios that emerge when a DP critical mode is coupled to a diffusive conserved field.

10.2.1 Multiple-species directed percolation

As it turns out, the directed percolation universality class extends even to generic multi-species systems. As a first illustration, let us return to the Lotka–Volterra predator–prey system discussed in Sections 9.1.1 and 9.5.2, but consider a variant where both predator and prey population growths are limited through binary fusion processes, $A + A \rightarrow A$ and $B + B \rightarrow B$ with rates τ and v, respectively.[11] In the associated shifted Doi–Peliti field theory (9.106), these reactions add the contributions

$$\Delta \mathcal{A}[\tilde{\psi}, \tilde{\varphi}; \psi, \varphi] = \int d^d x \int dt \left(\tau \, \tilde{\psi}(x, t)[1 + \tilde{\psi}(x, t)]\psi(x, t)^2 \right.$$

$$\left. + v \, \tilde{\varphi}(x, t)[1 + \tilde{\varphi}(x, t)]\varphi(x, t)^2 \right) \tag{10.44}$$

---

to the action (9.118). The equivalent Langevin description becomes

$$\frac{\partial \psi(x,t)}{\partial t} = (D\nabla^2 - \kappa)\,\psi(x,t) - \tau\,\psi(x,t)^2 + \lambda\,\psi(x,t)\varphi(x,t) + \zeta(x,t)\,,$$

(10.45)

$$\frac{\partial \varphi(x,t)}{\partial t} = (D\nabla^2 + \sigma)\,\varphi(x,t) - \nu\,\varphi(x,t)^2 - \lambda\,\psi(x,t)\varphi(x,t) + \eta(x,t)\,,$$

(10.46)

with Gaussian white noise $\langle \zeta(x,t)\rangle = 0 = \langle \eta(x,t)\rangle$, and (cross-)correlations

$$\langle \zeta(x,t)\,\zeta(x',t')\rangle = 2[\lambda\varphi(x,t) - \tau\psi(x,t)]\,\psi(x,t)\,\delta(x-x')\delta(t-t')\,, \quad (10.47)$$

$$\langle \zeta(x,t)\,\eta(x',t')\rangle = -\lambda\psi(x,t)\varphi(x,t)\,\delta(x-x')\delta(t-t')\,, \quad (10.48)$$

$$\langle \eta(x,t)\,\eta(x',t')\rangle = 2[\sigma - \nu\varphi(x,t)]\,\varphi(x,t)\,\delta(x-x')\delta(t-t')\,. \quad (10.49)$$

Aside from the entirely empty state $a = 0 = b$, the mean-field rate equations for this system allow for two different non-trivial stationary states, namely the predator–prey coexistence phase with $a_\infty = \langle \psi(x, t \to \infty)\rangle = (\sigma\lambda - \kappa\nu)/(\lambda^2 + \tau\nu)$, $b_\infty = \langle \varphi(x, t \to \infty)\rangle = (\kappa\lambda + \tau\sigma)/(\lambda^2 + \tau\nu)$, which exists and is stable for $\sigma\lambda > \kappa\nu$, whereas the predators are driven to extinction ($a_s = 0$) when $\sigma\lambda \leq \kappa\nu$, and the prey proliferate in the system to their maximum capacity $b_s = \sigma/\nu$. Equations (10.47) and (10.48) tell us that the multiplicative stochastic noise ζ completely vanishes in the predator extinction state, which is hence *absorbing*. We may thus suspect that the critical properties near the extinction threshold for the predator population A should be governed by the directed percolation universality class. Indeed, once the prey B are abundant and essentially uniformly distributed in the system, the predation reaction becomes effectively replaced with the branching process $A \to A + A$, and with the decay reactions $A \to \emptyset$ and $A + A \to A$ one arrives at the generic DP reactions discussed in Section 10.1.1.

In order to formally establish this assertion,[12] we introduce new fluctuating prey fields $\widetilde{\phi}(x,t) = -\sqrt{\sigma}\,\widetilde{\varphi}(x,t)$ and $\phi(x,t) = \sqrt{\sigma}\,[b_s - \varphi(x,t)]$, such that $\langle \phi\rangle = 0$. The Doi–Peliti field theory action then turns into

$$A[\widetilde{\psi}, \widetilde{\phi}; \psi, \phi] = \int d^d x \int dt \left[\widetilde{\psi}\left(\frac{\partial}{\partial t} - D\nabla^2 + \kappa - \lambda b_s\right)\psi - \lambda b_s\,\widetilde{\psi}^2\psi \right.$$

$$+ \tau\,\widetilde{\psi}(1 + \widetilde{\psi})\psi^2 + \frac{\widetilde{\phi}}{\sigma}\left(\frac{\partial}{\partial t} - D\nabla^2 + \sigma\right)\phi - \frac{\widetilde{\phi}^2\phi}{\sqrt{\sigma}} - \frac{\widetilde{\phi}\phi^2}{b_s\sqrt{\sigma}} + \frac{\widetilde{\phi}^2\phi^2}{b_s\sigma}$$

$$\left. - \frac{\lambda b_s}{\sqrt{\sigma}}\,\widetilde{\phi}\psi + \frac{\lambda}{\sqrt{\sigma}}\,\widetilde{\psi}(1 + \widetilde{\psi})\psi\phi - \frac{\lambda b_s}{\sqrt{\sigma}}\,\widetilde{\psi}\widetilde{\phi}\psi + \frac{\lambda}{\sigma}(1 + \widetilde{\psi})\widetilde{\phi}\psi\phi \right]. \quad (10.50)$$

---

[12] Mobilia, Georgiev, and Täuber (2007); Täuber (2012).

Near the extinction threshold at $r = (\kappa - \lambda b_s)/D \to 0$, the fluctuations of the prey fields ϕ are suppressed by their mass σ. Following the procedure in Section 5.4.2, we extract the asymptotic behavior by letting $\sigma \to \infty$, while keeping b_s fixed. This yields the much reduced effective action

$$\mathcal{A}_{\text{eff}}[\tilde{\psi}, \tilde{\phi}; \psi, \phi] = \int d^d x \int dt \left[\tilde{\psi}(x, t) \left[\frac{\partial}{\partial t} + D(r - \nabla^2) \right] \psi(x, t) \right.$$

$$- \lambda b_s \, \tilde{\psi}(x, t)^2 \psi(x, t)$$

$$\left. + \tau \, \tilde{\psi}(x, t)[1 + \tilde{\psi}(x, t)]\psi(x, t)^2 + \tilde{\phi}(x, t)\phi(x, t) \right]. \quad (10.51)$$

Integrating out the massive ϕ field, and rescaling to $\tilde{S}(x, t) = \sqrt{\lambda b_s/\tau} \, \tilde{\psi}(x, t)$ and $S(x, t) = \sqrt{\tau/\lambda b_s} \, \psi(x, t)$ leads directly to (10.6), where $u = \sqrt{\tau \lambda b_s}$ with scaling dimension $[u] = \mu^{2-d/2}$. Since $[\tau] = \mu^{2-d}$, the four-point vertex may again be dropped, and as anticipated we arrive at the Reggeon field theory effective action (10.9) that encapsulates the DP universality class.

After this specific example, let us now generalize to *multi-species* variants of directed percolation processes, which can be obtained in the discrete particle language by coupling the DP reactions $A_i \to \emptyset$, $A_i \rightleftharpoons A_i + A_i$ via stochastic processes of the form $A_i \rightleftharpoons A_j + A_j$ (with $j \neq i$). Alternatively, one may directly utilize the continuum Langevin representation with the stochastic equations of motion

$$\frac{\partial S_i(x, t)}{\partial t} = D_i \left(\nabla^2 - R_i[S_i(x, t)] \right) S_i(x, t) + \zeta_i(x, t), \quad (10.52)$$

$$\langle \zeta_i(x, t)\zeta_j(x', t') \rangle = 2S_i(x, t)N_i[S_i(x, t)] \delta(x - x')\delta(t - t') \delta_{ij}, \quad (10.53)$$

with the reaction functionals $R_i[S_i] = r_i + \sum_j u_{ij} S_j + \cdots$ and the noise correlations $\langle \zeta_i(x, t) \rangle = 0$, $N_i[S_i] = u_i + \cdots$. On first glance, this non-linear coupled system appears very rich. Remarkably, though, the ensuing renormalization constants turn out to be precisely those of single-species directed percolation, whence the *generic* critical properties of the associated active to absorbing phase transitions even in situations with multiple particle species are governed by the DP universality class.[13]

But through the consecutive processes $A_i \to A_i + A_i \to A_j$, the above reactions *generate* the linear transmutations $A_i \to A_j$, causing additional terms $\sum_{j \neq i} u_j S_j$ in Eq. (10.52). The resulting renormalization group flows typically produce asymptotically *unidirectional* inter-species couplings, which in turn allows for the appearance of special *multi-critical points* when several control parameters r_i

---

[13] Janssen (1997a, 2001).

vanish simultaneously.[14] To see how this comes about, we investigate a two-species system where both particle types A and B undergo the standard DP reactions (with rates $\kappa_{A/B}$, $\sigma_{A/B}$ and $\lambda_{A/B}$), supplemented with the unidirectional process $A \to B$, with rate μ_{AB}. On the mean-field level, the rate equation (9.15) remains intact,

$$\frac{\partial a(x,t)}{\partial t} = D(\nabla^2 - r_A)\,a(x,t) - \lambda_A\,a(x,t)^2 ,\qquad (10.54)$$

albeit with an increased effective decay rate $\kappa_A + \mu_{AB}$ and hence a modified control parameter $r_A = (\kappa_A + \mu_{AB} - \sigma_A)/D$. The rate equation for species B, on the other hand, contains a new gain term from the particle transmutation,

$$\frac{\partial b(x,t)}{\partial t} = D(\nabla^2 - r_B)\,b(x,t) - \lambda_B\,b(x,t)^2 + \mu_{AB}\,a(x,t) ,\qquad (10.55)$$

with the standard $r_B = (\kappa_B - \sigma_B)/D$.

As function of the asymptotic A particle density a_∞, the stationary, homogeneous solution of Eq. (10.55) is

$$b_\infty = \sqrt{\left(\frac{Dr_B}{2\lambda_B}\right)^2 + \frac{\mu_{AB}}{\lambda_B}\,a_\infty} - \frac{Dr_B}{2\lambda_B} .\qquad (10.56)$$

Thus, for $r_A > 0$, where $a_\infty = 0$, one finds $b_\infty = D(|r_B| - r_B)/2\lambda_B$, which is zero for $r_B > 0$, and becomes equal to $b_\infty = D|r_B|/\lambda_B$ for $r_B < 0$. When species A is in the inactive phase, the A and B hierarchy levels are effectively decoupled, and one would expect an ordinary DP active to absorbing phase transition for species B at $r_B = 0$, just as for $\mu_{AB} = 0$. For $r_A < 0$, $a_\infty = D|r_A|/\lambda_A$ and two cases need to be distinguished.

(i) For $(Dr_B/2\lambda_B)^2 \gg D|r_A|\mu_{AB}/\lambda_A\lambda_B$, one has approximately

$$b_\infty \approx \frac{D|r_B|}{2\lambda_B}\left[1 + \frac{D|r_A|\mu_{AB}}{2\lambda_A\lambda_B}\left(\frac{2\lambda_B}{Dr_B}\right)^2\right] - \frac{Dr_B}{2\lambda_B},\qquad (10.57)$$

whence $b_\infty > 0$ for $r_B < 0$. The B species is in its active state, and the DP transition at the half line ($r_A < 0$, $r_B = 0$) present in the uncoupled system has disappeared (see Fig. 10.6). Instead, for $r_B > 0$ the terms proportional to r_B cancel in Eq. (10.57), and $b_\infty \approx |r_A|\mu_{AB}/\lambda_A r_B$. Thus, the density of species B vanishes as the critical point $r_A = 0$ of species A is approached, and with the mean-field DP exponent $\beta = 1$. In effect, the DP critical half line ($r_A < 0$, $r_B = 0$) for species B has been rotated to ($r_A = 0$, $r_B > 0$) in the coupled system, as particles B are 'slaved' to the active/inactive behavior of the A species. The location of the critical lines for both species is shown in the phase diagram Fig. 10.6; the dotted parabola indicates the boundary curve separating the two different regimes for $r_A < 0$. A

---

[14] Täuber, Howard, and Hinrichsen (1998); Goldschmidt, Hinrichsen, Howard, and Täuber (1999).

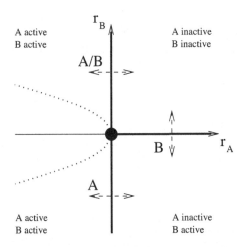

Fig. 10.6 Mean-field phase diagram for the two-species unidirectionally coupled DP process. The arrows mark continuous DP phase transitions for the A and B particle species. The dotted parabola denotes the boundary of the multi-critical regime, which includes $|r_A| = |r_B| = |r| \to 0$. [Figure reproduced with permission from: Y. Y. Goldschmidt, H. Hinrichsen, M. Howard, and U. C. Täuber, *Phys. Rev. E* **59**, 6381 (1999); DOI: 10.1103/PhysRevE.59.6381; copyright (1999) by The American Physical Society.]

detailed renormalization group analysis establishes that these critical lines are all characterized by DP scaling exponents, as expected.

(ii) For $(Dr_B/2\lambda_B)^2 \ll D|r_A|\mu_{AB}/\lambda_A\lambda_B$, i.e., in the interior of the dotted parabola in Fig. 10.6, which includes the special case when the critical points of both hierarchy levels are approached uniformly, $|r_A| = |r_B| = |r| \to 0$, we may neglect the terms $\propto r_B$ in Eq. (10.56), which yields

$$b_\infty \approx \sqrt{\frac{\mu_{AB}}{\lambda_B}} \, a_\infty = \sqrt{\frac{D|r_A|\mu_{AB}}{\lambda_A\lambda_B}} \, . \tag{10.58}$$

Consequently, the particle density critical exponents on each hierarchy level i are *different* in this regime, $a_i \sim |r_A|^{\beta_i}$, where $\beta_1 = \beta_A = 1$ and $\beta_2 = \beta_B = 1/2$ in the mean-field approximation. The other independent DP scaling exponents ν and z remain unaltered; but of course the critical density decay exponent α_i will differ for each hierarchy level i, namely $a_i(t) \sim t^{-\alpha_i}$ with $\alpha_i = \beta_i/z\nu$ according to the scaling relation (10.4).

A similar situation occurs in *tricritical directed percolation*, see Problem 10.2; indeed, here too *three* critical half lines converge at the *multi-critical* point $r_A = 0 = r_B$. When this special location in the phase diagram Fig. 10.6 is approached along a trajectory that crosses the dotted parabola, a *crossover* from ordinary directed percolation to multi-critical scaling ensues. These crossover features can be captured in the following scaling ansatz for the B density, valid in the limit $\tau \to 0$ or $\xi(\tau) \sim |\tau|^{-\nu} \to \infty$:

$$b(\tau, \mu_{AB}, x, t) = |\tau|^{\beta_1} \, \hat{a}\big(|\tau|^{-\phi} \mu_{AB}/D, x/\xi(\tau), Dt/\xi(\tau)^z\big) . \qquad (10.59)$$

According to Eq. (10.58), the mean-field *crossover exponent* is $\phi = 1$. It can be shown that this result persists to *all* orders in renormalized perturbation theory.[15] To lowest order in the dimensional expansion about the upper critical dimension $d_c = 4$, one obtains for the *hierarchy* of order parameter exponents β_i on the ith level of a unidirectional cascade

$$\beta_1 = 1 - \frac{\epsilon}{6} + O(\epsilon^2) , \quad \beta_2 = \frac{1}{2} - \frac{13\,\epsilon}{96} + O(\epsilon^2) , \dots, \quad \beta_i = \frac{1}{2^i} - O(\epsilon) .$$

$$(10.60)$$

There remains, however, an unresolved technical problem associated with the appearance of the relevant coupling $[\mu_{AB}] = \mu^2$ in the perturbation series, which renders the exponentiation of singular contributions ambiguous.

10.2.2 Directed percolation with long-range spreading

Hitherto we have assumed nearest-neighbor hopping or diffusive spreading of reacting particles or infectious agents. One may however also envision long-range interactions, e.g., Lévy flight propagation with hopping distances drawn from a probability distribution $P_L(x) \sim 1/|x|^{d+\sigma}$, with Lévy exponent $0 < \sigma \leq 2$ (see also Problem 9.5).[16] In Fourier space, this induces a non-analytic term $D_L|q|^\sigma$ in the propagator. Comparing with the critical diffusive spreading $\sim |q|^{2-\eta}$, one infers that Lévy flights do not alter the asymptotic scaling behavior of standard short-range directed percolation in the (small) parameter range $2 - \eta \leq \sigma \leq 2$. Technically, for $2 - \sigma = O(\epsilon)$ a systematic double expansion with respect to both $\epsilon = 4 - d$ and $2 - \sigma$ must be carried out; it yields an ultimate crossover to the DP critical exponents.

Yet for $\eta < 2 - \sigma = O(1)$, D_L should constitute a relevant perturbation, whereas ordinary diffusive spreading becomes irrelevant, and hence must be discarded in the perturbative RG analysis. The modified propagator reads

$$G_0(q, \omega) = \frac{1}{-i\omega + D_L(r + |q|^\sigma)} , \qquad (10.61)$$

---

[15] Janssen (2001).
[16] Janssen, Oerding, van Wijland, and Hilhorst (1999); Hinrichsen and Howard (1999); Janssen and Stenull (2008).

and instead of (10.8), one now finds the scaling dimensions

$$[D_L] = \mu^0, \quad [r] = \mu^\sigma, \quad [u] = \mu^{\sigma - d/2}, \tag{10.62}$$

indicating the new upper critical dimension $d_c(\sigma) = 2\sigma$. Fluctuation contributions are encoded by the identical Feynman graphs (Fig. 10.4) as for ordinary DP, and their analysis follows by straightforward replacements. For example, the fluctuation-induced percolation threshold shift is now

$$r_c = -\frac{u^2}{D_L^2} \int_k \frac{1}{r_c + |k|^\sigma} + O(u^4), \tag{10.63}$$

whence one obtains for the two-point vertex function to order u^2:

$$\Gamma^{(1,1)}(q, \omega) = i\omega + D_L (\tau + |q|^\sigma)$$

$$- \frac{u^2}{D_L} \int_k \frac{1}{|k|^\sigma} \frac{i\omega + D\left(2\tau + \left|\frac{q}{2} + k\right|^\sigma + \left|\frac{q}{2} - k\right|^\sigma - 2|k|^\sigma\right)}{i\omega + D\left(2\tau + \left|\frac{q}{2} + k\right|^\sigma + \left|\frac{q}{2} - k\right|^\sigma\right)} + O(u^4), \tag{10.64}$$

and similarly for the three-point vertex functions

$$\Gamma^{(1,2)}(\{0\}) = -\Gamma^{(2,1)}(\{0\}) = -2u\left[1 - \frac{2u^2}{D_L^2} \int_k \frac{1}{(\tau + |k|^\sigma)^2} + O(u^4)\right]; \tag{10.65}$$

note that rapidity reversal symmetry (10.10) is preserved even with Lévy flight propagation.

To one-loop order, there is but a single wavevector integral to be evaluated at the normalization point $\tau = \mu^2$,

$$\int_k \frac{1}{(\mu^2 + |k|^\sigma)^2} = \frac{A_{\sigma d} \, \mu^{d - 2\sigma}}{2\sigma - d}, \tag{10.66}$$

with appropriately chosen constant $A_{\sigma d}$ that is regular at $d_c(\sigma)$. Next we introduce renormalized parameters in analogy with (10.22) and (10.30),

$$D_{LR} = Z_{D_L} D_L, \quad \tau_R = Z_\tau \tau \mu^{-\sigma}, \tag{10.67}$$

$$u_R = Z_u u A_{\sigma d}^{1/2} \mu^{(d - 2\sigma)/2}, \quad v_R = \frac{Z_u^2}{Z_{D_L}^2} \frac{u^2}{D_L^2} A_{\sigma d} \mu^{d - 2\sigma}. \tag{10.68}$$

The crucial observation is that the long-range non-analytic term $\propto |q|^\sigma$ in the propagator cannot be renormalized through any perturbative correction; consequently $Z_S Z_{D_L} = 1$ to all orders in the perturbation expansion, and $\gamma_S = -\gamma_{D_L}$. This already implies the *exact* result

$$\eta = -\gamma_S^* - \gamma_{D_L}^* = 0. \tag{10.69}$$

Explicit evaluations of (10.64) and (10.65) result in the following one-loop Wilson flow functions:

$$\gamma_S = \frac{v_R}{2} + O(v_R^2), \quad \gamma_{D_L} = -\frac{v_R}{2} + O(v_R^2), \tag{10.70}$$

$$\gamma_\tau = -\sigma + v_R + O(v_R^2), \quad \gamma_u = \frac{d}{2} - \sigma + \frac{5v_R}{4} + O(v_R^2), \tag{10.71}$$

and the RG beta function for the effective coupling v_R of (10.68)

$$\beta_v = 2v_R(\gamma_u - \gamma_{D_L}) = v_R\left[-\epsilon' + \frac{7}{2}v_R + O(v_R^2)\right], \tag{10.72}$$

where $\epsilon' = 2\sigma - d$. The Gaussian fixed point $v_0^* = 0$ is stable for $d \geq d_c(\sigma) = 2\sigma$ and implies mean-field critical exponents $1/\nu = \sigma = z$. For $d < 2\sigma$, the non-trivial Lévy DP fixed point

$$v_{\text{LDP}}^* = \frac{2\epsilon'}{7} + O(\epsilon'^2) \tag{10.73}$$

becomes infrared-stable, and to first order in ϵ' yields the set of Lévy flight DP critical exponents

$$\frac{1}{\nu} = -\gamma_\tau^* = \sigma - \frac{2\epsilon'}{7} + O(\epsilon'^2), \tag{10.74}$$

$$\gamma = 2\nu = \frac{2}{\sigma}\left[1 + \frac{2\epsilon'}{7\sigma} + O(\epsilon'^2)\right], \tag{10.75}$$

$$z = \sigma + \gamma_{D_L}^* = \sigma - \frac{\epsilon'}{7} + O(\epsilon'^2), \tag{10.76}$$

$$\beta = \frac{\nu}{2}(d + z - \sigma) = 1 - \frac{2\epsilon'}{7\sigma} + O(\epsilon'^2), \tag{10.77}$$

$$\alpha = \frac{\beta}{z\nu} = 1 - \frac{3\epsilon'}{7\sigma} + O(\epsilon'^2), \tag{10.78}$$

$$\theta = \frac{\sigma}{z} - 1 = \frac{\epsilon'}{7\sigma} + O(\epsilon'^2). \tag{10.79}$$

Variants with built-in long-range *temporal* correlations that represent power-law distributions of 'incubation' times have also been investigated.

10.2.3 General epidemic processes, dynamic percolation

Let us next study a modification of the SEP rules (i) and (ii) in Section 10.1.2 that describe an 'epidemic with full recovery' of infected agents. Consider a situation where sick individuals either become immune or die, but in any rate cannot revert to a suseptible state; we also assume that in this 'epidemic with removal' all previously infected sites remain infectious to their neighbors, which induces temporal memory, as the 'debris' density is given by the accumulated decay products, $m(x, t) \propto \int_{-\infty}^{t} n(x, t') \, dt'$. For this *general epidemic process* (GEP),[17] the new stochastic rules become:

(i′) The susceptible medium becomes infected, depending on the densities $n(x, t)$ of sick individuals *and* $m(x, t)$ of the debris. After a brief time interval, the sick individuals decay into immune debris, which ultimately stops the disease locally by exhausting the supply of susceptible sites.

(ii′) States with $n = 0$ and any spatial distribution of m are *absorbing*, and describe the *extinction* of the epidemic.

After straightforward rescaling, the GEP is captured by the mesoscopic Langevin equation[18]

$$\frac{\partial S(x, t)}{\partial t} = -D(r - \nabla^2)S(x, t) - 2Du \, S(x, t) \int_{-\infty}^{t} S(x, t') \, dt' + \zeta(x, t) ,$$

(10.80)

with noise correlator (10.12). The associated response functional reads[19]

$$\mathcal{A}[\tilde{S}, S] = \int d^d x \int dt \left[\tilde{S}(x, t) \left[\frac{\partial}{\partial t} + D(r - \nabla^2) \right] S(x, t) \right.$$

$$\left. - u \, \tilde{S}(x, t)[\tilde{S}(x, t) - 2s(x, t)]S(x, t) \right] ,$$

(10.81)

where

$$s(x, t) = D \int_{-\infty}^{t} S(x, t') \, dt'$$

(10.82)

represents the locally accumulated debris density. The action (10.81) is invariant under the dynamical symmetry transformation

$$\tilde{S}(x, t) \leftrightarrow -s(x, -t) ,$$

(10.83)

the analog of rapidity reversal (10.10) for DP, see Problem 10.3(b). From the response functional (10.81), one readily infers the scaling dimensions

$$[\tilde{S}(x, t)] = \mu^{-1+d/2} = [s(x, t)] , \quad [S(x, t)] = \mu^{1+d/2} ,$$

(10.84)

$$[D] = \mu^0 , \quad [r] = \mu^2 , \quad [u] = \mu^{3-d/2} .$$

(10.85)

---

[17] Murray (2002).   [18] Grassberger (1983).
[19] Janssen (1985); Cardy and Grassberger (1985).

The coupling u becomes dimensionless at the upper critical dimension $d_c = 6$. The non-linear memory term in the Langevin equation (10.80) enhances the influence of correlations and consequently raises d_c as compared to DP.

One (of several) microscopic realization for the GEP is attained by adding the two-species reactions $A \to B$ and $A + B \to B$, with rates μ_{AB} and ν, respectively, to the standard DP reaction scheme. The B species here represents the immobile debris that suppresses the diffusively spreading active agents A. Constructing the corresponding Doi–Peliti action, and implementing the field shifts $\hat{\psi} = 1 + \tilde{\psi}$, $\hat{\varphi} = 1 + \tilde{\varphi}$, one arrives at

$$A[\tilde{\psi}, \tilde{\varphi}, \psi, \varphi] = \int d^d x \int dt \left[\tilde{\psi}(x, t) \left[\frac{\partial}{\partial t} + D(r - \nabla^2) \right] \psi(x, t) \right.$$
$$+ \tilde{\varphi}(x, t) \frac{\partial \varphi(x, t)}{\partial t} - \mu \tilde{\varphi}(x, t) \psi(x, t) - \sigma \tilde{\psi}(x, t)^2 \psi(x, t)$$
$$\left. + \nu \tilde{\psi}(x, t)[1 + \tilde{\varphi}(x, t)] \psi(x, t) \varphi(x, t) + \lambda \tilde{\psi}(x, t)^2 \psi(x, t)^2 \right],$$

(10.86)

where $r = (\kappa + \mu_{AB} - \sigma)/D$. Since the auxiliary field $\tilde{\varphi}$ only appears linearly in this action, it can be integrated out, yielding as constraint a non-linear deterministic differential equation for the debris field,

$$\frac{\partial \varphi(x, t)}{\partial t} = [\mu - \nu \tilde{\psi}(x, t) \varphi(x, t)] \psi(x, t) .$$

(10.87)

As before, power counting establishes that the four-point vertices in (10.86) are irrelevant near $d_c = 6$ and may be dropped; after rescaling one then arrives at the action (10.81), and the now linear Eq. (10.87) is easily integrated to yield (10.82).

For the theory defined by the Janssen–De Dominicis functional (10.81), one may take the *quasi-static limit* by introducing the asymptotic fields

$$\tilde{s}(x) = \tilde{S}(x, t \to \infty) , \quad s(x) = s(x, t \to \infty) = D \int_{-\infty}^{\infty} S(x, t') dt' .$$

(10.88)

In the long-time limit, the dynamic response functional (10.81) then becomes replaced by the quasi-static effective Hamiltonian, see Problem 10.3(c),

$$\mathcal{H}_{\mathrm{qst}}[\tilde{s}, s] = \int d^d x \left[\tilde{s}(x)(r - \nabla^2) s(x) - \frac{u}{D} \tilde{s}(x) [\tilde{s}(x) - s(x)] s(x) \right].$$

(10.89)

This functional of two independent fields $\tilde{s}$ and s assumes the very same form as Reggeon field theory (10.9) for directed percolation, with the reduced rapidity reversal invariance $\tilde{s}(x) \leftrightarrow -s(x)$, except that no special 'time' direction is singled out. Indeed, the field theory governed by (10.89) describes the universal

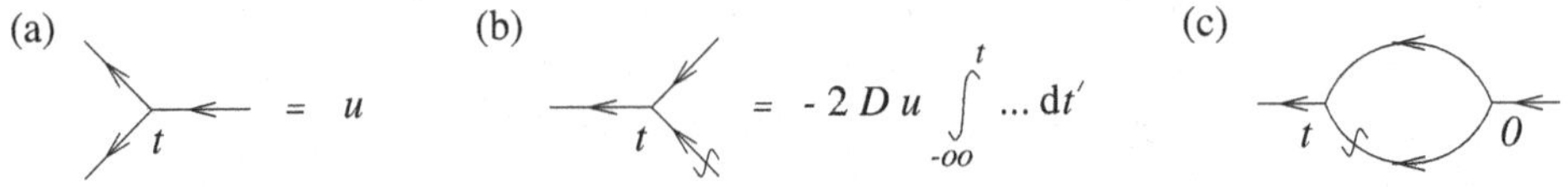

(a)   (b)   (c)

$$= u$$

$$= -2Du \int_{-\infty}^{t} \dots dt'$$

Fig. 10.7 Perturbation theory for dynamic isotropic percolation (DIP): (a) multiplicative noise vertex; (b) non-linear vertex with memory; and (c) one-loop Feynman graph for the two-point vertex function in the time domain.

scaling properties of *isotropic percolation*.[20] An isotropic percolation cluster is shown in Fig. 10.2(b), to be contrasted with the anisotropic DP structure visible in Fig. 10.2(a). The ensuing perturbation expansion with respect to the non-linear coupling u proceeds in strict analogy with DP, utilizing the vertices of Fig. 10.3 and to lowest order resulting in the same Feynman diagrams as depicted in Fig. 10.4, but involves the static propagator $G_0(q) = 1/(r + q^2)$. The explicit evaluation of the two- and three-point vertex functions and subsequent multiplicative renormalization of their UV divergences through our standard procedure is deferred to Problem 10.4. In an expansion about the upper critical dimension $d_c = 6$, one obtains the static critical exponents for isotropic percolation to first order in $\varepsilon = 6 - d$,

$$\eta = -\gamma_s^* = -\frac{\varepsilon}{21} + O(\varepsilon^2), \tag{10.90}$$

$$\frac{1}{\nu} = -\gamma_\tau^* = 2 - \frac{5\varepsilon}{21} + O(\varepsilon^2), \tag{10.91}$$

and therefrom

$$\beta = \frac{\nu}{2}(d - 2 + \eta) = 1 - \frac{\varepsilon}{7} + O(\varepsilon^2), \tag{10.92}$$

$$\gamma = \nu(2 - \eta) = 1 + \frac{\varepsilon}{7} + O(\varepsilon^2). \tag{10.93}$$

We now return to the full dynamical action (10.81), in order to calculate the dynamic critical exponents for this *dynamic isotropic percolation* (DIP) universality class. As shown in Fig. 10.7, one of the non-linear vertices becomes modified by the integration over previous times. In order to deal with the temporal non-locality, one best evaluates the fluctuation corrections in the time domain; subsequent Fourier transform yields

$$\Gamma^{(1,1)}(q, \omega) = i\omega + D(r + q^2)$$

$$+ 4Du^2 \int_k \frac{1}{i\omega + D[r + (\frac{q}{2} - k)^2]} \frac{1}{i\omega + 2D(r + \frac{q^2}{4} + k^2)} + O(u^4)$$

$$\tag{10.94}$$

[20] Benzoni and Cardy (1984); for a general introduction to percolation theory, see Stauffer and Aharony (1994).

(Problem 10.5). By means of our standard tools, we can now proceed with the additive and multiplicative renormalization program; note that the dynamical symmetry (10.83) implies the *exact* relations

$$Z_{\widetilde{S}} = Z_s = Z_S Z_D^2 \qquad (10.95)$$

for the field renormalization constants. In dimensions $d \leq d_c = 6$, there exists a non-trivial IR-stable RG fixed point, and the associated dynamic critical exponents for dynamic isotropic percolation become

$$z = 2 - \frac{\varepsilon}{6} + O(\varepsilon^2), \qquad (10.96)$$

$$\alpha = \frac{\beta}{z\nu} = 1 - \frac{5\varepsilon}{28} + O(\varepsilon^2), \qquad (10.97)$$

$$\theta = \frac{2 - \eta}{z} - 1 = \frac{3\varepsilon}{28} + O(\varepsilon^2) \qquad (10.98)$$

to first order in $\varepsilon = 6 - d$.[21]

It is also possible to describe the *crossover* from isotropic to directed percolation within the field-theoretic framework.[22] Imagine nearly isotropic percolation in $d + 1$ dimensions, where a bias $\propto g$ is implemented towards the 'positive' direction along one specific spatial component that we again label as 'time' t. In the continuum description, this biased percolation should be governed by the action

$$\mathcal{A}[\widetilde{S}, S] = \int d^d x \int dt \left[\widetilde{S}(x, t) \left[-\frac{\partial^2}{\partial t^2} + g \frac{\partial}{\partial t} + D(r - \nabla^2) \right] S(x, t) \right.$$

$$\left. - u \, \widetilde{S}(x, t)[\widetilde{S}(x, t) - S(x, t)]S(x, t) \right]. \qquad (10.99)$$

Power counting with $[t] = \mu^{-1}$ immediately establishes that $[g] = \mu$ constitutes a *relevant* parameter; the RG flow will take the action into the fully unidirectional (temporally irreversible) DP limit. Rescaling with the relevant parameter g yields an associated dimensional crossover whereupon the upper critical dimension changes from $5 + 1$ to $4 + 1$, similar to the quantum- to classical-crossover discussed for the non-linear sigma model in Section 7.4. One may thus construct an intriguing crossover sequence. Starting from directed percolation in $d + 1$ dimensions, the introduction of memory effects leads to dynamic isotropic percolation. In the quasi-static limit, DIP in turn is described by isotropic percolation in d spatial dimensions. Introducing directionality along one spatial coordinate finally yields DP again, but in $(d - 1) + 1$ dimensions.

---

[21] For an overview of the two-loop analysis, see Janssen and Täuber (2005).
[22] Frey, Täuber, and Schwabl (1994a, 1994b); Janssen and Stenull (2000).

Precisely as for directed percolation, multi-species generalizations of dynamic isotropic percolation processes can be defined and analyzed. They are indeed once again generically governed by the single-species DIP universality class, except for special multi-critical points in the phase diagram (Fig. 10.6), which are characterized by a set of order parameter exponents β_i that become markedly smaller (halved in the mean-field approximation) on each subsequent hierarchy level, and an exact crossover exponent $\phi = 1$.

As in DP, we may also replace diffusive spreading $\propto Dq^2$ in dynamic isotropic percolation with Lévy flight propagation $\propto D_L|q|^\sigma$. One then obtains the scaling dimensions

$$[\widetilde{S}(x,t)] = \mu^{(d-\sigma)/2}, \quad S(x,t)] = \mu^{(d+\sigma)/2}, \quad [u] = \mu^{(3\sigma-d)/2}, \quad (10.100)$$

indicating an upper critical dimension $d_c(\sigma) = 3\sigma$. Since the non-analytic Lévy flight term in the propagator is not renormalized, naturally $\eta = 0$ holds exactly. It is then a straightforward exercise (Problem 10.6) to determine the other static critical exponents utilizing the associated quasi-static long-range Hamiltonian, giving to first order in $\varepsilon' = 3\sigma - d$:

$$\frac{1}{\nu} = -\gamma_\tau^* = \sigma - \frac{\varepsilon'}{4} + O(\varepsilon'^2), \quad (10.101)$$

$$\gamma = 2\nu = \frac{2}{\sigma}\left[1 + \frac{\varepsilon'}{4\sigma} + O(\varepsilon'^2)\right], \quad (10.102)$$

$$\beta = \frac{\nu}{2}(d-\sigma) = 1 - \frac{\varepsilon'}{4\sigma} + O(\varepsilon'^2). \quad (10.103)$$

The dynamic critical exponents have to be calculated by means of the Janssen–De Dominicis response functional, see Problem 10.6(b):

$$z = \sigma + \gamma_{D_L}^* = \sigma - \frac{3\varepsilon'}{16} + O(\varepsilon'^2), \quad (10.104)$$

$$\alpha = \frac{\beta}{z\nu} = 1 - \frac{5\varepsilon'}{16\sigma} + O(\varepsilon'^2), \quad (10.105)$$

$$\theta = \frac{\sigma}{z} - 1 = \frac{3\varepsilon'}{16\sigma} + O(\varepsilon'^2). \quad (10.106)$$

10.2.4 Directed percolation coupled to a conserved field

In analogy with the equilibrium critical dynamics model C, where a scalar non-conserved order parameter is coupled to a conserved scalar field, see Sections 3.3.1 and 6.1, let us consider in a coarse-grained Langevin description an order parameter field $S(x,t)$ that undergoes an active to absorbing phase transition and interacts with a non-critical diffusive mode $\rho(x, t)$:

$$\frac{\partial S(x, t)}{\partial t} = -D[r - \nabla^2 + u\, S(x, t) + g\, \rho(x, t)]S(x, t) + \zeta(x, t), \qquad (10.107)$$

$$\frac{\partial \rho(x, t)}{\partial t} = \lambda\, \nabla^2[\rho(x, t) - v\, S(x, t)] + \eta(x, t), \qquad (10.108)$$

with $\langle \zeta \rangle = 0 = \langle \eta \rangle$ and noise correlators

$$\langle \zeta(x, t)\, \zeta(x', t') \rangle = 2\tilde{u}\, S(x, t)\, \delta(x - x')\delta(t - t'), \qquad (10.109)$$

$$\langle \eta(x, t)\, \eta(x', t') \rangle = -2\lambda\nabla_x^2 \delta(x - x')\delta(t - t'), \qquad (10.110)$$

$$\langle \zeta(x, t)\, \eta(x', t') \rangle = 0. \qquad (10.111)$$

One microscopic interpretation of this stochastic dynamics would be a diffusively spreading conserved pollutant C with density ρ that poisons the active A species, $A \leftrightarrow A + A$, $A \rightarrow \emptyset$, supplemented with $A + C \rightarrow C$;[23] in that case, $v = 0$ in (10.108). A feedback of the DP field S into the dynamics of the conserved density ρ is, e.g., introduced through the reaction scheme $A + B \rightarrow B + B, B \rightarrow A$, which leaves the total number of particles $c(t) = a(t) + b(t) = c_0$ fixed. Upon elimination of the A density and after appropriate rescaling, one arrives at Eqs. (10.107)–(10.111), where v is proportional to the difference of diffusivities for species A and B.[24]

Within the mean-field approximation, (10.108) implies $\rho(x) = \rho_0 + v\, S(x)$ in the stationary state. Equation (10.107) consequently yields

$$\nabla^2 S(x) = r'\, S(x) + u'\, S(x)^2, \qquad r' = r + g\rho_0, \qquad u' = u + gv. \qquad (10.112)$$

For $u' < 0$, the homogeneous solution $S(x) = 0$ becomes unstable even for $r' > 0$; in that case additional powers of the field $S(x, t)$ should be included in the effective coarse-grained description, and the phase transition is expected to become discontinuous. This instability can be understood as follows. Consider $g > 0$, when the pollutant inhibits the active field. The second term in (10.108) can be interpreted as a contribution to the diffusion current $j_c(x, t) = gv\nabla S(x, t)$. For $v > 0$, the inhibitor flows towards regions with enhanced density $S(x, t)$, and thereby suppresses the local fluctuation. If, however, $v < 0$, such inhomogeneities become amplified through the coupling to the conserved field. Indeed, a one-loop perturbative analysis yields runaway solutions for the RG flow equations in this situation, which indicates a fluctuation-induced first-order phase transition.

For positive v, the continuous character of the active to absorbing state transition remains preserved by the coupling to the conserved density, but the critical point

---

[23] Kree, Schaub, and Schmittmann (1989).
[24] van Wijland, Oerding, and Hilhorst (1998); Oerding, van Wijland, Leroy, and Hilhorst (2000).

at $\tau' = 0$ is governed by a new DPC universality class, distinct from DP, but still with upper critical dimension $d_c = 4$. Remarkably, dynamic symmetries of the associated response functional fully determine its critical exponents *exactly*, and one finds[25]

$$\nu = 2/d \,, \quad z = 2 \,, \quad \beta = 1 \,. \tag{10.113}$$

(Notice that ν reaches its mean-field value $1/2$ at $d_c = 4$.) For the special case $\upsilon = 0$, of these only the order parameter exponent picks up corrections of order $\epsilon = 4 - d$,

$$\beta = 1 - \frac{\epsilon}{32} + O(\epsilon^2) \,. \tag{10.114}$$

Finally, we briefly address the limit when the field $\rho(x)$ becomes static, i.e., a time-independent spatially random variable with short-range Gaussian correlator $\overline{\rho(x)\rho(x')} \propto \delta(x - x')$ which locally modifies the percolation threshold p_c. One can then directly perform the disorder average over the response functional; the ensuing RG flow equations, however, yield runaway flows.[26] This finding indicates that quenched randomness drastically changes the character of the DP active to absorbing non-equilibrium transition. Presumably spatial regions with rare extreme values of p_c dominate the system's physical properties. Indeed, a real-space RG approach specifically tailored to this type of problem has demonstrated the existence of a strong-disorder fixed point controlling the universal scaling of DP with quenched randomness, and predicts logarithmically slow kinetics.[27] For weaker disorder, the static critical exponents appear to vary with the disorder strength. The sensitivity of the directed percolation universality class to quenched disorder that describes a random spatial variation of the local percolation threshold may well be a decisive reason that DP scaling exponents do not feature quite as prominently in experimental studies as would otherwise be expected.

10.3 Branching and annihilating random walks

In Section 2.3.3, we mapped the kinetics of one-dimensional Ising spin systems in thermal equilibrium onto equivalent interacting particle dynamics, where the 'particles' represent the domain walls on the spin chain. Consider now a non-equilibrium setting, where zero-temperature Glauber spin flip dynamics is combined with Kawasaki spin exchange processes at $T = \infty$. In the domain wall picture, the ensuing stochastic dynamics is captured by unbiased particle diffusion, the pair annihilation reactions $A + A \to \emptyset$ from the Glauber kinetics (2.123) with

---

[25] For details, see Janssen and Täuber (2005).   [26] Janssen (1997b).
[27] Hooyberghs, Iglói, and Vanderzande (2003, 2004).

rate $\lambda = 2\Gamma_G$, and the branching processes $A \rightarrow A + A + A$ from the elementary spin exchanges (2.126) with rate $\sigma' = \Gamma_K$. In the following, we investigate a generalized version, namely *branching and annihilating random walks* (BAW) in d dimensions, where we combine the annihilation reactions $k A \rightarrow \emptyset$ ($k \geq 2$, see Section 9.3) with branching processes $A \rightarrow (m + 1) A$ (with $m \geq 1$).[28] Note that in contrast to directed percolation, spontaneous particle decay $A \rightarrow \emptyset$ is absent here. Following a brief discussion of the associated mean-field rate equations, we venture to explain why in the case $k = 2$ there exists a crucial distinction between the situations with odd and even numbers of offspring particles m. Outside the mean-field regime, the active to absorbing phase transition for BAW with odd m is in fact governed by the DP critical exponents. For even m and in low dimensions, fluctuations may cause the emergence of an inactive, absorbing phase characterized by the pair annihilation power laws. The emerging non-equilibrium phase transition is then described by a novel 'parity-conserving' (PC) universality class.

10.3.1 Rate equation and fluctuation corrections

It is always instructive to first study the mean-field rate equation for a given non-linear stochastic process. For BAW with kth order annihilation rate λ and branching rate σ_m, it reads

$$\frac{da(t)}{dt} = -k\lambda \, a(t)^k + m \, \sigma_m \, a(t) . \tag{10.115}$$

Irrespective of the initial conditions, as $t \rightarrow \infty$ the particle density $a(t)$ reaches a finite steady-state value

$$a_\infty = \left(\frac{m \, \sigma_m}{k\lambda}\right)^{1/(k-1)} , \tag{10.116}$$

provided $\sigma_m > 0$, whence the system always remains in an active state. Indeed, this saturation density is approached exponentially with a characteristic time scale $1/m \, \sigma_m$, as evident from the explicit solution of (10.115),

$$a(t) = a_\infty \big(1 + \big[(a_\infty/a_0)^{k-1} - 1\big]e^{-(k-1)m \, \sigma_m t}\big)^{-1/(k-1)} . \tag{10.117}$$

If the branching rate is set to zero, we are left with mere annihilation processes and the ensuing power law decay (9.8). One may thus view $\sigma_c = 0$ as a (degenerate) critical point; if diffusive propagation is added, it is characterized by the mean-field critical exponents $\nu = 1/2, z = 2, \beta = 1/(k - 1) = \alpha$. Since the scaling dimension for the rate of the non-linear annihilation reaction is known from (9.67), $[\lambda] = \mu^{2-(k-1)d}$, fluctuations are expected to play a decisive role only in low spatial

---

[28] Grassberger, Krause, and von der Twer (1984); Takayasu and Tretyakov (1992); the field-theoretic analysis presented here naturally relies heavily on Cardy and Täuber (1996, 1998).

dimensions $d < d_c(k) = 2/(k-1)$, the upper critical dimension of the annihilation fixed point. Since $d_c(3) = 1$ for triplet annihilation processes, we may henceforth restrict ourselves to BAW with pair annihilation reactions ($k = 2$).

Intriguingly, Monte Carlo simulations on one- and two-dimensional lattices do not display the simple mean-field picture at all. In one dimension, one invariably finds a critical point separating an active from an absorbing phase, but with different universal scaling behavior if the number m of offspring particles in the branching process is even or odd. For odd m, DP critical exponents are measured in one and two dimensions. In the case of even m, notice that the particle number *parity* remains conserved under the reaction kinetics, whence there are two disjoint absorbing states, namely (i) complete extinction if the initial particle number is even; and (ii) a single survivor if the system is initialized with an odd number of particles. In the following, we utilize our field theory tools to elucidate the origin and physical distinction between these quite distinct universality classes, and the mechanism through which the particle parity conservation law survives the thermodynamic and continuum limits.

The 'bulk' Doi–Peliti field theory action (9.52) for BAW with binary annihilation reads

$$A[\hat{\psi}, \psi] = \int d^d x \int dt \left[\hat{\psi}(x, t) \left(\frac{\partial}{\partial t} - D\nabla^2 \right) \psi(x, t) \right.$$

$$\left. - \sigma_m [\hat{\psi}(x, t)^m - 1] \hat{\psi}(x, t)\, \psi(x, t) + \lambda [\hat{\psi}(x, t)^2 - 1] \psi(x, t)^2 \right].$$

$$(10.118)$$

The branching rate with scaling dimension $[\sigma_m] = \mu^2$ of a mass term is obviously a relevant perturbation on the pure annihilation model. For even m, each term in (10.118) contains an even product of fields $\hat{\psi}$ and ψ, and the action is therefore invariant under simultaneous 'parity' transformations

$$\hat{\psi}(x, t) \to \hat{\psi}'(x, t) = -\hat{\psi}(x, t), \quad \psi(x, t) \to \psi'(x, t) = -\psi(x, t). \quad (10.119)$$

This special symmetry for even offspring number is of course still present but obscured when the usual field shift $\hat{\psi}(x, t) = 1 + \tilde{\psi}(x, t)$ is performed. Indeed, one would have to be very careful not to violate the requirement (10.119) when truncating the ensuing action to some low powers in the fields (which would in fact predict DP critical behavior for any integer m). It is therefore safer to work directly with the field theory (10.118). In addition to the propagator with mass σ_m/D,

$$G_0(q, \omega) = \frac{1}{-i\omega + \sigma_m + Dq^2}, \quad G_0(q, t) = e^{-(\sigma_m + Dq^2)t}\, \Theta(t), \quad (10.120)$$

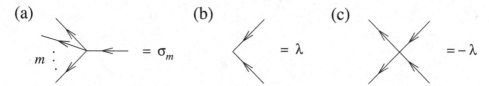

Fig. 10.8 Perturbation theory vertices for branching and annihilating random walks (BAW): (a) branching vertex; (b) pair annihilation vertex; and (c) associated 'scattering' vertex (see Fig. 9.4).

it is characterized by a branching vertex $\propto \sigma_m$ and the pair annihilation and associated 'scattering' vertices $\propto \lambda$, depicted in Fig. 10.8.

Let us now investigate fluctuation corrections to the mean-field scenario. The first, crucial observation is that consecutive branching and annihilation reactions *generate* the branching process with $m - 2$ offspring, as shown in Fig. 10.9(a). To higher orders in λ, consequently *all* offspring-producing reactions $A \to (m + 1)\, A, (m - 1)\, A, \ldots$ occur and should be represented in an apt coarse-grained description. Correspondingly, the original 'microscopic' Doi–Peliti action (10.118) needs to be complemented with a sum over branching terms with $m' = m - 2, m - 4, \ldots$. At this point the decisive distinction between even- and odd-offspring BAW emerges. For even m, one obtains the sequence of branching reactions $A \to (m + 1)\, A, (m - 1)\, A, \ldots, 3A, A$; at least a single particle remains even after several consecutive annihilation events. For *odd* m, on the other hand, $A \to (m + 1)\, A, (m - 1)\, A, \ldots, A + A$, and finally $A \to \emptyset$. Spontaneous particle decay is generated through stochastic fluctuations. This opens the possibility of adding a negative correction to the bare mass σ that would shift the critical point to $\sigma_c > 0$. The ensuing active to absorbing phase transition should clearly fall into the directed percolation universality class; and the emergent inactive phase would be governed by exponential temporal decay and correlations.

One may readily compute the renormalization constants and hence fluctuation corrections to the anomalous scaling dimensions of the rates σ_m and λ. Consider specifically the situation near the annihilation fixed point (9.76), where $\sigma_m = 0$. Defining the Z factors in the standard manner,

$$g_R = Z_g \frac{\lambda}{D} C_d \mu^{d-2}, \quad s_{mR} = Z_{s_m} \frac{\sigma_m}{D\mu^2}, \tag{10.121}$$

with the geometric factor C_d from (7.190), the one-loop Feynman diagrams in Fig. 10.9(b) and (c) immediately yield at the normalization point $\sigma_m = \mu^2$, $q = 0 = \omega$, or alternatively $\sigma_m = 0 = q$, $i\omega = D\mu^2$:

$$Z_{s_m} = 1 - \frac{m(m+1)}{2} \frac{\lambda}{D} \frac{C_d \mu^{d-2}}{2 - d}, \quad Z_g = 1 - \frac{\lambda}{D} \frac{C_d \mu^{d-2}}{2 - d}, \tag{10.122}$$

(a)   (b)   (c)

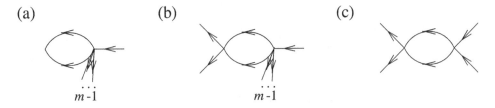

Fig. 10.9 One-loop Feynman diagrams for BAW fluctuation corrections: (a) gener-ation of the $A \to (m-1) A$ branching process; (b) renormalization of the branch-ing rate; and (c) renormalization of the annihilation rate.

where the combinatorial factor in Z_{s_m} simply reflects the number of possibilities $\binom{m+1}{2}$ of choosing two external legs from the branching vertex to form the loop in Fig. 10.9(b) through closure with the annihilation vertex. For the annihilation coupling, this results in the RG beta function (9.75), with $d_c(2) = 2$, non-trivial fixed point $g^* = 2 - d$, and

$$\gamma_{s_m} = \mu \frac{\partial}{\partial\mu}\Big|_0 \ln \frac{s_{mR}}{\sigma_m} = -2 + \frac{m(m+1)}{2} g_R + O(g_R^2) . \qquad (10.123)$$

Near two dimensions, all branching vertices remain relevant ($\gamma_m < 0$), whence the running parameters $\tilde{s}_m(\ell) \to \infty$ in the infrared limit $\ell \to 0$. In addition, we observe that the *most* relevant renormalized branching rates are those with *lowest* allowed value for m, i.e., $\sigma_2 = \sigma'$ for m even; and $\sigma_1 = \sigma$ for m odd.

10.3.2 BAW with odd offspring number and DP

For branching and annihilating random walks with an odd number m of offspring particles, we have just seen that the most relevant branching reaction is $A \to A + A$. As shown in Fig. 10.9(a), the combination of this process with pair annihilation $A + A \to \emptyset$ immediately generates spontaneous particle decay $A \to \emptyset$. Adding this process to the Doi–Peliti action (10.118) with $m = 1$ results in a coarse-grained effective action

$$\mathcal{A}_{\text{eff}}[\hat{\psi}, \psi] = \int d^d x \int dt \left[\hat{\psi}(x,t) \left(\frac{\partial}{\partial t} - D\nabla^2 \right) \psi(x,t) \right.$$

$$\left. + [\hat{\psi}(x,t) - 1][\kappa - \sigma\,\hat{\psi}(x,t) + \lambda\,(\hat{\psi}(x,t)+1)\,\psi(x,t)]\psi(x,t) \right]$$

$$(10.124)$$

that looks very similar to (10.5), differing only in the pair annihilation rather than binary fusion term. Proceeding just as in Section 10.1.1, one readily establishes that the continuous active to absorbing state transition at non-zero $\sigma_c > 0$ is described by the critical exponents of directed percolation as encoded by Reggeon field theory (10.9).

Yet it needs to be demonstrated that the fluctuations which both generate the single-particle decay process and renormalize the branching and annihilation vertices move the critical point away from the mean-field value $\sigma_c = 0$. To this end, we follow Section 9.3.1 to sum the perturbation expansion in λ for the most singular Feynman diagrams. Near two dimensions, they are given by a straightforward expansion of the one-loop contributions in Fig. 10.9 through a series of 'bubble' graphs, as depicted in Fig. 9.5. Since within this approximation there appears no renormalization of the diffusion constant D, one obtains for the resulting geometric sums in Fourier space:

$$\sigma_R = \frac{\sigma}{1 + \lambda\, I_d(\sigma)/D}\,, \qquad \kappa_R = \frac{\sigma \lambda\, I_d(\sigma)/D}{1 + \lambda\, I_d(\sigma)/D}\,. \tag{10.125}$$

This yields for the renormalized DP mass term

$$r_R = \frac{\kappa_R - \sigma_R}{D_R} = \frac{\sigma}{D}\,\frac{\lambda\, I_d(\sigma) - D}{\lambda\, I_d(\sigma) + D} \tag{10.126}$$

with the one-loop fluctuation integral

$$I_d(\sigma) = \int_k^\Lambda \frac{1}{\sigma + k^2} = \begin{cases} C_d\, \sigma^{1-d/2}/(2 - d), & d < 2, \\ \ln(\Lambda^2/\sigma), & d = 2, \end{cases} \tag{10.127}$$

where, as in (5.2), an ultraviolet cutoff Λ has been introduced in two dimensions. This simple calculation thus predicts the critical point to be shifted to a positive threshold value determined by $I_d(\sigma_c) = D/\lambda$, or

$$\sigma_c \approx \begin{cases} [\lambda\, C_d/D\,(2 - d)]^{2/(2-d)}, & d < 2, \\ \Lambda^2\, e^{-4\pi D/\lambda}, & d = 2, \end{cases} \tag{10.128}$$

in dimensions $d \leq 2$. The non-analytic exponential dependence on the annihilation rate λ at $d = 2$ can be confirmed through an analysis of the corresponding one-loop RG flow equations. In higher dimensions, σ_c remains zero within the perturbative approach, although a transition at large values of $\lambda\Lambda^{d-2}/D \gg 1$ cannot be ruled out. In fact, a non-perturbative RG treatment of BAW with odd m yields $\sigma_c > 0$ and consequently a DP phase transition in any dimension[29] (with mean-field critical exponents for $d > 4$).

---

[29] Canet, Delamotte, Deloubrière, and Wschebor (2004); Canet, Chaté, and Delamotte (2004).

10.3.3 Parity-conserving BAW with even offspring number

For BAW with even m that conserve the total particle number parity, fluctuations can never generate the process $A \to \emptyset$; the most relevant branching reaction is $A \to A + A + A$, to which we assign the rate σ'. Since the $\hat{\psi}$ fields appear to cubic order in the ensuing Doi–Peliti field theory (10.118) with $m = 2$, one cannot readily write down an equivalent stochastic differential equation for $\psi(x, t)$. However, nothing prevents us from interpreting $\psi(x, t)$ as the Martin–Siggia–Rose auxiliary response field instead, whence by means of the substitution $\phi(x, t) = \hat{\psi}(x, -t)$ one arrives at a 'dual' Langevin representation with Gaussian white noise $\langle \zeta \rangle = 0$ and

$$\frac{\partial \phi(x, t)}{\partial t} = D\nabla^2 \phi(x, t) - \sigma'[1 - \phi(x, t)^2]\phi(x, t) + \zeta(x, t), \qquad (10.129)$$

$$\langle \zeta(x, t)\zeta(x', t') \rangle = 2\lambda[1 - \phi(x, t)^2]\delta(x - x')\delta(t - t'). \qquad (10.130)$$

Interestingly, (10.129) constitutes a special case of a more general Langevin equation obeying a Z_2 inversion symmetry that has been proposed to capture a unified *voter model* universality class,[30]

$$\frac{\partial \phi(x, t)}{\partial t} = D\nabla^2 \phi(x, t) - [\sigma'\phi(x, t) - \rho\, \phi(x, t)^3][1 - \phi(x, t)^2] + \zeta(x, t),$$

$$(10.131)$$

with noise correlator (10.130). Notice that for $\rho > 0$ one may replace $1 - \phi(x, t)^2$ by 1, since the higher-order terms become irrelevant, whence (10.131) reduces to the scalar model A Langevin equation for a kinetic Ising model with Glauber dynamics. For $\rho < 0$, the stochastic differential equation (10.131) with multiplicative noise is designed to capture the genuine voter model phase transition. The PC universality class correspondingly emerges as a special bicritical point at $\rho = 0$.

Near two dimensions, i.e., for $\epsilon = d - 2 \ll 1$, the branching rate σ' constitutes a relevant parameter, since $\gamma_{s'}^* = -2 + 3\epsilon + O(\epsilon^2)$ at the annihilation fixed point $g^* = \epsilon$. Since $\sigma'_c = 0$, the *exact* scaling exponents at criticality are just those of the pure diffusion-limited pair annihilation process,

$$\eta = 0, \quad z = 2, \quad \alpha = d/2. \qquad (10.132)$$

---

[30] Al Hammal, Chaté, Dornic, and Muñoz (2005).

Only those exponents that describe divergences upon approaching the critical point from the active phase need to be computed perturbatively:

$$\frac{1}{\nu} = -\gamma_{s'}^* = 2 - 3\epsilon + O(\epsilon^2),$$ (10.133)

$$\beta = z\nu\alpha = d\nu = 1 + \epsilon + O(\epsilon^2).$$ (10.134)

A similar scenario pertains to a q-species generalization of BAW with $m = 2$ off-spring particles that can be analyzed exactly, see Problem 10.7. An ϵ expansion near the upper critical dimension $d_c = 2$ for pair annihilation cannot, however, capture the PC active to absorbing phase transition at a nonzero $\sigma_c' > 0$ that is observed in one dimension. The large positive fluctuation contribution to the anomalous dimension $\gamma_{s'}^*$ provides a hint that qualitatively different features emerge in low dimensions. Indeed, within the one-loop approximation, $\gamma_{s'}^*$ becomes positive, and hence the renormalized branching rate s' is *irrelevant* for $\epsilon = 2/3$, i.e., at $d < d_c' = 4/3$. Of course, a mere loop expansion is not to be trusted in this situation. To two-loop order, $d_c' \approx 1.1$ (Problem 10.8); and a self-consistent calculation to first order in σ' that, however, sums the perturbation series to all powers in λ finds $d_c' < 1$.[31]

Evaluating the renormalizations for the BAW annihilation and branching rates perturbatively in fixed dimension while keeping the full dependence on σ' in the propagator permits the construction of a RG flow scenario that at least qualitatively captures the Monte Carlo observations in low dimensions $d < d_c'$. Evaluating the one-loop diagrams in Fig. 10.9(b) and (c) at the normalization point $i\omega = D\mu^2$ thus yields the Z factors

$$Z_{s'} = 1 - \frac{3\,C_d}{2 - d} \frac{\lambda/D}{(\mu^2 + \sigma'/D)^{1-d/2}}, \quad Z_\lambda = 1 - \frac{C_d}{2 - d} \frac{\lambda/D}{(\mu^2 + \sigma'/D)^{1-d/2}},$$ (10.135)

which are functions of both σ'/D and λ/D, and for $\sigma' = 0$ reduce to (10.122) with $m = 2$. The resulting one-loop Wilson flow functions

$$\gamma_{s'} = -2 + f_R, \quad \gamma_g = d - 2 + f_R$$ (10.136)

are controlled by the *effective* coupling

$$f_R = \frac{g_R}{(1 + s_R')^{2-d/2}}.$$ (10.137)

---

[31] Benitez and Wschebor (2012, 2013).

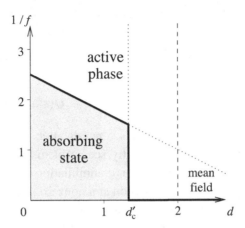

Fig. 10.10 Emergence of the inactive, absorbing state for even-offspring BAW in low dimensions $d < d'_c$; beyond $d_c = 2$, mean-field theory applies. [Figure adapted from Cardy and Täuber (1998).]

We expect $\tilde{s}'(\ell) \to \infty$ under the renormalization group flow; in that limit, the RG beta function for f_R becomes

$$\beta_f(f_R) \to f_R \left[\gamma_g - \left(2 - \frac{d}{2} \right) \gamma_{s'} \right] = f_R \left[2 - \frac{10 - 3d}{2} f_R + O(f_R^2) \right].$$

(10.138)

Aside from the Gaussian fixed point $f_0^* = 0$, (10.138) displays an unstable *critical fixed point*

$$f_c^* = \frac{4}{10 - 3d}$$

(10.139)

(provided $d < 10/3$). Yet since the bare reaction rate that corresponds to the pure annihilation fixed point is infinite already, we must demand on physical grounds that $f_c^* \leq g^* = 2 - d$, whence the critical fixed point comes into existence only for $d < d'_c = 4/3$. If $f_R < f_c^*$ initially, the RG flow for $\tilde{g}(\ell)$ tends to zero, consistent with $\tilde{s}'(\ell) \sim \ell^{-2} \to \infty$ as $\ell \to 0$. This Gaussian fixed point is of course characterized by naive scaling dimensions, and describes the *active* phase with exponential decay and correlations. On the other hand, for $f_R > f_c^*$, and provided that $d < d'_c$, $\tilde{s}'(\ell) \to 0$ and $\tilde{f}(\ell) \to g^*$, which represents the inactive state; note that the entire absorbing phase is governed by the power laws of the pure pair annihilation model (Section 9.3). The continuous phase transition in the PC universality class evidently has no true mean-field counterpart. For sufficiently low branching rate, fluctuations destroy the 'ordered' active state and give rise to a generically scale-invariant inactive phase below the new borderline dimension d'_c, which may thus be viewed as an 'inverted' lower critical dimension. The emerging phase diagram as a function of the spatial dimension d, as obtained within this (uncontrolled) one-loop approximation, is sketched in Fig. 10.10.

Although attempts to extend this approach to higher loop order have as yet remained futile, its general picture appears to be in accord with a non-perturbative numerical renormalization group study.[32] The above scenario is also compatible with an extension to branching and annihilating Lévy flights, see Sections 10.2.2 and 10.2.3. One may then stay in the physical dimension $d = 1$, whence the absorbing phase emerges only once the Lévy flight exponent exceeds a critical value σ_c. In the one-loop approximation, $\sigma_c = 3/2$, in good agreement with numerical simulations.[33]

As in our analysis of the non-linear sigma model (Section 7.4.3), we expand the beta function $\beta_f(f_R) \approx \beta'_f{}^*(f_R - f_c^*)$ near the PC critical fixed point f_c^* (for $d < d'_c$). Informed by the mean-field result for the particle density in the active phase, we write the solution of the associated RG equation as

$$a(t, s'_R, g_R, \mu) = \mu^d \frac{s'_R}{\lambda_R} \ell^{d+\gamma_{s'}^*-\gamma_g^*} \hat{a}\left(s'_R\mu^2 t \, \ell^{2+\gamma_{s'}^*}, (f_c^* - f_R)\ell^{\beta'_f{}^*}\right). \quad (10.140)$$

The matching condition $(f_c^* - f_R)\ell^{\beta'_f{}^*} = 1$ allows us to identify the critical exponents, and with (10.139) crudely estimate their numerical values,

$$z \approx 2, \quad \nu = \frac{2+\gamma_{s'}^*}{-\beta'_f{}^*} \approx \frac{3}{10-3d}, \quad \beta = \frac{d+\gamma_{s'}^*-\gamma_g^*}{-\beta'_f{}^*} \approx \frac{4}{10-3d}. \quad (10.141)$$

As should likely be expected, these values compare rather poorly with the PC critical exponents obtained through Monte Carlo simulations in one dimension, $\nu \approx 1.83$, $\nu_\parallel = z\nu \approx 3.22$, $z \approx 1.75$, $\beta \approx 0.92$, and $\alpha \approx 0.27$. As a final remark, we note that the presence of the dangerously irrelevant parameter $1/s'_R$ in fact precludes a direct calculation of the power laws precisely at the critical point, rather than approaching it from the active phase, and hence the derivation of 'hyperscaling' relations such as $\beta = z\nu\alpha$.

10.3.4 Higher-order processes

We first revisit BAW with triplet annihilation ($k = 3$) and m offspring particles. For $m = 3, 6, \ldots$ the particle number is conserved modulo 3 (even locally), and fluctuations cannot generate spontaneous decay $A \to \emptyset$. As for even-offspring BAW with pair annihilation, in this special situation potentially novel universal behavior might arise below some borderline dimension $d'_c < 1$. In one dimension

---

[32] Canet, Chaté, Delamotte, Dornic, and Muñoz (2005).   [33] Vernon and Howard (2001).

one should merely find logarithmic corrections to the mean-field scaling laws, e.g., the critical density decay at $\sigma_c = 0$ from pure triple annihilation (9.87): $a(t) \sim [(\ln Dt)/Dt]^{1/2}$. For all other values of m that are indivisible by $k = 3$, fluctuation loops induce processes such as $A \to A + A$, $A \to \emptyset$, $A + A \to \emptyset$, etc. Certainly for $d \leq d_c(3) = 1$, but probably in higher dimensions as well, one would expect the active to absorbing state phase transition to be shifted to $\sigma_c > 0$, and to fall into the directed percolation universality class.

Next we retain binary annihilation $A + A \to \emptyset$ or coagulation/fusion $A + A \to A$ (rate λ), but require a particle pair to meet in order to trigger a branching process, $A + A \to (m + 2) A$ (rate σ, $m \geq 1$). In the absence of any restrictions on the local particle density or the site occupation number on a lattice, obviously $a(t \to \infty) \to \infty$ in the active state ($\sigma > \lambda$), whereas the inactive, absorbing phase ($\sigma < \lambda$) is once again governed by the power laws of pure pair annihilation or fusion. The 'bosonic' version of this binary fission–annihilation reaction, now usually referred to as '*pair contact process with diffusion*' (PCPD), thus displays a strongly discontinuous non-equilibrium phase transition.[34] However, when site occupation restrictions are implemented or, equivalently, triplet annihilation reactions admitted, which renders the particle density in the active phase finite, the active to absorbing state transition becomes continuous. Owing to exceedingly long crossover time scales, its precise nature has been much debated in the literature.[35] Indeed, a direct and perhaps naive application of the Doi–Peliti formalism for this model yields runaway renormalization group flows, indicating the inadequacy of this 'microscopic' field theory approach.[36] The best effective coarse-grained description appears to be in terms of two separate fields that respectively represent the particle and particle pair densities. The order parameter field for the particle density then seems to undergo essentially a DP continuous phase transition, but interacts non-trivially with the pair density. Yet the corresponding continuum field theory entails an uncontrolled relevant parameter, reminiscent of the multi-critical point scenario in coupled multi-species DP or DIP processes (see Section 10.2.2).

This brief discussion underscores several important points. First, it is crucial to realize that formulating a microscopic model, say in the language of interacting and reacting particles, or even the associated Doi–Peliti continuum action that is based on the corresponding stochastic master equation, is *not* always sufficient to address large-scale and long-time emergent features, specifically to properly identify critical behavior and universality classes. This usually requires intermediate coarse-graining steps. Thus novel processes might be generated, others may turn out irrelevant for the asymptotic scaling; it may be possible to eliminate certain

---

[34] Howard and Täuber (1997).   [35] For a comprehensive overview, see Henkel and Hinrichsen (2004).
[36] Janssen, van Wijland, Deloubrière, and Täuber (2004).

particle species, but it is also conceivable that new types of agent must be introduced. Also, transforming to appropriate variables may point at conserved modes that are not immediately apparent, but that are obviously crucial. Mapping to field theory representations and applying the renormalization group program can be an invaluable aid in such investigations, but neither provides a self-correcting tool that automatically produces correct results. To date, proper coarse-graining and formulating apt mesoscopic descriptions of non-linear stochastic systems remains a skillful art, especially in non-equilibrium systems that are not restricted through detailed-balance constraints.

As a consequence, beyond the almost complete analysis of multi-species pair annihilation and directed as well as dynamic isotropic percolation processes, a full and satisfactory classification of universality classes that govern active to absorbing phase transitions and generic scale invariance in reaction-diffusion systems has remained elusive. Partial successes have certainly been achieved, based on insightful combinations of mean-field arguments, power counting, and the wealth of available numerical data and observations.[37] For reacting particle systems, the non-trivial structure of the zero-dimensional (non-spatial) pseudo-Hamiltonian trajectories (9.46) yields additional intriguing information[38] that complements the qualitative picture from straightforward rate equation analysis. Yet one must keep in mind that these trajectories still reside in the realm of a refined mean-field approximation. The truly difficult issue is to understand how non-linear stochastic fluctuations and spatio-temporal correlations might modify these trajectories quantitatively or perhaps even affect the phase diagram and character of the ensuing non-equilibrium phase transitions in a qualitative manner.

Problems

10.1 *Irrelevant terms in the continuum theory for simple epidemic processes*
Consider the density expansions $R[n] = Dr + \sum_{k \geq 1} u_k n^k$ and $L[n] = n \sum_{k \geq 0} v_k n^k$ for the reaction and noise functionals in Eqs. (10.13) and (10.14). Determine the scaling dimensions for the coefficients u_k and v_k, and show how rescaling for the leading terms recovers the Reggeon field theory effective action (10.9). Why do additional gradient expansions lead only to irrelevant contributions?

10.2 *Tricritical directed percolation: mean-field theory*
The mean-field theory for *tricritical directed percolation* is captured by the rate equation

$$\frac{\partial a(t)}{\partial t} = -r\, a(t) - u\, a(t)^2 - v\, a(t)^3 , \quad v > 0 .$$

---

[37] See, for example, Kockelkoren and Chaté (2003).   [38] Elgart and Kamenev (2006).

    (a) Demonstrate that for $u > 0$, there is a continuous active to absorbing (DP) phase transition at $r = 0$.

    (b) Find the mean-field tricritical DP exponents β_t, α_t at $u = 0$, $r \to 0$.

    (c) For $u < 0$, determine the line $r_s(u, v)$ below which there can be coexistence between the active and absorbing phases, and sketch the resulting phase diagram in the (r, u) plane. (Compare with Problem 1.3.)

10.3 *Dynamical symmetries for directed and dynamic isotropic percolation*

    (a) Confirm that the Reggeon field theory action (10.9) is invariant under rapidity reversal (10.10).

    (b) Check that the symmetry transformation (10.83) leaves the action (10.81) for dynamic isotropic percolation invariant.

    (c) Establish the emergence of the quasi-static Hamiltonian (10.89) from the Janssen–De Dominicis functional (10.81) for dynamic isotropic percolation with the fields (10.88) in the limit $t \to \infty$.

10.4 *Critical exponents for isotropic percolation*

Renormalize the one-loop contributions to the two- and three-point vertex functions for the quasi-static Hamiltonian (10.89), and compute the critical exponents for isotropic percolation to first order in $\varepsilon = 6 - d$.

10.5 *Dynamic critical exponents for dynamic isotropic percolation*

Evaluate the two-point vertex function $\Gamma^{(1,1)}(q, \omega)$ for dynamic isotropic percolation. First compute the loop integral in Fig. 10.7(c) in the time domain, and subsequently perform the Fourier transform. Employ the IR-stable RG fixed point and static critical exponents from Problem 10.4 to obtain the dynamic critical exponents z, α, and θ. Confirm (10.91) for the correlation length exponent ν, and check Eq. (10.95).

10.6 *Dynamic isotropic percolation with Lévy flight propagation*

Follow the procedure in Sections 10.2.2 and 10.2.3 to determine the critical exponents for dynamic isotropic percolation with Lévy flight spreading.

    (a) Employ the corresponding quasi-static Hamiltonian to confirm the static exponents (10.101)–(10.103), see Problem 10.4.

    (b) Use the full response functional and the results of Problem 10.5 to compute the associated dynamic critical exponents (10.104)–(10.106).

10.7 *Generalization of BAW with $m = 2$ offspring particles to q species*

Consider $q \geq 2$ interacting particle species that are subject to the branching processes $A_i \to A_i + A_i + A_i$ ($i = 1, \ldots, q$, rate σ') and $A_i \to A_i + A_j + A_j$ ($j \neq i$, rate $\bar{\sigma}$), while only particles of the same species may mutually annihilate: $A_i + A_i \to \emptyset$ (rate λ).

    (a) Use the one-loop fluctuation corrections to demonstrate that σ' is irrelevant compared to $\bar{\sigma}$.

(b) Show that after setting $\sigma' = 0$, the remaining perturbation series for both λ and σ can be summed to all orders.

(c) Thus derive the *exact* critical exponents at the degenerate critical point $\bar{\sigma}_c = 0$ in $d \leq 2$ dimensions: $\eta = 0, z = 2, \nu = 1/d, \alpha = d/2$, and $\beta = 1$, which are all *independent* of the number q of distinct species.

10.8 *Two-loop branching rate renormalization for BAW with two offspring*
Draw the Feynman graphs to two-loop order for the four-point vertex functions $\Gamma^{(2,2)}$ and $\Gamma^{(3,1)}$ for branching and annihilating random walks with two offspring particles, described by the field theory action (10.118) with $m = 2$. Therefrom obtain the renormalization constants

$$Z_g = 1 - \frac{\lambda}{D} \frac{C_d \, \mu^{-\epsilon}}{\epsilon} + \frac{\lambda^2}{D^2} \frac{C_d^2 \, \mu^{-2\epsilon}}{\epsilon^2} \, ,$$

$$Z_{s'} = 1 - 3 \frac{\lambda}{D} \frac{C_d \, \mu^{-\epsilon}}{\epsilon} + 6 \frac{\lambda^2}{D^2} \frac{C_d^2 \, \mu^{-2\epsilon}}{\epsilon^2} + \frac{3}{2} \ln \frac{4}{3} \frac{\lambda^2}{D^2} \frac{C_d^2 \, \mu^{-2\epsilon}}{\epsilon} \, ,$$

to compute the anomalous dimension of the branching rate σ' at the annihilation fixed point $g^* = \epsilon = 2 - d$:

$$\gamma_{s'}^* = 2 - 3\epsilon + 3 \ln \frac{4}{3} \epsilon^2 + O(\epsilon^3) \, .$$

Confirm that to this order the borderline dimension for the existence of an active phase is

$$d_c' = 2 - \left(1 - \sqrt{1 - \frac{8}{3} \ln \frac{4}{3}}\right) \Big/ 2 \ln \frac{4}{3} \approx 1.1 \, .$$

References

Al Hammal, O., H. Chaté, I. Dornic, and M.A. Muñoz, 2005, Langevin description of critical phenomena with two symmetric absorbing states, *Phys. Rev. Lett.* **94**, 230601-1–4.

Benitez, F. and N. Wschebor, 2012, Branching-rate expansion around annihilating random walks, *Phys. Rev. E* **86**, 010104(R)-1–5.

Benitez, F. and N. Wschebor, 2013, Branching and annihilating random walks: exact results at low branching rate, *Phys. Rev. E* **87**, 052132-1–18.

Benzoni, J. and J. L. Cardy, 1984, A hyperscaling relation in site-bond correlated percolation, *J. Phys. A: Math. Gen.* **17**, 179–196.

Canet, L., H. Chaté, and B. Delamotte, 2004, Quantitative phase diagrams of branching and annihilating random walks, *Phys. Rev. Lett.* **92**, 255703-1–4.

Canet, L., H. Chaté, B. Delamotte, I. Dornic, and M. A. Muñoz, 2005, Nonperturbative fixed point in a nonequilibrium phase transition, *Phys. Rev. Lett.* **95**, 100601, 1–4.

Canet, L., B. Delamotte, O. Deloubrière, and N. Wschebor, 2004, Nonperturbative renormalization-group study of reaction-diffusion processes, *Phys. Rev. Lett.* **92**, 195703-1–4.

Cardy, J. L. and P. Grassberger, 1985, Epidemic models and percolation, *J. Phys. A: Math. Gen.* **18**, L267–L272.

Cardy, J. L. and R. L. Sugar, 1980, Directed percolation and Reggeon field theory, *J. Phys. A: Math. Gen.* **13**, L423–L427.

Cardy, J. L. and U. C. Täuber, 1996, Theory of branching and annihilating random walks, *Phys. Rev. Lett.* **77**, 4780–4783.

Cardy, J. L. and U. C. Täuber, 1998, Field theory of branching and annihilating random walks, *J. Stat. Phys.* **90**, 1–56.

Elgart, V. and A. Kamenev, 2006, Classification of phase transitions in reaction-diffusion models, *Phys. Rev. E* **74**, 041101-1–16.

Frey, E., U. C. Täuber, and F. Schwabl, 1994a, Crossover from self-similar to self-affine structures in percolation, *Europhys. Lett.* **26**, 413–418.

Frey, E., U. C. Täuber, and F. Schwabl, 1994b, Crossover from isotropic to directed percolation, *Phys. Rev. E* **49**, 5058–5072.

Goldschmidt, Y. Y., H. Hinrichsen, M. Howard, and U. C. Täuber, 1999, Nonequilibrium critical behavior in unidirectionally coupled stochastic processes, *Phys. Rev. E* **59**, 6381–6408.

Grassberger, P., 1982, On phase transitions in Schlögl's second model, *Z. Phys. B Cond. Matt.* **47**, 365–374.

Grassberger, P., 1983, On the critical behavior of the general epidemic process and dynamical percolation, *Math. Biosci.* **63**, 157–172.

Grassberger, P., F. Krause, and T. von der Twer, 1984, A new type of kinetic critical phenomenon, *J. Phys. A: Math. Gen.* **17**, L105–110.

Henkel, M. and H. Hinrichsen, 2004, The non-equilibrium phase transition of the pair-contact process with diffusion, *J. Phys. A: Math. Gen.* **37**, R117–R159.

Henkel, M., H. Hinrichsen, and S. Lübeck, 2008, *Non-equilibrium Phase Transitions*, Vol. 1: *Absorbing phase transitions*, Dordrecht: Springer.

Hinrichsen, H., 2001, Nonequilibrium critical phenomena and phase transitions into absorbing states, *Adv. Phys.* **49**, 815–958.

Hinrichsen H. and M. Howard, 1999, A model for anomalous directed percolation, *Eur. Phys. J. B* **7**, 635–643.

Hooyberghs, J., F. Iglói, and C. Vanderzande, 2003, Strong disorder fixed point in absorbing-state phase transitions, *Phys. Rev. Lett.* **90**, 100601-1–4.

Hooyberghs, J., F. Iglói, and C. Vanderzande, 2004, Absorbing state phase transitions with quenched disorder, *Phys. Rev. E* **69**, 066140-1–16.

Howard, M. J. and U. C. Täuber, 1997, 'Real' vs 'imaginary' noise in diffusion-limited reactions, *J. Phys. A: Math. Gen.* **30**, 7721–7731.

Janssen, H. K., 1981, On the nonequilibrium phase transition in reaction-diffusion systems with an absorbing stationary state, *Z. Phys. B Cond. Matt.* **42**, 151–154.

Janssen, H. K., 1985, Renormalized field theory of dynamical percolation, *Z. Phys. B Cond. Matt.* **58**, 311–317.

Janssen, H. K., 1997a, Spontaneous symmetry breaking in directed percolation with many colors: differentiation of species in the Gribov process, *Phys. Rev. Lett.* **78**, 2890–2893.

Janssen, H. K., 1997b, Renormalized field theory of the Gribov process with quenched disorder, *Phys. Rev. E* **55**, 6253–6256.

Janssen, H. K., K. Oerding, F. van Wijland, and H. J. Hilhorst, 1999, Lévy-flight spreading of epidemic processes leading to percolating clusters, *Eur. Phys. J. B* **7**, 137–147.

Janssen H. K. and O. Stenull, 2000, Random resistor-diode networks and the crossover from isotropic to directed percolation, *Phys. Rev. E* **62**, 3173–3185.

Janssen, H. K., 2001, Directed percolation with colors and flavors, *J. Stat. Phys.* **103**, 801–839.

Janssen, H. K., F. van Wijland, O. Deloubrière, and U. C. Täuber, 2004, Pair contact process with diffusion: failure of master equation field theory, *Phys. Rev. E* **70**, 056114-1–7.

Janssen, H. K. and U. C. Täuber, 2005, The field theory approach to percolation processes, *Ann. Phys. (NY)* **315**, 147–192.

Janssen, H. K. and O. Stenull, 2008, Field theory of directed percolation with long-range spreading, *Phys. Rev. E* **78**, 061117-1–12.

Kinzel, W., 1983, Directed percolation, in: *Percolation Structures and Processes*, eds. G. Deutsch, R. Zallen, and J. Adler, Bristol: Adam Hilger.

Kockelkoren, J. and H. Chaté, 2003, Absorbing phase transitions of branching-annihilating random walks, *Phys. Rev. Lett.* **90**, 125701-1–4.

Kree, R., B. Schaub, and B. Schmittmann, 1989, Effects of pollution on critical population dynamics, *Phys. Rev. A* **39**, 2214–2221.

Mobilia, M., I. T. Georgiev, and U. C. Täuber, 2007, Phase transitions and spatio-temporal fluctuations in stochastic lattice Lotka–Volterra models, *J. Stat. Phys.* **128**, 447–483.

Moshe, M., 1978, Recent developments in Reggeon field theory, *Phys. Rep.* **37**, 255–345.

Murray, J. D., 2002, *Mathematical Biology*, Vols. I and II, New York: Springer, 3rd edn.

Obukhov, S. P., 1980, The problem of directed percolation, *Physica A* **101**, 145–155.

Ódor, G., 2004, Phase transition universality classes of classical, nonequilibrium systems, *Rev. Mod. Phys.* **76**, 663–724.

Oerding, K., F. van Wijland, J.-P. Leroy, and H. J. Hilhorst, 2000, Fluctuation-induced first-order transition in a nonequilibrium steady state, *J. Stat. Phys.* **99**, 1365–1395.

Rupp, P., R. Richter, and I. Rehberg, 2003, Critical exponents of directed percolation measured in spatiotemporal intermittency, *Phys. Rev. E* **67**, 036209-1–7.

Stauffer, D. and A. Aharony, 1994, *Introduction to Percolation Theory*, London: Taylor and Francis, 2nd edn.

Takayasu, H. and A. Yu. Tretyakov, 1992, Extinction, survival, and dynamical phase transition of branching annihilating random walk, *Phys. Rev. Lett.* **68**, 3060–3063.

Takeuchi, K. A., M. Kuroda, H. Chaté, and M. Sano, 2007, Directed percolation criticality in turbulent liquid crystals, *Phys. Rev. Lett.* **99**, 234503-1–5.

Takeuchi, K. A., M. Kuroda, H. Chaté, and M. Sano, 2009, Experimental realization of directed percolation criticality in turbulent liquid crystals, *Phys. Rev. E* **80**, 051116-1–12.

Täuber, U. C., 2012, Population oscillations in spatial stochastic Lotka–Volterra models: a field-theoretic perturbational analysis, *J. Phys. A: Math. Theor.* **45**, 405002-1–34.

Täuber, U. C., M. J. Howard, and H. Hinrichsen, 1998, Multicritical behavior in coupled directed percolation processes, *Phys. Rev. Lett.* **80**, 2165–2168.

Täuber, U. C., M. J. Howard, and B. P. Vollmayr-Lee, 2005, Applications of field-theoretic renormalization group methods to reaction-diffusion problems, *J. Phys. A: Math. Gen.* **38**, R79–R131.

van Wijland, F., K. Oerding, and H. J. Hilhorst, 1998, Wilson renormalization of a reaction-diffusion process, *Physica A* **251**, 179–201.

Vernon, D. and M. Howard, 2001, Branching and annihilating Lévy flights, *Phys. Rev. E* **63**, 041116-1–8.

Further reading

Al Hammal, O., J. A. Bonachela, and M. A. Muñoz, 2007, Absorbing state phase transitions with a non-accessible vacuum, *J. Stat. Mech.*, P12007-1–14.

Antonov, N. V., V. I. Iglovikov, and A. S. Kapustin, 2009, Effects of turbulent mixing on the nonequilibrium critical behaviour, *J. Phys. A: Math. Theor.* **42**, 135001-1–19.

Barkema, G. T. and E. Carlon, 2003, Universality in the pair contact process with diffusion, *Phys. Rev. E* **68**, 036113-1–7.

Cafiero, R., A. Gabrielli, and M. A. Muñoz, 1998, Disordered one-dimensional contact process, *Phys. Rev. E* **57**, 5060–5068.

Canet, L. and H. Hilhorst, 2006, Single-site approximation for reaction-diffusion processes, *J. Stat. Phys.* **125**, 517–531.

Canet, L., H. Chaté, and B. Delamotte, 2011, General framework of the non-perturbative renormalization group for non-equilibrium steady states, *J. Phys. A: Math. Theor.* **44**, 495001-1–26.

Ciafaloni, M., M. Le Bellac, and G. C. Rossi, 1977, Reggeon quantum mechanics: a critical discussion, *Nucl. Phys. B* **130**, 388–428.

Ciafaloni, M. and E. Onofri, 1979, Path integral formulation of Reggeon quantum mechanics, *Nucl. Phys. B* **151**, 118–146.

Dickman, R. and A. G. Moreira, 1998, Violation of scaling in the contact process with quenched disorder, *Phys. Rev. E* **57**, 1263–1268.

Grassberger, P., 2013, On the continuum time limit of reaction-diffusion systems, *EPL* **103**, 50009-1–3.

Janssen, H. K., Ü. Kutbay, and K. Oerding, 1999, Equation of state for directed percolation, *J. Phys. A: Math. Gen.* **32**, 1809–1818.

Janssen, H. K., S. Lübeck, and O, Stenull, 2007, Finite-size scaling of directed percolation in the steady state, *Phys. Rev. E* **76**, 041126-1–18.

Janssen, H. K., M. Müller, and O. Stenull, 2004, Generalized epidemic process and tricritical dynamic percolation, *Phys. Rev. E* **70**, 026114-1–20.

Moreira, A. G. and R. Dickman, 1996, Critical dynamics of the contact process with quenched disorder, *Phys. Rev. E* **54**, R3090–R3093.

Ohtsuki, T. and T. Keyes, 1987a, Nonequilibrium critical phenomena in one-component reaction-diffusion systems, *Phys. Rev. A* **35**, 2697–2703.

Ohtsuki, T. and T. Keyes, 1987b, Crossover in nonequilibrium multicritical phenomena of reaction-diffusion systems, *Phys. Rev. A* **36**, 4434–4438.

Sarkar, N. and A. Basu, 2012, Active-to-absorbing-state phase transition in the presence of fluctuating environments: weak and strong dynamic scaling, *Phys. Rev. E* **86**, 021122-1–13.

Schram, R. D. and G. T. Barkema, 2012, Critical exponents of the pair contact process with diffusion, *J. Stat. Mech.*, P03009-1–9.

Schram, R. D. and G. T. Barkema, 2013, Universality of the triplet contact process with diffusion, *J. Stat. Mech.*, P04020-1–7.

Smallenburg, F. and G. T. Barkema, 2008, Universality class of the pair contact process with diffusion, *Phys. Rev. E* **78**, 031129-1–8.

Vojta, T., 2004, Broadening of a nonequilibrium phase transition by extended structural defects, *Phys. Rev. E* **70**, 026108-1–4.

Vojta, T., 2012, Monte Carlo simulations of the clean and disordered contact process in three dimensions, *Phys. Rev. E* **86**, 051137-1–11.

Vojta, T. and M. Dickison, 2005, Critical behavior and Griffiths effects in the disordered contact process, *Phys. Rev. E* **72**, 036126-1–9.

Vojta, T. and M. Y. Lee, 2006, Nonequilibrium phase transition on a randomly diluted lattice, *Phys. Rev. Lett.* **96**, 035701-1–4.

Zhou, Z., J. Yang, R. M. Ziff, and Y. Deng, 2012, Crossover from isotropic to directed percolation, *Phys. Rev. E* **86**, 021102-1–8.

11

Driven diffusive systems and growing interfaces

The emergence of generic scale invariance, i.e., algebraic behavior without tuning to special critical points, appears to be remarkably common in systems that are settled in a non-equilibrium steady state. Prototypical examples are simple non-linear Langevin equations that describe driven diffusive systems and driven interfaces or growth models far from thermal equilibrium, whose distinct phases are characterized by non-trivial RG fixed points and hence universal scaling exponents. We start with driven lattice gases with particle exclusion that are described by generalizations of the one-dimensional noisy Burgers equation for fluid hydrodynamics. Symmetries and conservation laws completely determine the ensuing stationary power laws, as well as the intermediate aging scaling regime and even the large-deviation function for the particle current fluctuations. Next we address the non-equilibrium critical point for driven Ising lattice gases, whose critical exponents can again be computed exactly. We then turn our attention to the prominent Kardar–Parisi–Zhang equation, originally formulated to describe growing crystalline surfaces and the dynamics of driven interfaces, but also closely related to the noisy Burgers equation and even to the equilibrium statistical mechanics of directed lines in disordered environments. After introducing the scaling theory for interface fluctuations, we proceed to a renormalization group analysis at fixed dimension d. For $d > 2$, a non-trivial unstable RG fixed point separates a phase with Gaussian or Edwards–Wilkinson scaling exponents from a strong-coupling rough phase that is inaccessible by perturbative methods. Nevertheless, the critical exponents at this roughening transition can be obtained to all orders in $\epsilon = d - 2$, most conveniently in the directed-polymer representation by means of a stochastic Cole–Hopf transformation. Finally, we discuss the 'conserved KPZ' variant that does not display a similarly rich behavior, but apparently describes most of the experimentally studied growing surfaces.

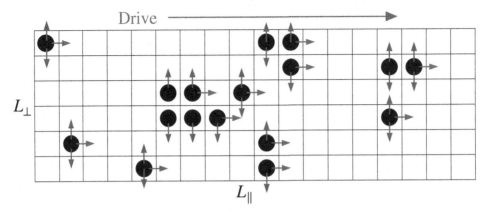

Fig. 11.1 Asymmetric exclusion process (ASEP) in two dimensions. Nearest-neighbor hops are only allowed onto empty lattice sites; particles move preferentially along the drive direction, setting up a stationary current. [Figure courtesy of G. L. Daquila (2011).]

11.1 Driven diffusive systems

Driven diffusive systems constitute perhaps the simplest paradigmatic examples for the emergence of generic scale invariance in non-equilibrium stationary states.[1] After introducing driven lattice gases and their continuum representation in terms of a non-linear stochastic differential equation, we take recourse to the associated Janssen–De Dominicis response functional and the dynamic renormalization group to derive the ensuing universal scaling properties. Remarkably, the fundamental conservation law in conjunction with Galilean invariance completely determine the asymptotic scaling exponents. Next we address the scaling laws in the non-equilibrium relaxation and physical aging regimes. It turns out that all these different scaling properties are governed by a *single* non-trivial exponent.

11.1.1 Driven lattice gases: continuum description

We wish to construct a coarse-grained continuum representation of the *asymmetric exclusion process* (ASEP), which constitutes a random walk of N particles on a d-dimensional lattice with L^d sites, illustrated in Fig. 11.1. Particle hopping is biased along one spatial direction, henceforth indicated as the '∥' direction. In addition, hard-core exclusion constraints are imposed on each lattice site, whence the particle occupation numbers may only assume the values $n_i = 0$ (empty) or 1 (filled); multiple site occupancies are not permitted. We shall consider periodic boundary conditions, and thus ignore any effects caused by the edges, including the intriguing

---

[1] For a comprehensive review, see Schmittmann and Zia (1995); also Marro and Dickman (1999).

boundary-induced phase transitions that have been widely studied for the ASEP.[2] As for the simple non-interacting biased diffusion discussed in Section 2.3.1, the bias induces a non-zero particle current along the direction of the drive. We describe this system in terms of the *conserved* particle density, whose fluctuations (in the comoving reference frame) we denote by $S(x, t) = \rho(x, t) - \langle \rho \rangle$, such that $\langle S \rangle = 0$. For a half-filled lattice with $\rho_0 = \langle \rho \rangle = 1/2$, one may also view $S(x, t) = \sigma(x, t)/2$ as the local magnetization in the equivalent spin description for the Ising lattice gas, see Eq. (1.30).

Let us start with heuristic considerations based on the fundamental symmetries of the problem. The density fluctuations must obey a *continuity equation*

$$\frac{\partial S(x, t)}{\partial t} + \nabla \cdot J(x, t) = 0 . \tag{11.1}$$

In the spatial sector transverse to the drive (with dimension $d_\perp = d - 1$), there is just a noisy *diffusion current*

$$J_\perp(x, t) = -D\nabla_\perp S(x, t) + \eta(x, t) , \tag{11.2}$$

$$\langle \eta_i(x, t) \rangle = 0 , \quad \langle \eta_i(x, t)\eta_j(x', t') \rangle = 2\tilde{D}\, \delta(x - x')\delta(t - t')\delta_{ij} , \tag{11.3}$$

where D and $\tilde{D}$ represent the transverse diffusivity and noise strength. In contrast, the current along the direction of the external drive (with non-zero mean) should be strongly affected by the hard-core particle interactions. For $\langle \rho \rangle = 1/2$, particle–hole symmetry dictates this non-linear contribution to lowest order in the density to be proportional to $\rho(1 - \rho) = \frac{1}{4} - S^2$. Denoting the longitudinal diffusion coefficient by cD, one arrives at

$$J_\parallel(x, t) = J_0 - cD\,\nabla_\parallel S(x, t) - \frac{Dg}{2} S(x, t)^2 + \eta_\parallel(x, t) , \tag{11.4}$$

$$\langle \eta_\parallel(x, t) \rangle = 0 , \quad \langle \eta_\parallel(x, t)\eta_\parallel(x', t') \rangle = 2\tilde{c}D\, \delta(x - x')\delta(t - t') , \tag{11.5}$$

where $\tilde{c}D$ measures the longitudinal noise strength. Transforming to a comoving reference frame, the mean current vanishes, $J_0 = \langle J_\parallel \rangle = 0$.

Since the system is driven away from thermal equilibrium, Einstein's relation need *not* be fulfilled. Yet one may of course rescale the scalar field $S(x, t) \to (D/\tilde{D})^{1/2} S(x, t)$ to satisfy it, say, in the transverse sector, see Section 8.3.1, but not simultaneously in the longitudinal one if $\tilde{c} \neq c$. The ratio

$$w = \tilde{c}/c \tag{11.6}$$

consequently measures the deviation from thermal equilibrium. These considerations yield the generic Langevin equation for the density fluctuations in *driven*

---

[2] For overviews, see, e.g., Derrida (1998); Schütz (2001).

diffusive systems (DDS)[3]

$$\frac{\partial S(x, t)}{\partial t} = D\big(c\,\nabla_\parallel^2 + \nabla_\perp^2\big) S(x, t) + \frac{Dg}{2}\,\nabla_\parallel S(x, t)^2 + \zeta(x, t)\,, \qquad (11.7)$$

with *conserved* noise

$$\zeta(x, t) = -\nabla_\parallel \eta_\parallel(x, t) - \nabla_\perp \cdot \eta(x, t)\,, \qquad (11.8)$$

satisfying $\langle \zeta \rangle = 0$ and, since $\langle \eta_\parallel(x, t)\eta_i(x', t') \rangle = 0$,

$$\langle \zeta(x, t)\zeta(x', t') \rangle = -2D\big(\tilde{c}\,\nabla_\parallel^2 + \nabla_\perp^2\big)\delta(x - x')\delta(t - t')\,. \qquad (11.9)$$

Notice that the drive term $\propto g$ breaks both the system's spatial inversion $x \to -x$ *and* the Ising Z_2 symmetry $S \to -S$.

A more formal, microscopic derivation utilizes the Fock space representation of the associated master equation, detailed in Section 9.2. Consider just the direction along the drive; the pseudo-Hamiltonian (9.37) for unbiased hopping is then readily generalized to

$$H_{\text{dif}} = \sum_i \left[\left(D + \frac{E}{2}\right)\big(a_i^\dagger - a_{i+1}^\dagger\big)a_i + \left(D - \frac{E}{2}\right)\big(a_i^\dagger - a_{i+1}^\dagger\big)a_i\right]\,, \qquad (11.10)$$

where we assume nearest-neighbor hopping with bias E along a one-dimensional chain. Next we take the continuum limit, retaining terms up to second order in the gradient expansion. Absorbing appropriate lattice constant factors into the rates, we arrive at the pseudo-Hamiltonian density

$$\mathcal{H}_{\text{dif}}(\hat{\psi}, \psi) = -\big[E\big(\nabla_\parallel\hat{\psi}\big) + D\big(\nabla_\parallel^2\hat{\psi}\big)\big]\psi\,f(\hat{\psi}\psi)\,. \qquad (11.11)$$

Here we have multiplied the drift and diffusion terms by a function of the density $\rho = \hat{\psi}\psi$ that phenomenologically accounts for the particle exclusion. The crucial step is to apply the Cole–Hopf transformation (9.64), which yields the Doi–Peliti action in terms of the density fields

$$\mathcal{A}[\tilde{\rho}, \rho] = \int dx_\parallel \int dt \left(\tilde{\rho}\frac{\partial\rho}{\partial t} - \big[E\,\nabla_\parallel\tilde{\rho} + D\,\nabla_\parallel^2\tilde{\rho} + D(\nabla_\parallel\tilde{\rho})^2\big]\rho f(\rho)\right)\,. \qquad (11.12)$$

Finally, we consider small density fluctuations $u(x, t) = \rho(x, t) - \rho_0$, and expand $\rho f(\rho) \approx \tilde{c} + c\,u(x, t) - \frac{Dg}{2}\,u(x, t)^2$, where $\tilde{c} = \rho_0 f(\rho_0)$. With $\tilde{u}(x, t) = \tilde{\rho}(x, t)$, this gives to leading order, and aside from boundary contributions:

$$\mathcal{A}[\tilde{u}, u] = \int dx_\parallel \int dt\,\tilde{u}\left(\frac{\partial u}{\partial t} + cE\,\nabla_\parallel u - \frac{Dg}{2}\,\nabla_\parallel u^2 - cD\,\nabla_\parallel^2 u + \tilde{c}D\,\nabla_\parallel^2\tilde{u}\right)\,, \qquad (11.13)$$

[3] Janssen and Schmittmann (1986a).

which in turn is equivalent to the Langevin equation

$$\frac{\partial u(x,t)}{\partial t} = -cE\,\nabla_{\|}u(x,t) + \frac{Dg}{2}\,\nabla_{\|}u(x,t)^2 + cD\,\nabla_{\|}^2 u(x,t) + \zeta(x,t)$$

with drift velocity cE, longitudinal diffusion coefficient cD, and noise correlation $\langle\zeta(x,t)\zeta(x',t')\rangle = -2\tilde{c}D\nabla_{\|}^2\,\delta(x-x')\delta(t-t')$. A Galilean transformation to the comoving reference frame $S(x,t) = u(x+cEt,t)$ removes the drift term $\propto \nabla_{\|}u$, and supplementing the mere diffusive transverse components at last yields Eqs. (11.7) and (11.9).

The corresponding Janssen–De Dominicis response functional (4.10) reads

$$\mathcal{A}[\tilde{S},S] = \int d^d x \int dt\,\tilde{S}(x,t)\left[\frac{\partial S(x,t)}{\partial t} - D(c\,\nabla_{\|}^2 + \nabla_{\perp}^2)S(x,t)\right.$$

$$\left. + D(\tilde{c}\,\nabla_{\|}^2 + \nabla_{\perp}^2)\tilde{S}(x,t) - \frac{Dg}{2}\,\nabla_{\|}S(x,t)^2\right], \tag{11.14}$$

compare with (11.13) for the longitudinal sector. It describes a *'massless'* field theory, whence we expect the system to be *generically scale-invariant*, without the need to tune it to a special point in parameter space. The non-linear drive term will induce anomalous scaling in the longitudinal (bias) direction, markedly distinct from ordinary diffusive behavior. The mesoscopic Langevin equation (11.7) and the equivalent action (11.14) display an *emergent* symmetry that is not explicitly manifest in the microscopic model, specifically, they are left invariant under (generalized) *Galilean transformations*

$$x_{\|} \to x_{\|}' = x_{\|} - Dgvt\,,\quad x_{\perp} \to x_{\perp}' = x_{\perp}\,,\quad t \to t' = t\,,$$

$$S(x_{\|},x_{\perp},t) \to S'(x_{\|}',x_{\perp}',t') = S(x_{\|} - Dgvt, x_{\perp},t) - v\,,$$

$$\tilde{S}(x_{\|},x_{\perp},t) \to \tilde{S}'(x_{\|}',x_{\perp}',t') = \tilde{S}(x_{\|} - Dgvt, x_{\perp},t)\,. \tag{11.15}$$

Since the arbitrary boost parameter v should not acquire any anomalous dimension under scale transformations, and the field S needs to scale as v to preserve the symmetry (11.15), one already deduces the absence of field renormalization for the response functional (11.14). This is confirmed by the explicit perturbative analysis in the following subsection.

The harmonic part of the action (11.14) yields the *anisotropic propagator*

$$G_0(q_{\|},q_{\perp},\omega) = \frac{1}{-i\omega + D(c\,q_{\|}^2 + q_{\perp}^2)}\,, \tag{11.16}$$

and an also anisotropic noise vertex $2D(\tilde{c}\,q_{\|}^2 + q_{\perp}^2)$, see Fig. 11.2(a). The Gaussian correlation function becomes

$$C_0(q_{\|},q_{\perp},\omega) = \frac{2D(\tilde{c}\,q_{\|}^2 + q_{\perp}^2)}{\omega^2 + D^2(c\,q_{\|}^2 + q_{\perp}^2)^2}\,, \tag{11.17}$$

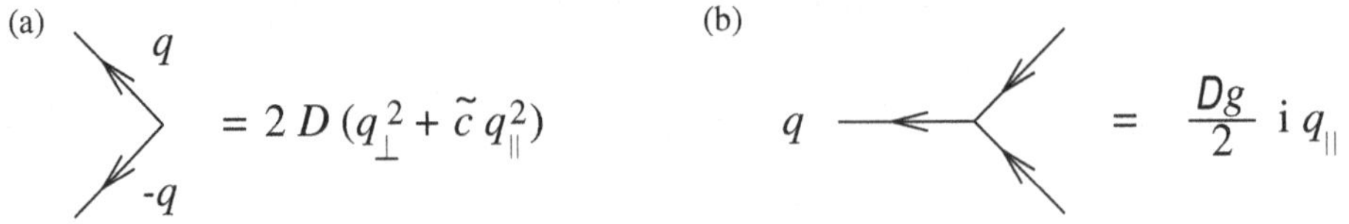

(a)

(b)

$$= 2D\,(q_\perp^2 + \tilde{c}\,q_\parallel^2)$$

$$q \longleftarrow \quad = \quad \frac{Dg}{2}\,i\,q_\parallel$$

Fig. 11.2 Vertices for driven diffusive systems (DDS): (a) noise source vertex; and (b) three-point vertex induced by the drive.

whence one obtains for the static structure factor

$$S_c(q_\parallel, q_\perp) = C_0(q_\parallel, q_\perp, t = 0) = \int \frac{d\omega}{2\pi}\, C_0(q_\parallel, q_\perp, \omega) = \frac{\tilde{c}\,q_\parallel^2 + q_\perp^2}{c\,q_\parallel^2 + q_\perp^2}\,. \qquad (11.18)$$

Owing to the different anisotropies in the noise and diffusion terms for $\tilde{c} \neq c$, the structure factor displays a discontinuity singularity at the origin of momentum space. Indeed, letting $q_\parallel = q\cos\theta$ and $q_\perp = q\sin\theta$, one finds

$$\lim_{q\to 0} S_c(q, \theta) = \frac{\tilde{c} + (\tan\theta)^2}{c + (\tan\theta)^2}\,, \qquad (11.19)$$

a non-trivial function of the approach angle θ. Specifically, consider

$$\frac{\lim_{q_\parallel\to 0} S_c(q_\parallel, q_\perp = 0)}{\lim_{q_\perp\to 0} S_c(q_\parallel = 0, q_\perp)} = \frac{\lim_{q\to 0} S_c(q, \theta = 0)}{\lim_{q\to 0} S_c(q, \theta = \pi/2)} = w\,, \qquad (11.20)$$

which demonstrates that the singularity disappears in thermal equilibrium ($w = 1$).

11.1.2 Renormalization and scaling exponents

Let us now proceed with the renormalization group analysis of the fluctuation corrections to the Gaussian model. The one-loop Feynman graphs contributing to the two-point vertex functions $\Gamma^{(1,1)}(q_\parallel, q_\perp, \omega)$ (propagator self-energy) and $\Gamma^{(2,0)}(q_\parallel, q_\perp, \omega)$ (noise vertex) are depicted in Fig. 11.3. Since the non-linear vertex induced by the drive is proportional to $iq_\parallel$, see Fig. 11.2(b), there are *no* loop contributions at all in the transverse sector, and we have to *all* orders in the perturbation expansion

$$\Gamma^{(1,1)}(q_\parallel = 0, q_\perp, \omega) = i\omega + Dq_\perp^2\,, \qquad (11.21)$$

$$\Gamma^{(2,0)}(q_\parallel = 0, q_\perp, \omega) = -2Dq_\perp^2\,. \qquad (11.22)$$

Consequently, there is no renormalization of the fields and the transverse diffusion coefficient,

$$Z_{\tilde{s}} = Z_S = Z_D = 1\,. \qquad (11.23)$$

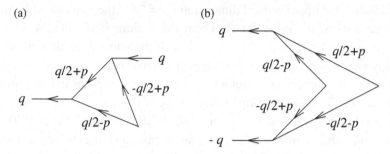

Fig. 11.3 One-loop Feynman diagrams for the DDS two-point vertex functions: (a) propagator self-energy $\Gamma^{(1,1)}(q_\|, q_\perp, \omega)$; and (b) noise strength $\Gamma^{(2,0)}(q_\|, q_\perp, \omega)$.

We define the scaling exponents for driven diffusive systems with similar conventions as for the randomly driven or two-temperature model B at its critical point, see Section 8.3.2. The scaling form for the Gaussian correlation function (11.17) is thus generalized to

$$C(q_\|, q_\perp, \omega) = |q_\perp|^{-2+\eta}\, \hat{C}\left(\frac{\sqrt{c}\,q_\|}{|q_\perp|^{1+\Delta}}, \frac{\omega}{D\,|q_\perp|^z}\right) \tag{11.24}$$

with anisotropy exponent Δ, compare with Eq. (8.89) for $\tau = 0$; note that here, however, $\Delta = 0$ in the mean-field approximation. Equations (11.23) then immediately yield the Gaussian values for the transverse scaling exponents

$$\eta = 0\,, \quad z = 2\,. \tag{11.25}$$

This leaves Δ as the sole non-trivial scaling exponent for DDS. If we write the last argument in (11.24) as $\omega/D|q_\||^{z_\|}$, which applies especially to one dimension, we identify the longitudinal dynamic exponent as $z_\| = 2/(1 + \Delta)$. The spatio-temporal Fourier transform of (11.24) yields with (11.25):

$$C(x_\|, x_\perp, t) = |x_\perp|^{-\lambda_C}\, \bar{C}\left(\frac{x_\|}{\sqrt{c}\,|x_\perp|^{1+\Delta}}, \frac{Dt}{|x_\perp|^2}\right) \tag{11.26}$$

$$= t^{-\zeta}\, \tilde{C}\left(\frac{x_\|}{\sqrt{c}\,|x_\perp|^{1+\Delta}}, \frac{Dt}{|x_\||^{z_\|}}\right), \tag{11.27}$$

to list two equivalent anisotropic scaling forms, where we have introduced

$$\lambda_C = d + \Delta\,, \quad \zeta = \frac{\lambda_C}{z} = \frac{d+\Delta}{2}\,, \tag{11.28}$$

compare with Eq. (8.31). Based on (5.3) with $a = 0$ and $[q_\perp] = \mu = [q_\|]$, power counting gives the scaling dimensions

$$[\tilde{S}(x, t)] = \mu^{d/2} = [S(x, t)]\,, \quad [\tilde{c}] = \mu^0 = [c]\,, \quad [g] = \mu^{1-d/2}\,, \tag{11.29}$$

which indicates the upper critical dimension $d_c = 2$ for the non-linear fluctuations to generate a non-zero value for Δ. From the scaling forms (11.24) or (11.26) one finally infers the relation $\Delta = -\gamma_c^*/2$ with the anomalous dimension of the anisotropy parameter c at an infrared-stable RG fixed point.

Moreover, the non-linear coupling g itself does not renormalize either as a consequence of *Galilean invariance*. According to (11.15), under renormalization the boost velocity v must scale as the field S, and since the product Dgv that constitutes the parameter for the Galilean transformation must also remain invariant under scale transformations, this leaves us with

$$Z_g^{1/2} = Z_D^{-1} Z_S^{-1/2} = 1 \,. \tag{11.30}$$

Formally, one may derive a *Ward identity* associated with Galilean invariance precisely as in Section 6.4.2, and thus arrive at the vertex function relation

$$- iq_{\parallel} Dg \, \frac{\partial \Gamma^{(1,1)}(q_{\parallel}, q_{\perp}, \omega)}{\partial(i\omega)} = \Gamma^{(1,2)}(-q_{\parallel}, -q_{\perp}, -\omega; q_{\parallel}, q_{\perp}, \omega; 0, 0, 0) \,,$$

$$\tag{11.31}$$

in direct correspondence with Eq. (6.166). Since this Ward identity must also hold for the renormalized vertex functions, (11.30) follows.

As we shall see shortly, the effective non-linear coupling governing the perturbation expansion in terms of loop diagrams turns out to be

$$v = g^2 c^{-3/2} \,, \quad [v] = \mu^{2-d} \,; \tag{11.32}$$

if we define its renormalized counterpart as

$$v_R = Z_c^{-3/2} v \, C_d \mu^{d-2} \,, \quad C_d = \frac{\Gamma(2 - d/2)}{2^{d-1}\pi^{d/2}} \,, \tag{11.33}$$

the associated renormalization group beta function becomes

$$\beta_v = \mu \left. \frac{\partial}{\partial \mu} \right|_0 v_R = v_R \left(d - 2 - \frac{3}{2}\gamma_c \right) \,. \tag{11.34}$$

At *any* non-trivial RG fixed point $0 < v^* < \infty$, therefore $\gamma_c^* = 2(d-2)/3$. We thus infer that while $\Delta = 0$, $z_{\parallel} = 2$, and $\zeta = d/2$ for $d > 2$, below the upper critical dimension $d_c = 2$ for DDS, the longitudinal scaling exponents are completely fixed by the conservation law and Galilean invariance:

$$\Delta = -\frac{\gamma_c^*}{2} = \frac{2-d}{3} \,, \quad z_{\parallel} = \frac{2}{1+\Delta} = \frac{6}{5-d} \,, \quad \zeta = \frac{\lambda_C}{2} = \frac{d+1}{3} \,. \tag{11.35}$$

In order to establish the existence of an infrared-stable RG fixed point, we require the explicit renormalization constants for the parameters c and $\tilde{c}$, as usual defined via $c_R = Z_c\, c$ and $\tilde{c}_R = Z_{\tilde{c}}\, \tilde{c}$. The Z factors follow from the fluctuation contributions to the two-point vertex functions $\Gamma^{(1,1)}(q_\parallel, q_\perp = 0, \omega)$ and $\Gamma^{(2,0)}(q_\parallel, q_\perp = 0, \omega)$, shown to one-loop order in Fig. 11.3. After performing the internal frequency integrals, the one-loop analytic expressions for these two-point vertex functions read

$$\Gamma^{(1,1)}(q_\parallel, q_\perp = 0, \omega) = i\omega + cD\,q_\parallel^2$$

$$+ \frac{Dg^2}{2} \int_k \frac{\tilde{c}\left(\frac{q_\parallel}{2} + k_\parallel\right)^2 + k_\perp^2}{c\left(\frac{q_\parallel}{2} + k_\parallel\right)^2 + k_\perp^2} \frac{q_\parallel\left(\frac{q_\parallel}{2} - k_\parallel\right)}{i\omega/2D + c\left(\frac{q_\parallel^2}{4} + k_\parallel^2\right) + k_\perp^2}, \quad (11.36)$$

$$\Gamma^{(2,0)}(q_\parallel, q_\perp = 0, \omega) = 2\tilde{c}D\,q_\parallel^2 + \frac{Dg^2}{2} q_\parallel^2$$

$$\times \int_k \frac{\tilde{c}\left(\frac{q_\parallel}{2} + k_\parallel\right)^2 + k_\perp^2}{c\left(\frac{q_\parallel}{2} + k_\parallel\right)^2 + k_\perp^2} \frac{\tilde{c}\left(\frac{q_\parallel}{2} - k_\parallel\right)^2 + k_\perp^2}{c\left(\frac{q_\parallel}{2} - k_\parallel\right)^2 + k_\perp^2} \operatorname{Re}\frac{1}{i\omega/2D + c\left(\frac{q_\parallel^2}{4} + k_\parallel^2\right) + k_\perp^2}.$$

$$(11.37)$$

As in Section 9.3.1, we choose the normalization point at finite frequency $i\omega = 2D\mu^2$ to avoid infrared singularities. Expanding (11.36) to $O(q_\parallel^2)$ and setting $q_\parallel = 0$ in the integrand of (11.37) yields

$$Z_c = \frac{g^2}{4} \int_k \frac{\tilde{c}\,k_\parallel^2 + k_\perp^2}{c\,k_\parallel^2 + k_\perp^2} \frac{1}{\mu^2 + c\,k_\parallel^2 + k_\perp^2} - \frac{g^2}{2} \int_k \frac{1}{c\,k_\parallel^2 + k_\perp^2} \frac{\tilde{c}\,k_\parallel^2}{\mu^2 + c\,k_\parallel^2 + k_\perp^2}$$

$$+ \frac{g^2}{2} \int_k \frac{\tilde{c}\,k_\parallel^2 + k_\perp^2}{\left(c\,k_\parallel^2 + k_\perp^2\right)^2} \frac{c\,k_\parallel^2}{\mu^2 + c\,k_\parallel^2 + k_\perp^2},$$

$$Z_{\tilde{c}} = 1 + \frac{g^2}{4\tilde{c}} \int_k \frac{\left(\tilde{c}\,k_\parallel^2 + k_\perp^2\right)^2}{c\,k_\parallel^2 + k_\perp^2} \frac{1}{\mu^2 + c\,k_\parallel^2 + k_\perp^2}.$$

Upon rescaling the internal longitudinal wavenumbers $\sqrt{c}\,k_\parallel \to k_\parallel$, with (11.6) and (11.32) one finally arrives at

$$Z_c = 1 + \frac{v}{4}\left[\int_k \frac{1}{\mu^2 + k^2} - (w-1)\int_k \frac{k_\parallel^2}{k^2(\mu^2 + k^2)} + 2(w-1)\right.$$

$$\left.\times \int_k \frac{k_\parallel^4}{k^4(\mu^2 + k^2)}\right] = 1 + \frac{v}{4}\left[1 - \frac{w-1}{d} + \frac{6(w-1)}{d(d+2)}\right]\int_k \frac{1}{\mu^2 + k^2}$$

$$\to 1 + \frac{v}{16}(3+w)\,\frac{C_d\mu^{d-2}}{2-d}, \quad (11.38)$$

$$Z_{\tilde{c}} = 1 + \frac{v}{4w} \left[\int_k \frac{1}{\mu^2 + k^2} + 2(w-1) \int_k \frac{k_\parallel^2}{k^2(\mu^2 + k^2)} + (w-1)^2 \right.$$

$$\times \left. \int_k \frac{k_\parallel^4}{k^4(\mu^2 + k^2)} \right] = 1 + \frac{v}{4w} \left[1 + \frac{2(w-1)}{d} + \frac{3(w-1)^2}{d(d+2)} \right] \int_k \frac{1}{\mu^2 + k^2}$$

$$\to 1 + \frac{v}{32} \left(\frac{3}{w} + 2 + 3w \right) \frac{C_d \mu^{d-2}}{2-d}, \tag{11.39}$$

where in the final steps we have replaced d with $d_c = 2$, following the minimal subtraction scheme.

Thus one obtains Wilson's flow functions

$$\gamma_c = \mu \left. \frac{\partial}{\partial \mu} \right|_0 \ln Z_c = -\frac{v_R}{16}(3 + w_R), \tag{11.40}$$

$$\gamma_{\tilde{c}} = \mu \left. \frac{\partial}{\partial \mu} \right|_0 \ln Z_{\tilde{c}} = -\frac{v_R}{32} \left(\frac{3}{w_R} + 2 + 3w_R \right), \tag{11.41}$$

whence

$$\beta_w = \mu \left. \frac{\partial}{\partial \mu} \right|_0 w_R = w_R(\gamma_{\tilde{c}} - \gamma_c) = -\frac{v_R}{32}(w_R - 1)(w_R - 3). \tag{11.42}$$

Provided $0 < v^* < \infty$, the *equilibrium* fixed point

$$w_{eq}^* = 1 \tag{11.43}$$

is infrared-stable, i.e., Einstein's relation is asymptotically recovered in the longitudinal sector as well. For $d < 2$, the DDS fixed point theory with

$$v_{eq}^* = \frac{8}{3}(2 - d) \tag{11.44}$$

and scaling exponents (11.35) hence represents in fact an equilibrium system.

In this context, it is instructive to make an intriguing connection with the *noisy Burgers equation*, describing simplified fluid hydrodynamics in terms of a vorticity-free velocity field $u(x, t)$:[4]

$$\frac{\partial u(x, t)}{\partial t} + \frac{Dg}{2} \nabla[u(x, t)^2] = D\nabla^2 u(x, t) + \zeta(x, t), \tag{11.45}$$

with the conserved thermal noise ($k_B T = 1$)

$$\langle \zeta_i(x, t) \rangle = 0, \quad \langle \zeta_i(x, t)\zeta_j(x', t') \rangle = -2D\nabla_i\nabla_j\delta(x - x')\delta(t - t'). \tag{11.46}$$

---

[4] Forster, Nelson, and Stephen (1977).

For $Dg = 1$, the non-linearity is just the usual fluid advection term. Naturally, Eq. (11.45) is invariant under a Galilean transformation

$$x \rightarrow x' = x - Dgvt , \quad t \rightarrow t' = t ,$$

$$u(x, t) \rightarrow u'(x', t') = u(x - Dgvt, t) + v . \tag{11.47}$$

Indeed, $\partial u'/\partial t = \partial u/\partial t - Dgv \cdot \nabla u = D\nabla^2 u - Dg\nabla u^2/2 - Dgv \cdot \nabla u + \zeta$, and because of the integrability condition $\partial_i u_j = \partial_j u_i$, the two terms $\propto g$ can be combined to $\partial u'/\partial t = D\nabla^2 u - Dg\nabla(u + v)^2/2 + \zeta = D\nabla^2 u' - Dg\nabla u'^2/2 + \zeta$. Neither is the noise correlator (11.46) modified by this transformation.

In one dimension, with $S(x, t) = -u(x, t)$ (or $g \rightarrow -g$) the Burgers equation (11.45) becomes *identical* with the DDS Langevin equation (11.7) at the equilibrium RG fixed point (11.43). Equation (11.35) consequently yields the anomalous dynamic scaling exponent $z_\| = 3/2$ for noisy Burgers kinetics. For $d = 1$ at least, the Langevin equation (11.45) with noise correlator (11.46) represents *equilibrium* stochastic dynamics, and the probability distribution for the velocity field $u(x, t)$ should asymptotically approach the canonical distribution with the fluid's kinetic energy serving as the Hamiltonian. By construction, Einstein's relation is satisfied; but we still need to check the equilibrium condition (3.58) for the reversible force term $F_{\text{rev}}[u] \propto \nabla u^2$ with $P_{\text{st}}[u] \propto \exp\left[-\frac{1}{2}\int u(x)^2 \, d^d x\right]$:

$$\int d^d x \, \frac{\delta}{\delta u(x, t)} \cdot [\nabla u(x, t)^2] e^{-\frac{1}{2}\int u(x',t)^2 \, d^d x'}$$

$$= \int [2\nabla \cdot u(x, t) - u(x, t) \cdot \nabla u(x, t)^2] d^d x \, e^{-\frac{1}{2}\int u(x',t)^2 \, d^d x'} .$$

With appropriate boundary conditions, the first term here vanishes, but the second one does so *only* for $d = 1$: $-\int u (du^2/dx) \, dx = \int u^2 (du/dx) \, dx = \frac{1}{3}\int (du^3/dx) \, dx = 0$. Driven diffusive systems in one dimension are therefore subject to a *'hidden'* fluctuation-dissipation theorem.

The predictions (11.35) and (11.28) for the DDS steady-state scaling exponents may be tested in computer simulations. To this end, based on Eq. (11.26), we postulate the finite-size scaling form

$$C(x_\|, x_\perp, t, L_\|, L_\perp) = L_\perp^{-\lambda_C} \, \bar{C}\left(\frac{x_\|}{L_\|}, \frac{x_\perp}{L_\perp}, \frac{Dt}{L_\perp^2}, \frac{L_\|}{\sqrt{c} \, L_\perp^{1+\Delta}}\right)$$

$$= L_\|^{-\lambda_C/(1+\Delta)} \, \tilde{C}\left(\frac{x_\|}{L_\|}, \frac{x_\perp}{L_\perp}, \frac{Dt}{L_\|^{z_\|}}, \frac{L_\|}{\sqrt{c} \, L_\perp^{1+\Delta}}\right) \tag{11.48}$$

for a system with dimension $L_\| \times L_\perp^{d-1}$. The last argument indicates that the scaling functions will in general depend on the aspect ratio $L_\|/L_\perp^{1+\Delta}$. In $d \geq 2$ dimensions, $\Delta = 0$ and the system is governed by mean-field scaling exponents,

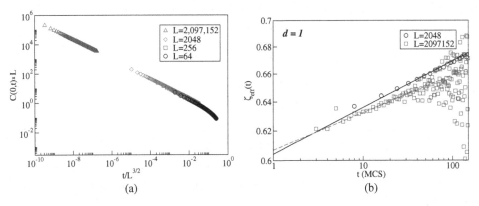

(a)   (b)

Fig. 11.4 Monte Carlo simulation results for the TASEP density autocorrelation function $C(0, t)$ in one dimension (time measured in Monte Carlo steps, MCS). (a) Finite-size scaling collapse for simulation data with (right to left) $L = 64, 256, 2048$, and 2097152 sites (averaged, respectively, over 80000, 10000, 40000, and 60 runs). (b) Time dependence of the local autocorrelation exponent $\zeta_{\text{eff}}(t)$, shown for different system sizes: $L = 2048$ (black circles, data averaged over 40000 realizations) and $L = 2097152$ (grey squares, 60 realizations). [Figures reproduced with permission from: G. L. Daquila and U. C. Täuber, *Phys. Rev. E* **83**, 051107 (2011); DOI: 10.1103/PhysRevE.83.051107; copyright (2011) by the American Physical Society.]

with logarithmic corrections at $d_c = 2$. In one dimension, according to (11.48) one finds for the density autocorrelation function at $x = 0$:

$$C(0, t, L) = L^{-1}\,\tilde{C}(Dt/L^{3/2})\,. \tag{11.49}$$

This scaling form is confirmed by Monte Carlo simulations: Fig. 11.4(a) shows convincing data collapse of $C(0, t, L)L$ plotted vs. $t/L^{3/2}$ for four vastly different system sizes, obtained for the totally asymmetric exclusion process (TASEP) in one dimension, where particle hops against the drive are prohibited (often referred to as DDS with 'infinite drive'), at half filling $\rho_0 = 1/2$.

The same numerical data also yield the asymptotic temporal decay of the density autocorrelation function $C(0, t) \sim t^{-\zeta}$ with $\zeta = 1/z_{\parallel} = 2/3$ for $d = 1$, see Eq. (11.35), whereas $\zeta = d/2$ for $d > 2$; at the critical dimension $d_c = 2$, one finds $C(0, t) \sim 1/t\,(\ln t)^{1/3}$ (Problem 11.1). However, a closer look at the *effective* ('local') autocorrelation exponent

$$\zeta_{\text{eff}}(t) = -\frac{d \ln C(0, t)}{d \ln t} \tag{11.50}$$

in the one-dimensional Monte Carlo simulations reveals that $\zeta_{\text{eff}}(t)$ initially approaches its asymptotic value from below, but in finite systems eventually *exceeds* it, as shown in Fig. 11.4(b). Better convergence to $\zeta = 2/3$ is observed only in very large TASEP chains, but the associated scaling function is manifestly governed by an exceedingly slow crossover. Since similarly unusual behavior is not observed in higher dimensions, it is clearly another consequence of the strong *temporal* correlations induced by the interplay of non-equilibrium drive and particle exclusion interactions that lead to anomalous features in low dimensions $d \leq d_c = 2$. The spatial correlations are in fact less interesting; for example in one dimension, the non-equilibrium stationary state is characterized by a *uniform* particle distribution, and hence vanishing correlations for $x \neq 0$.

11.1.3 Universal scaling in the physical aging regime

An alternative and often more effective means to numerically determine dynamic scaling exponents is to consider the universal aging and initial-slip regimes at intermediate and short times, respectively, rather than the scaling properties in the non-equilibrium stationary state. Let us therefore consider the two-time density autocorrelation function $C(x = 0, t, s)$ in the non-stationary regime where time translation invariance is still broken by the initial conditions. Note that in this chapter we denote the smaller 'waiting time' as $s < t$. The obvious generalization of the scaling ansatz (11.27) to two independent time arguments reads

$$C(0, t, s) = s^{-\zeta} \, \hat{C}(t/s) \,, \tag{11.51}$$

which in the literature is referred to as *simple aging scaling* form.[5] In order to access the aging time window, one must initially prepare the system in a configuration that differs drastically from its non-equilibrium steady state. A convenient choice in lattice simulations at half filling is to alternatingly fill the sites and leave them empty. The Monte Carlo simulation results for a one-dimensional TASEP shown in Fig. 11.5(a) display excellent data collapse with the simple aging scaling form (11.51) and the expected autocorrelation exponent $\zeta = 2/3$.

Additional information is contained in the initial-slip regime at very early times $s \ll t$ that independently probes the anomalous scaling dimension γ_c^*. Following the analysis for the relaxational models A and B in Section 8.1, one first realizes that as a consequence of the wavenumber dependence of the non-linear DDS vertex $\propto iq_{\parallel}$, there emerges no new singularity on the initial-time sheet, hence $Z_0 = 1$ and $\gamma_0 = 0$ to all orders in the perturbation expansion, just as for model B with conserved order parameter. Second, in direct analogy with (8.18), within correlation functions we may identify

$$\left. \frac{\partial S(x, t)}{\partial t} \right|_{t=0} \simeq 2cD \, \tilde{S}(x, 0) \,, \tag{11.52}$$

[5] Henkel and Pleimling (2010).

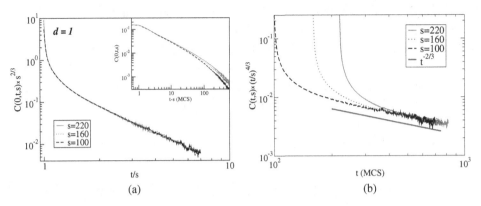

Fig. 11.5 Monte Carlo simulation results for the decay of the two-time density autocorrelation function $C(x = 0, t, s)$ in a one-dimensional driven lattice gas (TASEP) with 1000 lattice sites at half filling $\rho_0 = 1/2$. The particles were initally placed at alternating sites, and measurements taken at waiting times $s = 100$ (dashed), 160 (dotted), and 220 MCS (full line); the data were averaged over 60 000 simulation runs. (a) Simple aging data collapse according to Eq. (11.51) with autocorrelation exponent $\zeta = 2/3$; the inset shows the same data plotted against $t - s$, demonstrating breaking of time translation invariance; (b) initial-slip scaling plot, confirming Eq. (11.53) with $1 - \theta = 4/3$. [Figures reproduced with permission from: G. L. Daquila and U. C. Täuber, *Phys. Rev. E* **83**, 051107 (2011); DOI: 10.1103/PhysRevE.83.051107; copyright (2011) by the American Physical Society.]

which enters the short-time expansion (8.37). Since non-trivial scaling can only emerge for $d \leq d_c = 2$, we focus on the longitudinal sector here, in the limit $s \to 0$. The initial field $S_0(x, s)$ thus picks up the anomalous dimension $\theta = \gamma_c^*/2$ from the longitudinal diffusivity cD. For $d > d_c = 2$, $\theta = 0$, but non-trivial initial-slip scaling ensues for $d < 2$, compare with Eq. (8.5):

$$C(0, t, s/t \to 0) \sim t^{-\zeta} \left(\frac{s}{t}\right)^{1-\theta} , \qquad 1 - \theta = \frac{5 - d}{3} \qquad (11.53)$$

(for the logarithmic corrections at $d_c = 2$, see Problem 11.1). In one dimension, therefore[6] $C(0, t, s/t \to 0) = t^{-2/3}(s/t)^{4/3}$, as is indeed nicely confirmed by TASEP Monte Carlo simulations, evidenced by the data in Fig. 11.5(b).

11.2 Critical dynamics of driven Ising lattice gases

This section addresses the continuous non-equilibrium phase transition in driven lattice gases with both on-site exclusion and attractive nearest-neighbor Ising interactions. We construct a coarse-grained Langevin representation for the microscopic Katz–Lebowitz–Spohn model, and formulate its scaling theory akin to

---

[6] This result was first derived by Krech (1997) for the (in one dimension) equivalent Kardar–Parisi–Zhang equation, see Section 11.3.2.

the two-temperature model B. Once again, the conservation law in conjunction with Galilean invariance allow an *exact* determination of the associated critical exponents, both for the steady-state and critical aging regimes. In addition, our field-theoretic tools allow us to access the large-deviation function of the particle current for both non-critical and critical driven diffusive systems; more precisely, to analyze the scaling properties of the generating function for the current cumulants.

11.2.1 Katz–Lebowitz–Spohn model: Langevin equation

Even richer scaling behavior ensues if in addition to the mutual exclusion, the particles in a driven lattice gas (see Fig. 11.1) are subject to attractive nearest-neighbor Ising interactions. The dynamics with conserved total particle number or net magnetization (which is zero at precisely half-filling) is generated through Kawasaki exchange processes at a fixed temperature T, cf. Eq. (2.126) with (2.128). The drive is then implemented through an appropriate directed bias in the spin exchanges or particle hops. In this model originally designed by Katz, Lebowitz, and Spohn (KLS), the Ising interactions induce a genuine non-equilibrium second-order phase transition in dimensions $d \geq 2$, where at sufficiently low temperatures phase separation into ultimately two domains of respectively filled and empty sites takes place, which are always oriented *parallel* to the drive.[7] At zero bias, one of course recovers the equilibrium Ising phase transition. With the drive, the critical temperature is raised relative to the equilibrium situation; e.g., in two dimensions and with maximum bias ('infinite drive'), $T_c^\infty = 1.41\, T_c^{eq}$.

In order to describe the phase transition in the KLS model through a continuum Langevin equation, we add the drive term $\propto \nabla_\parallel S(x, t)^2$ to model B critical dynamics for a conserved scalar field. Naturally, we must allow for anisotropic scaling, especially since only the $d_\perp$-dimensional *transverse* spatial sector becomes critical. Far away from the critical point in the high-temperature phase ($r \propto T - T_c^0 > 0$), this recovers the stochastic differential equation (11.7), with $q_\perp^2 \to r q_\perp^2$ in the propagator, and hence the static structure factor

$$S_c(r, q_\parallel, q_\perp) = \frac{\tilde{c}\, q_\parallel^2 + q_\perp^2}{c\, q_\parallel^2 + r\, q_\perp^2}.\tag{11.54}$$

This implies an inevitable singularity at the origin of wavevector space as the phase transition is approached. Indeed, the ratio (11.20) now assumes the temperature-dependent value $rw \to 0$ as $T \to T_c^0$.

---

[7] Katz, Lebowitz, and Spohn (1983, 1984); for more recent reviews, see Schmittmann and Zia (1995); Marro and Dickman (1999).

In the critical region, and generalizing to $d_\| = d - d_\perp \geq 1$ longitudinal directions subject to a hopping bias, we may subsequently proceed precisely as for the two-temperature or randomly driven model B introduced in Section 8.3.2. Applying the anisotropic scaling (8.80), which corresponds to $\Delta = 1$ in the mean-field approximation, and (8.81), we are then left with the scalar anisotropic model B dynamics (8.83), augmented with the DDS non-linearity. As coarse-grained representation of the KLS model we thus obtain the *driven model B* or *critical DDS* Langevin equation[8]

$$\frac{\partial S(x,t)}{\partial t} = D\big[c\nabla_\|^2 + \nabla_\perp^2(r - \nabla_\perp^2)\big]S(x,t) + \frac{Dg}{2}\nabla_\| S(x,t)^2$$

$$+ \frac{D\tilde{u}}{6}\nabla_\perp^2 S(x,t)^3 + \zeta(x,t), \tag{11.55}$$

with the (scalar) conserved noise specified in (8.84),

$$\langle \zeta(x,t)\zeta(x',t')\rangle = -2D\nabla_\perp^2\,\delta(x - x')\delta(t - t'). \tag{11.56}$$

The corresponding Janssen–De Dominics response functional is

$$\mathcal{A}[\tilde{S}, S] = \int d^d x \int dt\, \tilde{S}(x,t)\left(\frac{\partial S(x,t)}{\partial t} - D\big[c\nabla_\|^2 + \nabla_\perp^2(r - \nabla_\perp^2)\big]S(x,t)\right.$$

$$\left. + D\left[\nabla_\perp^2\,\tilde{S}(x,t) - \frac{g}{2}\nabla_\| S(x,t)^2 - \frac{\tilde{u}}{6}\nabla_\perp^2 S(x,t)^3\right]\right), \tag{11.57}$$

and its Gaussian propagator and dynamic correlation functions are just those of the two-temperature model B,

$$G_0(q_\|, q_\perp, \omega) = \frac{1}{-i\omega + D\big[c\,q_\|^2 + q_\perp^2(r + q_\perp^2)\big]}, \tag{11.58}$$

$$C_0(q_\|, q_\perp, \omega) = \frac{2Dq_\perp^2}{\omega^2 + D^2\big[c\,q_\|^2 + q_\perp^2(r + q_\perp^2)\big]^2}, \tag{11.59}$$

resulting in the static density correlation or structure factor (8.87). The general scaling forms can also be transferred from the two-temperature model B, e.g., Eq. (8.89) for the dynamic susceptibility.

The essential difference from the two-temperature model B resides in the non-linearity $\propto g$ originating jointly from the non-equilibrium drive and particle exclusion, and generating the vertex $\propto iq_\|$ in Fig. 11.2(b). Consequently, the critical DDS model always contains non-vanishing *three-point* correlations, which are absent in the high-temperature phase of the randomly driven model B. Moreover,

---

[8] Janssen and Schmittmann (1986b); Leung and Cardy (1986).

the associated upper critical dimensions differ; with

$$[\tilde{S}(x,t)] = \mu^{1+d_\parallel+d_\perp/2}, \quad [S(x,t)] = \mu^{-1+d_\parallel+d_\perp/2}, \quad [c] = \mu^0 \tag{11.60}$$

as in (8.88), one obtains for the scaling dimensions of the two competing non-linear couplings

$$[\tilde{u}] = \mu^{4-2d_\parallel-d_\perp}, \quad [g] = \mu^{3-d_\parallel-d_\perp/2}. \tag{11.61}$$

Hence the critical dimensions associated with the static and dynamic couplings are, respectively, $d'_c = 4 - d_\parallel$ and $d_c = 6 - d_\parallel$. Since $[\tilde{u}/g^2] = \mu^{-2}$, the static non-linearity $\propto \tilde{u}$ is thus *irrelevant* relative to the drive $\propto g$, and may be omitted if we wish to determine the asymptotic universal scaling laws. It of course represents a *dangerously irrelevant* parameter, since it is responsible for the occurrence of the phase transition in the system. The two-temperature model B is, moreover, characterized by non-trivial static critical exponents, while the kinetics is purely diffusive along the drive direction, $z_\parallel = 2$. Conversely, for the driven model B or critical DDS, only the longitudinal scaling exponents differ from their mean-field values.

11.2.2 Renormalization and critical exponents

The renormalization procedure closely follows the analysis for non-critical driven diffusive systems in Section 11.1.2. Since the asymptotically remaining drive vertex is proportional to $iq_\parallel$, Eqs. (11.21), (11.22), and therefore (11.23) hold for critical DDS as well, whence the transverse critical exponents are *exactly* those of the Gaussian model B,

$$\eta = 0, \quad \nu = \frac{1}{2}, \quad z = 4. \tag{11.62}$$

In addition, Galilean invariance with respect to the transformation (11.15) and therefore the Ward identity (11.31) and relation (11.30) are also valid as before. With $d_\parallel$ biased directions, the effective coupling becomes

$$v = g^2/c^{1+d_\parallel/2}, \quad [v] = \mu^{6-d-d_\parallel}, \tag{11.63}$$

and with its renormalized counterpart defined similarly to (11.33),

$$v_R = Z_c^{-(1+d_\parallel/2)} v \, B_{d_\parallel d} \mu^{d+d_\parallel-6}, \quad B_{d_\parallel d} = \frac{\Gamma(4-d/2-d_\parallel/2)\Gamma(d/2)}{2^d \pi^{d/2}(d-2)\Gamma(d/2-d_\parallel/2)}, \tag{11.64}$$

one obtains the associated renormalization group beta function

$$\beta_v = v_R \left[d + d_\parallel - 6 - \left(1 + \frac{d_\parallel}{2}\right)\gamma_c \right]. \tag{11.65}$$

Consequently, $\gamma_c^* = 2(d + d_\parallel - 6)/(2 + d_\parallel)$ at *any* non-trivial RG fixed point $0 < v^* < \infty$, which again allows us to determine the longitudinal exponents for critical DDS to *all* orders in the perturbation expansion. Whereas the mean-field values $\Delta = 1$ and $z_\parallel = 2$ apply in high dimensions $d > d_c = 6 - d_\parallel$, we predict for $d < d_c$:

$$\Delta = 1 - \frac{\gamma_c^*}{2} = \frac{8 - d}{2 + d_\parallel}, \qquad z_\parallel = \frac{4}{1 + \Delta} = \frac{4(2 + d_\parallel)}{10 + d_\parallel - d}. \tag{11.66}$$

The only non-zero anomalous dimension γ_c for the critical DDS system follows from the renormalization of the vertex function $\Gamma^{(1,1)}(q_\parallel, q_\perp = 0, \omega)$. The one-loop Feynman graph is identical to Fig. 11.3(a), and the corresponding analytic expression follows immediately from Eq. (11.36) by setting $\tilde{c} = 0$ in the noise vertex and replacing $k_\perp^2$ by $k_\perp^2(r + k_\perp^2)$ in the propagators:

$$\Gamma^{(1,1)}(q_\parallel, q_\perp = 0, \omega) = i\omega + cD\,q_\parallel^2$$

$$+ \frac{Dg^2}{2} \int_k \frac{k_\perp^2}{c\left(\frac{q_\parallel}{2} + k_\parallel\right)^2 + k_\perp^2\left(r + k_\perp^2\right)} \frac{q_\parallel\left(\frac{q_\parallel}{2} - k_\parallel\right)}{i\omega/2D + c\left(\frac{q_\parallel^2}{4} + k_\parallel^2\right) + k_\perp^2\left(r + k_\perp^2\right)}.$$

$$\tag{11.67}$$

Expanding to $O(q_\parallel^2)$, rescaling $\sqrt{c}\,k_\parallel$ to $k_\parallel$, and choosing $q = 0 = \omega$, but $r = \mu^2$ as the normalization point, yields

$$Z_c = 1 + \frac{v}{4} \int_k \frac{k_\perp^2}{\left[k_\parallel^2 + k_\perp^2\left(\mu^2 + k_\perp^2\right)\right]^2} + \frac{v}{2d_\parallel} \int_k \frac{k_\parallel^2 k_\perp^2}{\left[k_\parallel^2 + k_\perp^2\left(\mu^2 + k_\perp^2\right)\right]^3}$$

$$= 1 + \frac{3v}{2} \frac{B_{d_\parallel d}}{6 - d_\parallel - d} \mu^{d + d_\parallel - 6} \tag{11.68}$$

after first evaluating the $k_\parallel$, then the $k_\perp$ integrals (Problem 11.2). Therefore

$$\gamma_c = \mu \left.\frac{\partial}{\partial\mu}\right|_0 \ln Z_c = -\frac{3}{2} v_R, \tag{11.69}$$

whence (11.65) gives a non-zero, infrared-stable RG fixed point

$$v^* = \frac{4(6 - d_\parallel - d)}{3(2 + d_\parallel)} \tag{11.70}$$

in dimensions $d < d_c = 6 - d_\parallel$, with the driven model B critical exponents (11.62) and (11.66).

For the critical DDS dynamic correlation function, the general scaling forms analogous to Eqs. (11.26) and (11.27) become

$$C(\tau, x_\parallel, x_\perp, t) = |x_\perp|^{-\lambda_C} \, \bar{C}\left(\tau |x_\perp|^{1/\nu}, \frac{x_\parallel}{\sqrt{c}\,|x_\perp|^{1+\Delta}}, \frac{Dt}{|x_\perp|^z}\right) \qquad (11.71)$$

$$= t^{-\zeta} \, \tilde{C}\left(\tau |x_\parallel|^{1/\nu_\parallel}, \frac{x_\parallel}{\sqrt{c}\,|x_\perp|^{1+\Delta}}, \frac{Dt}{|x_\parallel|^{z_\parallel}}\right), \qquad (11.72)$$

where $\tau \propto T - T_c$ measures the distance from the true critical point, and

$$\lambda_C = d - 2 + \eta + \Delta, \quad \nu_\parallel = \nu(1 + \Delta), \quad \zeta = \frac{\lambda_C}{z} = \frac{d + 4 + (d-2)d_\parallel}{4(2 + d_\parallel)}, \qquad (11.73)$$

with the Gaussian exponents (11.62). Finite-size scaling renders (11.72) into

$$C(\tau, x_\parallel, x_\perp, t, L_\parallel, L_\perp) = L_\parallel^{-\lambda_C/(1+\Delta)} \, \tilde{C}\left(\tau |L_\parallel|^{1/\nu_\parallel}, \frac{x_\parallel}{L_\parallel}, \frac{x_\perp}{L_\perp}, \frac{Dt}{L_\parallel^{z_\parallel}}, \frac{L_\parallel}{\sqrt{c}\,L_\perp^{1+\Delta}}\right); \qquad (11.74)$$

in two dimensions and with $d_\parallel = 1$, one obtains with $\lambda_C = \Delta = 2$, $\nu_\parallel = 3/2$, and $z_\parallel = 4/3$ for the density autocorrelation function at the critical point:

$$C(0, 0, t, L_\parallel, L_\perp) = L_\parallel^{-2/3} \, \tilde{C}\left(Dt/L_\parallel^{4/3}, L_\parallel/\sqrt{c}\,L_\perp^3\right). \qquad (11.75)$$

In order to test the finite-size scaling law (11.75) for a driven Ising lattice gas (KLS model), one needs to run the numerical simulations on rectangular lattices with fixed aspect ratio $L_\parallel/L_\perp^{1+\Delta}$. For each finite system, the corresponding 'critical temperature' must be determined separately, e.g., by locating the variance maximum of the order parameter fluctuations. Figure 11.6 displays Monte Carlo data obtained for three different systems with $L_\perp^3/L_\parallel = 256$, which yield apparent but not truly satisfactory scaling collapse for the larger system sizes. These results exemplify the difficulties of measuring the DDS critical exponents accurately in the non-equilibrium steady state: observables such as the order parameter or correlation functions are typically characterized by extended crossover regions, with the asymptotic scaling regime often beyond the range of accessible simulations.[9]

11.2.3 Critical aging and order parameter growth

The critical exponents of the driven Ising lattice gas can also be accessed in the non-equilibrium relaxation regime, following a quench from a randomized

---

[9] See, e.g., Caracciolo, Gambassi, Gubinelli, and Pelissetto (2004a).

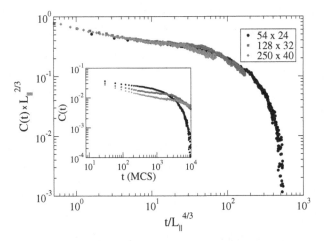

Fig. 11.6 Finite-size scaling for the KLS model density autocorrelation function in the non-equilibrium stationary state according to Eq. (11.75), obtained from Monte Carlo simulations on rectangular lattices with $(L_\parallel, L_\perp) = (54, 24)$, $(128, 32)$, and $(250, 40)$ at their numerically determined 'critical temperatures'. The data were respectively averaged over 2000, 550, and 200 realizations; the inset depicts the unscaled curves. [Figure reproduced with permission from: G. L. Daquila and U. C. Täuber, *Phys. Rev. Lett.* **108**, 110602 (2012); DOI: 10.1103/PhysRevLett.108.110602; copyright (2012) by the American Physical Society.]

high-temperature initial state to the critical point. Indeed, this critical aging regime permits a precise measurement of the DDS critical exponents.[10] At $T = T_c$ ($\tau = 0$), the density autocorrelation function should obey the simple scaling form (11.51) with $\zeta = 1/2$ at $d = 2$, $d_\parallel = 1$, see (11.66). This is nicely confirmed by the Monte Carlo simulation data shown in Fig. 11.7, which yield convincing scaling collapse for waiting times s ranging from 200 to 2000 Monte Carlo steps (MCS). The best data fit is in fact obtained with an autocorrelation exponent value of $\zeta \approx 0.48$. For the associated initial-slip scaling, we may follow Section 11.1.3 to derive the scaling law (11.53), with $\theta = \gamma_c^*/2$. Thus $1 - \theta = \Delta = (8 - d)/(2 + d_\parallel)$.

Independent confirmation may be gathered from an investigation of the critical order parameter growth laws. To be properly sensitive to the density modulations transverse to the drive, we define an anisotropic order parameter as follows (for $d_\parallel = 1 = d_\perp$), utilizing the spin representation:[11]

$$m(\tau, t, L_\parallel, L_\perp) = \frac{\sin(\pi/L_\perp)}{2L_\parallel} \left| \sum_{i=1}^{L_\parallel} \sum_{j=1}^{L_\perp} \sigma(x_i, x_j, t) \, e^{2\pi i x_j/L_\perp} \right|, \qquad (11.76)$$

[10] Daquila and Täuber (2012).   [11] Leung (1991); Wang (1996).

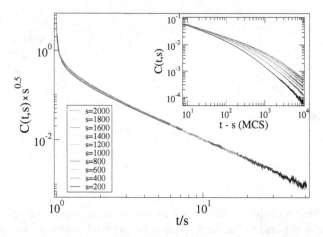

Fig. 11.7 Monte Carlo simulation results for the two-time density autocorrelation function of the KLS model on a two-dimensional rectangular lattice ($L_\parallel = 125\,000$, $L_\perp = 50$) following a quench to the critical point from a high-temperature initial state. Inset: data plotted vs. $t - s$, demonstrating the breaking of time translation invariance; main panel: simple aging scaling data collapse according to Eq. (11.51) with $\zeta = 1/2$. Each curve was averaged over 200 realizations. [Figure reproduced with permission from: G. L. Daquila, Ph.D. Dissertation, Virginia Tech (2011).]

and focus on its temporal evolution starting from a highly correlated checkerboard initial state with sites alternating occupied and empty, and $m_0 = 0$. In the vicinity of the critical point and past mere microscopic time scales, this quantity should be governed by the universal finite-size scaling law[12]

$$m(\tau, t, L_\parallel, L_\perp) = |\tau|^\beta \, \tilde{m}\left(\tau |L_\parallel|^{1/\nu_\parallel}, \frac{Dt}{L_\parallel^{z_\parallel}}, \frac{L_\parallel}{\sqrt{c}\, L_\perp^{1+\Delta}}\right) \qquad (11.77)$$

with order parameter critical exponent

$$\beta = \frac{\nu}{2}\,(d - 2 + \Delta - \eta) = \frac{d - 2 + \Delta}{4}\,; \qquad (11.78)$$

for $d = 2$ and $d_\parallel = 1$, $\beta = 1/2$. At T_c therefore

$$m(0, t, L_\parallel, L_\perp) = L_\parallel^{-\beta/\nu_\parallel}\, \hat{m}\left(Dt/L_\parallel^{z_\parallel}, L_\parallel/\sqrt{c}\, L_\perp^{1+\Delta}\right), \qquad (11.79)$$

which can be readily tested in numerical simulations for various system sizes with fixed aspect ratio $L_\parallel/L_\perp^{1+\Delta}$. The Monte Carlo simulation results for the two-dimensional anisotropic KLS model on reactangular lattices with $L_\parallel/L_\perp^3 = 1$ shown in Fig. 11.8(a) indeed display data collapse with the DDS critical exponents $z_\parallel = 4/3$ and $\beta/\nu_\parallel = 1/3$.

---

[12] Caracciolo, Gambassi, Gubinelli, and Pelissetto (2004b).

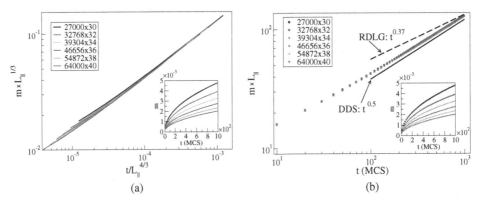

Fig. 11.8 Monte Carlo simulation results for the order parameter growth in the critical initial-slip regime for the KLS model on a two-dimensional rectangle with $L_\parallel \times L_\perp$ sites; each curve was averaged over 500 realizations. (a) Finite-size scaling following Eq. (11.79); (b) initial-slip order parameter growth according to (11.81). The insets display the unscaled growth data. [Figures reproduced with permission from: (left) G. L. Daquila, Ph.D. Dissertation, Virginia Tech (2011); (right) G. L. Daquila and U. C. Täuber, *Phys. Rev. Lett.* **108**, 110602 (2012); DOI: 10.1103/PhysRevLett.108.110602; copyright (2012) by the American Physical Society.]

To furthermore extract the temporal order parameter growth in the initial-slip scaling regime, we rewrite (11.79) as

$$m(0, t, L_\parallel, P_\perp) = t^{-\beta/z_\parallel \nu_\parallel} \, \bar{m}\big((Dt)^{1/z_\parallel}/L_\parallel, L_\parallel/\sqrt{c}\, L_\perp^{1+\Delta}\big) . \qquad (11.80)$$

Holding the aspect ratio fixed, this becomes a function of $y = (Dt)^{1/z_\parallel}/L_\parallel$ only. Expanding the regular scaling function $\bar{m}$ for small arguments $y \ll 1$, $\bar{m}(y) = \bar{m}_0 + \tilde{m}_1 y + \cdots$, where $\bar{m}_0 = 0$ for the checkerboard initial configuration, one arrives at the order parameter growth law at short times,

$$m(t, L_\parallel) L_\parallel \sim t^\kappa , \qquad \kappa = (1 - \beta/\nu_\parallel)/z_\parallel . \qquad (11.81)$$

Figure 11.8(b) demonstrates convincing scaling collapse with this initial-slip scaling law, using the same Monte Carlo data as in Fig. 11.8(a). A fit to the largest lattice with $64\,000 \times 40$ sites gives $\kappa \approx 0.487$, close to the expected value 0.5 for the critical KLS model in two dimensions, labeled 'DDS' here. For comparison, the different power law for a randomly driven Ising lattice gas ('RDLG') or two-temperature model B with $\kappa \approx 0.37$ at $d = 2$ is indicated as well.

11.2.4 Large-deviation function for the particle current

To conclude our discussion of the scaling behavior in driven diffusive systems, we address the asymptotic properties of the large-deviation function for the integrated particle current

$$Q(t_f) = \int d^d x \int_0^{t_f} dt \, J_\parallel(x, t) \tag{11.82}$$

along the drive direction that crucially distinguishes the ensuing non-equilibrium stationary state from thermal equilibrium.[13] We are interested in its time-dependent probability distribution $\mathcal{P}(Q, t_f)$ for a statistical sampling over noise histories up to time t_f. Specifically, we aim for the associated *large-deviation function* $\pi(q)$ defined as

$$\pi(q) = \lim_{t_f \to \infty} \frac{1}{t_f} \ln \mathcal{P}(Q(t_f) = q \, t_f, t_f) \,, \tag{11.83}$$

where q represents the fluctuating time-averaged current. For finite systems $\pi(q)$ will be peaked around its stationary maximum $q_{st} = \lim_{t_f \to \infty} \langle Q \rangle / t_f$, which sharpens as the system size increases. Even in equilibrium, $\pi(q)$ typically is not given by a simple Gaussian form; it may therefore potentially serve as a tool to distinguish characteristic fluctuation features in non-equilibrium stationary states. It satisfies a *symmetry relation* in analogy to Crooks' theorem (8.110): since driving the particles along with the longitudinal current is associated with a work contribution $\propto Q$, one generically obtains $\pi(q) - \pi(-q) \propto q$, connecting the probabilities of observing lasting current deviations both close to and far from its stationary value.[14]

The longitudinal current fluctuations are also fully encoded in the moments of Q, accessible through the *generating function*

$$Z(s, t_f) = \left\langle e^{-sQ(t_f)} \right\rangle, \quad \left\langle Q(t_f)^n \right\rangle = (-1)^n \left. \frac{d^n Z(s, t_f)}{ds^n} \right|_{s=0}. \tag{11.84}$$

One expects the 'dynamical partition function' $Z(s, t_f)$ to typically grow exponentially with time t_f, whence it is appropriate to introduce a time-intensive 'dynamical free energy'

$$f(s) = \lim_{t_f \to \infty} \frac{1}{t_f} \ln Z(s, t_f) \,, \tag{11.85}$$

which is formally related to the current large deviation function $\pi(q)$ defined in (11.83) through a Legendre transform, $\pi(q) = \max_s \{ f(s) + sq \}$, and $f(s)$ is just the generating function for the *cumulants* of the integrated longitudinal current $Q(t_f)$ in the asymptotic long-time limit:

$$(-1)^n \left. \frac{d^n f(s)}{ds^n} \right|_{s=0} = \lim_{t_f \to \infty} \frac{\langle Q(t_f)^n \rangle_c}{t_f} \,. \tag{11.86}$$

[13] Lecomte, Täuber, and van Wijland (2007).
[14] Bertini, De Sole, Gabrielli, Jona-Lasinio, and Landim (2005).

The relationship between $\mathcal{P}(Q, t_f)$ and $Z(s, t_f)$ is analogous to switching between the micro-canonical and canonical ensembles in equilibrium statistical mechanics, and both quantities ultimately contain the same information.

With the explicit form (11.4) for the longitudinal current in driven diffusive systems, and reinstating the mean current J_0, we may use the procedures of Section 4.1.2 to write the generating function (11.84) in the form of a functional integral

$$Z(s, t_f) = \int \mathcal{D}[i\tilde{S}] \int \mathcal{D}[S]\, e^{-\tilde{A}[\tilde{S}, S, s]} . \tag{11.87}$$

Observing the periodic boundary conditions, the first three contributions to the longitudinal current simply yield terms linear in s,

$$\tilde{A}_J[S, s] = J_0 V t_f s - \frac{Dgs}{2} \int d^d x \int_0^{t_f} dt\, S(x, t)^2 \tag{11.88}$$

in a system with total volume V. When the average over the Gaussian white noise is carried out as in Eq. (4.8), the effect of the stochastic current $\eta_\parallel$ is a shift $\nabla_\parallel \tilde{S}(x, t) \rightarrow \nabla_\parallel \tilde{S}(x, t) + s$ in the Janssen–De Dominicis functional. Again taking the periodic boundary conditions into account, one thus obtains the modified action

$$\tilde{A}[\tilde{S}, S, s] = A[\tilde{S}, S] + \tilde{A}_J[S, s] - \tilde{c}DVt_f s^2 , \tag{11.89}$$

which aside from the two deterministic contributions $\propto V t_f$ differs from the DDS response functionals $A[\tilde{S}, S]$ in (11.14) or (11.57) through a new quadratic term in the field $S(x, t)$ from (11.88). For non-interacting particles ($g = 0$), each of the cumulants $\langle Q(t_f)^n \rangle_c$ is extensive in the volume V *and* the elapsed time t_f, and the associated dynamical free energy assumes a simple scaling form,

$$\hat{f}(s) = \frac{f_0(s)}{V} = -J_0 s + \tilde{c}Ds^2 , \tag{11.90}$$

which entails a regular polynomial expansion around $s = 0$. In contrast, particle interactions along with the non-equilibrium drive tend to generate terms involving non-integer powers of $|s|$ in the analogous expansion of $\hat{f}(s)$ (in the infinite-volume thermodynamic limit); this non-analytic behavior at $s = 0$ reflects the existence of an infinite current cumulant.

For $s \neq 0$, the dynamical action (11.89) does not represent a stochastic process, and $Z(s \neq 0, t_f) \neq 1$ carries non-trivial information. We proceed to evaluate it in the tree-level approximation. To this end, we write the Gaussian action in Fourier space in the form

$$\tilde{A}_0[\tilde{S}, S, s] = J_0 V t_f s - \tilde{c}DVt_f s^2$$
$$+ \frac{1}{2} \int \frac{d^d q}{(2\pi)^d} \int \frac{d\omega}{2\pi} \left(\tilde{S}(-q, -\omega)\ S(-q, -\omega) \right) \tilde{A}(q, \omega, s) \begin{pmatrix} \tilde{S}(q, \omega) \\ S(q, \omega) \end{pmatrix} . \tag{11.91}$$

For non-critical driven diffusive systems described by the Langevin equation (11.7) with noise correlator (11.9), one has explicitly

$$\bar{A}(q, \omega, s) = \begin{pmatrix} -2D(\tilde{c}\, q_\parallel^2 + q_\perp^2) & -i\omega + D(c\, q_\parallel^2 + q_\perp^2) \\ i\omega + D(c\, q_\parallel^2 + q_\perp^2) & -Dgs \end{pmatrix},$$

with determinant

$$\det \bar{A}(q, \omega, s) = -\omega^2 - D^2(c\, q_\parallel^2 + q_\perp^2)^2 + 2D^2 gs (\tilde{c}\, q_\parallel^2 + q_\perp^2).$$

With proper normalization, the Gaussian integrals over the fields $\widetilde{S}$ and S in (11.91) result in (compare with Section 8.5 and Problem 4.1):

$$\ln Z_0(s, t_f) + J_0 V t_f s - \tilde{c} D V t_f s^2$$

$$= -\frac{V t_f}{2} \int \frac{d^d q}{(2\pi)^d} \int \frac{d\omega}{2\pi} \ln \left| \frac{\det \bar{A}(q, \omega, s)}{\det \bar{A}(q, \omega, 0)} \right|$$

$$= -\frac{V t_f}{2} \int \frac{d^d q}{(2\pi)^d} \int \frac{d\omega}{2\pi} \ln \left[1 - \frac{2D^2 gs\, (\tilde{c}\, q_\parallel^2 + q_\perp^2)}{\omega^2 + D^2\,(c\, q_\parallel^2 + q_\perp^2)^2} \right]$$

$$= -\frac{V t_f D}{2} \int \frac{d^d q}{(2\pi)^d} \left[\sqrt{(c\, q_\parallel^2 + q_\perp^2)^2 - 2gs(\tilde{c}\, q_\parallel^2 + q_\perp^2)} - (c\, q_\parallel^2 + q_\perp^2) \right].$$

$$(11.92)$$

At the renormalization group fixed point (11.43), $\tilde{c}_R = c_R$; we therefore identify $\tilde{c} = c$ and proceed to evaluate the wavevector integral. Separating its leading term in a small s expansion (with ultraviolet cutoff Λ) isolates the infrared divergence for $s \to 0$ of the dynamical free energy:

$$\hat{f}_0(s) = -J_0 s + c D s^2 + \frac{Dgs}{2} \Lambda^d + \frac{\Gamma(1 - d/2)}{4(4\pi)^{d/2}} \frac{D(-gs)^{1+d/2}}{\sqrt{c}}. \qquad (11.93)$$

With minimal subtraction, one finds

$$[\hat{f}_0(s) + J_0 s]_{\text{sing.}} = c D s^2 \left[1 + \frac{C_d\, v}{4(2 - d)} (-gs)^{(d-2)/2} \right]; \qquad (11.94)$$

choosing the normalization point $-gs = \mu^2$, the expression in the square brackets reduces to Eq. (11.38) for Z_c (at $w_{\text{eq}}^* = 1$), whence the divergence is properly absorbed into the renormalized parameter c_R,

$$\hat{f}_0(s) = -J_0 s + c_R D s^2. \qquad (11.95)$$

Through the associated RG equation, the renormalized parameter c_R becomes scale-dependent, and near the infrared-stable fixed point $w_{\text{eq}}^* = 1$ it follows the power law $\tilde{c}(\ell) \sim c_R \ell^{\gamma_c^*} = c_R \ell^{-2(2-d)/3}$ in dimensions $d < d_c = 2$. Upon matching

$\mu^2 \ell^2 = -gs$, this leads to the asymptotic scaling

$$\hat{f}(s) = -J_0 s + a_d\, s^\delta\,, \quad \delta = \frac{d+4}{3} \tag{11.96}$$

as $s \to 0$, with a constant, non-universal amplitude a_d. Note that the universal scaling exponent δ has again been determined to *all* orders in perturbation theory. For the totally asymmetric exclusion process in one dimension, $J_0 = \rho(1-\rho)$, and it is possible to compute $\delta = 5/3$ exactly.[15] At the critical dimension $d_c = 2$, one finds

$$\hat{f}(s) = -J_0 s + a_2\, s^2(-\ln|s|)^{2/3}\,, \tag{11.97}$$

see Problem 11.1(c). It is crucial to realize that the exponent δ follows directly from the longitudinal noise renormalization and could hence have been inferred through straightforward dynamic scaling.

We may take this shortcut for the driven model B or critical DDS described by the Langevin equation (11.55). In analogy to Eq. (11.92), the Gaussian contribution to its dynamical partition function becomes

$$\ln Z_0(s, t_f) + J_0 V t_f s - \tilde{c} D V t_f s^2$$

$$= -\frac{V t_f}{2} \int \frac{d^d q}{(2\pi)^d} \int \frac{d\omega}{2\pi} \ln\left[1 - \frac{2D^2 gs\,(\tilde{c}\,q_\parallel^2 + q_\perp^2)}{\omega^2 + D^2 \left[c\,q_\parallel^2 + q_\perp^2(r + q_\perp^2)\right]^2}\right], \tag{11.98}$$

and we once more require the anomalous scaling dimension of the noise strength $\tilde{c}$, which is an *irrelevant* parameter, in the RG sense, for this model. In dimensions $d < d_c = 6 - d_\parallel$ one finds (Problem 11.2)

$$\gamma_{\tilde{c}}^* = 2 - \frac{2d_\parallel(6 - d_\parallel - d)}{3(2 + d_\parallel)} \tag{11.99}$$

at the infrared-stable RG fixed point (11.70). Since time now scales according to $[t] = \mu^{-4}$, the appropriate matching condition becomes $\mu^4 \ell^4 = -gs$, which implies

$$\hat{f}(s) = -J_0 s + a_d'\, s^\delta\,, \quad \delta = 2 - \frac{d_\parallel(6 - d_\parallel - d)}{6(2 + d_\parallel)}\,. \tag{11.100}$$

Note that this result holds only to first order in ε in the dimensional expansion for $d = d_c - \varepsilon$. At the upper critical dimension $d_c = 6 - d_\parallel$,

$$\tilde{v}(\ell) = \frac{v_R}{1 - \frac{3}{4}(d_\parallel + 2)v_R \ln \ell}\,, \quad \tilde{c}(\ell) \sim \tilde{c}\,\ell^2\,|\ln \ell|^{2d_\parallel/3(2+d_\parallel)}\,, \tag{11.101}$$

[15] Derrida and Lebowitz (1998); Derrida and Appert (1999).

wherefrom we obtain the logarithmic correction

$$\hat{f}(s) = -J_0 s + a'_{6-d_\parallel} s^2 (-\ln|s|)^{2d_\parallel/3(2+d_\parallel)} . \qquad (11.102)$$

11.3 Driven interfaces and the KPZ equation

The *Kardar–Parisi–Zhang (KPZ) equation* represents one of the most prominent examples for generic scale invariance in non-equilibrium systems, and in addition displays a dynamic phase transition between two distinct scaling regimes. It was originally formulated to describe growing crystalline surfaces, but has found a variety of diverse applications in other contexts, most notably for the dynamics of driven interfaces, and the equilibrium statistical mechanics of directed lines in disordered environments.[16] In one dimension, it is also intimately connected with the noisy Burgers equation for the hydrodynamics of potential fluids. We set out with derivations of the KPZ equation for the interface fluctuations of a driven domain wall, and for the height fluctuations of a crystalline surface, growing under non-equilibrium conditions. Next we consider the elementary scaling theory for the KPZ problem, and analyze its linearized version, the *Edwards–Wilkinson model*. Finally, we utilize the mapping to the noisy Burgers equation encountered in Section 11.1.2 to exactly determine the scaling exponents in one dimension.

11.3.1 Driven domain walls and surface growth

Let us first consider an interface (domain wall) in a $(d + 1)$-dimensional Ising ferromagnet, separating regions of positive and negative magnetization. When a positive magnetic field of strength $\bar{g}$ along the z direction is applied, the interface is driven into the negative magnetization domain, as depicted schematically in Fig. 11.9. Under conditions of strong drive, it is reasonable to assume that there appear no 'overhangs' in the interface contour. Choosing an appropriate d-dimensional base hyperplane, we can hence describe the interface at time t by the vector $X(t) = (x, z = h(x, t))$, where $h(x, t)$ is a single-valued scalar function (Monge gauge). In this parametrization, $dX = (1, \ldots, 1, \nabla h(x, t)) \, d^d x$, and the surface element becomes $ds = \sqrt{1 + [\nabla h(x, t)]^2} \, d^d x$.

The equilibrium Landau–Ginzburg Hamiltonian for an Ising model with scalar order parameter $\bar{S}(x, z)$ in $d + 1$ dimensions reads

$$\mathcal{H}_0[\bar{S}] = \int \left[\frac{1}{2} [\nabla \bar{S}(x, z)]^2 + \frac{1}{2} \left(\frac{\partial \bar{S}(x, z)}{\partial z} \right)^2 + V(\bar{S}(x, z)) \right] d^d x \, dz , \quad (11.103)$$

---

[16] For overviews, see Krug and Spohn (1992); Halpin-Healy and Zhang (1995); Barabási and Stanley (1995); Krug (1997); and Lässig (1998).

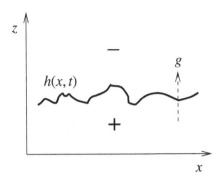

Fig. 11.9 Schematic picture of an interface in a $(d + 1)$-dimensional Ising model, characterized by a height function $z = h(x, t)$, which is driven along the z direction by a uniform magnetic field of strength $\bar{g}$.

where ∇ denotes the d-dimensional gradient with respect to the transverse coordinates x, and $V(\bar{S})$ is an even function of $\bar{S}$, e.g., the potential of the scalar Φ^4 model, $V(\bar{S}) = r\bar{S}^2/2 + u\bar{S}^4/4!$ (see Section 1.2.1). The interface profile is then determined by the solution of the Landau–Ginzburg stationarity equation

$$0 = \frac{\delta \mathcal{H}_0[\bar{S}]}{\delta \bar{S}(x, z)} = -\nabla^2 \bar{S}(x, z) - \frac{\partial^2 \bar{S}(x, z)}{\partial z^2} + V'(\bar{S}(x, z)) \,. \qquad (11.104)$$

With the boundary conditions $\bar{S}(x, z \to \pm\infty) = \mp|\bar{S}_\infty|$, the minimum free energy interface becomes independent of x, and is given by $\bar{S}(x, z) = \bar{f}(z)$, satisfying

$$\frac{d^2 \bar{f}(z)}{dz^2} = V'(\bar{f}(z)) \,. \qquad (11.105)$$

We impose purely relaxational model A dynamics for the time-dependent non-conserved order parameter field $\bar{S}(x, z, t)$,

$$\frac{\partial \bar{S}(x, z, t)}{\partial t} = -D \frac{\delta \mathcal{H}_0[\bar{S}]}{\delta \bar{S}(x, z, t)} + \eta(x, z, t) \,, \qquad (11.106)$$

with $\langle \eta(x, z, t) \rangle = 0$, and the Einstein relation

$$\langle \eta(x, z, t)\eta(x', z', t') \rangle = 2D\, \delta(x - x')\delta(z - z')\delta(t - t') \qquad (11.107)$$

for the noise correlations, as required by the condition that in the absence of the drive the system should relax to an equilibrium thermodynamic state with the Boltzmann weight $\exp(-\mathcal{H}_0[\bar{S}])$ (where $k_B T = 1$).

We now switch on the magnetic field $\bar{g}$, and thus enter a genuinely non-equilibrium regime, which, however, can be described by replacing the free energy

$\mathcal{H}_0$ with the modified Hamiltonian

$$\mathcal{H}[\bar{S}] = \mathcal{H}_0[\bar{S}] - \bar{g} \int \bar{S}(x, z) \, d^d x \, dz . \qquad (11.108)$$

Note that $\bar{g}$ measures the strength of the driving field relative to the coefficient of the gradient terms in Eq. (11.103), which in turn determine the interfacial surface tension. Differential geometry dictates the general form for the interface shape function in the Monge gauge,

$$\bar{S}(x, z, t) = \bar{f} \left(\frac{h(x, t) - z}{\sqrt{1 + [\nabla h(x, t)]^2}} \right) . \qquad (11.109)$$

Next, we replace $\mathcal{H}_0$ with $\mathcal{H}$ in the model A Langevin equation (11.106), and insert the ansatz (11.109). Retaining only the lowest-order derivatives, this leads to

$$\frac{\dot{h} \, \bar{f}'}{\sqrt{1 + (\nabla h)^2}} = D \left[\frac{\nabla^2 h}{\sqrt{1 + (\nabla h)^2}} \bar{f}' + \bar{f}'' - V'(\bar{f}) + \bar{g} \right] + \eta .$$

In the spirit of an expansion near equilibrium, we may use Eq. (11.105) to eliminate $V'(\bar{f})$, and then divide by $\bar{f}'(\tilde{z})$ and average over its argument $\tilde{z} = (h - z)/\sqrt{1 + (\nabla h)^2}$ in an interval of length L_z. This gives

$$\frac{\partial h}{\partial t} = D \nabla^2 h + \lambda \sqrt{1 + (\nabla h)^2} + \sqrt{1 + (\nabla h)^2} \, \bar{\zeta} , \qquad (11.110)$$

where $\lambda = D\bar{g} \int_0^{L_z} f'(\tilde{z})^{-1} d\tilde{z} / L_z$ and $\bar{\zeta}(x, t) = \int_0^{L_z} \eta(x, z, t) f'(\tilde{z})^{-1} d\tilde{z} / L_z$, and consequently with $\tilde{D} = D \int_0^{L_z} f'(\tilde{z})^{-2} d\tilde{z} / L_z^2$,

$$\langle \bar{\zeta}(x, t) \bar{\zeta}(x', t') \rangle = 2\tilde{D} \, \delta(x - x') \delta(t - t') . \qquad (11.111)$$

For small interface height fluctuations $(\nabla h)^2 \ll 1$, we may expand

$$\frac{\partial h(x, t)}{\partial t} = D \nabla^2 h(x, t) + \lambda \left(1 + \frac{1}{2} [\nabla h(x, t)]^2 + \cdots \right) + \bar{\zeta} \, (1 + \cdots) .$$

$$(11.112)$$

Naturally the drive leads to an average drift velocity $v_d = Dg$ along the z direction. In order to describe the interface properties in a comoving frame, we define new height fields $\bar{h}(x, t)$ via the simple shift $\bar{h}(x, t) = h(x, t) - Dgt$, and arrive at the *Kardar–Parisi–Zhang (KPZ) equation*[17]

$$\frac{\partial \bar{h}(x, t)}{\partial t} = D \nabla^2 \bar{h}(x, t) + \frac{\lambda}{2} [\nabla \bar{h}(x, t)]^2 + \bar{\zeta}(x, t) . \qquad (11.113)$$

---

[17] Kardar, Parisi, and Zhang (1986).

Here, the non-linearity $\propto \lambda$ in effect describes curvature-driven growth perpendicular to the surface, counteracted by diffusive smoothening $\propto D$. For convenience we rescale the fields according to $S(x, t) = (D/\tilde{D})^{1/2}\bar{h}(x, t)$, $\zeta(x, t) = (D/\tilde{D})^{1/2}\bar{\zeta}(x, t)$, and define $g = \lambda(\tilde{D}/D^3)^{1/2}$, which at last yields the non-linear stochastic differential equation

$$\frac{\partial S(x, t)}{\partial t} = D\nabla^2 S(x, t) + \frac{Dg}{2}[\nabla S(x, t)]^2 + \zeta(x, t), \tag{11.114}$$

with the white noise correlator

$$\langle\zeta(x, t)\zeta(x', t')\rangle = 2D\,\delta(x - x')\delta(t - t'). \tag{11.115}$$

It is important to realize that, although in this parametrization Einstein's relation between the relaxation constant and the noise strength is superficially fulfilled, the KPZ equation describes a genuine *non-equilibrium* situation, because the reversible, non-linear force term $\propto g$ does in general *not* satisfy the potential condition (3.58).

Ending this discussion of a driven Ising domain wall, we remark that alternatively we could have directly started from an effective *interfacial* free energy, given by the integral of the surface element $ds = \sqrt{1 + [\nabla h(x)]^2}\,d^d x$ over the d-dimensional 'substrate' space,

$$\mathcal{H}_0[h] = \int \sqrt{1 + [\nabla h(x)]^2}\,d^d x\,. \tag{11.116}$$

Here we have omitted an overall energy scale, as determined by the ratio of the surface tension and temperature, i.e., we set $\sigma/k_B T = 1$. Imposing again purely relaxational dynamics directly for the non-conserved height variable,

$$\frac{\partial h(x, t)}{\partial t} = -L[h]\frac{\delta\mathcal{H}_0[h]}{\delta h(x, t)} + \zeta(x, t), \tag{11.117}$$

accompanied by the corresponding Einstein relation, we then need to specify the Onsager coefficient L. With the additional assumption that physically the relaxation rate should be proportional to the magnitude of the surface element, we write $L[h] = D\sqrt{1 + (\nabla h)^2}$. One may then define the Hamiltonian for the driven system (with driving force $\propto g$) in analogy with Eq. (11.108), whereupon the model A Langevin equation (11.117) becomes

$$\frac{\partial h}{\partial t} = -D\sqrt{1 + (\nabla h)^2}\left[-\sum_{i=1}^{d}\nabla_i\left(\frac{\nabla_i h}{\sqrt{1 + (\nabla h)^2}}\right) - g\right] + \zeta$$

$$= D\left[\nabla^2 h - \frac{\sum_{ij}(\nabla_i h)(\nabla_i\nabla_j h)(\nabla_j h)}{1 + (\nabla h)^2}\right] + Dg\sqrt{1 + (\nabla h)^2} + \zeta\,.$$

After expanding for $(\nabla h)^2 \ll 1$ and shifting $S(x, t) = h(x, t) - Dgt$, the KPZ equation (11.114) follows, with noise correlator (11.115).

Originally, the non-linear Langevin equation (11.113), with the noise specified by Eq. (11.111), was derived for the height fluctuations of crystalline surfaces under non-equilibrium growth conditions. The basic ansatz here is that the growth velocity should be a smooth function of the height profile gradient,

$$\frac{\partial h(x, t)}{\partial t} = v(\nabla h(x, t)) . \tag{11.118}$$

In addition, we allow for random particle deposition, which we model as a stochastic force ζ with zero mean and Gaussian correlator (11.111). Furthermore, we introduce some effective surface smoothing process (restructuring), described by a diffusive relaxational term $D\nabla^2 h$ (as the growing crystal is manifestly not in thermal equilibrium, the diffusion coefficient D and the noise strength $\widetilde{D}$ are *not* related). Next we expand $v(u = \nabla h) = v(0) + \nabla v|_{u=0} \cdot \nabla h + \frac{1}{2}\sum_{ij}(\partial^2 v/\partial u_i \partial u_j)|_{u=0} \nabla_i h \nabla_j h + \cdots \approx v(0) + \nabla v|_{u=0} \cdot \nabla h + \frac{\lambda}{2}(\nabla h)^2$, if the growth process itself is isotropic. Again, we can transform to a comoving frame by means of the shift $\bar{h}(x, t) = h(x - \nabla v|_{u=0} t, t) - v(0)t$, and end up with the KPZ equation (11.113). Upon rescaling the fields as before, this reduces to our previous Eq. (11.114).

A functional relationship of the form (11.118) is rather general, as is hence the appearance of the leading non-linearity $\propto (\nabla h)^2$. It is therefore not surprising that the Kardar–Parisi–Zhang problem, and simple variants thereof, appears in a variety of different physical circumstances, especially in one dimension, where spatial anisotropies are absent. Prominent examples include the Kuramoto–Sivashinski equation, flame front propagation, asymmetric exclusion processes (see Section 11.1), multiplicative noise problems, the dynamics of sine-Gordon chains, and the kinetics of two-dimensional planar ferromagnets subject to (two-dimensional) dipolar forces. Identifying t with a spatial dimension, one may also map the one-dimensional KPZ problem to the thermodynamics of two-dimensional smectic-A liquid crystals. We shall explore two important equivalent representations of the KPZ equation in more detail below, namely its mapping to the noisy Burgers equation and hence to driven diffusive systems in one dimension, cf. Section 11.1.2, and its relation to the equilibrium statistical mechanics of directed lines in random media (Section 11.3.4).

11.3.2 Scaling theory of interface fluctuations

As in our previously studied models for the critical dynamics in the vicinity of equilibrium phase transitions, reaction-diffusion models, and driven diffusive systems, the difficulties in analyzing the KPZ equation (11.114) originate from the *combined* non-linearity (here $\propto [\nabla S]^2$) and stochasticity. By means of a Cole–Hopf transformation (see Section 11.4.2) the deterministic problem is easily integrated for any given initial condition (for an explicit integration of the one-dimensional deterministic KPZ equation, see Problem 11.3). As the driven lattice gas studied in Section 11.1, the KPZ equation is *generically scale-invariant*, and does not require tuning to a special critical point in parameter space in order to display algebraic scaling behavior. This is already evident if we explore the linear version of Eq. (11.114),

$$\frac{\partial S(x,t)}{\partial t} = D\nabla^2 S(x,t) + \zeta(x,t), \tag{11.119}$$

which is of course nothing but a noisy diffusion equation. In the surface growth context, (11.119) is known as the *Edwards–Wilkinson* model.[18] With the noise correlator (11.115), detailed balance holds (by setting $g = 0$, we have switched off the non-equilibrium driving term), and therefore the stationary probability distribution is given by the canonical distribution with Hamiltonian (11.116), expanded to quadratic order in $\nabla h = \nabla S$ as in Eq. (11.112):

$$\mathcal{P}_{\mathrm{st}}[S] \propto e^{-\frac{1}{2}\int (\nabla S)^2 \, d^d x}. \tag{11.120}$$

The dynamical Edwards–Wilkinson correlation function naturally coincides with that of the Gaussian model A for non-conserved relaxational kinetics at the critical point $T = T_c$ (see Section 3.2.2),

$$C_0(q,\omega) = \frac{2D}{\omega^2 + (Dq^2)^2}, \tag{11.121}$$

which is consistent with the general scaling form

$$C(x,t) = |x|^{-(d-2+\eta)} \hat{C}(Dt/|x|^z), \tag{11.122}$$

where $\hat{C}(y) \to$ const. in the limit $y \to 0$, and $\hat{C}(y) \to y^{(2-d-\eta)/z}$ as $y \to \infty$. In Fourier space, this becomes

$$C(q,\omega) = |q|^{-z-2-\eta} \hat{C}(\omega/D|q|^z). \tag{11.123}$$

Thus, as to be expected from purely diffusive linear dynamics, $z = 2$ and $\eta = 0$ in the Edwards–Wilkinson model.

The scaling properties in the KPZ problem are characterized by the two exponents η and z. In the terminology of non-equilibrium growth processes, usually the so-called *roughness* (or *wandering*) exponent χ is introduced instead of η, via

$$C(x,t) = |x|^{2\chi} \hat{C}(Dt/|x|^z), \quad \chi = \frac{2-d-\eta}{2}. \tag{11.124}$$

---

[18] Edwards and Wilkinson (1982).

Physically, χ describes the interface height fluctuations in the stationary state, $\Delta S(x)^2 = \lim_{t \to \infty} \langle [S(x, t) - S(0, t)]^2 \rangle \sim |x|^{2\chi}$. The associated temporal fluctuations scale as $\Delta S(t)^2 = \lim_{x \to \infty} \langle [S(x, t) - S(x, 0)]^2 \rangle \sim t^{2\chi/z}$. For $\chi < 0$ the height fluctuations die out on long length scales, and the surface is smooth. On the other hand, if $0 < \chi < 1$, the spatial fluctuations grow with the distance $|x|$ between the reference points, and there will be no long-range positional order while orientational order is maintained; the interface is then called *rough*. If one starts with an initially flat interface, and lets it evolve under the non-linear KPZ dynamics, then after the elapsed time t the height correlations begin to display the roughness power law $\sim |x|^{2\chi}$ up to the characteristic length scale $L_x(t) \sim t^{1/z}$, beyond which the surface still looks smooth. At any finite time $t < \infty$, these correlations therefore do not grow indefinitely with $|x|$, despite the positivity of the roughness exponent χ. When the roughness exponent exceeds the upper limit $\chi \to 1$, the large-scale surface orientation differs from that of the substrate, since $\lim_{|x| \to \infty} \Delta S(x)/|x| \neq 0$ in this case, and the above assumption of a single-valued height function breaks down. In the marginal case $\chi = 0$, one usually expects logarithmic roughness. Thus the linear Edwards–Wilkinson model, for which $\chi = 1 - \frac{d}{2}$, predicts that the interface should be rough for $d < 2$, logarithmically rough at $d_c = 2$, and smooth in large dimensions $d > 2$.

In fact, $d_c = 2$ turns out to be a special dimension in the non-linear growth equation as well, but in a very different sense. In order to see this, we write down the response functional (4.10) for the KPZ problem,

$$A[\tilde{S}, S] = \int d^d x \int dt \, \tilde{S}(x, t) \left[\frac{\partial S(x, t)}{\partial t} - D \nabla^2 S(x, t) \right.$$

$$\left. - D \tilde{S}(x, t) - \frac{Dg}{2} [\nabla S(x, t)]^2 \right]. \quad (11.125)$$

Measuring lengths according to $[x] = \mu^{-1}$ and introducing the corresponding time scale $[t] = \mu^{-2}$ ($[D] = \mu^0$), one readily obtains the scaling dimensions

$$[\tilde{S}(x, t)] = \mu^{(d+2)/2}, \quad [S(x, t)] = \mu^{(d-2)/2}, \quad [g] = \mu^{(2-d)/2}. \quad (11.126)$$

Therefore we conclude at the Gaussian Edwards–Wilkinson fixed point $g_0^* = 0$ that g constitutes a *relevant* coupling for $d < 2$. Compared to the KPZ non-linearity $\propto (\nabla S)^2$, higher-order terms in the gradient expansion are irrelevant in the renormalization group sense, which justifies our omissions in the above derivations of the KPZ Langevin equation. The scaling exponents are then governed by a non-trivial infrared-*stable* fixed point $g^* > 0$, and we shall find that the interface is always rough in this situation. For $d > d_c = 2$, on the other hand, the non-linearity is *irrelevant* for sufficiently small g. In this 'weak-coupling' regime,

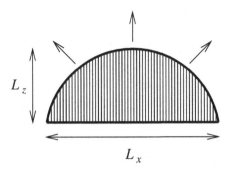

Fig. 11.10 Interface fluctuation, widening through the curvature-driven growth process normal to its surface.

asymptotically one expects a smooth interface characterized by the Edwards–Wilkinson correlation function (11.121). However, we shall in addition find a non-trivial *unstable* fixed point g_c for $d \geq d_c = 2$, separating this smooth phase and a rough strong-coupling regime, where the perturbational coupling g diverges formally (see Fig. 11.14 below). The unstable fixed point g_c thus describes a continuous non-equilibrium *'roughening'* transition, which is to be distinguished from temperature-driven roughening transitions that occur in thermal equilibrium. For isotropic two-dimensional surfaces, these typically fall into the Berezinskii–Kosterlitz–Thouless universality class.

For any non-trivial and finite fixed point $0 < g^* < \infty$, we shall soon derive a scaling relation for the KPZ roughness and dynamic exponents,

$$\chi + z = 2 , \qquad (11.127)$$

which is technically a consequence of the Galilean invariance in the equivalent Burgers representation. This identity, which reduces the number of independent exponents to merely one, can be illustrated by means of the following scaling argument. Consider a growing interface element of width L_x and height L_z, as shown in Fig. 11.10. According to the definition of the roughness exponent, its typical height grows with the lateral scale as $L_z \sim L_x^{\chi}$. Furthermore, as the growth rate is proportional to the local surface gradient, we have $dL_x/dt \sim L_z/L_x \sim L_x^{\chi-1}$, whence $L_x(t) \sim t^{1/(2-\chi)}$. According to the definition of the dynamic exponent, on the other hand, the characteristic size of the interface fluctuations should scale as $L_x \sim t^{1/z}$, and consequently Eq. (11.127) results in the non-linear curvature-driven regime. Note that the Edwards–Wilkinson exponents, governed by the Gaussian fixed point $g_0^* = 0$, satisfy $\chi + z = 3 - \frac{d}{2}$, which conincides with Eq. (11.127) only in two dimensions.

Upon setting $u(x,t) = -\nabla S(x,t)$, the KPZ equation (11.114) turns into the noisy Burgers equation (11.45) with noise correlations (11.46). Remarkably, the KPZ problem is thus equivalent to the stochastic hydrodynamics of randomly stirred fluids, for which the velocity field u can be written as a gradient of a scalar potential S, and the only non-linearity taken into account is the convective term. In terms of the interface height fluctuations, the invariance with respect to a Galilean transformation can be interpreted as a symmetry with respect to tilting the surface infinitesimally, such that $S'(x',t')$ remains a single-valued function without overhangs,

$$S(x,t) \rightarrow S'(x',t') = S(x - Dgvt, t) - v \cdot x . \tag{11.128}$$

Since the product Dg appears as the structure constant of the Galilean symmetry group, there should be no independent renormalization constant associated with g (see Section 11.1 and Problems 11.4, 11.5). Heuristically, one may understand the origin of the exponent identity (11.127) by applying the scale transformation $x \rightarrow bx$, $t \rightarrow b^z t$, to Eq. (11.128). As Dg is scale-invariant, v scales as x/t, $v \rightarrow b^{1-z}v$. In addition we require the scaling hypothesis

$$S(x,t) \rightarrow S(bx, b^z t) = b^\chi S(x,t), \tag{11.129}$$

which is meant to hold in a *statistical* sense, i.e., for averages and correlation functions of $S(x,t)$. Thus we find that the invariance under infinitesimal tilt implies $b^\chi S \rightarrow b^\chi S - b^{2-z}v \cdot x$. Both terms on the right-hand side must now scale in the same manner, which leads directly to Eq. (11.127).

11.3.3 KPZ scaling exponents in one dimension

In the special case of substrate dimension $d = 1$, through the mapping to the equivalent noisy Burgers equation and the ensuing 'hidden' fluctuation-dissipation relation, the KPZ stationary probability distribution turns out to be identical to the simple Gaussian (11.120) of the linear theory, and consequently the KPZ scaling exponents can be determined *exactly*. One immediately infers that the 'static' scaling exponent for the Gaussian model is $\eta = 0$. The identification (11.124) and the scaling relation (11.127) then imply the exact results

$$d = 1 : \quad \chi = \frac{1}{2}, \quad z = \frac{3}{2}, \tag{11.130}$$

which of course coincides with the DDS dynamic exponent $z_{\parallel}$ in one dimension. We finally remark that z can also be understood by investigating the dynamical properties of the soliton excitations (see Problem 11.3) in the one-dimensional

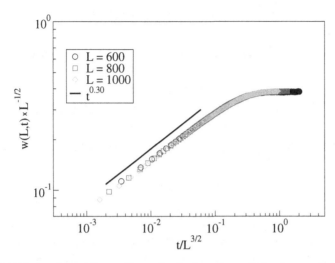

Fig. 11.11 Finite-size scaling collapse for the interface width of a one-dimensional growth model in the KPZ universality class (equivalent to the TASEP) with scaling exponents $\chi = 1/2$ and $z = 3/2$. The data were taken for systems with $L = 600$, 800, and 1000 lattice sites, and each curve is averaged over 50 000 realizations. [Figure reproduced with permission from: G. L. Daquila and U. C. Täuber, *Phys. Rev. E* **83**, 051107 (2011); DOI: 10.1103/PhysRevE.83.051107; copyright (2011) by the American Physical Society.]

Burgers equation.[19] A direct mapping of the totally asymmetric exclusion process (TASEP) to a microscopic growth model in the KPZ universality class is afforded by respectively identifying occupied/empty lattice sites for the driven lattice gas with upward/downward steps for the growing one-dimensional interface. Figure 11.11 shows numerical finite-size scaling for such a TASEP growth model; convincing data collapse is indeed achieved by means of the one-dimensional KPZ exponents. The graph also clearly shows that the growing interface is rough up to the scale $L_x(t) \sim t^{1/z}$.

There are a few experimental realizations for surface growth in $1 + 1$ dimensions under non-equilibrium conditions that clearly reveal the KPZ scaling exponents. For example, Fig. 11.12 shows data obtained from an experiment with silver electrodeposition under ionic mass transport control, where, following an initial diffusion-limited aggregation (DLA) regime, KPZ scaling ensues in the ballistic deposition region.[20] Interestingly, the one-dimensional flame fronts in slow combustion of sheets of paper also unambiguously display KPZ scaling properties in the kinetic roughening regime.[21] Striking evidence for the

[19] Fogedby, Eriksson, and Mikheev (1995).
[20] Schilardi, Azzaroni, Salvarezza, and Arvia (1999).
[21] Myllys, Maunuksela, Alava, Ala-Nissila, Merikoski, and Timonen (2001).

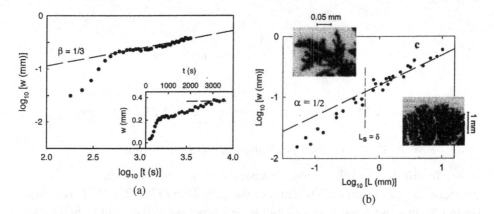

Fig. 11.12 Experimental data for the interface width scaling as function of: (a) time t, and (b) length L for Ag electrodeposition from an aqueous solution. The insets in (b) depict the emerging patterns at different length scales. [Figures reproduced with permission from: P. L. Schilardi, O. Azzaroni, R. C. Salvarezza, and A. J. Arvia, *Phys. Rev. B* **59**, 4638 (1999); DOI: 10.1103/PhysRevB.59.4638; copyright (1999) by the American Physical Society.]

$(1 + 1)$-dimensional KPZ scaling exponents has recently been observed in the turbulent dynamics of topological defects in the electroconvection of nematic liquid crystals.[22]

11.4 Renormalization group analysis of the KPZ equation

A renormalization group analysis at fixed dimension of the Kardar–Parisi–Zhang Langevin equation (11.114) or equivalent field theory (11.125) reveals that two quite different dimensional regimes must be distinguished. For dimensions $d < 2$, an infrared-stable RG fixed point governs the universal interface scaling properties. For $d > 2$, a non-trivial unstable critical fixed point separates a phase with Edwards–Wilkinson scaling exponents from a strong-coupling rough phase, which is not accessible by perturbational means. Thus, $d_{lc} = 2$ constitutes a *lower critical dimension* for this non-equilibrium *roughening transition*, and similarly to the non-linear sigma model analyzed in Section 7.4, the associated critical exponents can be accessed in a dimensional expansion in powers of $\epsilon = d - 2$. Indeed, this may be carried out to *all* orders in the directed-polymer representation, as is achieved through a stochastic Cole–Hopf transformation. Finally, we turn our attention to an important variant, the '*conserved* KPZ' equation, which has turned out relevant for the description of many experiments on growing surfaces.

---

[22] Takeuchi and Sano (2010, 2012).

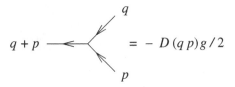

Fig. 11.13 Three-point vertex stemming from the KPZ non-linearity.

11.4.1 Renormalization at fixed dimension

Before turning to an explicit loop expansion for the KPZ equation, we explore the consequences of the general structure of the field theory (11.125) for its renormalization. To this end, we define multiplicatively renormalized quantities in the usual manner according to Eqs. (5.19) and

$$D_R = Z_D D , \qquad g_R^2 = Z_g \, g^2 C_d \, \mu^{d-2} , \qquad (11.131)$$

and it is understood that the UV divergences are to be absorbed into the Z factors. The elements of the perturbation expansion consist of the massless diffusion propagator (9.66), the noise vertex $\Gamma_0^{(2,0)} = -2D$, and the non-linear vertex $\Gamma^{(1,2)}$ shown in Fig. 11.13. Because of the momentum scalar product in the vertices, any loop contribution to the vertex function $\Gamma^{(1,1)}(q, \omega)$ vanishes $\propto q^2$ as $q \to 0$, see Fig. 11.3(a); hence, as in the equilibrium models with conserved order parameter (Section 5.2 and Chapter 6), and in driven diffusive systems (Sections 11.1 and 11.2):

$$\Gamma^{(1,1)}(q = 0, \omega) = i\omega \qquad (11.132)$$

to all orders in perturbation theory, and consequently

$$Z_{\tilde{S}} Z_S = 1 . \qquad (11.133)$$

Furthermore, we noted above that Dg appears as a structure constant in the 'Galilean' tilt transformation (11.128). Since v scales as the field S, one expects

$$Z_D (Z_g Z_S)^{1/2} = 1 . \qquad (11.134)$$

Equation (11.134) can also be directly confirmed via deriving a Ward identity associated with the infinitesimal interface tilt invariance, and an ensuing relation between derivatives of $\Gamma^{(1,1)}$ and $\Gamma^{(1,2)}$, see Problem 11.4; or through explicit evaluation of the three-point vertex function, see Problem 11.5.

This leaves us with only *two* independent renormalization constants Z_S and Z_D, which can be inferred from $\partial \Gamma^{(1,1)}(q, \omega)/\partial q^2$ and $\Gamma^{(2,0)}(q, \omega)$, respectively, evaluated at an appropriate normalization point (NP) outside the IR-singular region. As the KPZ model constitutes a massless field theory, one must take care not to set both momentum and frequency to zero simultaneously, which would inevitably lead to severe infrared problems. Instead one may either use $q^2 = 4\mu^2$ and $\omega = 0$, say, or, equivalently, $q = 0$ and $i\omega = 2D_R\mu^2$, which we shall choose for convenience. The one-loop diagrams for $\Gamma^{(1,1)}(q, \omega)$ and $\Gamma^{(2,0)}(q, \omega)$ are of the same form as for driven diffusive systems, depicted in Fig. 11.3. The corresponding analytic expressions, after performing the internal frequency integration via the residue theorem, read

$$\Gamma^{(1,1)}(q, \omega) = i\omega + Dq^2 + \frac{Dg^2}{2} \int_k \frac{\frac{q^2}{2} - q \cdot k}{\left(\frac{q}{2} - k\right)^2} \frac{\frac{q^2}{4} - k^2}{i\omega/2D + \frac{q^2}{4} + k^2}, \qquad (11.135)$$

$$\Gamma^{(2,0)}(q, \omega) = -2D - \frac{Dg^2}{2} \int_k \frac{\left(\frac{q^2}{4} - k^2\right)^2}{\left(\frac{q}{2} + k\right)^2 \left(\frac{q}{2} - k\right)^2} \text{Re} \frac{1}{i\omega/2D + \frac{q^2}{4} + k^2}, \qquad (11.136)$$

see Problem 11.6. Using $\int_p (q \cdot k)^2 f(k) = (q^2/d) \int_k k^2 f(k)$, we compute at the normalization point $q = 0$, $i\omega = 2D\mu^2$,

$$\frac{\partial}{\partial q^2} \Gamma^{(1,1)}(q, \omega)\Big|_{NP} = D\left[1 - \frac{d-2}{d} \frac{g^2}{4} \int_k \frac{1}{\mu^2 + k^2}\right], \qquad (11.137)$$

$$\Gamma^{(2,0)}(q, \omega)\Big|_{NP} = -2D\left[1 + \frac{g^2}{4} \int_k \frac{1}{\mu^2 + k^2}\right]. \qquad (11.138)$$

The brackets on the right-hand sides of these equations become identical for $d = 1$, reflecting the 'hidden' fluctuation-dissipation theorem and the absence of field renormalization. If we want to include the one-dimensional case in our general RG analysis, we must be careful not to lose this specific feature of Eq. (11.137), which originates in the angular part of the d-dimensional integral (11.135) that leads to the prefactor $d - 2$. For, in dimensional regularization this integral becomes, according to Eq. (5.10):

$$\int_k \frac{1}{\mu^2 + k^2} = \frac{C_d\mu^{d-2}}{2 - d}.$$

In the *minimal* subtraction scheme, therefore, the factor $d - 2$ would simply cancel, and one would infer $Z_D = 1$, inevitably leading to $z = 2$. Thus, if the exactly known KPZ scaling exponents in $d = 1$, with $z = 3/2$, are to be captured as well, the 'geometric' factor $d - 2$ in (11.137) needs to be preserved. Consequently, we apply a *non-minimal* regularization procedure here, and identify the right-hand sides of Eqs. (11.137) and (11.138) with the appropriate Z factors, i.e., using the identity (11.133),

$$Z_D = 1 - \frac{d-2}{d} \frac{g^2}{4} \frac{C_d \mu^{d-2}}{2-d} , \qquad (11.139)$$

while Eq. (11.138) yields the product $Z_{\tilde{s}} Z_D$, and therefrom

$$Z_S = 1 - \frac{d-1}{d} \frac{g^2}{2} \frac{C_d \mu^{d-2}}{2-d} . \qquad (11.140)$$

As required, this field renormalization vanishes in one dimension.

Proceeding to the associated renormalization group equation, which is once again set up and solved in precisely the same manner as for model A critical dynamics at $T = T_c$ (see Section 5.3), we compute the RG flow functions at fixed dimension d,

$$\gamma_S = \mu \left. \frac{\partial}{\partial \mu} \right|_0 \ln Z_S = \frac{d-1}{d} \frac{g_R^2}{2} , \qquad (11.141)$$

$$\gamma_D = \mu \left. \frac{\partial}{\partial \mu} \right|_0 \ln Z_D = \frac{d-2}{d} \frac{g_R^2}{4} . \qquad (11.142)$$

Galilei invariance of the Burgers equation, or tilt invariance of the corresponding growth problem, had implied the additional relation $Z_g = Z_S^{-1} Z_D^{-2}$, which means that the RG beta function for the renormalized non-linear coupling g_R^2 is given by

$$\beta_g = \mu \left. \frac{\partial}{\partial \mu} \right|_0 g_R^2 = g_R^2 (d - 2 - \gamma_S - 2\gamma_D) . \qquad (11.143)$$

This immediately leads us to the scaling relation (11.127) which reduces the number of independent scaling exponents to one, *provided* there exists a non-trivial finite fixed point $0 < g^* < \infty$. In this case, the fixed point condition $\beta_g(g^*) = 0$ becomes $\gamma_S^* + 2\gamma_D^* = d - 2$. From the analogy with the equilibrium critical dynamics of model A, we infer

$$\eta = -\gamma_S^* , \quad \chi = \frac{2-d+\gamma_S^*}{2} , \quad z = 2 + \gamma_D^* , \qquad (11.144)$$

and consequently $\chi + z = 2 + 1 - \frac{d}{2} + \frac{1}{2}(\gamma_S^* + 2\gamma_D^*) = 2$.

To one-loop order in our non-minimal prescription at fixed dimension,

$$\beta_g = g_R^2 \left(d - 2 - \frac{2d-3}{2d} g_R^2 \right) , \qquad (11.145)$$

yielding the Edwards–Wilkinson fixed point $g^* = 0$, and the non-trivial KPZ fixed point

$$(g^*)^2 = \frac{2d(d-2)}{2d-3} , \qquad (11.146)$$

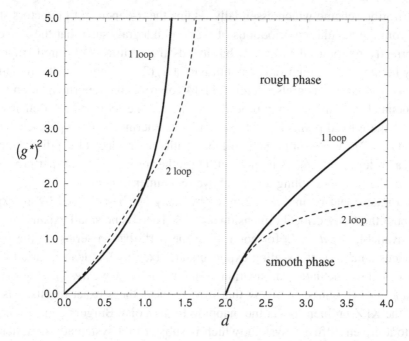

Fig. 11.14 KPZ fixed point value $(g^*)^2$ as function of (fixed) dimension d in one-loop (solid line) and two-loop approximation (dashed). For $d < 2$, the non-trivial fixed point g^* is stable. For $d \geq 2$, it becomes unstable, and controls a continuous non-equilibrium roughening transition between two distinct dynamic phases. [Figure adapted from Frey and Täuber (1994).]

which is plotted in Fig. 11.14 as a function of d. Clearly, the regimes $d < 2$ and $d > 2$ must be distinguished. For $d < 2$, $\beta_g'^* = \partial \beta_g / \partial g_R^2 \big|_{g^*} = 2 - d > 0$, and the non-trivial fixed point (11.146) is infrared-*stable*. Upon inserting into Eqs. (11.141), (11.142), and (11.144), one finds

$$\eta = -\frac{(d-1)(2-d)}{3-2d} \ , \qquad \chi = \frac{(2-d)^2}{2(3-2d)} \ , \qquad z = 2 - \frac{(2-d)^2}{2(3-2d)} \ . \qquad (11.147)$$

Thus, $\chi > 0$ and the interface is rough for $d < 3/2$. In $d = 1$, we recover the exact results $\eta = 0$, $\chi = 1/2$, and $z = 3/2$. Notice the divergence of g^* as $d \uparrow 3/2$, which is merely a consequence of the fact that the one-loop contribution to β_g happens to vanish at this point. In the range $3/2 < d < 2$, the fixed point $(g^*)^2$ is negative, which is of course unphysical; the one-loop approximation is obviously insufficient here.

In a massless theory a loop expansion at fixed dimension generates more and more troublesome *infrared* divergences with increasing order in perturbation theory. Yet, at least to two-loop order, it is possible to devise a modified dimensional regularization scheme which essentially retains the d-dependent prefactors stemming from the angular contributions of the loop integrals such that the Z factors are correctly reproduced in $d = 1$, but avoids the additional infrared singularities by taking into account only the ultraviolet $1/(2 - d)$ poles of the remaining one-dimensional $|k|$ integrals.[23] Figure 11.14 compares the renormalization group fixed points from this two-loop procedure with our one-loop results. Remarkably, in $d = 1$ these fixed points actually coincide. Furthermore, the unphysical singularity at $d = 3/2$ is absent in the two-loop result, and replaced by a divergence as $d \uparrow 2$; as it turns out, this is no artifact, but rather reflects the non-perturbational nature of the strong-coupling regime above two dimensions.

The dimensional regime $d < 2$ may obviously not be accessed by an expansion about the *lower* critical dimension $d_{lc} = 2$. There is no small expansion parameter available for $d < d_{lc}$ for controlling the perturbation series. In the physical dimension $d = 1$, the scaling exponents are fixed by Galilean invariance and the 'accidental' fluctuation-dissipation relation, and we have *designed* our renormalization scheme in such a way that it reproduces these exact results. Also, in $d = 1$ the KPZ problem is via the mapping to the noisy Burgers equation equivalent to a driven diffusive system, which is subject to a systematic treatment by means of a $2 - \epsilon$ dimensional expansion, cf. Section 11.1. On the other hand, as the exponents that characterize the leading power laws are known exactly, a self-consistent theory for the scaling functions promises to produce very accurate results. Indeed, one may show that all the singular two-loop contributions to the two-point functions stemming from vertex renormalizations vanish in $d = 1$. Consequently, the self-consistent one-loop 'mode-coupling' approximation (see Section 6.2.2 and Problem 11.6) works very well, and the numerical solution of the ensuing coupled integral equations matches simulation results quite accurately.[24] Finally, in the trivial zero-dimensional case, the height variable simply performs a random walk in time, $\Delta S(t)^2 \sim t^{2\chi/z}$, which in terms of the KPZ exponents means $\chi/z = 1/2$. Although formally $g^* \to 0$ as $d \downarrow 0$ (there exist no gradients in zero dimension), see Fig. 11.14, and neither χ nor z have a physical meaning for themselves, Eq. (11.147) yields $\chi = 2/3$ and $z = 4/3$, which reproduces the correct ratio.

11.4.2 Directed polymers and the roughening transition

In dimensions $d > 2$, on the other hand, $\beta_g'^* = 2 - d < 0$, and the KPZ fixed point (11.146) becomes *unstable*. Indeed, the RG beta function (11.143) now looks just as in Fig. 7.7 for the non-linear sigma model for $d > d_{lc}$. For initial values

---

[23] Frey and Täuber (1994).   [24] Frey, Täuber, and Hwa (1996).

$g_R < g_c = g^*$, the renormalization group flow leads into the (Gaussian) Edwards–Wilkinson fixed point $g_0^* = 0$, which is stable above two dimensions, and describes a smooth interface. For $g_R > g_c$, the non-linear coupling will tend to infinity; this *strong-coupling* regime is expected to be characterized by a positive roughness exponent. Hence the unstable fixed point $g_c = g^*$ defines a separatrix for the RG flow trajectories in parameter space, and is to be interpreted as indicating a dynamic phase transition between a smooth Edwards–Wilkinson phase, and a rough strong-coupling regime (Fig. 11.14). Since $g_c \to 0$ as $d \downarrow 2$, we may indeed identify $d_{lc} = 2$ as the *lower critical dimension* for the roughening transition. We may actually further exploit the analogy with the non-linear sigma model discussed in Section 7.4, which described quantum or classical critical behavior at an equilibrium phase transition in the vicinity of its lower critical dimension. Despite the apparent non-renormalizability of the field theory above $d_{lc} = 2$, its critical properties can be systematically accessed by means of a $d = 2 + \epsilon$ expansion, because the ultraviolet regime is controlled by the appearance of an infrared-unstable, ultraviolet-stable fixed point g_c. The role of the rotation symmetry in order parameter space which ensured the renormalizability of the theory to all orders in ϵ for the non-linear sigma model is assumed by the Galilean/tilt invariance in the KPZ problem. Within the one-loop approximation, or, equivalently, to first order in $\epsilon = d - 2$, we find at the roughening transition

$$g_c^2 = 4\epsilon + O(\epsilon^2) \,, \tag{11.148}$$

and $\gamma_S^* = \epsilon + O(\epsilon^2)$, $\gamma_D^* = O(\epsilon^2)$, resulting in the *critical exponents*

$$\eta_c = -\epsilon + O(\epsilon^2) \,, \quad \chi_c = 0 + O(\epsilon^2) \,, \quad z_c = 2 + O(\epsilon^2) \,. \tag{11.149}$$

In addition, we define a crossover length scale and associated correlation length exponent ν, again in complete analogy with our earlier analysis of the non-linear sigma model. Expanding β_g in the vicinity of the critical point, $\beta_g(g_R^2) \approx (g_R^2 - g_c^2) \beta_g'^* \approx 2g_c (g_R - g_c) \beta_g'^*$, the flow equation $\beta_g = \ell \, d\tilde{g}(\ell)^2/d\ell \approx 2g_c \ell \, d\tilde{g}(\ell)/d\ell$ becomes

$$\ell \frac{d\tilde{g}(\ell)}{d\ell} \approx \beta_g'^* [\tilde{g}(\ell) - g_c] \,, \tag{11.150}$$

with the solution $\tilde{g}(\ell) - g_c \sim (g_R - g_c) \ell^{\beta_g'^*}$. If we then identify the crossover length ξ via $\ell \sim \xi^{-1} \sim |g_R - g_c|^\nu$, and integrate Eq. (11.150) until $\tilde{g}(\ell) - g_c = O(1)$, we obtain

$$\nu^{-1} = -\beta_g'^* = \epsilon + O(\epsilon^2) \,, \tag{11.151}$$

precisely as in Eq. (7.205) for the non-linear sigma model. As is usually the case, the correlation length exponent ν diverges upon approaching the lower critical

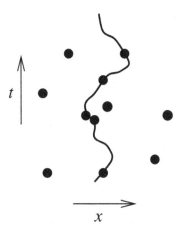

Fig. 11.15 Directed line interacting with randomly distributed point defects.

dimension $d_{lc} = 2$. An explicit two-loop calculation in the KPZ representation yields, after extensive algebra, that the above results (11.149) and (11.151) remain unchanged to order ϵ^3. Indeed, as we shall see next, they actually hold to *all* orders in the perturbation series expansion in $\epsilon = d - 2$.

To this end, consider an apparently unrelated equilibrium statistical mechanics problem, namely that of a single *directed* line in $d + 1$ dimensions, which interacts with randomly distributed pinning centers,[25] depicted schematically in Fig. 11.15. Physically, this represents, for example, the dilute limit of magnetic flux lines in type-II superconductors subject to quenched point defects (such as oxygen vacancies in ceramic high-T_c compounds) that act as attractive binding sites.[26] If we label the space direction along which the idealized directed 'polymer' with line tension ε is oriented as t, and model the disorder by a Gaussian-distributed, short-range, local attractive potential $V(x, t)$, then the restricted partition function $Z(x, t)$ for the line trajectory starting at the point $(y, s) = (0, 0)$ and terminating at (x, t) is given by the functional integral

$$Z(x, t) = \int_{y(0)=0}^{y(t)=x} \mathcal{D}[y] \exp\left(-\frac{1}{k_B T} \int_0^t \left[\frac{\varepsilon}{2}\left(\frac{dy(s)}{ds}\right)^2 + V(y(s), s)\right] ds\right),$$

$$(11.152)$$

where one has to sum over all possible intermediate paths $y(s)$ connecting the start and end points. This expression resembles an imaginary-time Feynman path integral for quantum particles of 'mass' ε, subject to the external potential $V(x, t)$, see Section 7.1. In an equivalent formulation of the problem, we can therefore

[25] See, e.g., Kardar and Zhang (1987); Fisher and Huse (1991); Hwa and Fisher (1994).
[26] See, e.g., Nattermann and Scheidl (2000).

search for the solution of the imaginary-time Schrödinger equation

$$k_B T \frac{\partial Z(x,t)}{\partial t} = \frac{(k_B T)^2}{2\varepsilon} \nabla^2 Z(x,t) + V(x,t) Z(x,t), \quad (11.153)$$

where the temperature $k_B T$ plays the role of Planck's constant $\hbar$, and measures the strength of thermal (rather than quantum) fluctuations. Bypassing the quantum-mechanical analogy, we could of course just have stated that the diffusion equation (11.153) with multiplicative noise is formally solved by means of the 'propagator' (11.152).

The linear diffusion equation with multiplicative noise (11.153) can be mapped onto the non-linear KPZ equation with additive noise via the *Cole–Hopf transformation*

$$S(x,t) = \frac{k_B T}{Dg\varepsilon} \ln Z(x,t), \quad (11.154)$$

which, upon identifying

$$k_B T/2\varepsilon \to D, \quad V/Dg\varepsilon \to \zeta, \quad (11.155)$$

yields Eq. (11.114). Specifically, this transformation implies that the deterministic KPZ and Burgers equations are equivalent to a simple diffusion equation for the variable $Z(x,t)$, and may therefore be readily integrated for any given initial condition. Translating the KPZ renormalization group results to the directed-line picture, we find that $d_{lc} + 1 = 2 + 1$ represents the lower critical dimension for a second-order *equilibrium* phase transition between a phase where the line tension dominates over the pinning energy, and the polymer remains essentially straight, and a low-temperature (notice that $g > g_c$ corresponds to $T < T_c$) disorder-dominated 'glassy' phase, in which the line minimizes its free energy by finding an energetically most favorable path amongst the point defects. In $d = 1 + 1$ dimensions, the binding potential always prevails, and as a consequence the pinned line fluctuates strongly, $\Delta x \sim (\Delta t)^{\chi/z} = (\Delta t)^{1/3}$, as it traverses the sample in the t direction.

For the roughening transition itself, the directed-polymer representation allows the formulation of a simple (hyper-)scaling argument[27] which determines the critical exponents χ_c and z_c. To this end, consider rescaling lengths x by a factor b, and correspondingly 'time' t by a factor b^z in Eq. (11.152). In the low-temperature glassy phase, the random contribution to the free energy should scale as b^Θ, with a positive *stiffness exponent* $\Theta \geq 0$, see Section 1.3.2, which should become marginal precisely at the critical point, $\Theta_c = 0$. The elastic term in Eq. (11.152), on the other hand, scales as b^{z-2}, for the line tension ε should not be affected by the random potential V, as a consequence of Galilean invariance. The hyperscaling argument now postulates that at the critical point the elastic and random contributions to

---

[27] Doty and Kosterlitz (1992).

the free energy should balance each other, and therefore $z_c - 2 = \Theta_c = 0$. Thus, using the scaling relations (11.127) and (11.124), one arrives at the *exact* critical exponent values

$$z_c = 2, \quad \chi_c = 0, \quad \eta_c = 2 - d. \tag{11.156}$$

In the field theory framework, we can actually demonstrate the validity of Eq. (11.156) at the roughening transition to *all* orders in perturbation theory, thus generalizing our previous one-loop results, (11.149) and (11.151), considerably.[28] Motivated by the Cole–Hopf transformation (11.154), we apply the non-linear canonical stochastic transformation

$$n(x, t) = \frac{2}{g} e^{g\, S(x,t)/2}, \quad \tilde{n}(x, t) = \tilde{S}(x, t) e^{-g\, S(x,t)/2} \tag{11.157}$$

to the KPZ Janssen–De Dominicis action (11.125). Because the associated functional determinant is one, this leads directly to the response functional for a linear diffusion process subject to multiplicative noise,[29]

$$A[\tilde{n}, n] = \int d^d x \int dt \left[\tilde{n}(x, t) \left(\frac{\partial}{\partial t} - D\nabla^2 \right) n(x, t) \right.$$

$$\left. - \frac{Dg^2}{4} \tilde{n}(x, t)^2\, n(x, t)^2 \right]. \tag{11.158}$$

Notice that this ensuing field theory is of precisely the same form as that for the diffusion-limited pair annihilation reactions, cf. Section 9.3. It is hence impossible to construct any loop diagrams which contribute to the renormalization of the propagator or $\Gamma^{(1,1)}(q, \omega)$, which means

$$Z_{\tilde{n}} Z_n = 1, \quad Z_D = 1, \tag{11.159}$$

and therefore immediately implies $z = 2$ in the entire *perturbational* regime. Provided there exists a non-trivial RG fixed point $0 < g_c < \infty$, then $\eta_c = 2 - d$ and $\chi_c = 0$ at the critical point as well, which follows from Eqs. (11.143) and (11.144) with $\gamma_D^* = 0$.

In order to establish the existence of an unstable critical fixed point describing the roughening transition for $d \geq d_{lc} = 2$, we just have to consider the renormalization of the four-point vertex function $\Gamma^{(2,2)}$. As for the pair annihilation processes, one can sum the entire perturbation expansion for $\Gamma^{(2,2)}$ via a geometric series in Fourier–Laplace space, cf. Fig. 9.5. Using Eq. (11.159), this yields at a suitable

---

[28] Lässig (1995); Wiese (1998).

[29] We are slightly sloppy here in treating the fields S and $\tilde{S}$ as continuous differentiable functions. To be more precise, one needs to take recourse to an apt discretization of the functional integral; this procedure leads eventually to the action (11.158); see Janssen, Täuber, and Frey (1999).

normalization point

$$Z_g^{-1} = 1 + \frac{g^2}{4} \frac{C_d \mu^{d-2}}{d-2} \,, \tag{11.160}$$

from which we obtain the *exact* RG beta function

$$\beta_g = g_R^2 \left(d - 2 - \frac{g_R^2}{4} \right) . \tag{11.161}$$

This coincides with the one-loop result (11.145), if we expand the latter to first order in $\epsilon = d - 2$. To all orders in perturbation theory, we thus find the unstable fixed point $g_c^2 = 4(d-2)$, which would be represented in Fig. 11.14 by the tangent to the curves originating at the point $(d = 2, g^* = 0)$. The associated critical correlation length exponent reads

$$\nu^{-1} = -\beta_g'^* = d - 2 \,, \tag{11.162}$$

compare with Eqs. (11.148) and (11.151). For disordered systems, the Chayes–Chayes–Fisher–Spencer inequality[30] states that an appropriately defined correlation length exponent is bound by $\nu \geq 2/d$. Under the assumption that the corresponding premises hold here, and also apply to our exponent ν which characterizes the divergence of the crossover length scale near the phase transition, the exact result (11.162) predicts that ν reaches the lower bound of the Chayes–Chayes–Fisher–Spencer inequality at $d = d_c' = 4$. Above this *upper critical dimension* for the KPZ non-equilibrium roughening transition, the correlation length critical exponent should remain equal to its 'mean-field' value $\nu = 1/2$.

While the non-linear coupling flows to zero for $g < g_c$, even the summation of the perturbation series to all orders does not suffice to identify a finite fixed point in the strong-coupling regime $g > g_c$. We are therefore led to the conclusion that the rough phase is *inaccessible* through perturbational methods. Consequently, little is known about the strong-coupling KPZ regime, and even the simulation and direct numerical integration results have remained somewhat controversial. Monte Carlo simulations for discrete growth models have given values for the roughness exponent in the range $\chi \approx 0.3$–0.4 in $d = 2$, and $\chi \approx 0.2$–0.3 in $d = 3$; the dynamic exponent has been estimated to be $z \approx 1.6$ and $z \approx 1.7$ in two and three dimensions, respectively. Remarkably, the scaling relation $\chi + z = 2$ appears to be fulfilled (at least within the numerical accuracy) even in the strong-coupling rough phase, where the perturbational coupling $g \to \infty$. By allowing for spatially long-range noise correlations in the KPZ problem (see Problem 11.7), whereby the lower critical dimension d_{lc} is shifted upwards, it can be established that the short-range KPZ scaling behavior *below* d_{lc} cannot exist beyond $d = 4$ dimensions.

---

[30] Chayes, Chayes, Fisher, and Spencer (1986).

However, this scaling regime with $\chi > 0$ is conceptually distinct from the strong-coupling phase beyond the roughening transition above d_{1c}, and another, hitherto unsolved problem is the question whether there exists an upper critical dimension for this rough phase as well.

In addition to the perturbational renormalization group and numerical simulations, a variety of non-perturbative methods have been applied to the KPZ problem, e.g., mode-coupling theory and functional renormalization, but to date no fully coherent and generally accepted picture for the strong-coupling rough phase has emerged. There are even some indications that universality might break down in this regime, in the sense that the scaling exponents could depend on lattice details and discretization schemes. In this context, one should bear in mind that our 'derivations' of the KPZ equation in Section 11.3 fundamentally and explicitly assumed small height fluctuations $(\nabla S)^2 \ll 1$, a condition that is evidently violated in the rough phase. Notice also that our earlier arguments regarding the irrelevance of higher non-linear and gradient terms were merely based on the scaling dimensions at the Gaussian (Edwards–Wilkinson) fixed point. Perhaps the divergence of the non-linear coupling is but a signature of the inadequacy of the simple KPZ continuum description in the strong-coupling regime. Nevertheless, non-perturbative RG approaches starting directly from a continuum field theory representation have had success in reaching and characterizing a KPZ strong-coupling scaling regime.[31]

11.4.3 The 'conserved KPZ' equation

In many experiments on growing surfaces under non-equilibrium conditions, surface diffusion plays an important role. The ensuing interface fluctuations are therefore usually not described by the KPZ universality class, but can be modeled by certain variants, subsumed under the name 'conserved KPZ' equation,[32]

$$\frac{\partial S(x,t)}{\partial t} = -D\nabla^2\left[\nabla^2 S(x,t) + \frac{g}{2}[\nabla S(x,t)]^2\right] + \zeta(x,t). \qquad (11.163)$$

The relation of the Langevin equation (11.163) to the original KPZ equation (11.114) is analogous to that of the equilibrium conserved and non-conserved relaxational models B and A of Section 3.1. Yet there exists no detailed balance constraint here that fixes the noise correlations, and one may legitimately study the two distinct cases of

$$\langle \zeta(x,t)\zeta(x',t')\rangle = 2D(i\nabla)^a\delta(x-x')\delta(t-t'), \qquad (11.164)$$

[31] Canet, Chaté, Delamotte, and Wschebor (2010, 2011); Kloss, Canet, and Wschebor (2012).
[32] Sun, Guo, and Grant (1989); Wolf and Villain (1990); for the field-theoretical RG analysis of these models, see Janssen (1997).

with (i) fully conserved noise with $a = 2$ (*Sun–Guo–Grant model*), and (ii) shot-noise $a = 0$ (*Wolf–Villain model*), which should, for example, describe ideal molecular beam epitaxy with random particle deposition.

Equations (11.163) and (11.164) are equivalent to the Janssen–De Dominicis functional

$$A[\widetilde{S}, S] = \int d^d x \int dt \, \widetilde{S}(x, t) \left(\frac{\partial S(x, t)}{\partial t} - D(i\nabla)^a \widetilde{S}(x, t) \right.$$

$$\left. + D\nabla^2 \left[\nabla^2 S(x, t) + \frac{g}{2} [\nabla S(x, t)]^2 \right] \right). \quad (11.165)$$

The propagator is now the one of model B at the critical point, $G_0(q, \omega) = (-i\omega + Dq^4)^{-1}$, the noise vertex reads $\Gamma_0^{(2,0)} = -2Dq^a$, and the non-linear three-point vertex $\Gamma^{(1,2)}$ has the same form as in Fig. 11.13, albeit supplemented by an additional external momentum factor $(q + p)^2$. Measuring lengths and times according to $[x] = \mu^{-1}$ and $[t] = \mu^{-4}$, respectively, such that again $[D] = \mu^0$, we obtain the scaling dimensions

$$[\widetilde{S}(x, t)] = \mu^{2+(d-a)/2}, \quad [S(x, t)] = \mu^{-2+(d+a)/2}, \quad [g] = \mu^{(4-d-a)/2}.$$
$$(11.166)$$

We therefore expect a critical dimension $d_c = 4 - a = 2$ in the fully conserved case, and $d_c = 4$ for the Wolf–Villain variant. As opposed to the KPZ problem, however, these are the *upper* critical dimensions for the conserved KPZ equation, above which mean-field theory holds with (obviously) $\eta = 0$, $z = 4$, and $\chi = (d_c - d)/2$, see Eq. (11.171) below, in analogy with the Edwards–Wilkinson case. For $d < d_c$, strong fluctuations lead to non-trivial scaling exponents, and the growing surface is rough. However, *no* kinetic roughening transition exists in either of these conserved KPZ variants.

In fact, there is once more only *one* independent scaling exponent. This follows from Eq. (11.132), which of course holds again, and thus $Z_{\widetilde{S}} Z_S = 1$, and the observation that as a consequence of the additional momentum factors in the vertices all the loop contributions to $\Gamma^{(2,0)}(q, \omega)$ are proportional to q^4. Hence

$$\left. \frac{\partial}{\partial q^a} \Gamma^{(2,0)}(q, \omega) \right|_{q=0} = -2D, \quad (11.167)$$

which, using the definitions of Eq. (11.131), leads to

$$Z_D = Z_{\widetilde{S}}^{-1} = Z_S, \quad (11.168)$$

whence for the conserved KPZ variants

$$z = 4 + \gamma_D^* = 4 + \gamma_S^* = 4 - \eta, \quad (11.169)$$

precisely as for the equilibrium model B with conserved order parameter, see Section 5.3. Furthermore, Eqs. (11.132) and (11.167) imply

$$C(q, \omega) = -\frac{\Gamma^{(2,0)}(q, \omega)}{|\Gamma^{(1,1)}(q, \omega)|^2} = |q|^{a-2z}\, \hat{C}(\omega/D|q|^z)\,. \tag{11.170}$$

Thus $C(q, t = 0) \sim |q|^{a-z}$ and $C(x, t = 0) \sim 1/|x|^{d+a-z}$, giving the roughness exponent

$$\chi = \frac{z - d - a}{2} = \frac{4 - a - d - \eta}{2}\,. \tag{11.171}$$

These scaling exponents are easily computed to first order in $\epsilon = 4 - a - d$. To this end, we define a renormalized coupling strength

$$g_R^2 = Z_g\, g^2 C_{ad}\mu^{d+a-4}\,, \qquad C_{ad} = \frac{\Gamma(3 - a/2 - d/2)}{2^{d-1}\pi^{d/2}}\,. \tag{11.172}$$

Within the one-loop approximation (but *not* to higher orders in perturbation theory), the three-point vertex function $\Gamma^{(1,2)}$ turns out to be non-singular in the ultraviolet (see Problem 11.8), as in the non-conserved KPZ problem, where this relation followed as a consequence of the underlying Galilean invariance. Thus, $Z_g = Z_S^{-1}Z_D^{-2} = Z_S^{-3}$ to one-loop order, and

$$\beta_g = \mu \left.\frac{\partial}{\partial\mu}\right|_0 g_R^2 = g_R^2\big[d + a - 4 - 3\gamma_S^* + O(g_R^4)\big]\,. \tag{11.173}$$

At any non-zero finite fixed point $0 < g^* < \infty$, the scaling exponents are then necessarily fixed to $O(\epsilon)$ by the condition $3\gamma_S^* = d + a - 4 = -\epsilon$, or

$$\eta = \chi = \frac{\epsilon}{3} + O(\epsilon^2)\,, \qquad z = 4 - \frac{\epsilon}{3} + O(\epsilon^2)\,. \tag{11.174}$$

Indeed, by explicitly computing the renormalization of $\Gamma^{(1,1)}(q, \omega)$ in the one-loop approximation (Problem 11.8), one finds the $O(\epsilon)$ RG fixed point

$$(g_a^*)^2 = \frac{2^{3-a/2}}{3}\epsilon + O(\epsilon^2)\,, \tag{11.175}$$

which is infrared-stable for $d < d_c = 4 - a$. The two-loop corrections are in fact numerically rather tiny, and the one-loop exponents (11.174) describe the scaling behavior of the conserved KPZ equation variants remarkably well.

Problems

11.1 *Logarithmic corrections for driven diffusive systems*
Solve the RG flow equations for the renormalized non-linear coupling v_R and the parameter c_R for driven diffusive systems at the upper critical dimension $d_c = 2$. Therefrom derive the logarithmic corrections to

(a) the temporal decay of the density autocorrelations, see Eq. (11.27),

(b) the initial-slip scaling law (11.53), and

(c) the generating function for the current cumulants.

11.2 *Anomalous scaling dimensions for critical driven diffusive systems*

(a) Confirm the renormalization factor (11.68) for the anisotropy parameter c of the critical driven Ising lattice gas (KLS model).

(b) Compute the anomalous scaling dimension $\gamma_{\tilde{c}} = 2 - d_\| v_R / 2$ and hence (11.99) for the irrelevant longitudinal noise strength $\tilde{c}D$.

11.3 *Solitary waves in the one-dimensional Burgers equation*

Show that the one-dimensional deterministic Burgers equation has the stationary kink solution

$$u(x) = -u_0 \tanh\left[\frac{g u_0}{2}(x - x_0)\right] ,$$

centered at x_0. By applying a Galilean boost, obtain the moving solitary wave

$$u(x, t) = v - u_0 \tanh\left[\frac{g u_0}{2}(x - x_0 - Dgvt)\right] ,$$

and, correspondingly for the one-dimensional KPZ equation,

$$S(x, t) = \frac{2}{g} \ln \cosh\left[\frac{g u_0}{2}(x - x_0 - Dgvt)\right] - vx .$$

11.4 *Ward identity from tilt invariance*

(a) Use the tilt invariance of the Kardar–Parisi–Zhang surface profile,

$$S(x, t) \to S'(x', t') = S(x - Dgvt, t) - v \cdot x ,$$

$$\tilde{S}(x, t) \to \tilde{S}'(x', t') = \tilde{S}(x - Dgvt, t) ,$$

to derive the following Ward identity for the generating functional Γ of the vertex functions:

$$\int d^d x \int dt \left(Dgt \left[\frac{\delta \Gamma}{\delta \tilde{S}(x, t)} \nabla \tilde{S}(x, t) + \frac{\delta \Gamma}{\delta S(x, t)} \nabla S(x, t) \right] + x \frac{\delta \Gamma}{\delta S(x, t)} \right) = 0 .$$

(b) Taking appropriate functional derivatives, show that

$$Dg q \frac{\partial \Gamma^{(1,1)}(q, \omega)}{\partial (i\omega)} = \left. \frac{\partial \Gamma^{(1,2)}(-q - p, -\omega; q, \omega; p, 0)}{\partial p} \right|_{p=0} ,$$

compare with Eq. (11.31), and consequently $Z_D(Z_g Z_S)^{1/2} = 1$.

11.5 *Renormalization of the KPZ three-point function*

By explicitly computing the three-point vertex function $\Gamma^{(1,2)}$ for the Kardar–Parisi–Zhang equation in the one-loop approximation, demonstrate that $Z_D(Z_g Z_S)^{1/2} = 1$. Using Eq. (11.135), check the Ward identity of Problem 11.4 to this order.

11.6 *Two-point functions and Burgers/KPZ exponents in one dimension*
Confirm the one-loop results (11.135) and (11.136) for the KPZ two-point
vertex functions. In one dimension, show that

$$\text{Re } \Gamma^{(1,1)}(q, \omega) = -\frac{q^2}{2} \Gamma^{(2,0)}(q, \omega) ,$$

and consequently $Z_S = Z_{\tilde{s}} = 1$, implying $\eta = 0$ and $\chi = 1/2$. In $d = 1$,
write down the self-consistent one-loop ('mode-coupling') equation for the
line width $\Delta(q) = \Gamma^{(1,1)}(q, \omega = 0)$ of the dynamic correlation function in the
Lorentzian approximation (see Section 6.2.2), and show that it is solved by
$\Delta(q) \propto |q|^z$ with $z = 3/2$.

11.7 *KPZ equation with spatially long-range correlated noise*
Consider the KPZ equation (11.114) with long-range noise correlator

$$\langle \zeta(q, t)\zeta(q', t') \rangle = 2D(1 + w |q|^{-2\rho})(2\pi)^d \delta(q + q')\delta(t - t') .$$

Convince yourself that the non-analytic noise contribution $\propto |q|^{-2\rho}$ cannot
renormalize perturbatively. As a consequence, deduce the RG beta function
for the additional long-range effective coupling $f = g^2 w$:

$$\beta_f = f(d - 2 - 2\rho - 3\gamma_D) .$$

In conjunction with Eq. (11.143), this already fixes the dynamic and roughness
exponents to the values

$$z = \frac{4 + d - 2\rho}{3} , \qquad \chi = \frac{2 - d + 2\rho}{3} ,$$

provided both $0 < g^* < \infty$ and $0 < f^* < \infty$. The long-range noise fixed
point with these scaling exponents competes with the short-range noise fixed
point ($w^* = f^* = 0$); show that in $d = 1$, the latter remains infrared-stable
for $\rho < \rho_c = 1/4$.

11.8 *'Conserved KPZ' equation (SGG and Wolf–Villain models)*
For the conserved KPZ problem defined by Eqs. (11.163) and (11.164),
compute the two-point vertex function $\Gamma^{(1,1)}(q, \omega)$ to one-loop order, see
Fig. 11.3(a) and compare with (11.135),

$$\Gamma^{(1,1)}(q, \omega) = i\omega + Dq^4$$

$$+ (Dg)^2 q^2 \int_k \frac{\frac{q^2}{2} - q \cdot k}{\left(\frac{q}{2} - k\right)^{4-a}} \frac{\left(\frac{q}{2} + k\right)^2 \left(\frac{q^2}{4} - k^2\right)}{i\omega + D\left(\frac{q}{2} + k\right)^4 + D\left(\frac{q}{2} - k\right)^4} .$$

Deduce $Z_D = Z_S$ from this result, and therefrom compute the fixed point
(11.175) and scaling exponents (11.174).

References

Barabási, A.-L. and H. E. Stanley, 1995, *Fractal Concepts in Surface Growth*, Cambridge: Cambridge University Press.

Bertini, L., A. De Sole, D. Gabrielli, G. Jona-Lasinio, and C. Landim, 2005, Current fluctuations in stochastic lattice gases, *Phys. Rev. Lett.* **94**, 030601-1–4.

Canet, L., H. Chaté, B. Delamotte, and N. Wschebor, 2010, Nonperturbative renormalization group for the Kardar–Parisi–Zhang equation, *Phys. Rev. Lett.* **104**, 150601-1–4.

Canet, L., H. Chaté, B. Delamotte, and N. Wschebor, 2011, Nonperturbative renormalization group for the Kardar–Parisi–Zhang equation: general framework and first applications, *Phys. Rev. E* **84**, 061128-1–18.

Caracciolo, S., A. Gambassi, M. Gubinelli, and A. Pelissetto, 2004a, Finite-size scaling in the driven lattice gas, *J. Stat. Phys.* **115**, 281–322.

Caracciolo, S., A. Gambassi, M. Gubinelli, and A. Pelissetto, 2004b, Comment on "Dynamic behavior of anisotropic nonequilibrium driving lattice gases", *Phys. Rev. Lett.* **92**, 029601-1.

Chayes, J. T., L. Chayes, D. S. Fisher, and T. Spencer, 1986, Finite-size scaling and correlation lengths for disordered systems, *Phys. Rev. Lett.* **57**, 2999–3002.

Daquila, G. L., 2011, *Monte Carlo Analysis of Non-equilibrium Steady States and Relaxation Kinetics in Driven Lattice Gases*, Ph.D. dissertation, Virginia Polytechnic Institute and State University.

Daquila, G. L. and U. C. Täuber, 2011, Slow relaxation and aging kinetics for the driven lattice gas, *Phys. Rev. E* **83**, 051107-1–11.

Daquila, G. L. and U. C. Täuber, 2012, Nonequilibrium relaxation and critical aging for driven Ising lattice gases, *Phys. Rev. Lett.* **108**, 110602-1–5.

Derrida, B., 1998, An exactly soluble non-equilibrium system: the asymmetric simple exclusion process, *Phys. Rep.* **301**, 65–83.

Derrida, B. and C. Appert, 1999, Universal large deviation function of the Kardar–Parisi–Zhang equation in one dimension, *J. Stat. Phys.* **94**, 1–30.

Derrida, B. and J. L. Lebowitz, 1998, Exact large deviation function in the asymmetric exclusion process, *Phys. Rev. Lett.* **80**, 209–213.

Doty, C. A. and J. M. Kosterlitz, 1992, Exact dynamical exponent at the Kardar–Parisi–Zhang roughening transition, *Phys. Rev. Lett.* **69**, 1979–1981.

Edwards, S. F. and D. R. Wilkinson, 1982, The surface statistics of a granular aggregate, *Proc. R. Soc. London A* **381**, 17–31.

Fisher, D. S. and D. A. Huse, 1991, Directed paths in a random potential, *Phys. Rev. B* **43**, 10 728–10 742.

Fogedby, H. C., A. B. Eriksson, and L. V. Mikheev, 1995, Continuum limit, Galilean invariance, and solitons in the quantum equivalent of the noisy Burgers equation, *Phys. Rev. Lett.* **75**, 1883–1886.

Forster, D., D. R. Nelson, and M. J. Stephen, 1977, Large-distance and long-time properties of a randomly stirred fluid, *Phys. Rev. A* **16**, 732–749.

Frey, E. and U. C. Täuber, 1994, Two-loop renormalization-group analysis of the Burgers–Kardar–Parisi–Zhang equation, *Phys. Rev. E* **50**, 1024–1045.

Frey, E., U. C. Täuber, and T. Hwa, 1996, Mode-coupling and renormalization group results for the noisy Burgers equation, *Phys. Rev. E* **53**, 4424–4438.

Halpin-Healy, T. and Y.-C. Zhang, 1995, Kinetic roughening phenomena, stochastic growth, directed polymers and all that, *Phys. Rep.* **254**, 215–414.

Henkel, M. and M. Pleimling, 2010, *Nonequilibrium Phase Transitions*, Vol. 2: *Ageing and Dynamical Scaling Far from Equilibrium*, Dordrecht: Springer.

Hwa, T. and D. S. Fisher, 1994, Anomalous fluctuations of directed polymers in random media, *Phys. Rev. B* **49**, 3136–3154.

Janssen, H. K., 1997, On critical exponents and the renormalization of the coupling constant in growth models with surface diffusion, *Phys. Rev. Lett.* **78**, 1082–1085.

Janssen, H. K. and B. Schmittmann, 1986a, Field theory of long time behaviour in driven diffusive systems, *Z. Phys. B Cond. Matt.* **63**, 517–520.

Janssen, H. K. and B. Schmittmann, 1986b, Field theory of critical behaviour in driven diffusive systems, *Z. Phys. B Cond. Matt.* **64**, 503–514.

Janssen, H. K., U. C. Täuber, and E. Frey, 1999, Exact results for the Kardar–Parisi–Zhang equation with spatially correlated noise, *Eur. Phys. J. B* **9**, 491–511.

Kardar, M., G. Parisi, and Y.-C. Zhang, 1986, Dynamic scaling of growing interfaces, *Phys. Rev. Lett.* **56**, 889–892.

Kardar, M. and Y.-C. Zhang, 1987, Scaling of directed polymers in random media, *Phys. Rev. Lett.* **58**, 2087–2090.

Katz, S., J. L. Lebowitz, and H. Spohn, 1983, Phase transitions in stationary nonequilibrium states of model lattice systems, *Phys. Rev. B* **28**, 1655–1658.

Katz, S., J. L. Lebowitz, and H. Spohn, 1984, Nonequilibrium steady states of stochastic lattice gas models of fast ionic conductors, *J. Stat. Phys.* **34**, 497–537.

Kloss, T., L. Canet, and N. Wschebor, 2012, Nonperturbative renormalization group for the Kardar–Parisi–Zhang equation: scaling functions and amplitude ratios in $1 + 1$, $2 + 1$, and $3 + 1$ dimensions, *Phys. Rev. E* **86**, 051124-1–19.

Krech, M., 1997, Short-time scaling behavior of growing interfaces, *Phys. Rev. E* **55**, 668–679; err. *Phys. Rev. E* **56**, 1285.

Krug, J., 1997, Origins of scale invariance in growth processes, *Adv. Phys.* **46**, 139–282.

Krug, J. and H. Spohn, 1992, Kinetic roughening of growing surfaces, in: *Solids Far from Equilibrium*, ed. C. Godrèche, Cambridge: Cambridge University Press, 479–582.

Lässig, M., 1995, On the renormalization of the Kardar–Parisi–Zhang equation, *Nucl. Phys. B* **448**, 559–574.

Lässig, M., 1998, On growth, disorder, and field theory, *J. Phys. Cond. Matt.* **10**, 9905–9950.

Lecomte, V., U. C. Täuber, and F. van Wijland, 2007, Current distribution in systems with anomalous diffusion: renormalization group approach, *J. Phys. A: Math. Theor.* **40**, 1447–1465.

Leung, K.-t., 1991, Finite-size scaling of driven diffusive systems: theory and Monte Carlo studies, *Phys. Rev. Lett.* **66**, 453–456.

Leung, K.-t. and J. L. Cardy, 1986, Field theory of critical behavior in a driven diffusive system, *J. Stat. Phys.* **44**, 567–588.

Marro, J. and R. Dickman, 1999, *Nonequilibrium Phase Transitions in Lattice Models*, Cambridge: Cambridge University Press.

Myllys, M., J. Maunuksela, M. Alava, T. Ala-Nissila, J. Merikoski, and J. Timonen, 2001, Kinetic roughening in slow combustion of paper, *Phys. Rev. E* **64**, 036101-1–12.

Nattermann, T. and S. Scheidl, 2000, Vortex-glass phases in type-II superconductors, *Adv. Phys.* **49**, 607–704.

Schilardi, P. L., O. Azzaroni, R. C. Salvarezza, and A. J. Arvia, 1999, Validity of the Kardar–Parisi–Zhang equation in the asymptotic limit of metal electrodeposition, *Phys. Rev. B* **59**, 4638–4641.

Schmittmann, B. and R. K. P. Zia, 1995, Statistical mechanics of driven diffusive systems, in: *Phase Transitions and Critical Phenomena*, Vol. 17, eds. C. Domb and J. L. Lebowitz, London: Academic Press.

Schütz, G. M., 2001, Exactly solvable models for many-body systems far from equilibrium, in: *Phase Transitions and Critical Phenomena*, Vol. 19, eds. C. Domb and J. L. Lebowitz, London: Academic Press.

Sun, T., H. Guo, and M. Grant, 1989, Dynamics of driven interfaces with a conservation law, *Phys. Rev. A* **40**, 6763–6766.

Takeuchi, K. A. and M. Sano, 2010, Universal fluctuations of growing interfaces: evidence in turbulent liquid crystals, *Phys. Rev. Lett.* **104**, 230601-1–4.

Takeuchi, K. A. and M. Sano, 2012, Evidence for geometry-dependent universal fluctuations of the Kardar–Parisi–Zhang interfaces in liquid-crystal turbulence, *J. Stat. Phys.* **147**, 853–890.

Wang, J.-S., 1996, Anisotropic finite-size scaling analysis of a two-dimensional driven diffusive system, *J. Stat. Phys.* **82**, 1409–1427.

Wiese, K. J., 1998, On the perturbation expansion of the KPZ equation, *J. Stat. Phys.* **93**, 143–154.

Wolf, D. E. and J. Villain, 1990, Growth with surface diffusion, *Europhys. Lett.* **13**, 389–394.

Further reading

Bouchaud, J. P. and M. E. Cates, 1993, Self-consistent approach to the Kardar–Parisi–Zhang equation, *Phys. Rev. E* **47**, R1455–R1458.

Chou, T., K. Mallick, and R. K. P. Zia, 2011, Non-equilibrium statistical mechanics: from a paradigmatic model to biological transport, *Rep. Prog. Phys.* **74**, 116601-1–41.

Colaiori, F. and M. A. Moore, 2001a, Upper critical dimension, dynamic exponent, and scaling functions in the mode-coupling theory for the Kardar–Parisi–Zhang equation, *Phys. Rev. Lett.* **86**, 3946–3949.

Colaiori, F. and M. A. Moore, 2001b, Stretched exponential relaxation in the mode-coupling theory for the Kardar–Parisi–Zhang equation, *Phys. Rev. E* **63**, 057103-1–4.

Colaiori, F. and M. A. Moore, 2001c, Numerical solution of the mode-coupling equations for the Kardar–Parisi–Zhang equation in one dimension, *Phys. Rev. E* **65**, 017105-1–3.

Derrida, B. and H. Spohn, 1988, Polymers on disordered trees, spin glasses, and traveling waves, *J. Stat. Phys.* **51**, 817–840.

Fogedby, H. C., 2001, Scaling function for the noisy Burgers equation in the soliton approximation, *Europhys. Lett.* **56**, 492–498.

Fogedby, H. C., 2005, Localized growth modes, dynamic textures, and upper critical dimension for the Kardar–Parisi–Zhang equation in the weak-noise limit, *Phys. Rev. Lett.* **94**, 195702-1–4.

Fogedby, H. C., 2006, Kardar–Parisi–Zhang equation in the weak noise limit: pattern formation and upper critical dimension, *Phys. Rev. E* **73**, 031104-1–26.

Frey, E., U. C. Täuber, and H. K. Janssen, 1999, Scaling regimes and critical dimensions in the Kardar–Parisi–Zhang problem, *Europhys. Lett.* **47**, 14–20.

Golubović, L. and Z.-G. Wang, 1994, Kardar-Parisi-Zhang model and anomalous elasticity of two- and three-dimensional smectic-A liquid crystals, *Phys. Rev. E* **49**, 2567–2578.

Gueudré, T., P. Le Doussal, A. Rosso, A. Henry, and P. Calabrese, 2012, Short-time growth of a Kardar–Parisi–Zhang interface with flat initial conditions, *Phys. Rev. E* **86**, 041151-1–8.

Halpin-Healy, T., 2012, $(2 + 1)$-dimensional directed polymer in a random medium: scaling phenomena and universal distributions, *Phys. Rev. Lett.* **109**, 170602-1–5.

Hwa, T. and E. Frey, 1991, Exact scaling function of interface growth dynamics, *Phys. Rev. A* **44**, R7873–R7876.

Katzav, E. and M. Schwartz, 2004, Numerical evidence for stretched exponential relaxations in the Kardar–Parisi–Zhang equation, *Phys. Rev. E* **69**, 052603-1–4.

Kelling, J. and G. Ódor, 2011, Extremely large-scale simulation of a Kardar–Parisi–Zhang model using graphics cards, *Phys. Rev. E* **84**, 061150-1–7.

Kim, D., 1995, Bethe ansatz solution for crossover scaling functions of the asymmetric XXZ chain and the Kardar–Parisi–Zhang-type growth model, *Phys. Rev. E* **52**, 3512–3524.

Krapivsky, P. L. and B. Meerson, 2012, Fluctuations of current in nonstationary diffusive lattice gases, *Phys. Rev. E* **86**, 031106-1–11.

Lässig, M. and H. Kinzelbach, 1997, Upper critical dimension of the Kardar–Parisi–Zhang equation, *Phys. Rev. Lett.* **78**, 903–906.

Marinari, E., A. Pagnani, G. Parisi, and Z. Rácz, 2002, Width distributions and the upper critical dimension of Kardar–Parisi–Zhang interfaces, *Phys. Rev. E* **65**, 026136-1–4.

Medina, E., T. Hwa, M. Kardar, and Y.-C. Zhang, 1989, Burgers equation with correlated noise: renormalization-group analysis and applications to directed polymers and interface growth, *Phys. Rev. A* **39**, 3053–3075.

Moore, M. A., T. Blum, J. P. Doherty, M. Marsili, J.-P. Bouchaud, and P. Claudin, 1995, Glassy solutions of the Kardar–Parisi–Zhang equation, *Phys. Rev. Lett.* **74**, 4257–4260.

Nattermann, T. and L.-H. Tang, 1992, Kinetic surface roughening. I. The Kardar–Parisi–Zhang equation in the weak-coupling regime, *Phys. Rev. A* **45**, 7156–7161.

Nicoli, M., R. Cuerno, and M. Castro, 2013, Dimensional fragility of the Kardar–Parisi–Zhang universality class, *J. Stat. Mech.*, P11001, 1–11.

Ódor, G., B. Liedke, and K.-H. Heinig, 2009, Mapping of $(2+1)$-dimensional Kardar–Parisi–Zhang growth onto a driven lattice gas model of dimers, *Phys. Rev. E* **79**, 021125-1–5.

Prähofer, M. and H. Spohn, 2004, Exact scaling functions for one-dimensional stationary KPZ growth, *J. Stat. Phys.* **115**, 255–279.

Schehr, G., 2012, Extremes of N vicious walkers for large N: application to the directed polymer and KPZ interfaces, *J. Stat. Phys.* **149**, 385–410.

Schmittmann, B. and R. K. P. Zia, 1998, Driven diffusive systems: an introduction and recent developments, *Phys. Rep.* **301**, 45–64.

Schwartz, M. and S. F. Edwards, 1992, Nonlinear deposition: a new approach, *Europhys. Lett.* **20**, 301–306.

Schwartz, M. and E. Katzav, 2008, The ideas behind self-consistent expansion, *J. Stat. Mech.*, P04023-1–12.

Täuber, U. C., B. Schmittmann, and R. K. P. Zia, 2001, Critical behaviour of driven bilayer systems: a field-theoretic renormalisation group study, *J. Phys. A: Math. Gen.* **34**, L583–L589.

Täuber, U. C. and E. Frey, 2002, Universality classes in the anisotropic Kardar–Parisi–Zhang model, *Europhys. Lett.* **59**, 655–661.

Index